Lecture Notes in Computer Science 7562

Commenced Publication in 1973
Founding and Former Series Editors:
Gerhard Goos, Juris Hartmanis, and Jan van Leeuwen

Hartmut Ehrig   Gregor Engels
Hans-Jörg Kreowski   Grzegorz Rozenberg (Eds.)

Graph Transformations

6th International Conference, ICGT 2012
Bremen, Germany, September 24-29, 2012
Proceedings

 Springer

Volume Editors

Hartmut Ehrig
Technical University of Berlin
Franklinstr. 28/29
10587 Berlin, Germany
E-mail: ehrig@cs.tu-berlin.de

Gregor Engels
University of Paderborn
Zukunftsmeile 1
33102 Paderborn, Germany
E-mail: engels@upb.de

Hans-Jörg Kreowski
University of Bremen
P.O. Box 33 04 40
28334 Bremen, Germany
E-mail: kreo@informatik.uni-bremen.de

Grzegorz Rozenberg
Leiden University, LCNC - LIACS
Niels Bohrweg 1
2333 CA Leiden, The Netherlands
E-mail: rozenber@liacs.nl

ISSN 0302-9743   e-ISSN 1611-3349
ISBN 978-3-642-33653-9   e-ISBN 978-3-642-33654-6
DOI 10.1007/978-3-642-33654-6
Springer Heidelberg Dordrecht London New York

Library of Congress Control Number: 2012947426

CR Subject Classification (1998): G.2, D.2.4, D.2, E.1, F.3, F.1, F.4

LNCS Sublibrary: SL 1 – Theoretical Computer Science and General Issues

Typesetting: Camera-ready by author, data conversion by Scientific Publishing Services, Chennai, India

Printed on acid-free paper

Springer is part of Springer Science+Business Media (www.springer.com)

Preface

ICGT 2012 was the sixth International Conference on Graph Transformation, held at the University of Bremen in September 2012 under the auspices of the European Association of Theoretical Computer Science (EATCS), the European Association of Software Science and Technology (EASST), and the IFIP Working Group 1.3, Foundations of Systems Specification. ICGT 2012 continued the series of conferences previously held in Barcelona (Spain) in 2002, Rome (Italy) in 2004, Natal (Brazil) in 2006, Leicester (UK) in 2008, and in Enschede (The Netherlands) in 2010 following a series of six International Workshops on Graph Grammars and Their Application to Computer Science from 1978 to 1998.

The conference motto was "Modeling and Analysis of Dynamic Structures". Dynamic structures are the predominant concept for modeling and understanding complex problem situations. They consist of elements and interrelations which either may be added or removed, or may change their state. Dynamic structures are used in many computer science disciplines as the fundamental modeling approach. Examples are software architectures, software models, program structures, database structures, communication and network structures, or artifact versions and configurations. These structures are dynamic as they may be changed at design time or at runtime. These changes are known as, e.g., architectural refactorings, model transformations, or artifact evolutions. In the case of executable descriptions, dynamic structures are also used as semantic domains or as computational models for formal specification approaches, providing in this way means to formally analyze dynamic structures for certain predefined or user-defined properties.

All these approaches rely on the same uniform structure of graphs as well as on graph transformations to describe their dynamic behavior. Both aspects of graphs and graph transformations have been studied for more than 40 years by the graph grammar and graph transformation community. The conference aims at fostering this community as well as attracting researchers from other research areas to join the community. This could happen by contributing to the theory of graph transformation or by applying graph transformations to already known or novel application areas. Examples are self-adaptive systems, virtual structures in cloud computing, or advanced computational models such as models for DNA computing.

The conference program was split into the foundations track and the applications track with separate program committees, in order to yield a high-quality conference program covering all aspects of graph transformations. The proceedings of ICGT 2012 consist of three invited papers and 24 contributions, which were selected following a thorough reviewing process. Moreover, the three presentations accepted for the Doctoral Symposium are documented by extended abstracts.

The volume starts with the invited papers. The further papers are divided into the foundations track and the applications track. The foundations track consists of 15 papers subdivided into the thematic topics behavioral analysis, high-level graph transformation, revisited approaches, general transformation models, and structuring and verification while the applications track consists of nine papers subdivided into the thematic topics graph transformations in use, (meta-)model evolution, and incremental approaches. The volume ends with the abstracts of the presentations given at the Doctoral Symposium.

We are grateful to the University of Bremen for hosting ICGT 2012, and would like to thank the members of the organization committee and of the two program committees as well as the subreviewers. Particular thanks go to Andrea Corradini and Gabriele Taentzer for organizing the Doctoral Symposium as part of the conference. Moreover, according to the tradition of the ICGT series, three satellite workshops were organized:

- 7th International Workshop on Graph Based Tools (GraBaTs 2012) organized by Christian Krause and Bernhard Westfechtel,
- 4th International Workshop on Graph Computation Models (GCM 2012)
  organized by Rachid Echahed, Annegret Habel, and Mohamed Mosbah, and
- 5th International Workshop on Petri Nets, Graph Transformation and other
  Concurrency Formalisms (PNGT 2012) organized by Kathrin Hoffmann and
  Julia Padberg.

We are also grateful to Marcus Ermler, Melanie Luderer, and Caroline von Totth for their help in editing this volume. Finally, we would like to acknowledge the excellent support throughout the publishing process by Alfred Hofmann and his team at Springer, and the helpful use of the EasyChair and ConfTool conference management systems.

July 2012

Hartmut Ehrig
Gregor Engels
Hans-Jörg Kreowski
Grzegorz Rozenberg

Program Committee

Foundations Track

Paolo Baldan	Padua (Italy)
Michel Bauderon	Bordeaux (France)
Paolo Bottoni	Rome (Italy)
Andrea Corradini	Pisa (Italy)
Hartmut Ehrig	Berlin (Germany)
Annegret Habel	Oldenburg (Germany)
Reiko Heckel	Leicester (UK)
Berthold Hoffmann	Bremen (Germany)
Dirk Janssens	Antwerp (Belgium)
Barbara König	Duisburg-Essen (Germany)
Hans-Jörg Kreowski (Chair)	Bremen (Germany)
Sabine Kuske	Bremen (Germany)
Leen Lambers	Potsdam (Germany)
Ugo Montanari	Pisa (Italy)
Mohamed Mosbah	Bordeaux (France)
Fernando Orejas	Barcelona (Spain)
Francesco Parisi-Presicce	Rome (Italy)
Detlef Plump	York (UK)
Arend Rensink	Twente (The Netherlands)

Applications Track

Luciano Baresi	Milan (Italy)
Artur Boronat	Leicester (UK)
Juan de Lara	Madrid (Spain)
Gregor Engels (Chair)	Paderborn (Germany)
Claudia Ermel	Berlin (Germany)
Holger Giese	Potsdam (Germany)
Esther Guerra	Madrid (Spain)
Ralf Lämmel	Koblenz-Landau (Germany)
Mark Minas	Munich (Germany)
Manfred Nagl	Aachen (Germany)
Rinus Plasmeijer	Nijmegen (The Netherlands)
Leila Ribeiro	Porto Alegre (Brazil)
Wilhelm Schäfer	Paderborn (Germany)
Andy Schürr	Darmstadt (Germany)
Gabriele Taentzer	Marburg (Germany)
Pieter Van Gorp	Eindhoven (The Netherlands)

Dániel Varró	Budapest (Hungary)
Gergely Varró	Darmstadt (Germany)
Jens H. Weber	Victoria, BC (Canada)
Bernhard Westfechtel	Bayreuth (Germany)
Albert Zündorf	Kassel (Germany)

Subreviewers

Christopher Bak	Bartek Klin	Christopher Poskitt
Marcus Ermler	Andreas Koch	Jan Rieke
Fabio Gadducci	Christian Krause	Andreas Scharf
László Gönczy	Marius Lauder	Jan Stückrath
Jonathan Hayman	Jerome Leroux	Caroline von Totth
Tobias Heindel	Melanie Luderer	Sebastian Wätzoldt
Frank Hermann	Tim Molderez	

Local Organization

Marcus Ermler	Sabine Kuske	Helga Reinermann
Berthold Hoffmann	Melanie Luderer	Caroline von Totth
Hans-Jörg Kreowski	Sylvie Rauer	

Table of Contents

Foundations 3: Revisited Approaches

Foundations 4: General Transformation Models

Foundations 5: Structuring and Verification

Applications 1: Graph Transformations in Use

Applications 2: (Meta-)Model Evolution

Applications 3: Incremental Approaches

Doctoral Symposium

A Graph-Based Design Framework for Services

Antónia Lopes[1] and José Luiz Fiadeiro[2]

[1] Faculty of Sciences, University of Lisbon
Campo Grande, 1749–016 Lisboa, Portugal
mal@di.fc.ul.pt
[2] Department of Computer Science, Royal Holloway University of London
Egham TW20 0EX, UK
Jose.Fiadeiro@rhul.ac.uk

Abstract. Service-oriented systems rely on software applications that offer services through the orchestration of activities performed by external services procured on the fly when they are needed. This paper presents an overview of a graph-based framework developed around the notions of service and activity module for supporting the design of service-oriented systems in a way that is independent of execution languages and deployment platforms. The framework supports both behaviour and quality-of-service constraints for the discovery, ranking and selection of external services. Service instantiation and binding are captured as algebraic operations on configuration graphs.

1 Introduction

Service-oriented systems are developed to run on global computers and respond to business needs by interacting with services and resources that are globally available. The development of these systems relies on software applications that offer services through the orchestration of activities performed by other services procured on the fly, subject to a negotiation of service level agreements, in a dynamic market of service provision. The binding between the requester and the provider is established at run time at the instance level, i.e., each time the need for the service arises. Over the last few years, our research has addressed some challenges raised by this computing paradigm, namely:

(i) to understand the impact of service-oriented computing (SOC) on software engineering methodology;

(ii) to characterise the fundamental structures that support SOC independently of the specific languages or platforms that may be adopted to develop or deploy services;

(iii) the need for concepts and mechanisms that support the design of service-oriented applications from business requirements;

(iv) the need for mathematical models that offer a layer of abstraction at which we can capture the nature of the transformations that, in SOC, are operated on configurations of global computers;

(v) the need for an interface theory for service-oriented design.

H. Ehrig et al.(Eds.): ICGT 2012, LNCS 7562, pp. 1–19, 2012.

As a result of (i) above, we identified two types of abstractions that are useful for designing service-oriented systems: business *activities* and *services*. Activities correspond to applications developed by business IT teams according to requirements provided by their organisation, e.g., the applications that, in a bank, implement the financial products that are made available to the customers. The implementation of activities may resort to direct invocation of components and can also rely on services that will be procured on the fly. Services differ from activities in that they are applications that are not developed to satisfy specific business requirements of an organisation; instead they are developed to be published in ways that they can be discovered by activities.

Taking into account this distinction, we developed a graph-based framework for the design of service-oriented systems at a level of abstraction that supports this "business-oriented" perspective. In this framework, services and activities are defined through *activity modules* and *service modules*, respectively. These modules differ in the type of interface and binding they provide to their clients, which in the case of activities is for direct invocation or static binding (e.g., human-computer interaction or system-to-system interconnections established at configuration time) and, in the case of services, for dynamic discovery and binding. Activity and service modules are graph-based primitives that define a workflow and the external services that may need to be procured and bound to in order to fulfil business goals. Behaviour and service-quality constraints can be imposed over the external services to be procured. These constraints are taken into account in the processes of discovery, ranking and selection.

The proposed design framework is equipped with a layered graph-based model for state configurations of global computers. Configurations are made to be *business reflective* through an explicit representation of the types of business activities that are active in the current state. This model captures the transformations that occur in the configuration of global computers when the discovery of a service is triggered, which results in the instantiation and binding of the selected service.

In this paper, we present an overview of this framework. In Sec. 2, we present the notions of service and activity modules, the cornerstone of our framework, grounded on an interface theory for service-oriented design. In Sec. 3, we present a model for state configurations of global computers that in Sec. 4 is used to provide the operational semantics of discovery, instantiation and binding. We conclude in Sec. 5 by pointing to other aspects of SOC that have been investigated within the framework.

2   Design Primitives for Service-Oriented Systems

The design primitives we propose for service-oriented systems were inspired by the Service Component Architecture (SCA) [22]. As in SCA, we view SOC as providing an architectural layer that can be superposed over a component infrastructure – what is sometimes referred to as a service overlay. More concretely, we adopt the view that services are delivered by ensembles of components (orchestrations) that are able to bind dynamically to other services discovered at run time. For the purposes of this paper, the model that is used for defining orchestrations is not relevant. As discussed in Sec. 2.1, it is enough to know that we have a component algebra in the sense of [9] that makes explicit the structure of the component ensembles.

We illustrate our framework with a simplified credit service: after evaluating the risk of a credit request, the service either proposes a deal to the customer or denies the request; in the first case and if the proposal is accepted, the service takes out the credit and informs the customer of the expected transfer date. This activity relies on an external risk evaluator that is able to evaluate the risk of the transaction.

2.1   The Component Algebra

We see the ensembles of components that orchestrate services as networks in which *components* are connected through *wires*. For the purpose at hand, the nature of these components and the communication model is not relevant. The design framework is defined in terms of a set *COMP* of components, a set *PORT* of ports that components make available for communication with their environment, a set *WIRE* of wires for interconnecting pairs of ports, and a component algebra built around those elements. In the sequel we use *ports(c)* and *ports(w)* to denote, respectively, the set of ports of a component c and the pair of ports interconnected by a wire w.

Definition 1 (Component Net). *A component net α is a tuple $\langle C, W, \gamma, \mu \rangle$ where:*

- *$\langle C, W \rangle$ is a simple finite graph: C is a set of nodes and W is a set of edges. Each edge is an unordered pair $\{c_1, c_2\}$ of nodes.*
- *γ is a function assigning $\gamma_c \in COMP$ to every $c \in C$ and $\gamma_w \in WIRE$ to every $w \in W$.*
- *μ is a W-indexed family of bijections μ_w establishing a correspondence between $ports(\gamma_w)$ and the components $\{c_1, c_2\}$ interconnected by w, such that:*
  1. *For every $P \in ports(\gamma_w)$, $P \in ports(\mu_w(P))$.*
  2. *If $w' = \{c_1, c_3\}$ is an edge with $c_2 \neq c_3$, then $\mu_w(c_1) \neq \mu_{w'}(c_1)$.*

This definition reflects component-and-connector architectural configurations where the mapping μ defines the attachments between component ports and connector roles. Because in SOC communication is essentially peer-to-peer, we take all connectors to be binary. The fact that the graph is simple means that all interactions between two components are supported by a single wire and that no component can interact with itself. Through (2), ports of a component cannot be used in more than one connection.

The ports of a component net that are still available for establishing further interconnections, i.e., not connected to any other port, are called interaction-points:

Definition 2 (Interaction-point). *An interaction-point of a component net* $\alpha = \langle C, W, \gamma, \mu \rangle$ *is a pair* $\langle c, P \rangle$ *where* $c \in C$ *and* $P \in ports(\gamma_c)$ *such that there is no edge* $\{c, c'\} \in W$ *such that* $\mu_{\{c,c'\}}(c) = P$. *We denote by* I_α *the set of interaction-points of* α.

Component nets can be composed through their interaction points via wires that interconnect the corresponding ports.

Definition 3 (Composition of Component Nets). *Let* $\alpha_1 = \langle C_1, W_1, \gamma_1, \mu_1 \rangle$ *and* $\alpha_2 = \langle C_2, W_2, \gamma_2, \mu_2 \rangle$ *be component nets such that* C_1 *and* C_2 *are disjoint,* $(w^i)_{i=1...n}$ *a family of wires, and* $\langle c_1^i, P_1^i \rangle_{i=1...n}$ *and* $\langle c_2^i, P_2^i \rangle_{i=1...n}$ *families of interaction points of, respectively,* α_1 *and* α_2, *such that: (1) each* w^i *is a wire connecting* $\{P_1^i, P_2^i\}$, *(2) if* $c_1^i = c_1^j$ *and* $c_2^i = c_2^j$ *then* $i = j$, *(3) if* $c_1^i = c_1^j$ *with* $i \neq j$, *then* $P_1^i \neq P_1^j$ *and (4) if* $c_2^i = c_2^j$ *with* $i \neq j$, *then* $P_2^i \neq P_2^j$. *The composition*

$$\alpha_1 \left\|_{\langle c_1^i, P_1^i \rangle, w^i, \langle c_2^i, P_2^i \rangle}^{i=1...n} \alpha_2 \right.$$

is the component net defined as follows:

- *Its graph is* $\langle C_1 \cup C_2, W_1 \cup W_2 \cup \bigcup_{i=1...n} \{c_1^i, c_2^i\} \rangle$.
- *Its functions* γ *and* μ *coincide with that of* α_1 *and* α_2 *on the corresponding subgraphs. For the new edges,* $\gamma_{\{c_1^i, c_2^i\}} = w^i$ *and* $\mu_{\{c_1^i, c_2^i\}}(P_j^i) = c_j^i$.

In order to illustrate the notions just introduced, we take the algebra of *Asynchronous Relational Nets* (ARN) defined in [13]. In that algebra, components interact asynchronously through the exchange of messages transmitted through channels. Ports are sets of messages classified as incoming or outgoing. A component consists of a finite collection of mutually disjoint ports and a set of infinite sequences of sets of actions (traces), each action being the publication of an outgoing message or the reception of an incoming message (for simplicity, the data that messages may carry is ignored). Interconnection of components is established through channels – a set P of messages and a set of traces. A wire consists of a channel and a pair of injections $\mu_i : P \to P_i$ that uniquely establishes connections between incoming and outgoing messages.

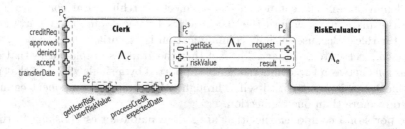

Fig. 1. An example of an ARN with two components connected through a wire

Fig. 1 presents an example of an ARN with two components connected through a wire, which support part of the activities involved in the request of a credit. The net has two nodes $\{c{:}Clerk, e{:}RiskEvaluator\}$ and a single edge $\{c, e\}{:}w_{ce}$.

The component *Clerk* has four ports. Its behaviour Λ_c is as follows: after the delivery of the first *creditReq* message on port P_c^1, it publishes *getUserRisk* on port P_c^2 and waits for the delivery of *userRiskValue* in the same port; if the credit request comes from a known user, this may be enough for making a decision on the request and sending *approved* or *denied*; if not, it publishes *getRisk* on P_c^3 and waits for the delivery of *riskValue* for making the decision; after sending *approved* (if ever), *Clerk* waits for the delivery of *accept*, upon which it publishes *processCredit* on P_c^4 and waits for *expectedDate*; when this happens, it sends *transferDate*.

The component *RiskEvaluator* has a single port and its behaviour is quite simple: every time *request* is delivered, it publishes *result*. The wire w_{ce} interconnects ports P_c^3 and P_e and establishes that the publication of *getRisk* in P_c^3 will be delivered in P_e under the name *request* and the publication of *request* in P_e will be delivered in P_c^3 under the name *riskValue*.

The example presented in Fig. 1 can also be used to illustrate the composition of ARNs: this ARN is the composition of the two single-component ARNs defined by *Clerk* and *RiskEvaluator* via the wire w_{ce}.

2.2   The Interface Algebra

As discussed in the introduction, the interfaces of services and business activities need to specify the functionality that customers can expect as well as the dependencies that they may have on external services.

In our approach, a service interface identifies a port through which the service is provided (*provides-point*), a number of ports through which external services are required (*requires-points*) and a number of ports for those persistent components of the underlying configuration that the service will need to use once instantiated (*uses-points*). Activity interfaces are similar except that they have a *serves-point* instead of a provides-point. The differences are that the binding of provides and requires-points is performed by the runtime infrastructure whereas the binding of uses and serves-points has to be provided by developers.

In addition, interfaces describe the behavioural constraints imposed over the external services to be procured and quality-of-service constraints through which service-level agreements can be negotiated with these external services during matchmaking. The first are defined in terms of a logic while the latter are expressed through constraint systems defined in terms of c-semirings [5].

More concretely, we consider that sentences of a specification logic *SPEC* are used for specifying the properties offered or required. The particular choice of the specification logic – logic operators, their semantics and proof-theory – can be abstracted away. For the purpose of this paper, it is enough to know that the logic satisfies some structural properties, namely that we have available an entailment system (or π- institution) $\langle SIGN, gram, \vdash \rangle$ for *SPEC* [17,11]. In this structure, *SIGN* is the category of signatures of the logic: signatures are sets of actions (e.g., the actions of sending and receiving a message m, which we denote, respectively, by $m!$ and m_{i}) and signature morphisms are maps that preserve the structure of actions (e.g., their type, their parameters, etc). The grammar functor $gram{:}SIGN{\rightarrow}SET$ generates the language used for describing properties of the interactions in every signature. Notice that, given a signature morphism $\sigma{:}\Sigma{\rightarrow}\Sigma'$, $gram(\sigma)$ translates properties in the language of Σ to the language of Σ'. Translations induced by isomorphims (i.e., bijections between sets of actions) are required to be conservative.

We also assume that $PORT$ is equipped with a notion of morphism that defines a category related to $SIGN$ through a functor $A{:}PORT{\rightarrow}SIGN$. The idea is that each port defines a signature that allows to express properties over what happens in that port. We use A_P to denote the signature corresponding to port P. Moreover, we consider that every port P has a *dual* port P^{op} (e.g., the dual of a set of messages classified as incoming or outgoing is the same set of messages but with the dual classification).

For quality-of-service constraints, we adopt so-called soft constraints, which map each valuation of a set of variables into a space of degrees of satisfaction A. The particular soft-constraint formalism used for expressing constraints is not relevant for the approach that we propose. It is enough to know that we consider constraint systems defined in terms of a fixed c-semiring S, as defined in [5]:

- A *c-semiring* S is a semiring of the form $\langle A, +, \times, 0, 1 \rangle$ in which A represents a space of degrees of satisfaction. The operations $\times$ and $+$ are used for composition and choice, respectively. Composition is commutative, choice is idempotent and 1 is an absorbing element (i.e., there is no better choice than 1). S induces a partial order $\leq_S$ (of satisfaction) over A: $a \leq_S b$ iff $a + b = b$.
- A *constraint system* defined in terms of c-semiring S is a pair $\langle D, V \rangle$ where V is a totally ordered set (of variables), and D is a finite set (domain of possible values taken by the variables).
- A *constraint* consists of a subset *con* of V and a mapping $def{:}D^{|con|}{\rightarrow}A$ assigning a degree of satisfaction to each assignment of values to the variables in *con*.
- The *projection* of a constraint c over $I{\subseteq}V$, denoted by $c \Downarrow_I$, is $\langle def', con' \rangle$ with $con' = con \cap I$ and $def'(t') = \sum_{\{t \in D^{|con|}: t\downarrow_{con'}^{con} = t'\}} def(t)$, where $t\downarrow_X^Y$ denotes the projection of Y-tuple t over X.

We start by defining a notion of service and activity interface.

Definition 4 (Service and Activity Interface). *An interface i consists of:*

- *A set I (of interface-points) partitioned into a set $I^{\rightarrow}$ with at most one element, which (if it exists) is called the* provides-point *and denoted by $i^{\rightarrow}$, a set $I^{\leftarrow}$ the member of which are called the* requires-points, *a set $I^{\uparrow}$ with at most one element, which (if it exists) is called the* serves-point *and denoted by $i^{\uparrow}$, a set $I^{\downarrow}$ the member of which are called the* uses-points, *such that either $I^{\rightarrow}$ or $I^{\uparrow}$ is empty.*
- *For every $r{\in}I$, a port P_r and a consistent set of formulas Φ_r over A_{P_r}.*

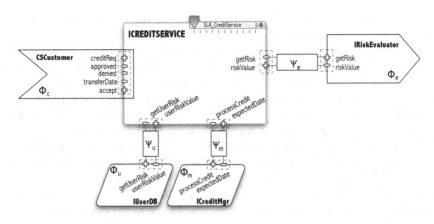

Fig. 2. An example of a service interface

- *For every* $r \in I^{\leftarrow} \cup I^{\downarrow}$, *a consistent set of formulas* Ψ_r *over the amalgamated union of* A_{P_r} *and* $A_{P_r^{op}}$.
- *A pair* $C = \langle C_{cs}, C_{sla} \rangle$ *where* C_{cs} *is a constraint system* $\langle C_D, C_V \rangle$ *and* C_{sla} *is a set of constraints over* C_{cs}.

A service interface i *is an interface such that* $I^{\uparrow}$ *is empty and* $I^{\rightarrow}$ *is a singleton. Conversely, an* activity interface i *is an interface such that* $I^{\rightarrow}$ *is empty and* $I^{\uparrow}$ *is a singleton.*

The formulas Φ_r at each interface-point r specify the protocols that the element requires from external services (in the case of requires-points) or from other components (in the case of uses-points) and those that it offers to customer services (in the case of the provides-point) or to users (in the case of the serves-point). The formulas Ψ_r express requirements on the wire through which the element expects to interact with r. C defines the constraints through which SLAs can be negotiated with external services during discovery and selection.

In Fig. 2, we present an example of an interface for a credit service using a graphical notation similar to that of SCA. On the left, we have a provides-point *CSCustomer* through which the service is provided; on the right, a requires-point *IRiskEvaluator* through which an external service is required; and, on the bottom, two uses-points through which the service connects to persistent components (a database that stores information about users and a manager of approved credit requests).

In this example, the specification of behavioural properties is defined in linear temporal logic. For instance, Φ_c includes $\Box(creditReq_\mathsf{i} \supset \Diamond(approved! \lor denied!))$ specifying that the service offers, in reaction to the delivery of the message *creditReq*, to reply by publishing either *approved* or *denied*. Moroever, Φ_c also specifies that if *accept* was received after the publication of *approved*, then *transferDate* will eventually be published. On the other hand, Φ_e only includes $\Box(getRisk_\mathsf{i} \supset \Diamond riskValue!)$ specifying that the required service is asked to react to the delivery of *getRisk* by eventually publishing *riskValue*. Ψ_e specifies that the wire used to connect the requires-point with the external service has to ensure that the transmission of both messages is reliable.

The quality-of-service constraints are defined in terms of the c-semiring $\langle [0, 1],$ $max, min, 0, 1 \rangle$ of soft fuzzy constraints. $SLA\_CreditService$ declares three configuration variables – $c.amount$ (the amount conceded to the customer), $e.fee$ (the fee to be paid by the credit service to the risk evaluator service) and $e.cfd$ (the confidence level of the risk evaluator) – and has two constraints: (1) the credit service targets only credit requests between 1000 and 10 000; (2) the fee f to be paid to the risk evaluator must be less than 50, the confidence level c must be greater than 0.9 and, if these conditions are met, the preference level is given by $\frac{c-0.9}{0.2} + \frac{50-f}{100}$.

Composition of interfaces is an essential ingredient of any interface algebra. As discussed before, activities and services differ in the form of composition they offer to their customers. In this paper, we focus on the notion of composition that is specific to SOC, which captures the binding of a service or activity with a required service.

Definition 5 (Interface Match). *Let i be an interface, $r \in I^{\leftarrow}$ and j a service interface. An* interface match *from $\langle i, r \rangle$ to j consists of a port morphism $\delta : P_r^i \to P_{j \to}^j$ such that $\Phi_{j \to}^j \vdash \delta(\Phi_r^i)$ and a partial injective function $\rho : C_V^i \to C_V^j$. Interface i is said to be* compatible *with service interface j w.r.t. requires-point r if (1) I and J are disjoint, (2) $blevel(C_{sla}^i \oplus_\rho C_{sla}^j) >_S 0$ and (3) there exists a match from $\langle i, r \rangle$ to j.*

An interface match defines a relation between the port of the requires-point r of interface i and the port of the provides-point of j in such a way that the required properties are entailed by the provided ones. Moreover, the function ρ identifies the configuration variables in the constraint systems of the two interfaces that are shared. The formulation of condition (2) above relies on a composition operator $\oplus_\rho$ that performs amalgamated unions of constraint systems and constraints, taking into account the shared configuration variables. These operations are defined as follows.

Definition 6 (Amalgamation of Constraints). *Let $S_1 = \langle D_1, V_1 \rangle$ and $S_2 = \langle D_2, V_2 \rangle$ be two constraint systems and $\rho : V_1 \to V_2$ a partial injective function.*

- *$S_1 \oplus_\rho S_2$ is $\langle D, V \rangle$ where D is $D_1 \cup D_2$ and V is $V_1 \oplus_\rho V_2$, the amalgamated union of V_1 and V_2. We use ι_i to denote the injection from V_i into $V_1 \oplus_\rho V_2$.*
- *Let $c = \langle con, def \rangle$ be a constraint in S_i. $\rho(c)$ is the constraint $\langle \iota_i(con), def' \rangle$ in $S_1 \oplus_\rho S_2$ where $def'(t)$ is $def(t)$ for $t \in D_i^{|con|}$ and 0 otherwise.*
- *Let C_1, C_2 be sets of constraints in, respectively, S_1 and S_2. $C_1 \oplus_\rho C_2$ is $\rho(C_1) \cup \rho(C_2)$.*

The consistency of a set of constraints C in $S = \langle D, V \rangle$ is defined in terms of the notion of best level of consistency as follows:

$$blevel(C) = \sum_{t \in D^{|V|}} \prod_{c \in C} def_c(t \downarrow_{con_c}^V)$$

Intuitively, this notion gives us the degree of satisfaction that we can expect for C. We choose (through the sum) the best among all possible combinations (product) of all constraints in C (for more details see [5]). C is said to be consistent iff $blevel(C) >_S 0$. If a set of constraints C is consistent, a valuation for the variables used in C is said to be a *solution* for C and can be also regarded as a constraint.

Definition 7 (Composition of Interfaces). *Given an interface i compatible with a service interface j w.r.t. r, a match $\mu=\langle\delta,\rho\rangle$ between $\langle i,r\rangle$ and j, and a solution Δ for $(C^i_{cs}\oplus_\rho C^j_{cs})\Downarrow_{\iota_j\circ\rho(C^i_V)}$, the composition $i\parallel_{r:\mu,\Delta} j$ is $\langle K^\rightarrow,K^\leftarrow,K^\uparrow,K^\downarrow,$ $P,\Phi,\Psi,C\rangle$ where:*

- $K^\rightarrow = I^\rightarrow$, $K^\leftarrow = J^\leftarrow \cup (I^\leftarrow \setminus \{r\})$, $K^\uparrow = I^\uparrow$ and $K^\downarrow = I^\downarrow \cup J^\downarrow$.
- P,Φ,Ψ *coincides with* P^i,Φ^i,Ψ^i *and* P^j,Φ^j,Ψ^j *on the corresponding points.*
- $C = \langle C^i_{cs}\oplus_\rho C^j_{cs},(C^i_{sla}\oplus_\rho C^j_{sla}) \cup \{\Delta\}\rangle$.

Notice that the composition of interfaces is not commutative: the interface on the left plays the role of client and the one on the right plays the role of supplier of services.

2.3 Service and Activity Modules

A component net orchestrates a service interface by assigning interaction-points to interface-points in such a way that the behaviour of the component net validates the specifications of the provides-points on the assumption that it is interconnected to component nets that validate the specifications of the requires- and uses-points through wires that validate the corresponding specifications.

In order to reason about the behaviour of component nets we take the behaviour of components $c \in COMP$ and wires $w \in WIRE$ to be captured, respectively, by specifications $\langle A_c,\Phi_c\rangle$ and $\langle A_w,\Phi_w\rangle$ in $SPEC$ defining the language of components and wires to be the amalgamated union of the languages associated with their ports. Given a pair of port morphisms $\theta_1:P_1\rightarrow P'_1$ and $\theta_2:P_2\rightarrow P'_2$, we denote by $\langle\theta_1,\theta_2\rangle$ the unique mapping from the amalgamated sum of A_{P_1} and A_{P_2} to the amalgamated sum of $A_{P'_1}$ and $A_{P'_2}$ that commutes with θ_1 and θ_2.

Notice that nodes and edges denote *instances* of components and wires, respectively. Different nodes (resp. edges) can be labelled with the same component (resp. wire). Therefore, in order to reason about the properties of the component net as a whole we need to translate the properties of the components and wires involved to a language in which we can distinguish between the corresponding instances. We take the translation that uses the node as a prefix for the elements in their language. Given a set A and a symbol p, we denote by $(p._-)$ the function that prefixes the elements of A with 'p.'. Note that prefixing defines a bijection between A and its image $p.A$.

Definition 8 (Component Net Properties). *Let $\alpha = \langle C,W,\gamma,\mu\rangle$ be a component net. $A_\alpha = \bigcup_{c\in C} c.(A_{\gamma_c})$ is the language associated with α and Φ_α is the union of, for every $c \in C$, the prefix-translation of Φ_{γ_c} by $(c._-)$ and, for every $w \in W$, the translation of Φ_{γ_w} by μ_w, where, for $a \in A_P$, $\mu_w(a) = \mu_w(P).a$.*

The set Φ_α consists on the translations of all the specifications of the components and wires using the nodes as prefixes for their language. Notice that because we are using bijections, these translations are conservative, i.e. neither components nor wires gain additional properties because of the translations. However, by taking the union of all such descriptions, new properties may emerge, i.e., Φ_α is not necessarily a conservative extension of the individual descriptions.

Definition 9 (Orchestration). *An orchestration of an interface i consists of:*

- *a component net $\alpha = \langle C, W, \gamma, \mu \rangle$ where C and I are disjoint;*
- *an injective function $\theta : I \to I_\alpha$ that assigns a different interaction-point to each interface-point; we write $r \xrightarrow{\theta} c$ to indicate that $\theta(r) = \langle c, P_c \rangle$ for some $P_c \in ports(\gamma_c)$;*
- *for every r of $I^\to \cup I^\uparrow$, a port morphism $\theta_r : P_r \to P_{c_r}$ where $r \xrightarrow{\theta} c_r$;*
- *for every r of $I^\leftarrow \cup I^\downarrow$, a port morphism $\theta_r : P_r^{op} \to P_{c_r}$ where $r \xrightarrow{\theta} c_r$;*

If i is a service interface, we require that

$$\bigcup_{r \in I^\leftarrow \cup I^\downarrow} (\, r.\Phi_r \cup \mu_r(\Psi_r) \,) \cup \Phi_\alpha \vdash c_{i \to}.(\theta_{i \to}(\Phi_{i \to}))$$

where $\mu_r(a) = r.a$ for $a \in A_{P_r}$ and $\mu_r(a) = c_r.\theta_r(a)$ for $a \in A_{P_r^{op}}$. We use $\alpha \lhd_\theta i$ to denote an orchestrated interface.

Consider again the single-component ARN defined by *Clerk*. This ARN, together with the correspondences $CSCustomer \mapsto \langle Clerk, P_c^1 \rangle$, $IRiskEvaluator \mapsto \langle Clerk, P_c^3 \rangle$, $IUserDB \mapsto \langle Clerk, P_c^2 \rangle$ and $ICreditMgr \mapsto \langle Clerk, P_c^4 \rangle$, defines an orchestration for the service interface ICREDITSERVICE. The port morphisms involved are identity functions. The traces in Λ_c are such that they validate Φ_c on the assumption that *Clerk* is interconnected through *IRiskEvaluator*, *IUserDB* and *ICreditMgr* to component nets that validate, respectively, Φ_e, Φ_u and Φ_m via wires that validate Ψ_e, Ψ_u and Ψ_m.

Services are designed through service modules. These modules define an orchestrated service interface, the initialisation conditions for the components and the triggers for the requires-points (stating when external services need to be discovered). The proposed framework is independent of the language used for specifying initialisation conditions and triggers: we assume that we have available a set STC of conditions and a set TRG of triggers.

Definition 10 (Service and Activity Module). *A service (resp. activity) module consists of an orchestrated service (resp. activity) interface $\alpha \lhd_\theta i$ and a pair of mappings $\langle trigger, init \rangle$ such that trigger assigns a condition in STC to each $r \in I^\leftarrow$ and init assigns a condition in STC to each c in the nodes of α.*

We use $interface(M)$, $orch(M)$ and θ^M to denote, respectively, i, α and θ; $M^\to$ and $M^\leftarrow$ to denote, respectively, $i^\to$ and $I^\leftarrow$; $\mathcal{C}_{cs}(M)$ and $\mathcal{C}_{sla}(M)$ to denote, respectively, the constraint system and the set of constraints of i.

The service interface ICREDITSERVICE orchestrated by *Clerk* together with a initialization condition for *Clerk* and a trigger condition for *IRiskEvaluator* define the service module CREDITSERVICE. We do not illustrate these conditions because their formulation depends on the formalism used for specifying the behaviour of components and wires (e.g., state machines, process calculi, Petrinets). See [15,16] for examples in SRML, a modelling language for SOC that we defined in the SENSORIA project.

Definition 11 (Service Match). *Let M be a module and $r \in M^{\leftarrow}$. A service match for M w.r.t. r is a triple $\langle S, \mu, w \rangle$ where*

- *S is a service module such that the set of nodes of $orch(S)$ is disjoint from that of $orch(M)$ and $interface(M)$ is compatible with $interface(S)$ w.r.t. r,*
- *μ is an interface match from $\langle interface(M), r \rangle$ to $interface(S)$,*
- *w is a wire connecting ports $\{P, P'\}$ such that $\Phi_w \vdash \langle \theta_r^M, \theta_{S\rightarrow}^S \circ \rho \rangle(\Psi_r^M)$, assuming that $\theta^M(r) = \langle c, P \rangle$ and $\theta^S(S^{\rightarrow}) = \langle c', P' \rangle$.*

Proposition and Definition 12 (Module Composition). *Let M be a service (resp. activity) module with interface i and $r \in M^{\leftarrow}$; $\langle S, \mu, w \rangle$ a service match for M w.r.t. r with $\mu = \langle \delta, \rho \rangle$ and $j = interface(S)$; Δ a solution for $(\mathcal{C}_{cs}^i \oplus_\rho \mathcal{C}_{cs}^j) \Downarrow_{\iota_j \circ \rho(\mathcal{C}_V^i)}$. The composition $M \oplus_{r:\mu,w,\Delta} S$ is the service (resp. activity) module with:*

- *$(i \parallel_{r:\mu,\Delta} j) \rhd_\theta (orch(M) \parallel_{\theta^i(r),w,\theta^j(j^{\rightarrow})} orch(S))$, where θ coincides with θ^i on the interface-points inherited from i and with θ^j on those inherited from j*
- *trigger and init have the conditions that are inherited from M and S.*

$M \oplus_{r:\mu,w} S$ *is the composition in which no additional constraints are imposed on the external services, i.e., $M \oplus_{r:\mu,w,\emptyset} S$.*

Fig. 3 illustrates the elements involved in the composition of CREDITSERVICE (presented before) and RISKEVALSERVICE. The interface of this new service has the provides-point *RECustomer*, with Φ_r including $\square(request_i \supset \bigcirc result!)$, and its constraint system includes the configuration variables $r.fee$ and $r.cfd$ constrained by $r.fee = -3 + \frac{3}{(1-r.cfd)}$. The orchestration of this service is provided by the single-component ARN with the component *RiskEvaluator* involved in the ARN presented before with $\kappa_{RECustomer}$ being the identity.

The match between the two services is given by the mappings δ: $getRisk \mapsto request$, $riskValue \mapsto result$ and ρ: $e.fee \mapsto r.fee$, $e.cfd \mapsto r.cfd$. The required property included in Φ_e translated by δ is $\square(request_i \supset \Diamond result!)$ which is trivially entailed by Φ_r. For the composition of the two services, we take the wire w_{ce} also used in the ARN presented in Fig. 1 (its properties entail $\delta(\Psi_e)$) and the constraint $cfd = 0.9095$. This confidence level implies that the *fee* is approximately 30. This pair of values is one that provides the best level of consistency among the solutions for $(\mathcal{C}_{sla}(\text{CREDITSERVICE}) \oplus_\rho \mathcal{C}_{sla}(\text{RISKEVALSERVICE})) \Downarrow_{\{fee,cfd\}}$.

The result of this composition is a service module whose interface has two uses- and two provides-points (inherited from CREDITSERVICE), the variables

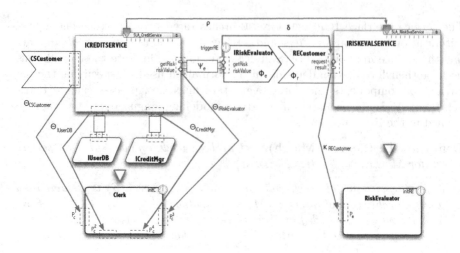

Fig. 3. Example of a service match

c.amount, *fee* and *cfd* subject to the constraints inherited from the two interfaces and also $cfd = 0.9095 \land fee = 30.14917$. The service is orchestrated by the ARN presented in Fig. 1.

This example illustrates how the proposed notion of composition of services can be used for obtaining more complex services from simpler ones. The service provider of *IRiskEvaluator* was chosen at design-time as well as the SLA and the result of this choice was made available in a new service CREDITSERVICE ⊕ RISKEVALSERVICE.

3   Business-Reflective Configurations

As mentioned before, component nets define configurations of global computers. In order to account for the way configurations evolve, it is necessary to consider the states of the configuration elements and the steps that they can execute. For this purpose, we take that every component $c \in COMP$ and wire $w \in WIRE$ of a component net may be in a number of states, the set of which is denoted by $STATE_c$ and $STATE_w$, respectively.

Definition 13 (State Configuration). *A state configuration $\mathcal{F}$ is a pair $\langle \alpha, \mathcal{S} \rangle$, where $\alpha = \langle C, W, \gamma, \mu \rangle$ is a component net and $\mathcal{S}$ is a configuration state, i.e., a mapping that assigns an element of $STATE_c$ to each $c \in C$ and of $STATE_w$ to each $w \in W$.*

A state configuration $\langle \alpha, \mathcal{S} \rangle$ may change in two different ways: (1) A state transition from $\mathcal{S}$ to $\mathcal{S}'$ can take place within α – we call such transitions *execution steps*. An execution step involves a local transition at the level of each component and wire, though some may be idle; (2) Both a state transition from $\mathcal{S}$ to $\mathcal{S}'$ and a change from α to another component net α' can take place – we call such transitions *reconfiguration steps*. In this paper, we are interested in the *reconfigurations steps* that happen when the execution of business activities triggers the discovery and binding to other services. In order to determine how state configurations of global computers evolve, we need a more sophisticated typing mechanism that goes beyond the typing of the individual components and wires: we need to capture the business activities that perform in a state configuration. We achieve this by typing the sub-configurations that, in a given state, execute the activities with activity module, thus making the configurations reflective.

Business configurations need also to include information about the services that are available in a given state (those that can be subject to procurement). We consider a space $\mathcal{U}$ of service and activity identifiers (e.g., URIs) to be given and, for each service and activity that is available, the configuration has information about its module and for each uses-point u: (i) the component c_u in the configuration to which u must be connected and (ii) a set of pairs of ports and wires available for establishing a connection with c_u. This information about uses-points of modules captures a 'direct binding' between the need of u and a given provider c_u, reflecting the fact that composition at uses-points is integration-oriented. The multiplicity of pairs of ports and wires opens the possibility of having a provider c_u serving different instances of a service at the same time.

We also consider a space $\mathcal{A}$ of business activities to be given, which can be seen to consist of reference numbers (or some other kind of identifier) such as the ones that organisations automatically assign when a service request arrives.

Definition 14 (Business Configuration). *A business configuration is* $\langle \mathcal{F}, \mathcal{P}, \mathcal{B}, \mathcal{C} \rangle$ *where*

- *$\mathcal{F}$ is a state configuration*
- *$\mathcal{P}$ is a partial mapping that assigns to services $s \in \mathcal{U}$, a pair*
  $$\langle \mathcal{P}_M(s), \{\mathcal{P}^u(s) : u \in \mathcal{P}_M(s)^{\downarrow}\} \rangle$$
  where $\mathcal{P}_M(s)$ is a service module and $\mathcal{P}^u(s)$ consists of a node c^u in $\mathcal{F}$ (i.e., a component instance) and a set of pairs $\langle P_i^u, w_i^u \rangle$, where each P_i^u is a port of γ_{c^u} distinct from the others and w_i^u is a wire connecting P_i^u and the port of u that satisfies the properties Ψ^u imposed by $\mathcal{P}_M(s)$. The services and activities in the domain of this mapping are those that are available in that state.
- *$\mathcal{B}$ is a partial mapping that assigns an activity module $\mathcal{B}(a)$ to each activity $a \in \mathcal{A}$ (the workflow being executed by a in $\mathcal{F}$). We say that the activities in the domain of this mapping are those that are active in that state.*
- *$\mathcal{T}$ is a mapping that assigns an homomorphism $\mathcal{T}(a)$ of graphs $\mathrm{orch}(\mathcal{B}(a)) \rightarrow \mathcal{F}$ to every activity $a \in \mathcal{A}$ that is active in $\mathcal{F}$. We denote by $\mathcal{F}(a)$ the image of $\mathcal{T}(a)$ – the sub-configuration of $\mathcal{F}$ that corresponds to the activity a.*

Let us consider a configuration in which CREDITSERVICE (presented before) and CREDITACTIVITY are available. The latter is an activity that the same provider makes available in order to serve requests that are placed, not by other services,

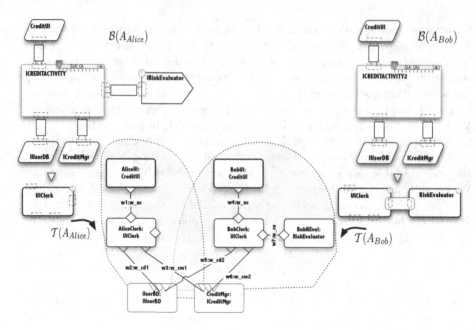

Fig. 4. Excerpt of a business configuration

but by applications that interact with users (e.g., a web application that supports online credit requests). Suppose that the configuration also defines that the uses-points *IUserDB* and *ICreditMgr* of both interfaces should be connected, respectively, to *UserDB* and *CreditMgr* – a database of users and a manager of approved credit requests that are shared by all instances of this service and activity. The other elements of the business configuration are (partially) described in Fig. 4. It is not difficult to recognise that there are currently two active business activities – A_{Alice} and A_{Bob}. Intuitively, both correspond to two instances of the same business logic (two customers requesting a credit using the same business activity) but at different stages of their workflow: one (launched by *BobUI*) is already connected to a risk evaluator (*BobREval*) while the other (launched by *AliceUI*) has still to discovery and bind to a risk evaluator service. The active computational ensemble of component instances that collectively pursue the business goal of each activity in the current state are highlighted through a dotted line.

4   Service Discovery and Binding

Every activity module declares a triggering condition for each requires-point, which determines when a service needs to be discovered and bound to the current configuration through that point. Let $\mathcal{L}=\langle\mathcal{F},\mathcal{P},\mathcal{B},\mathcal{C}\rangle$ be the current business configuration. The discovery of a service for a given activity a and requires-point r of $\mathcal{B}(a)$ consists of several steps. First, it is necessary to find, among the services that are available in $\mathcal{L}$, those that are able to guarantee the properties associated with r in $\mathcal{B}(a)$ and with which it is possible to reach a service-level agreement. Then, it is necessary to rank the services thus obtained, i.e., to calculate the most favourable service-level agreement that can be achieved with each S – the contract that will be established between the two parties if S is selected. The last step is the selection of one of the services that maximises the level of satisfaction offered by the corresponding contract.

Definition 15 (discover($M, r, \mathcal{P}$)). *Let $\mathcal{P}$ be a mapping as in Def. 14, M an activity module and $r \in M^{\leftarrow}$.* **discover($M, r, \mathcal{P}$)** *is the set of tuples $\langle s, \langle \delta, \rho \rangle, w, \Delta \rangle$ such that:*

1. *$s \in \mathcal{U}$ and $S = \mathcal{P}_M(s)$ is defined;*
2. *$\langle S, \langle \delta, \rho \rangle, w \rangle$ is a service match for M w.r.t. r;*
3. *Δ is a solution for $(\mathcal{C}_{sla}(M) \oplus_\rho \mathcal{C}_{sla}(S)) \Downarrow_{\iota_S \circ \rho(\mathcal{C}_V^M)}$ and $blevel(\mathcal{C}_{sla}(M) \oplus_\rho \mathcal{C}_{sla}(S) \cup \{\Delta\})$ is greater than or equal to the value obtained for any other solution of that set of constraints;*
4. *$blevel(\mathcal{C}_{sla}(M) \oplus_\rho \mathcal{C}_{sla}(S) \cup \{\Delta\})$ is greater than or equal to the value obtained for any other tuple $\langle s', \delta', \rho', \Delta' \rangle$ satisfying the conditions 1-3, above.*

The discovery process for an activity module and one of its requires-points r also provides us with a wire to connect r with the provides-point of the discovered service. By Def. 11, this wire guarantees the properties associated with r in $\mathcal{B}(a)$.

The process of binding an activity to a discovered service for one of its requires-points can now be defined:

Definition 16 (Service Binding). *Let $\mathcal{L} = \langle \mathcal{F}, \mathcal{P}, \mathcal{B}, \mathcal{T} \rangle$ be a business configuration with $\mathcal{F} = \langle \alpha, \mathcal{S} \rangle$, a an active business activity in $\mathcal{L}$ and $r \in \mathcal{B}(a)^{\leftarrow}$.*

- *If $\mathcal{F}(a) \models trigger_{\mathcal{B}(a)}(r)$ and* **discover($\mathcal{B}(a), r, \mathcal{L}$)** $\neq \emptyset$, *then binding $\mathcal{B}(a)$ to r using any of the elements in* **discover($\mathcal{B}(a), r, \mathcal{P}$)** *is enabled in $\mathcal{L}$.*
- *Binding $\mathcal{B}(a)$ to r using $\langle s, \langle \delta, \rho \rangle, w, \Delta \rangle \in$* **discover($\mathcal{B}(a), r, \mathcal{L}$)**, *and assuming that $\mathcal{P}_M(s) = S$, induces a business configuration $\langle \langle \alpha', \mathcal{S}' \rangle, \mathcal{P}, \mathcal{B}', \mathcal{T}' \rangle$ such that:*
  - *$\mathcal{B}'(x) = \mathcal{B}(x)$, if $x \neq a$ and $\mathcal{B}'(a)$ is the activity module $\mathcal{B}(a) \oplus_{r:\mu, w, \Delta} S$*
  - *if $\theta^{\mathcal{B}(a)}(r) = \langle c, P \rangle$ and $\theta^S(S^{\rightarrow}) = \langle c', P' \rangle$, α' is $\alpha \parallel_{\Xi} \alpha_S$ where*
    - *α_S is a component net obtained by renaming the nodes in $orch(S)$ in such a way this set becomes disjoint from the set of nodes of $orch(\mathcal{B}(a))$,*
    - *c'' is the node in α_S corresponding to c',*
    - *$\Xi = \{\langle \langle \langle \mathcal{T}(c), P \rangle, w, \langle c'', P' \rangle \rangle, \langle \langle c^u, P_i^u \rangle, w_i^u, \theta^S(u) \rangle \rangle : u \in S^{\uparrow} \}$ where c^u is the component identified by $\mathcal{P}^u(s)$, P_i^u is a port in $\mathcal{P}^u(s)$ that is still available for connection and w_i^u is the corresponding wire, as defined by $\mathcal{P}^u(s)$.*
  - *$\mathcal{S}'$ coincides with $\mathcal{S}$ in the nodes of α and assigns, to every node c in α_S, a state that satisfies $init_S(c)$.*
  - *$\mathcal{T}'$ is the homomorphism that results from updating $\mathcal{T}$ with the renaming from $orch(S)$ to α_S.*

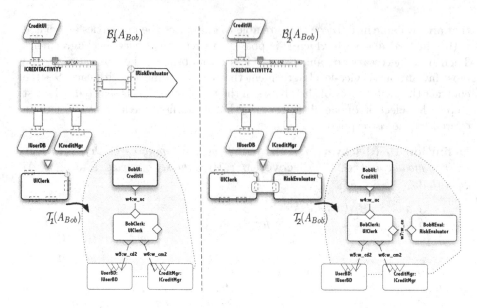

Fig. 5. Reconfiguration step induced by a service binding

That is to say, new instances of components and wires of S are added to the configuration while existing components are used for uses-interfaces, according to the direct bindings defined in the configuration.

Figure 5 illustrates the binding process. On the left side is depicted part of the business configuration $\mathcal{L}_1$, before the activity A_{Bob} has bound to a risk evaluator service (activity A_{Alice} was not yet active at this time). On the right side is depicted the result of the binding of *IRiskEvaluator* to the service RISKEVALSERVICE using the service match presented in Fig. 3. This means that this service is, among all the services available in $\mathcal{P}_1$ that fit the purpose, one of the services that best fits the quality-of-service constraints. According to what discussed in Sec. 2.3, the contract established between the two parties is a confidence level of 0.9095 and a fee of 30.14917.

5   Conclusions

In this paper, we presented an overview of a graph-based framework for the design of service-oriented systems developed around the notion of service and activity module. Service modules, introduced in [14], were originally inspired by concepts proposed in SCA [22]. They provide formal abstractions for composite services whose execution involves a number of external parties that derive from the logic of the business domain. Our approach has also been influenced by algebraic component frameworks for system modelling [10] and architectural modelling [2]. In those frameworks, components are, like services, self-contained modelling units with interfaces describing what they require from the environment and what they themselves provide. However, the underlying composition model is quite different as, unlike components, services are not assembled but, instead, dynamically discovered and bound through QoS-aware composition mechanisms. Another architectural framework inspired by SCA is presented in [25]. This framework is also language independent but its purpose is simply to offer a meta-model that covers service-oriented modelling aspects such as interfaces, wires, processes and data.

Our notion of service module also builds on the theory of interfaces for service-oriented design proposed in [13], itself inspired by the work reported in [9] for component based design. Henzinger and colleagues also proposed a notion of interface for web-services [4] but the underlying notion of composition, as in component-based approaches, is for integration.

We presented a mathematical model that accounts for the evolutionary process that SOC induces over software systems and used it to provide an operational semantics of discovery, instantiation and binding. This model relies on the mechanism of reflection, by which configurations are typed with models of business activities and service models. Reflection has been often used as a means of making systems adaptable through dynamic reconfiguration (e.g. [8,19,20]). A more detailed account of the algebraic properties of this model can be found in [15].

The presented framework was defined in terms of abstractions like *COMP*, *PORT* and *WIRE* so that the result was independent of the nature of components, ports and wires and of the underlying computation and communication model. In the same way, we have considered that the behavioural constraints imposed over the interface-points were defined in terms of a specification logic *SPEC*. A large number of formalisms have been proposed for describing each of these concepts in the context of SOC – e.g., process-calculi [7,18,26], automata-based models [3,23] and models based on Petri-nets [21,24]. An example of the instantiation of the framework is provided by the language SRML [12,16], a modelling language of service-oriented systems we have developed in the context of SENSORIA project. This modelling language is equipped with a logic for specifying stateful, conversational interactions, and a language and semantic model for the orchestration of such interactions. Examples of quantitative and qualitative analysis techniques of service modules modelled in SRML can be found in [1,6].

References

1. Abreu, J., Mazzanti, F., Fiadeiro, J.L., Gnesi, S.: A Model-Checking Approach for Service Component Architectures. In: Lee, D., Lopes, A., Poetzsch-Heffter, A. (eds.) FMOODS/FORTE 2009. LNCS, vol. 5522, pp. 219–224. Springer, Heidelberg (2009)
2. Allen, R., Garlan, D.: A formal basis for architectural connection. ACM Trans. Softw. Eng. Methodol. 6(3), 213–249 (1998)
3. Benatallah, B., Casati, F., Toumani, F.: Web service conversation modeling: A cornerstone for e-business automation. IEEE Internet Computing 8(1), 46–54 (2004)
4. Beyer, D., Chakrabarti, A., Henzinger, T.A.: Web service interfaces. In: Ellis, A., Hagino, T. (eds.) WWW, pp. 148–159. ACM (2005)

5. Bistarelli, S., Montanari, U., Rossi, F.: Semiring-based constraint satisfaction and optimization. J. ACM 44(2), 201–236 (1997)

6. Bocchi, L., Fiadeiro, J.L., Gilmore, S., Abreu, J., Solanki, M., Vankayala, V.: A formal approach to modelling time properties of service oriented systems. In: Handbook of Research on Non-Functional Properties for Service-Oriented Systems: Future Directions. Advances in Knowledge Management Book Series. IGI Global (in print)

7. Carbone, M., Honda, K., Yoshida, N.: Structured Communication-Centred Programming for Web Services. In: De Nicola, R. (ed.) ESOP 2007. LNCS, vol. 4421, pp. 2–17. Springer, Heidelberg (2007)

8. Coulson, G., Blair, G.S., Grace, P., Taïani, F., Joolia, A., Lee, K., Ueyama, J., Sivaharan, T.: A generic component model for building systems software. ACM Trans. Comput. Syst. 26(1) (2008)

9. de Alfaro, L., Henzinger, T.A.: Interface Theories for Component-Based Design. In: Henzinger, T.A., Kirsch, C.M. (eds.) EMSOFT 2001. LNCS, vol. 2211, pp. 148–165. Springer, Heidelberg (2001)

10. Ehrig, H., Orejas, F., Braatz, B., Klein, M., Piirainen, M.: A component framework for system modeling based on high-level replacement systems. Software and System Modeling 3(2), 114–135 (2004)

11. Fiadeiro, J.L.: Categories for Software Engineering. Springer (2004)

12. Fiadeiro, J.L., Lopes, A.: A Model for Dynamic Reconfiguration in Service-Oriented Architectures. In: Babar, M.A., Gorton, I. (eds.) ECSA 2010. LNCS, vol. 6285, pp. 70–85. Springer, Heidelberg (2010)

13. Fiadeiro, J.L., Lopes, A.: An Interface Theory for Service-Oriented Design. In: Giannakopoulou, D., Orejas, F. (eds.) FASE 2011. LNCS, vol. 6603, pp. 18–33. Springer, Heidelberg (2011)

14. Fiadeiro, J.L., Lopes, A., Bocchi, L.: A Formal Approach to Service Component Architecture. In: Bravetti, M., Núñez, M., Zavattaro, G. (eds.) WS-FM 2006. LNCS, vol. 4184, pp. 193–213. Springer, Heidelberg (2006)

15. Fiadeiro, J.L., Lopes, A., Bocchi, L.: An abstract model of service discovery and binding. Formal Asp. Comput. 23(4), 433–463 (2011)

16. Fiadeiro, J., Lopes, A., Bocchi, L., Abreu, J.: The SENSORIA Reference Modelling Language. In: Wirsing, M., Hölzl, M. (eds.) SENSORIA Project. LNCS, vol. 6582, pp. 61–114. Springer, Heidelberg (2011)

17. Fiadeiro, J., Sernadas, A.: Structuring Theories on Consequence. In: Sannella, D., Tarlecki, A. (eds.) Abstract Data Types 1987. LNCS, vol. 332, pp. 44–72. Springer, Heidelberg (1988)

18. Kitchin, D., Quark, A., Cook, W., Misra, J.: The Orc Programming Language. In: Lee, D., Lopes, A., Poetzsch-Heffter, A. (eds.) FMOODS/FORTE 2009. LNCS, vol. 5522, pp. 1–25. Springer, Heidelberg (2009)

19. Kon, F., Costa, F.M., Blair, G.S., Campbell, R.H.: The case for reflective middleware. Commun. ACM 45(6), 33–38 (2002)

20. Léger, M., Ledoux, T., Coupaye, T.: Reliable Dynamic Reconfigurations in a Reflective Component Model. In: Grunske, L., Reussner, R., Plasil, F. (eds.) CBSE 2010. LNCS, vol. 6092, pp. 74–92. Springer, Heidelberg (2010)

21. Martens, A.: Analyzing Web Service Based Business Processes. In: Cerioli, M. (ed.) FASE 2005. LNCS, vol. 3442, pp. 19–33. Springer, Heidelberg (2005)

22. OSOA. Service component architecture: Building systems using a service oriented architecture (2005), White paper available from http://www.osoa.org

23. Ponge, J., Benatallah, B., Casati, F., Toumani, F.: Analysis and applications of timed service protocols. ACM Trans. Softw. Eng. Methodol. 19(4), 11:1–11:38 (2010)
24. Reisig, W.: Towards a Theory of Services. In: Kaschek, R., Kop, C., Steinberger, C., Fliedl, G. (eds.) UNISCON 2008. LNBIP, vol. 5, pp. 271–281. Springer, Heidelberg (2008)
25. van der Aalst, W., Beisiegel, M., van Hee, K., Konig, D.: An SOA-based architecture framework. Journal of Business Process Integration and Management 2(2), 91–101 (2007)
26. Vieira, H.T., Caires, L., Seco, J.C.: The Conversation Calculus: A Model of Service-Oriented Computation. In: Drossopoulou, S. (ed.) ESOP 2008. LNCS, vol. 4960, pp. 269–283. Springer, Heidelberg (2008)

Evolutionary Togetherness: How to Manage Coupled Evolution in Metamodeling Ecosystems

Davide Di Ruscio, Ludovico Iovino, and Alfonso Pierantonio

Dipartimento di Ingegneria e Scienze dell'Informazione e Matematica
Università degli Studi dell'Aquila
Via Vetoio, L'Aquila, Italy
{davide.diruscio,ludovico.iovino,alfonso.pierantonio}@univaq.it

Abstract. In Model-Driven Engineering (MDE) metamodels are cornerstones for defining a wide range of related artifacts interlaced with explicit or implicit correspondences. According to this view, models, transformations, editors, and supporting tools can be regarded as a whole pursuing a common scope and therefore constituting an *ecosystem*. Analogously to software, metamodels are subject to evolutionary pressures too. However, changing a metamodel might compromise the validity of the artifacts in the ecosystem which therefore require to *co-evolve* as well in order to restore their validity.

Different approaches have been proposed to support at different extent the adaptation of artifacts according to the changes operated on the corresponding metamodels. Each technique is specialized in the adaptation of specific kind of artifact (e.g., models, or transformations) by forcing modelers to learn different technologies and languages. This paper discusses the different relations occurring in a typical metamodeling ecosystem among the metamodel and the related artifacts, and identifies the commonalities which can be leveraged to define a unifying and comprehensive adaptation process. A language and corresponding supporting tools are also proposed for the management of metamodel evolution and the corresponding *togetherness* with the related artifacts.

1 Introduction

Model-Driven Engineering (MDE) [1] is increasingly emerging as a software discipline which employs metamodels to engineer domains. They permit to formalize *problems* in a given domain by means of 'upstream' models, which can be regarded as approximation of reality [2]: each problem can then be step-wise refined into 'downstream' models corresponding to the *solution*.

Admittedly, metamodels play a precise and formal role in the development of models and transformations because of the conformance and domain conformance relations [3], respectively. However, the entities which are defined upon metamodels are numerous and include editors, models for concrete syntaxes, model differencing and versioning, and many more. The nature of the dependencies existing among metamodels and such components is often explicit, formal, and well-known, other times depending on the artifact it is implicit, intricate, or blurred, if not obscured. All these constituent elements

H. Ehrig et al.(Eds.): ICGT 2012, LNCS 7562, pp. 20–37, 2012.

and their interlaced correspondences form what we like to call a *metamodeling ecosystem*, i.e., a metamodel-centered environment whose entities are traditionally subject to distinct evolutionary pressures but cannot have independent life-cycles[1]. In essence, the way the ecosystem can evolve is not always cooperative and coordinated since metamodels have two different categories of competing *clients*:

- the modelers which are continuously attempting to rework the metamodel in order to accommodate additional requirements deriving from new insights emerged from the domain [5], and
- the artifacts whose definition is implicitly or explicitly depending on the metamodel and therefore rely on its immutability [3].

Clearly, the second of the two might represent a serious impediment to possible changes in the metamodel as any modification in the metamodel tend to render the artifacts not valid any longer. As a consequence, this can easily culminate in having the metamodel locked in the ecosystem. In order to let the artifacts remain valid after a metamodel undergoes modifications, they need to be consistently *adapted*. Naturally, the modeler and the implementors can always adapt the models, the transformations and any other tool in the ecosystem by inspecting the artifacts, detecting the necessary refactorings, and finally applying them with manual operations. However, carrying on this activity without specialized tools and techniques presents intrinsic difficulties. The intricacy of metamodel changes on one hand, but also the size and diversity of the artifacts on the other hand, can rapidly affect the accuracy and precision of the adaptations [6]. In fact, if the adaptation is based on spontaneous and individual skills and pursued without any automated support, it can easily give place to inconsistencies and lead to irremediable information erosion [7].

Over the last year, different proposals have been made to mitigate this problem by supporting the automated adaptation of modeling artifacts. However, each approach is specialized in adapting a specific kind of artifact (e.g., models, or transformations) by forcing modelers to learn different languages and technologies. In order to let the metamodels freely evolve without - at a certain extent - compromising the related artifacts and tools, we aim at realizing an *evolutionary togetherness* by endowing the metamodel and consequently the ecosystem with a migration infrastructure capable of preserving the validity of the components. To this end, the paper discusses the different relationships occurring in a typical metamodeling ecosystem, and identifies the commonalities which can be exploited to define a systematic adaptation process. By leveraging such commonalities, a language and corresponding supporting tools are proposed for a comprehensive and uniform management of metamodel evolution. Migration programs are therefore used to let the artifacts uniformly co-evolve throughout the ecosystem.

The paper is structured as follows: Sect. 2 presents the problem of coupled evolution in MDE. Section 3 is an overview of existing approaches which have been conceived to support the adaptation of different modeling artifacts, and identify the commonalities of both the problems and adaptations which can be leveraged to define EMFMigrate,

---

[1] A more comprehensive definition of software ecosystem considers it as a set of software solutions that enable, support and automate the activities and transactions by the actors in the associated social or business ecosystem and the organizations that provide these solution [4].

a uniform approach to coupled evolution presented in Sect. 4. The supporting tools for such a new approach are presented in Sect. 5. Conclusions and future work are discussed in Sect. 6.

2   Coupled Evolution in MDE

As aforementioned, metamodels are fundamental ingredients of an ecosystem and they rarely exist in isolation. Resolving the pragmatic issues of their evolutionary nature is core to improving the overall maintainability. Evolution may happen at different levels as once the metamodel changes it causes a ripple of adaptations to be operated throughout the ecosystem. Such adaptations are different and must be individually but coherently designed as each of them is traditionally not unequivocal [3]. On the contrary, if the problem is inaccurately or wrongly handled, it can easily let the system slide towards a lock-in situation where the metamodel becomes easily immutable and resilient to variations [8]. The adaptation process can be regarded as a three-steps process as

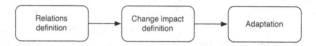

Fig. 1. The adaptation process

illustrated in Fig. 1:

1. *Relations definition*. A set of relations between the metamodel and the other modeling artifacts are identified. Intuitively, those relations can be considered as dependencies between artifacts, and they play a role similar to that of tracing information between source and target models of a model-to-model transformation.
2. *Change impact detection*. In this step the relationships defined in step 1 can be considered in order to assess the impact on the related artifacts of the changes made in the metamodel.
3. *Adaptation*. In this step the developer apply some adaptation actions on the (possibly corrupted) artifacts. This step can imply the use of very different adaptation policies, depending on the types of artifacts to be adapted.

Current practices usually end up with blurring the distinction between impact assessment and adaptation semantics. Thus, it is important to clarify the nature of the relation which exists between the metamodel and the other artifacts. Indeed, dependencies emerge at different stages during the metamodel life-cycle, and with different degrees of causality depending on the nature of the considered artifact. For instance, by referring to Fig. 2, there may be a *transformation* that takes an *input* model and produces an *output* model, each conforming to a metamodel; also a graphical or textual *editor* to properly edit models or other kinds of artifacts may be utilized, and all of them are typically related to the metamodel at different extent.

Being more precise, metamodels may evolve in different ways: some changes may be additive and independent from the other elements, thus requiring no or little co-changes. However, in other cases metamodel manipulations introduce incompatibilities

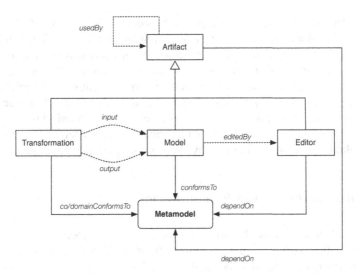

Fig. 2. A sample of an ecosystem with some artifacts and relations

and inconsistencies which can not be always easily (and automatically) resolved. In particular, the co-changes which are required to adapt modeling artifacts according to the changes executed on the corresponding metamodel, depend on the relation which couples together the modeling artifact and the metamodel. According to our experience and to the literature (e.g., [2,9,10,11,12]), at least the following relations are involved in the metamodel co-evolution problem:

- *conformsTo:* it holds between a model and a metamodel, it can be considered similar to a typing relation. A model *conforms to* a metamodel, when the metamodel specifies every concept used in the model definition, and the models uses the metamodel concepts according to the rules specified by the metamodel [2];
- *domainConformsTo*: it is the relation between the definition of a transformation and the metamodels it operates on. For instance, a sample domain conformance constraint might state that the source elements of every transformation rule must correspond to a metaclass in the source metamodel [10] and same for target metamodel elements;
- *dependsOn*: it is a generic and likely the most complex relation, since it occurs between a metamodel and modeling artifacts, which do not have a direct and a well-established dependence with the metamodel elements. For instance, in case of GMF [12] models, some of them do not refer directly to the elements specified in the metamodel, even though some form of consistency has to be maintained in order to do not generate GMF editors with limited functionalities.

In particular, the *dependsOn* relation denotes a large class of possible dependencies between metamodels and artifacts. In Fig. 2, for instance, the correspondence between the editor and the metamodel is one of such cases, it is clearly a dependency as the editor is specifically design for editing models conforming to that given metamodel but it is also a looser relation if compared with the conformance. As a result, its management is more challenging and requires a deeper understanding of the established relations. Additionally, dotted arrows are denoting a usage relation, which describes how a given artifact can be consumed by other artifacts, as for instance a transformation consuming a model in order to produce another one, or an editor designed to edit models.

The different coupled evolution scenarios occurring because of the relations described above are discussed in the next section. We are aiming at identifying and leveraging the commonalities of the approaches in order to devise a unifying paradigm to manage the evolutionary togetherness of metamodeling ecosystems. Therefore, each coupled evolution problem is discussed by outlining a representative solution.

3   Existing Coupled Evolution Scenarios

As said, each different artifact has to be *ad-hoc* adapted in response to a given metamodel modification. Hence, models, transformations, and any other metamodel-based tools require a different support to their automated adaptation. In this section, we present different coupled evolution problems in terms of the available techniques. In particular, this section is organized to discuss separately the metamodel/model (Sect. 3.1), metamodel/transformation (Sect. 3.2), and metamodel/editor (Sect. 3.3) coupled evolution problems.

During the discussions the typical explanatory PetriNet metamodel evolution shown in Fig. 3 is considered. In the evolved version of the metamodel shown in Fig. 3.b the metaclasses *Arc*, *PlaceToTransition*, and *TransitionToPlace* have been added, and other changes have been executed, i.e., the merging of the references *places* and *transitions* into the new *elements*, and the renaming of the metaclass *Net* as *PetriNet*. Such metamodel evolution might affect different existing artifacts defined on the initial version of the PetriNet metamodel, thus proper adaptation techniques have to be employed as discussed in the sequel.

3.1   Metamodel/Model Coupled Evolution

As mentioned in the previous section, when a metamodel evolves, existing models may no longer conform to the newer version of the metamodel. For instance, Fig. 4 shows a simple model conforming to the first version of the PetriNet metamodel in Fig. 3. Such model represents a PetriNet (see the *Net* element named *net1*) consisting of three places and one transition. Because of the performed metamodel modifications, the model in Fig. 4 has to be migrated in order to adapt the element named *net1* (since the metaclass *Net* has been renamed in *PetriNet*), and to add *PlaceToTransition* and *TransitionToPlace* elements to connect the existing places and transition elements, which cannot be directly related according to the new version of the metamodel.

Over the last years, different approaches have been defined to automate the migration of models, and all of them can be classified according to the following categories [13]:

- *manual specification* approaches (e.g., [14,15]) provide transformation languages to manually specify model migrations;

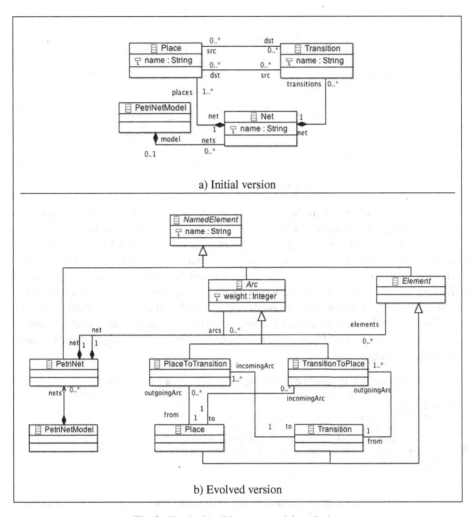

Fig. 3. Simple PetriNet metamodel evolution

- *operator-based* approaches (e.g., [16,7]) provide coupled operators that permit to specify metamodel changes together with the corresponding migrations to be operated on the existing models;
- *metamodel matching* approaches (e.g., [9,17]) are able to generate model transformations from difference models representing the changes between subsequent versions of the same metamodel.

As discussed in [13] it is impossible to identify the best tool for supporting the coupled evolution of metamodels and models. Each tool has strengths and weaknesses and depending on the particular situation at stake (e.g., frequent, and incremental coupled evolution, minimal guidance from user, and unavailability of the metamodel change history) some approaches can be preferred with respect to others.

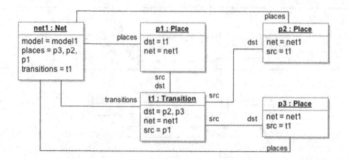

Fig. 4. Petrinet model sample

Just to give an example, in the following we consider Flock [14], one of the approaches belonging to the *manual specification* category previously mentioned. It is an EMF-based textual language that permits the specification of model migration strategies. When a migration strategy is executed, some parts of the migrated model are derived directly from the original model, other parts are derived from user-defined rules. In Listing 1.1 a simple Flock migration program is specified in order to adapt models which have been affected by the metamodel evolution shown in Fig. 3. In particular, for each *Net* element in the source model, a target *PetriNet* element is generated (see lines 3-4). Moreover, the source *Place* and *Transition* elements will be simply copied to the new model (see lines 5-6).

Listing 1.1. Fragment of Flock model migration for Petrinet example

```
1 ...
2 migrate Nets{
3     for (net in petrinetmodel.nets){
4         var petrinet = new Migrated!PetriNet;
5         petrinet.places := net.places.equivalent();
6         petrinet.transitions := net.transitions.equivalent();
7     }
8 }
9 ...
```

Flock has been applied on different metamodel/model coupled evolution situations. However, more efforts are still required to provide some native mechanisms for supporting the reuse of already developed migrations, and to support user's intervention when required.

3.2 Metamodel/Transformation Coupled Evolution

Because of changes operated on a given metamodel, model transformation inconsistencies can occur and are those elements in the transformation, which do not longer satisfy the *domain conformance* relation discussed in the previous section. The metamodel/-transformation coupled evolution problem is less investigated than the metamodel/model one, partly because is intrinsically more difficult, and because the metamodel/model problem has been considered first by natural choice. In [11] the authors propose a set of tasks that should be performed in order to re-establish the transformation consistency after metamodel evolution. In particular, the authors claim that the transformation migration should be performed by means of three phases [11]:

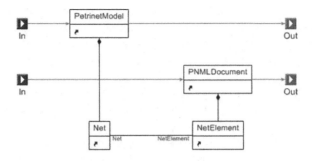

Fig. 5. GReAT transformation example Petrinet to PNML

– *impact detection*, to identify the transformation inconsistencies caused by meta-model evolution;
– *impact analysis*, to obtain - possibly by using human assistance - the set of transformation updates to re-establish the domain conformance;
– *transformation adaptation*, the updates identified in the previous phase are really applied.

Besides such an exploratory work, another attempt has been proposed by Levendovszky et al. in [18]. They propose an approach based on higher-order transformations (HOTs) able to automatically migrate, when possible, existing transformations according to occurred metamodel changes. If automatic adaptations can not be performed, user interventions are demanded. The approach is specifically tailored for managing transformations developed in the GME/GReAT toolset[2]. The approach is able to partially automate the adaptation, and the developed algorithms alert the user about missing information, which can then be provided manually after the execution of the automatic part of the evolution. A simple GReAT transformation is shown in Fig. 5 which specifies how to create a PNML[3] document (conforming to the metamodel shown in Fig. 6) out of a source PetriNet model.

Because of the metamodel evolution shown in Fig. 3, the simple GReAT transformation needs to be adapted since the *Net* element is no longer existing in the new version of the metamodel. In [18] the Model Change Language (MCL) is proposed to specify metamodel evolutions and automatically generate the corresponding migration. Thus, MCL rules, like the one in Fig. 7, are the input for the generation of the HOT able to adapt the transformations which have to be migrated. Essentially, the simple MCL rule shows how the *Net* metaclass in the original transformation has to be mapped into the adapted one.

As said above, the outlined metamodel/transformation coupled evolution approach has been a first attempt to support such complex coupled evolution scenario. However, the approach misses adequate support for reusing and customizing developed higher-order transformations. Moreover, the presented adaptations are limited to not too complex cases and leaved the responsibility of more complex adaptations to the designer.

---

[2] www.isis.vanderbilt.edu/Projects/gme/
[3] Petrinet Markup Language

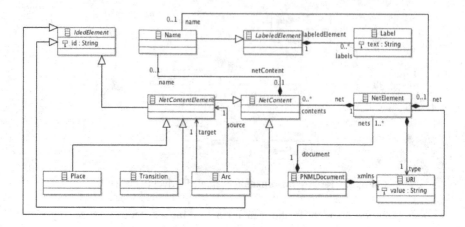

Fig. 6. PNML metamodel

3.3   Metamodel/Editor Coupled Evolution

As already discussed, metamodels underpin the development of a wide number of related artifacts in addition to models and transformations. For instance, in the definition
of domain specific modeling languages (DSMLs), metamodels play a central role since
they define the abstract syntaxes of the languages being developed. A number of other
related artifacts are produced in order to define the concrete syntaxes and possibly further aspects related to semantics or requirements of a particular DSML tool. In such
cases, specific techniques are required to propagate any abstract-syntax changes to the
dependent artifacts, e.g., models of the graphical or concrete syntaxes.

In the Eclipse Modeling Framework (EMF) [19] different approaches have been proposed to define concrete syntaxes of modeling languages, e.g., GMF [12] for developing
graphical editors, EMFText [20], TCS [21], and XText [22] for producing textual editors. Essentially, all of them are generative approaches able to generate working editors
starting from source specifications that at different extent are related to the abstract
syntax of the considered DSML. The relation between the metamodel of the language
and the editor models is *weaker* than the *conformance* and *domain conformance*. Thus,
even the detection of the inconsistencies between the editor models and the new version
of the metamodel is more difficult to be determined. Besides, even if we could detect
such inconsistencies, re-establishing consistency requires a deep knowledge of the used
editor technology in order to properly propagate the metamodel changes to the editor
models, and in case also to the generated code.

Concerning the adaptation of GMF editors, in [8] the authors introduce the approach
shown in Fig. 8. The approach relies on three adapters able to automate the propagation

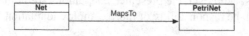

Fig. 7. Migration rule for Net

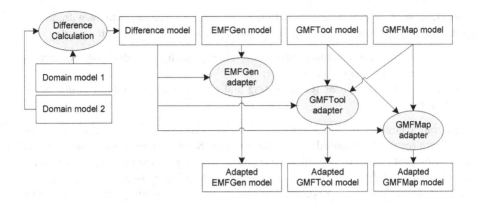

Fig. 8. Automated adaptation of GMF editor models

of domain-model changes (i.e., metamodel changes) to the *EMFGen*, *GMFTool*, and *GMFMap* models required by GMF to generate the graphical editor. In particular, *EMFGen* is a model used by the EMF generator to produce Java code required to manage models conforming to the metamodel of the considered modeling language. *GMFTool* defines toolbars and other periphery of the editor to facilitate the management of diagram content. *GMFMap* links together all the other GMF models. The adapters in Fig. 8 are special model-to-model transformations that are driven by a difference model which represents the differences between two subsequent version of the considered domain-model.

The approach is a first attempt the deal with the co-evolution challenge of GMF and more investigations are required. In particular, more efforts are required to achieve a full coverage of Ecore, the metamodeling language of EMF, and full understanding of the implicit semantics of GMF model dependencies and tools.

4   A Unifying Paradigm for Metamodeling Ecosystems Adaptation

Managing the coupled evolution problem is a complex task. However, with the acceptance of Model-Driven Engineering has come the urgent need to manage the complex change evolution within the modeling artifact in a more comprehensive and satisfactory manner. Different attempts have been proposed to cope with the intricacies raised by the adaptation of different kinds of artifacts. Thus, modelers have to learn different techniques, and notations depending on the artifacts that have to be adapted.

By taking into account the coupled evolution scenarios previously discussed, we believe that the metamodel co-evolution problem has to be managed in a more uniform way, i.e., with a unique technique, notation, and corresponding paradigm for any kind of migration problem regardless of the artifact type. In particular, we aim at a *declarative* approach as migration strategies are problems in a restricted domain and thus a programmatic approach based on a general-purpose language is not always necessary in our opinion. Since migration patterns occur repeatedly, the notation must allow the specification of default behaviors with the possibility to customize them with an emphasis on usability and readability. Being more precise, we would like to specify

- *relation libraries*, each formalizing a relation (like the ones discussed in Sect. 2) and the default adaptation behavior;
- *custom migration rules* for extending and customizing the default migration libraries.

A migration engine can then adapt existing artifacts by executing the proper migration rules with respect to occurred metamodel changes. Such changes can be automatically calculated (with state- or operation-based methods) or specified by the modeler assuring enough flexibility in choosing the most adequate solution. In any case, metamodel changes, regardless whether specified or calculated, should be properly represented by means of difference models (e.g., as in [23]) which are amenable to automatic manipulations. To satisfy such requirements, in Sect. 4.1 we describe the EMFMigrate approach. The provided support for the definition of migration libraries and corresponding customizations is discussed in Sect. 4.2

4.1   EMFMigrate Overview

EMFMigrate [3,24] is an attempt aiming at supporting the coupled evolution in a uniform and comprehensive way, in the sense that it is not restricted to specific kinds of artifacts. As shown in Fig. 9 the approach consists of a DSL which provides modelers with dedicated constructs for *i)* specifying migration libraries, which aim to embody and enable the reuse of recurrent artifact adaptations; *ii)* customizing migrations already available in libraries; and *iii)* managing those migrations which are not fully automated and that require user intervention. In other words, metamodel refactorings originate different adaptations depending on the kind of artifact to be kept consistent, with each adaptation formalized in a library. Recurrent adaptations are specified in default libraries which can be in turn customized in order to address ad-hoc needs.

A sample EMFMigrate migration program is shown in Fig. 10. It consists of a header (see lines 1-4) which defines the name of the migration (line 1), specifies the migration

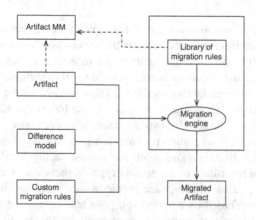

Fig. 9. Overview of EMFMigrate

library to be imported (line 2), refers the artifact to be adapted together with its meta-model, and the difference model representing the changes which have been operated on the modified metamodel (see lines 3-4). After the header section, a number of migration rules are specified (see lines 6-23). Each migration rule consists of a guard (e.g., see lines 7-11) and a body (e.g., see lines 12-15). The former filters the application of the latter. For instance, the migration rule RenameClass will be applied for all the occurrences of the metamodel change specified in lines 7-11, which represents meta-class renamings. For all renamed metaclasses, the body in lines 12-15 will be applied on the considered petrinet2PNML.atl ATL transformation. Essentially, the body of a migration rule consists of rewriting rules defined as follows

$$s \rightarrow t_1[assign_1]; t_2[assign_2]; \ldots t_n[assign_n]$$

where s, t_1, ..., t_n refer to metaclasses of the metamodel of the artifact to be adapted (in the example in Fig. 10, the ATL metamodel is considered). After the specification of the rewriting rules, a *where* condition can be specified (e.g., see line 15). It is a boolean expression which has to be *true* in order to execute all the specified rewriting rules. It is possible to specify the values of the target term properties by means of assignment operations (see $assign_i$ above). For instance, in case of the RenameClass rule in Fig. 10, the considered ATL transformation is migrated by rewriting its OclModelElement instances which have the same name of the matched class c involved in the renaming operation (see line 8). In particular, each OclModelElement instance is rewritten with a new OclModelElement having the property name set with the new name of the matched class c (see name <- %{newName} in line 13).

The migration in Fig. 10 is able to adapt the transformation in Fig. 11 with respect to the renaming operation performed on the initial metaclass Net in Fig. 3.a. In fact, the guard of the migration rule in lines 7-11 in Fig. 10 matches with the renaming operation performed on the metaclass Net to obtain the final PetriNet in Fig. 3.b.

```
  *ATLExample.emig 
 1  migration test;
 2  import "lib_atl.emig";
 3  metamodel ATL : 'http://www.eclipse.org/gmt/2005/ATL';
 4  migrate "petrinet2PNML.atl" : ATL with 'PN1_PN2.edelta' ;
 5
 6 rule RenameClass
 7  [    -- filter
 8      class c = changeClass {
 9          set name <- var newName;
10      }
11  ]
12  {    --migration actions
13      o1 : ATL!OclModelElement -> o1 : ATL!OclModelElement [ name <- %{newName} ];
14  }
15  where [ o1.name = c.name ]
16
17 rule CustomAddMetaClass replaces addMetaclass
18  [
19      super.guard
20  ]
21  {
22
23  } where[c.name= 'PlaceToTransition' or c.name= 'TransitionToPlace' ]
```

Fig. 10. Sample migration program in EMFMigrate

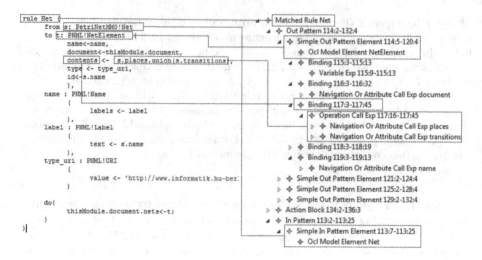

Fig. 11. Sample ATL transformation rule and its abstract syntax

In case of metaclass renamings the considered ATL transformation can be adapted by replacing all the occurrences of the old metaelement with the new one (see line 13). By considering the transformation in Fig. 11, the execution of the migration program in Fig. 10 adapts the input pattern `PetriNetMM0!Net` of the rule `Net` by replacing it with the new `PetriNetMM0!PetriNet`.

4.2   EMFMigrate Migration Libraries and Customization Support

As said, there are metamodel changes which require the intervention of the designer since it is not possible to fully automatize the migration of the affected artifacts. However, in such situations it is still feasible to implement default migration policies which can be refined/completed or even fully replaced by the user. EMFMigrate provides the modeler with the `library` construct to specify default migration rules for a given metamodel. For instance, Fig. 12 shows migration rules which are defined in the library named `libATL` specifically conceived for managing the adaptation of ATL transformations. The shown rules are related to the merge of references (see lines 5-21) and to the addition of new metaclasses (see lines 24-44).

The `mergeReference` rule in Fig. 12 is able to adapt ATL transformations when two references in a given metaclass of the source metamodel are merged together by giving place to a new one. For instance, this is the case of the references `places` and `transitions` in the sample PetriNet metamodel in Fig. 3.a, which are merged in the new reference `elements` as shown in the new version of the metamodel in Fig. 3.b. Such a complex difference is specified in the guard in lines 6-15. The adaptation implemented by the body in lines 16-21 rewrites all the occurrences of the matched references `ref1` and `ref2` with target ATL `select` operations which properly filter the new reference `newName` by selecting elements of type `ref1.type` and `ref2.type`. For instance, in case of the reference `place` of the running example, all the instances

```
  *lib_atl.emig ⊠
 1  library libATL;
 2
 3  metamodel ATL : 'http://www.eclipse.org/gmt/2005/ATL';
 4
 5⊖ rule mergeReferences
 6  [
 7      class c = changeClass{
 8          reference ref1 = deleteReference{}
 9          reference ref2 = deleteReference{}
10          reference ref_merged=addReference{
11              set name<-var newName;
12              set eType <- var commonsupertype;
13          }
14      }
15  ]
16  {
17  s1:ATL!NavigationorAttributeCallExp ->[[%{newName}->select(ele.oclIsKindOf(%{ref1.type}))]]
18  ;
19  s2:ATL!NavigationorAttributeCallExp ->[[%{newName}->select(ele.oclIsKindOf(%{ref2.type}))]]
20  ;
21  }where[s1.name=ref1.name or s2.name=ref2.name ]
22
23
24⊖ rule addMetaclass
25  [
26      class c= addClass{
27          set name<-var className;
28          set abstract<-var isAbstract;
29      }
30  ]
31  {
32  m: ATL!Module -> m: ATL!Module[
33      m.matchedRules<-
34          [[
35          rule copy%{className}{
36              from s: %{package}!%{className}
37
38          to
39              --Insert your ATL output pattern code here
40          }
41          ]]
42  ]
43
44  }where[isAbstract=false]
```

Fig. 12. Sample EMFMigrate migration library

of NavigationOrAttributeCallExp[4] named place will be rewritten with the expression elements->select(e|e.oclIsKindOf(Place)) (see line 17 in Fig. 12). To simplify the specification or rewriting rules, EMFMigrate permits to specify terms by using the concrete syntax of artifact to be adapted instead of its abstract syntax (see lines 17 and 19 in Fig. 10). Concerning the addMetaclass migration rule in lines 24-44 of Fig. 12, the implemented default migration consists of adding a new matched rule that has the input pattern which refers to the new added class. In the output pattern,

---

[4] NavigationOrAttributeCallExp is the metaclass of the ATL metamodel which is used to refer to structural features of a given element. For instance, on the right-hand side of Fig. 11, there are two NavigationOrAttributeCallExp instances since the references places, and transitions of the source metaclass Net are used to set the value of the target contents reference.

an ATL comment is generated since there is no information which can be exploited to generate a corresponding manipulation.

EMFMigrate permits to customize migration rules defined in libraries as in the case of the migration program in Fig. 10. In particular, the shown migration program uses the library shown in Fig. 12 and customizes the addMetaclass migration rule. EMFMigrate provides developers with two different constructs named refines, and replaces. The former enables the refinement of an existing rule in the default library. In particular, after the execution of the considered migration rule in the library, the rewriting rules in the customization are executed. Migration rules can be also completely overridden whenever the default migration policy does not reflect the intentions of the developers. In this respect, the construct replaces can be used to overwrite the body of an existing migration rule. For instance, CustomAddMetaClass replaces the body of the existing migration rule addMetaclass. In particular, CustomAddMetaClass is applied if the guard of addMetaclass holds (see super.guard in line 19 in Fig. 12) and if the name of the added class is TPArc or PTArc. The intention of the developer is to overwrite the addMetaclass migration rule only for added metaclasses which are abstract, or that are named TPArc or PTArc. In this case, since the body of of the CustomAddMetaClass migration rule is empty, the effect of such a customization is disabling the application of the default addMetaclass migration rule when the guard in lines 19 and 23 in Fig. 10 holds.

5   Tool Support

This section outlines the existing supporting tools for specifying and executing EMFMigrate migration programs. As with other approaches, it is of great importance not to neglect the role of dedicated environments in the management of the general coupled evolution problem as usability and accuracy are hardly achievable without any automated support.

As described in the previous section, migration rules are executed according to the metamodel changes which can be generated or manually specified. In this respect, a textual editor for defining metamodel changes (in the EDelta language [3]) is provided. Moreover, an additional editor is available for specifying migration strategies that use EDelta specifications as input. The editors are based on *EMFText* [5] and are freely available[6]. The editors provide syntax coloring and error identification as shown in Fig. 10 and Fig. 12.

The semantics of EMFMigrate is defined in transformational terms by a semantic anchoring, which targets the EMF Transformation Virtual Machine (EMFTVM) [25] – a reusable runtime engine for the model transformation domain [24]. EMFTVM was conceived as a common runtime engine for heterogeneous model transformation languages, and provides a high-level bytecode metamodel to express transformations. Domain-specific language primitives are part of the bytecode metamodel, such as explicit transformation modules and rules, including their composition mechanisms, and model manipulation instructions. Since the bytecode of EMFTVM is represented as a

---

[5] http://www.emftext.org/index.php/EMFText
[6] http://code.google.com/a/eclipselabs.org/p/emfmigrate/

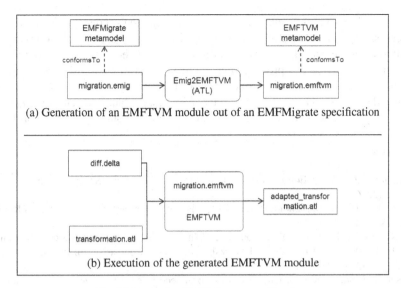

(a) Generation of an EMFTVM module out of an EMFMigrate specification

(b) Execution of the generated EMFTVM module

Fig. 13. Executing EMFMigrate specifications

metamodel, the semantics of EMFMigrate has been given in terms of a model transformation as shown in Fig. 13. In particular, an ATL transformation module translates EMFMigrate migrations into EMFTVM specifications (see Fig. 13.a). The generated EMFTVM module is then executed on top of the existing runtime engine infrastructure to obtain a migrated version of the modelling artefact (see Fig. 13.b). The infrastructure shown in Fig. 13 has been fully implemented and is freely available[7].

6   Conclusions

Similarly to any software artifact, metamodels are entities whose nature is steadily leaning towards evolution. Since they are pivotal ingredients in metamodeling ecosystems, any modification in a metamodel must be accurately assessed and managed because of the ripple effect which causes over the rest of the corresponding ecosystem. This requires proper co-evolution techniques able to propagate the changes over those modeling artifacts which are defined on top of the metamodels. If the adaptation is operated with spontaneous or individual skills without automated support, it might easily lead to inconsistencies and irremediable information erosion on the refactored artifacts.

In recent years, different approaches have been proposed to cope with this intrinsically difficult problem. However, each of them is specifically focussing on a particular kind of modeling artifact requiring the modelers to become familiar with different languages and systems in order to deal with the overall re-alignment. For these reasons, we are interested in dealing with the metamodel-based artifacts as a whole and considering their adaptation as something which should be inherently belonging to the metamodel definition. Migration facilities must be related to generic metamodel refactoring patterns which can be applied off-the-shelf throughout the ecosystem, whenever

---

[7] http://tinyurl.com/emig2EMFTVM-atl

the metamodel undergoes an evolution. This should lead to the concept of evolutionary togetherness: metamodels are related to *all* artifacts, as metamodel evolution is related to a comprehensive and uniform adaptation of the *whole* ecosystem. Without such a holistic attitude towards this problem, metamodels sooner or later will become always more resilient to change as the consequences of their modifications would be too expensive and complex to be socially and financially sustainable. In essence, the key lies in resolving pragmatic issues related to the intricacy of the coupled evolution problem.

References

1. Schmidt, D.C.: Guest Editor's Introduction: Model-Driven Engineering. Computer 39(2), 25–31 (2006)
2. Bézivin, J.: On the Unification Power of Models. Journal on Software and Systems Modeling 4(2), 171–188 (2005)
3. Di Ruscio, D., Iovino, L., Pierantonio, A.: What is needed for managing co-evolution in MDE? In: Procs. 2nd International Workshop on Model Comparison in Practice, IWMCP 2011, pp. 30–38. ACM (2011)
4. Bosch, J.: From software product lines to software ecosystems. In: Procs. 13th International Software Product Line Conference, SPLC 2009, pp. 111–119. Carnegie Mellon University Press (2009)
5. Favre, J.M.: Meta-Model and Model Co-evolution within the 3D Software Space. In: Procs. International Workshop on Evolution of Large-scale Industrial Software Applications, ELISA 2003 (2003)
6. Beck, K., Fowler, M.: Bad smells in code, pp. 75–88. Addison Wesley (1999)
7. Wachsmuth, G.: Metamodel Adaptation and Model Co-adaptation. In: Ernst, E. (ed.) ECOOP 2007. LNCS, vol. 4609, pp. 600–624. Springer, Heidelberg (2007)
8. Di Ruscio, D., Lämmel, R., Pierantonio, A.: Automated Co-evolution of GMF Editor Models. In: Malloy, B., Staab, S., van den Brand, M. (eds.) SLE 2010. LNCS, vol. 6563, pp. 143–162. Springer, Heidelberg (2011)
9. Cicchetti, A., Di Ruscio, D., Eramo, R., Pierantonio, A.: Automating Co-evolution in Model-Driven Engineering. In: Procs. 12th International IEEE Enterprise Distributed Object Computing Conference, EDOC 2008, pp. 222–231. IEEE Computer Society (2008)
10. Rose, L., Etien, A., Méndez, D., Kolovos, D., Paige, R., Polack, F.: Comparing Model-Metamodel and Transformation-Metamodel Coevolution. In: Procs. International Workshop on Models and Evolution, ME 2010 (2010)
11. Mendez, D., Etien, A., Muller, A., Casallas, R.: Towards Transformation Migration After Metamodel Evolution. In: Procs. International Workshop on Models and Evolution, ME 2010 (2010)
12. Eclipse project: GMF - Graphical Modeling Framework,
    http://www.eclipse.org/gmf/
13. Rose, L.M., Herrmannsdoerfer, M., Williams, J.R., Kolovos, D.S., Garcés, K., Paige, R.F., Polack, F.A.C.: A Comparison of Model Migration Tools. In: Petriu, D.C., Rouquette, N., Haugen, Ø. (eds.) MODELS 2010, Part I. LNCS, vol. 6394, pp. 61–75. Springer, Heidelberg (2010)
14. Rose, L.M., Kolovos, D.S., Paige, R.F., Polack, F.A.C.: Model Migration with Epsilon Flock. In: Tratt, L., Gogolla, M. (eds.) ICMT 2010. LNCS, vol. 6142, pp. 184–198. Springer, Heidelberg (2010)
15. Narayanan, A., Levendovszky, T., Balasubramanian, D., Karsai, G.: Automatic Domain Model Migration to Manage Metamodel Evolution. In: Schürr, A., Selic, B. (eds.) MODELS 2009. LNCS, vol. 5795, pp. 706–711. Springer, Heidelberg (2009)

16. Herrmannsdoerfer, M., Benz, S., Juergens, E.: COPE - Automating Coupled Evolution of Metamodels and Models. In: Drossopoulou, S. (ed.) ECOOP 2009. LNCS, vol. 5653, pp. 52–76. Springer, Heidelberg (2009)

17. Garcés, K., Jouault, F., Cointe, P., Bézivin, J.: Managing Model Adaptation by Precise Detection of Metamodel Changes. In: Paige, R.F., Hartman, A., Rensink, A. (eds.) ECMDA-FA 2009. LNCS, vol. 5562, pp. 34–49. Springer, Heidelberg (2009)

18. Levendovszky, T., Balasubramanian, D., Narayanan, A., Karsai, G.: A Novel Approach to Semi-automated Evolution of DSML Model Transformation. In: van den Brand, M., Gašević, D., Gray, J. (eds.) SLE 2009. LNCS, vol. 5969, pp. 23–41. Springer, Heidelberg (2010)

19. Steinberg, D., Budinsky, F., Paternostro, M., Merks, E.: EMF: Eclipse Modeling Framework. Addison-Wesley (2009)

20. Reuseware Team: EMFText, http://www.emftext.org/index.php/EMFText

21. Jouault, F., Bézivin, J., Kurtev, I.: TCS: a DSL for the specification of textual concrete syntaxes in model engineering. In: Procs. 5th International Conference on Generative Programming and Component Engineering, pp. 249–254. ACM Press, New York (2006)

22. Efftinge, S., Völter, M.: oAW xText: A framework for textual DSLs. In: Eclipsecon Summit Europe (2006)

23. Cicchetti, A., Di Ruscio, D., Pierantonio, A.: A Metamodel Independent Approach to Difference Representation. Journal of Object Technology 6(9), 165–185 (2007)

24. Wagelaar, D., Iovino, L., Di Ruscio, D., Pierantonio, A.: Translational Semantics of a Coevolution Specific Language with the EMF Transformation Virtual Machine. In: Hu, Z., de Lara, J. (eds.) ICMT 2012. LNCS, vol. 7307, pp. 192–207. Springer, Heidelberg (2012)

25. Wagelaar, D., Tisi, M., Cabot, J., Jouault, F.: Towards a General Composition Semantics for Rule-Based Model Transformation. In: Whittle, J., Clark, T., Kühne, T. (eds.) MODELS 2011. LNCS, vol. 6981, pp. 623–637. Springer, Heidelberg (2011)

Completeness-Driven Development[*]

Rolf Drechsler[1,2], Melanie Diepenbeck[1], Daniel Große[1], Ulrich Kühne[1],
Hoang M. Le[1], Julia Seiter[1], Mathias Soeken[1,2], and Robert Wille[1]

[1] Institute of Computer Science, University of Bremen, 28359 Bremen, Germany
[2] Cyber-Physical Systems, DFKI GmbH, 28359 Bremen, Germany
drechsle@informatik.uni-bremen.de

Abstract. Due to the steadily increasing complexity, the design of embedded systems faces serious challenges. To meet these challenges additional abstraction levels have been added to the conventional design flow resulting in *Electronic System Level* (ESL) design. Besides abstraction, the focus in ESL during the development of a system moves from design to verification, i.e. checking whether or not the system works as intended becomes more and more important. However, at each abstraction level only the validity of certain properties is checked. Completeness, i.e. checking whether or not the entire behavior of the design has been verified, is usually not continuously checked. As a result, bugs may be found very late causing expensive iterations across several abstraction levels. This delays the finalization of the embedded system significantly. In this work, we present the concept of *Completeness-Driven Development* (CDD). Based on suitable completeness measures, CDD ensures that the next step in the design process can only be entered if completeness at the current abstraction level has been achieved. This leads to an early detection of bugs and accelerates the whole design process. The application of CDD is illustrated by means of an example.

1 Introduction

Although embedded systems have witnessed a reduction of their development time and life time in the past decades, their complexity has been increasing steadily. To keep up with the (customer) requirements, design reuse is common and, hence, more and more complex *Intellectual Property* (IP) is integrated. According to a recent study [1], the external IP adoption increased by 69% from 2007 to 2010. In 2010, 76% of all designs included at least one embedded processor. As a result, the development of embedded systems moves from design to verification, i.e. more time is spent in checking whether the developed design is correct or not. In fact, in the above mentioned time period, there has been a 4% increase of designers compared to an alarming 58% increase of verification engineers.

To face the respective verification challenges, significant effort has been put into clever verification methodologies and new flows have been investigated.

---

[*] This work was supported in part by the German Research Foundation (DFG) within the Reinhart Koselleck project DR 287/23-1.

H. Ehrig et al.(Eds.): ICGT 2012, LNCS 7562, pp. 38–50, 2012.

A major milestone for the development and verification of embedded systems has become the so-called *Electronic System Level* (ESL) design which is state-of-the-art today [2]. Here, the idea is to start designing a complex system at a high level of abstraction – typically using an algorithm specification of the design. At this level, the functionality of the system is realized and evaluated in an abstract fashion ignoring e.g. which parts might become hardware or software later.

The next level of abstraction is based on *Transaction Level Modeling* (TLM) [3, 4]. As modeling language typically SystemC [5–7] is used which offers the TLM-2.0 standard [8]. A TLM model consists of modules communicating over channels, i.e. data is transferred in terms of transactions. Within TLM, different levels of timing accuracy are available such as untimed, loosely-timed, approximately-timed, and cycle-accurate. The respective levels allow e.g. for early software development, performance evaluation, as well as HW/SW partitioning and, thus, enable a further refinement of the system.

Finally, the hardware part of the TLM model is refined to the *Register Transfer Level* (RTL), i.e. a description based on precise hardware building blocks which can subsequently be mapped to the physical level. Here, the resulting chip is eventually prepared for manufacturing.

While this flow is established in industry today, ESL-based design focuses on the implementation and verification of the system. However, although the validity of certain properties of the implementation is checked at the various abstraction levels, often the behavior is not completely considered in these stages. Completeness, i.e. checking whether or not certain behavior of the resulting design has been verified, is usually not continuously checked. This typically causes expensive iterations across several abstraction levels and delays the finalization of the embedded system significantly.

In this work, we present the concept of *Completeness-Driven Development* (CDD). The idea of CDD ensures that the next step in the design process can only be entered if completeness at the current abstraction level has been achieved. For this purpose, suitable completeness measures are needed for each abstraction level in a CDD flow. With CDD, the focus moves from *implementation* to *completeness* while completeness is targeted immediately. Overall, CDD has the following advantages:

- **In-place verification:** New details are verified when they are added.
- **No bug propagation:** Bugs are found as soon as possible since completeness ensures verification quality. As a consequence, bugs are not propagated to lower levels.
- **Long loop minimization:** Loops over several abstraction levels may only occur due to design exploration or unsatisfied non-functional requirements.
- **Correctness and efficient iterations:** The essential criterion is design correctness which is ensured via completeness along each design step. Thus, iterations are only necessary at the current abstraction level.

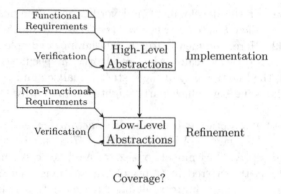

Fig. 1. Conventional flow

CDD is illustrated by means of an example at two representative abstraction levels (Sect. 3). Before, the addressed problem and the proposed CDD flow is described in more detail in the next section.

2 Completeness-Driven Development

2.1 Established ESL Flow

Figure 1 shows a rough sketch of the established ESL flow for embedded systems. To cope with the increasing complexity, requirements for the system are not incorporated at once, but subsequently added leading to a continuous refinement. Usually, the functional requirements are considered first at higher levels of abstraction. Non-functional requirements are added afterwards in the lower levels of the design flow. This allows designers to concentrate on the behavior of the system first. This procedure is sufficient for early simulation (through an executable specification) as well as analysis of the correctness of the functional aspects.

As can be seen in Fig. 1, newly incorporated requirements are verified against prior design states and the specification. However, although a positive verification outcome ensures the correctness of the system with respect to the properties that are checked, full correctness cannot be ensured as it is unclear whether enough properties have been considered. For this task, completeness checks have to be applied. However, today completeness checkers are typically not used continuously, i.e. coverage checks are performed after several design steps and, even worse, mainly at lower levels of abstraction – too late in the overall design process.

As a consequence, behavior that has not been verified at the current abstraction level is not considered until the lower stages of the design flow. If it turns out that this unconsidered part contains bugs, a large portion of the design flow needs to be repeated, leading to long and expensive verification and debugging loops.

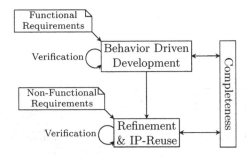

Fig. 2. Overview CDD flow

2.2   Envisioned Solution

To solve the problem described above, we propose the concept of CDD. CDD shifts the focus from implementation to completeness, i.e. completeness is added orthogonally to the state-of-the-art flow. By this, completeness checks are performed at a high level of abstraction and during all refinement steps. As a consequence in each situation, the next step in the design process can only be entered if completeness at the current abstraction level has been achieved.

A major requirement for this flow is that suitable coverage measures must be available for each abstraction level. For the lower levels of abstractions different approaches already exist, see e.g. [9–13]. Also solutions of industrial strength are available, for instance [14]. In contrast, on higher level of abstraction only a few approaches have been proposed. Most of them are based on simulation, e.g. [15–18], and, hence, are not sufficient since the identification of uncovered behavior is not guaranteed. On the formal side, initial approaches have been devised for instance in [19, 20]. If, within a specific abstraction level, an implementation step can be adequately formalized as a *model transformation*, then completeness results can be propagated through several transformations, as long as their correctness is ensured. As an example, in [21], behavior preserving transformations are used to refine the communication model of a system.

In the following, we demonstrate CDD at two representative abstraction levels. The design is composed through *Behavior Driven Development* (BDD) [22] and subsequent refinement/IP-reuse. BDD is a recent development approach which has its roots in software *Test Driven Development* (TDD) [23]. Essentially, in TDD testing and writing code is interleaved while the test cases are written before the code. In doing so, testing is no longer a post-development process, which in practice is often omitted due to strict time constraints. BDD extends TDD in the sense that the test cases are written in natural language, easing the communication between engineers and stakeholders. In BDD, all test cases are called *acceptance tests*. To summarize, in contrast to the current flow from Fig. 1, with CDD completeness is considered additionally at each abstraction level as shown on the right hand side in Fig. 2. This is demonstrated using a concrete example in the next section.

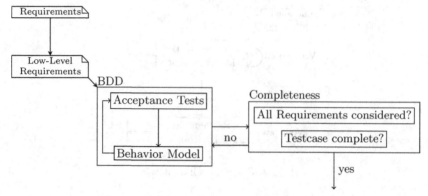

Fig. 3. CDD flow on high-level abstractions

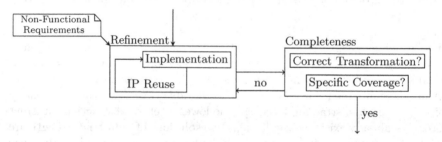

Fig. 4. CDD flow on lower-level abstractions

3   Completeness-Driven Development in Action

In this section, we present an example to demonstrate the proposed CDD flow. We first review the applied abstraction levels and the respective flows at a glance. The details are then explained in the following subsections.

In the example, the development of a calculator is considered. In the process, we use two abstraction levels: the *behavioral level* and the *register transfer level*. The overall flow from Fig. 2 is partitioned into two subflows for each respective abstraction level as depicted in Figs. 3 and 4, respectively. As can be seen, in conjunction with BDD at the behavioral level and the refinement/IP-reuse approach at RTL, completeness analysis techniques tailored for each respective method are employed (see right hand side of both figures).

The development process starts with a document of requirements (see Fig. 3). The translation of these initial requirements to acceptance tests requires an intermediate step that derives low-level requirements. The following BDD process first produces the acceptance tests from these requirements and generates a set of corresponding testcases written in SystemC. Afterwards, the SystemC behavioral model is incrementally developed to pass the defined testcases one at a time. The defined behavior has been fully implemented, if the model passes all testcases. As a consequence, we perform a completeness check to ensure that all requirements have already been considered and the testcases are complete. The result of this step is a complete set of properties that have been generalized from the tests.

After the completeness at behavioral level has been achieved, we proceed to the lower abstraction level at RTL (see Fig. 4). The SystemC model is refined to an RTL model in Verilog. In this refinement step, IP components are integrated. The functionality and the completeness of the RTL model are subsequently assured by using property checking and property-based coverage analysis, respectively.

In the remainder of this section, we first briefly describe the path from the initial requirements to the acceptance tests of BDD. Afterwards, the model and the development process at the behavior level and the RTL are presented in necessary detail focusing on the completeness analysis at both levels of abstractions.

3.1   From Requirements to Acceptance Tests

An excerpt of a list of requirements describing the functionality of the considered calculator reads as follows:

REQ1. The system shall be able to perform calculation with two given numbers. At least addition, subtraction, and multiplication shall be supported.

REQ2. The system shall be able to store the last calculated result and perform calculation with this number and another given number.

REQ3. A given number:
1. can have a positive integer value;
2. can have a negative integer value;
3. shall have up to 3 digits;
4. can be 0.

REQ4. If the result of a calculation has more than 3 digits, the system shall report an error.

REQ...

These requirements are then translated to low-level requirements that capture precisely the expected behavior of the calculator in each specific case. For example, the first two low-level requirements (LLR1 and LLR2 in the following) specify the addition operation of the calculator. Note that the relation to the initial high-level requirements is maintained when specifying each low-level requirement.

LLR1. The system shall be able to add two given numbers in the range $[-999, 999]$. If the sum of the given numbers fits in the range $[-999, 999]$, the system shall return this value. This requirement corresponds to REQ1, REQ3 and REQ4.

LLR2. The system shall be able to add two given numbers in the range $[-999, 999]$. If the sum of the given numbers does not fit in the range $[-999, 999]$, the system shall report an error. This requirement corresponds to REQ1, REQ3 and REQ4.

LLR3. The system shall be able to subtract the second given number from the first one ...

LLR...

At the beginning of the BDD process, the low-level requirements are compiled into acceptance tests which are provided in a very close form to a testcase and also contain precise values of the numbers given to the calculator. For example, the acceptance test that corresponds to LLR1 is as follows:

When the numbers $< a >$ and $< b >$ are given
And I want to add $< a >$ and $< b >$
Then the result should be $< c >$ (where $< c >=< a > + < b >$)
Examples:

a	0	7	20	1	...
b	0	4	17	997	...
c	0	11	37	998	...

Through BDD, these acceptance tests can now be used to determine an according SystemC description.

3.2 CDD at High Level of Abstraction

Generating the Behavioral Model in SystemC. First, the BDD process generates a system description following a TLM modeling style. That is, the data transported to and from the calculator is modeled as a payload shown in Fig. 5. It contains the requested operator, two given numbers, and also the status and the result of the calculation. The functionality of the calculator shall be fully captured in a function *calculate* which receives a payload, performs the requested calculation, and writes back the result into the payload.

After this basic structure has been defined, the BDD process continues with the translation of the acceptance tests to executable testcases in SystemC. Figure 6 exemplarily shows a testcase which corresponds to one of the precise cases in the acceptance test for the low-level requirement LLR1 shown earlier. Line 2 declares a SystemC port to which the calculator will be connected later. In Lines 11–16, a payload with a request operator as well as numbers is initialized and sent to the calculator through the port, while afterwards the received results are checked.

After all testcases have been written, the SystemC model (essentially the function *calculate*) is developed step-by-step to gradually pass all testcases. The final version of *calculate* is depicted in Fig. 7. For example, the first development step has added Line 5 and Lines 14–21 to satisfy the testcases defined for the addition of two given numbers. Lines 14–21 check the intermediate result, then raise the error status flag or write the valid result back, respectively. This code lines are also common for the other operations, so that only Line 6 and Line 7 had to be added to make the testcases for subtraction and multiplication pass. The SystemC model has been successfully tested against all defined testcases.

```
1   struct calc_payload {
2       Operator op;
3       int number1;
4       int number2;
5       CalcStatus calc_status;
6       int result;
7   };
```

Fig. 5. Calculator payload

```
1   struct testcase : public sc_module {
2       sc_port<calculator_if> calc_port;
3
4       SC_HAS_PROCESS(testcase);
5
6       testcase(sc_module_name name) : sc_module(name) {
7           SC_THREAD(main);
8       }
9
10      void main() {
11          calc_payload p;
12          p.op = ADD;
13          p.number1 = 7;
14          p.number2 = 4;
15          calc_port->calculate(p);
16          assert(p.calc_status == CALC_OKAY && p.result == 7+4);
17      }
18  };
```

Fig. 6. A testcase for the calculator

Checking the Completeness. Since the relation between the initial requirements, the low-level requirements, the acceptance tests, and the testcases in SystemC have always been maintained in each translation step, it is very easy to trace back and check whether all requirements have been considered.

To check the completeness of the testcases, they are first generalized into formal properties. As the whole functionality of the calculator is captured in the function *calculate*, the properties only need to reason about the behavior at the start and the end of each *calculate* transaction. Most of the generalization process can be automated, however, human assistance is still required in providing adequate invariants.

To illustrate the concept, Fig. 8 shows two generalized properties for the addition of two given numbers. Both properties are written in a flavor of the *Property Specification Language* (PSL) extended for SystemC TLM [24]. As mentioned earlier, we only need to sample at the start (*:entry*) and the end (*:exit*) of the function/transaction *calculate*. Property P1 covers the case that the sum of two

```
1   void calculate(calc_payload& p) {
2       p.calc_status = CALC_OKAY;
3       switch (p.op) {
4           case NOP : break;
5           case ADD : acc = p.number1 + p.number2; break;
6           case SUB : acc = p.number1 - p.number2; break;
7           case MULT : acc = p.number1 * p.number2; break;
8           case ACC_ADD :
9               ...
10          default :
11              // unknown op -> error response
12              p.calc_status = CALC_ERROR;
13      }
14      if (p.calc_status == CALC_OKAY) {
15          if (acc > MAX_VAL || acc < MIN_VAL) {
16              p.calc_status = CALC_ERROR;
17              acc_out_of_range = true;
18          } else {
19              p.result = acc;
20          }
21      }
22  }
```

Fig. 7. Function *calculate*

given numbers fits in the range so that the calculation will be successful and the sum will be returned, while P2 specifies the calculation in the other case, i.e. the sum is out of range. In both cases, the valid range had to be provided manually as an invariant. Both properties then represent the generalized behavior which is partly considered by the testcases. This generalized behavior is also proven by the high-level property checking method in [25].

However, as determined by the completeness check, behavior remained uncovered. In fact, the invariant for P1 is insufficient. More precisely, the completeness check has detected an uncovered testcase:

$$p.number1 == 0 \text{ and } p.number2 == 999.$$

This is representative for the general forgotten case of

$$p.number1 + p.number2 == 999.$$

The result in this case is not defined by neither P1 nor P2. If this uncovered testcase would have been included in the set of testcases from the beginning, it would have been impossible to provide the insufficient invariant for P1, since a generalized property must be compliant with the testcases it covers. This demonstrates clearly the usefulness and necessity of completeness at this high level of abstraction.

*P1: default clock = calculate:**entry** || calculate:**exit**;*
always *(calculate:**entry** && p.op == ADD && (−999 < p.number1 +*
     p.number2 && p.number1 + p.number2 < 999))
     *−> **next** (calculate:**exit** && p.calc_status == CALC_OKAY &&*
     p.result == p.number1 + p.number2)

*P2: default clock = calculate:**entry** || calculate:**exit**;*
always *(calculate:**entry** && p.op == ADD && ((p.number1 +*
     p.number2 >= 1000) || (p.number1 + p.number2 <= −1000))
     *−> **next** (calculate:**exit** && p.calc_status == CALC_ERROR)*

Fig. 8. Generalized properties for addition

3.3 CDD at RTL

Generating the RTL Model in Verilog. The RTL model is created in a re-finement process starting with the behavioral SystemC model. First, the payload (see Fig. 5) is refined to inputs and outputs of the overall design: both numbers and the operator become inputs, while the result and the calculation status become outputs. Subsequently, the sufficient bit-width for each input and output has to be determined based on the values it has to represent. Both number inputs and the result output are in the range $[−999, 999]$ and thus each of them needs 11 bits. The calculation status contains two states that can be represented using only one bit. For the operator input, three bits are required since its value can either be reset or one of the six supported arithmetic operators.

After the inputs and the outputs have been identified, we proceed to the translation of the algorithmic behavior. Some parts of the algorithmic behavior can be translated one-to-one, for example, the range check of the numbers. Before any computation, the respective inputs are checked if their values are within the valid range. The function *calculate* of the SystemC model is refined to two additional modules: the module CALCULATE to perform the actual calculation, and the module SELECT that stores the last calculated result and delivers it to CALCULATE when an accumulative operation is chosen.

To speed up the development, we integrate two existing IP components into the module CALCULATE: an *Arithmetic Logic Unit* (ALU) – for the addition and subtraction – and a multiplier. Both IPs are taken from the M1 Core [26]. The ALU itself has 15 different operation modes, including the arithmetic functions addition and subtraction, some shift and some Boolean operations. Both units can handle numbers up to 32 bits. Thus, an additional check has to be added to ensure that either the calculated result is in the allowed range or the status output is set.

The overall structure of the RTL design is depicted in Fig. 9. As can be seen, the two number inputs, the operator input, the result output, and the calculation status are denoted as *a*, *b*, *op*, *results*, and *status*, respectively. In each calculation, the inputs are checked first in the unit CHK_1 whether they are within the valid range. Then, they are forwarded to the CALCULATE module.

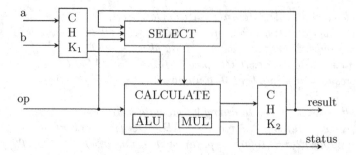

Fig. 9. RTL model overview

The CALCULATE module calculates the result using either the ALU or the multiplier depending on the value of the input *op*. This result is then checked again in unit CHK_2.

Checking the Completeness. After the RTL model has been completely implemented, its correctness has to be verified. For this task, the complete set of properties at the behavioral model is also refined to a set of RTL properties. Essentially, timing needs to be added to the properties while adjusting the syntax of the PSL properties.

After all refined properties have been proven, we perfórm the completeness check at RTL using the method proposed in [13]. The check detects uncovered behavior of the RTL model for the value 111_2 of the 3-bit operator input *op*. The other seven values correspond to the defined operations of the calculator and hence the behavior in these cases is fully specified by the property set. In the case of 111_2, the ALU performs an unintended operation (a shift operation). This mismatch is possible because the ALU has not been specifically developed for the calculator (in fact as mentioned above the ALU is an external IP component). This shows clearly that completeness checks are necessary, in particular, since integrated IPs may have additional but unintended behavior.

4   Conclusions

In this paper, we have presented the concept of *Completeness-Driven Develop-ment* (CDD). With CDD, completeness checks are added orthogonally to the state-of-the-art design flow. As a result, completeness is ensured already at the highest level of abstraction and during all refinement steps. Hence, bugs are found as soon as possible and are not propagated to lower levels. As a result, expensive design loops are avoided. We have demonstrated the advantages of CDD for an example. For two abstraction levels (behavioral level and RTL) we have shown that completeness is essential for correctness and efficient development.

Going forward, to implement the concept of CDD, high-level and continu-ous completeness measures are necessary. Furthermore, innovative methods to support correct transformation as well as property refinement need to be inves-tigated.

References

1. Wilson Research Group and Mentor Graphics: 2010-2011 Functional Verification Study (2011)
2. Bailey, B., Martin, G., Piziali, A.: ESL Design and Verification: A Prescription for Electronic System Level Methodology. Morgan Kaufmann/Elsevier (2007)
3. Cai, L., Gajski, D.: Transaction level modeling: an overview. In: IEEE/ACM/IFIP International Conference on Hardware/Software Codesign and System Synthesis, pp. 19–24 (2003)
4. Ghenassia, F.: Transaction-Level Modeling with SystemC: TLM Concepts and Applications for Embedded Systems. Springer (2006)
5. Accellera Systems Initiative: SystemC (2012), http://www.systemc.org
6. Black, D.C., Donovan, J.: SystemC: From the Ground Up. Springer-Verlag New York, Inc. (2005)
7. Große, D., Drechsler, R.: Quality-Driven SystemC Design. Springer (2010)
8. Aynsley, J.: OSCI TLM-2.0 Language Reference Manual. Open SystemC Initiative (OSCI) (2009)
9. Chockler, H., Kupferman, O., Vardi, M.Y.: Coverage Metrics for Temporal Logic Model Checking. In: Margaria, T., Yi, W. (eds.) TACAS 2001. LNCS, vol. 2031, pp. 528–542. Springer, Heidelberg (2001)
10. Claessen, K.: A coverage analysis for safety property lists. In: Int'l Conf. on Formal Methods in CAD, pp. 139–145 (2007)
11. Große, D., Kühne, U., Drechsler, R.: Analyzing functional coverage in bounded model checking. IEEE Trans. on CAD 27(7), 1305–1314 (2008)
12. Chockler, H., Kroening, D., Purandare, M.: Coverage in interpolation-based model checking. In: Design Automation Conf., pp. 182–187 (2010)
13. Haedicke, F., Große, D., Drechsler, R.: A guiding coverage metric for formal verification. In: Design, Automation and Test in Europe, pp. 617–622 (2012)
14. Bormann, J., Beyer, S., Maggiore, A., Siegel, M., Skalberg, S., Blackmore, T., Bruno, F.: Complete formal verification of Tricore2 and other processors. In: Design and Verification Conference, DVCon (2007)
15. Helmstetter, C., Maraninchi, F., Maillet-Contoz, L.: Full simulation coverage for SystemC transaction-level models of systems-on-a-chip. Formal Methods in System Design 35(2), 152–189 (2009)
16. Heckeler, P., Behrend, J., Kropf, T., Ruf, J., Weiss, R., Rosenstiel, W.: State-based coverage analysis and UML-driven equivalence checking for C++ state machines. In: FM+AM 2010. Lecture Notes in Informatics, vol. P-179, pp. 49–62 (September 2010)
17. Bombieri, N., Fummi, F., Pravadelli, G., Hampton, M., Letombe, F.: Functional qualification of TLM verification. In: Design, Automation and Test in Europe, pp. 190–195 (2009)
18. Sen, A.: Concurrency-oriented verification and coverage of system-level designs. ACM Trans. Design Autom. Electr. Syst. 16(4), 37 (2011)
19. Apvrille, L.: Ttool for diplodocus: an environment for design space exploration. In: Proceedings of the 8th International Conference on New Technologies in Distributed Systems, NOTERE 2008 (2008)
20. Le, H.M., Große, D., Drechsler, R.: Towards analyzing functional coverage in SystemC TLM property checking. In: IEEE International High Level Design Validation and Test Workshop, pp. 67–74 (2010)

21. Andova, S., van den Brand, M.G.J., Engelen, L.: Reusable and Correct Endogenous Model Transformations. In: Hu, Z., de Lara, J. (eds.) ICMT 2012. LNCS, vol. 7307, pp. 72–88. Springer, Heidelberg (2012)
22. North, D.: Behavior Modification: The evolution of behavior-driven development. Better Software 8(3) (2006)
23. Beck, K.: Test Driven Development: By Example. Addison-Wesley Longman Publishing Co., Inc., Boston (2002)
24. Tabakov, D., Vardi, M., Kamhi, G., Singerman, E.: A temporal language for SystemC. In: Int'l Conf. on Formal Methods in CAD, pp. 1–9 (2008)
25. Große, D., Le, H.M., Drechsler, R.: Proving transaction and system-level properties of untimed SystemC TLM designs. In: ACM & IEEE International Conference on Formal Methods and Models for Codesign, pp. 113–122 (2010)
26. Fazzino, F., Watson, A.: M1 core (2012), http://opencores.org/project,m1_core

Exploiting Over- and Under-Approximations for Infinite-State Counterpart Models*

Fabio Gadducci[1], Alberto Lluch Lafuente[2], and Andrea Vandin[2]

[1] Department of Computer Science, University of Pisa, Italy
{gadducci}@di.unipi.it
[2] IMT Institute for Advanced Studies Lucca, Italy
{alberto.lluch,andrea.vandin}@imtlucca.it

Abstract. Software systems with dynamic topology are often infinite-state. Paradigmatic examples are those modeled as graph transformation systems (GTSs) with rewrite rules that allow an unbounded creation of items. For such systems, verification can become intractable, thus calling for the development of approximation techniques that may ease the verification at the cost of losing in preciseness and completeness. Both over- and under-approximations have been considered in the literature, respectively offering more and less behaviors than the original system. At the same time, properties of the system may be either preserved or reflected by a given approximation. In this paper we propose a general notion of approximation that captures some of the existing approaches for GTSs. Formulae are specified by a generic quantified modal logic that generalizes many specification logics adopted in the literature for GTSs. We also propose a type system to denote part of the formulae as either reflected or preserved, together with a technique that exploits under- and over-approximations to reason about typed as well as untyped formulae.

Keywords: model checking, graph transition systems, abstraction, graph logics, approximations, simulations.

1 Introduction

Various approaches have been proposed to equip visual specification formalisms with tools and techniques for verification. Recently, quite some attention has been devoted to those proposals that have adapted traditional approaches (such as model checking) to the area of graph transformation. Among others, we mention here two research lines that have integrated the techniques they argue for into suitable verification tools: GROOVE [5, 9, 12–16] and AUGUR [1–4, 10][1].

A main ingredient in those works is the adoption of a suitable language for property specification. The language is in form of a modal logic capturing very

---

* Partly supported by the EU FP7-ICT IP ASCEns and by the MIUR PRIN SisteR.
[1] See groove.cs.utwente.nl and www.ti.inf.uni-due.de/research/tools/augur2.

H. Ehrig et al.(Eds.): ICGT 2012, LNCS 7562, pp. 51–65, 2012.
© Springer-Verlag Berlin Heidelberg 2012

often two essential dimensions of the state space of graph transformation systems (GTSs): the *topological* structure of states (i.e. graphs) and the *temporal* structure of transitions (i.e. graph rewrites). The topological dimension is usually handled by variants of monadic second-order (MSO) logics [6], spatial logics [7] or regular expressions [12], while the temporal dimension is typically tackled with standard modal logics from the model checking tradition like LTL, CTL or the modal μ-calculus. Our own contribution [8] to this field follows the tradition of [2] and is based on a quantified version of the μ-calculus that mixes temporal modalities and graph expressions in MSO-style.

The model checking problem for GTSs is in general not decidable for such logics, since GTSs are Turing complete languages. Pragmatically, the state space of GTSs (usually called *graph transition system*) is often infinite and it is well known that only some infinite-state model checking problems are decidable. Paradigmatic examples are GTSs with rewrite rules allowing an unbounded creation of items. Verification becomes then intractable and calls for appropriate state space reduction techniques. For example, many efforts have been devoted to the definition of approximation techniques inspired by abstract interpretation. The main idea is to consider a *finite-state* abstract system that approximates (the properties of) an *infinite-state* one, so that verification becomes feasible (at the acceptable cost of losing preciseness in the verification results). Approximated systems represent either more or less behaviours than the original one, resulting respectively in over- or under-approximations. In general, in order to consider meaningful approximations, it is necessary to relate them with the original systems via behavioural relations, like *simulation* ones. Such approximation techniques have been developed in both the above mentioned research lines: namely *neighbourhood abstractions* [5] and *unfoldings* [1, 3, 4].

Contribution. Even if such techniques have been shown to be very effective, we do believe that there is still space for pushing forward their exploitation in the verification of GTSs. In this paper we propose a general formalization of similarity-based approximations, and a verification technique exploiting them. We focus on the type system of [4] proposed within the unfolding technique to classify formulae as preserved or reflected by a given approximation. We extend and generalize such type system in several directions: (i) our type system is *technique-agnostic*, meaning that it does not require the approximated systems to be obtained with a particular mechanism (e.g. the unfolding one); (ii) we consider counterpart models, a generalization of graph transition systems; (iii) our type system is parametric with respect to a given simulation relation (while the original one considers only simulations with certain properties); (iv) we use the type system to reason on all formulae (rather than just on closed ones); and (v) we propose a technique that exploits over- and under-approximations to estimate properties more precisely, handling also part of the untyped formulae.

Synopsis. Sect. 2 provides the necessary background. Sect. 3 defines simulation relations between counterpart models. Sect. 4 provides a type system to classify formulae as *preserved*, *reflected* or *strongly preserved*, exploited in Sect. 5 to define the approximated evaluation of formulae. Finally, Sect. 6 discusses related works, and concludes the paper outlining future research avenues.

2 Background

We summarize here the basic machinery of our approach: essentially, the notion of *counterpart models* (which generalize graph transition systems) and a logic to reason about such models. A detailed presentation can be found in [8].

2.1 Counterpart Models

While graph transition systems have graphs associated to states, counterpart models use many-sorted algebras to denote the structure of states (*worlds*).

Recall that a *(many-sorted) signature* Σ is a pair (S_Σ, F_Σ) composed by a set of sorts $S_\Sigma = \{\tau_1, \cdots, \tau_m\}$ and by a set of function symbols $F_\Sigma = \{f_\Sigma : \tau_1 \times \ldots \times \tau_n \to \tau \mid \tau_i, \tau \in S_\Sigma\}$ typed over S_Σ^*, and that a *(many-sorted) algebra* $\mathbf{A}$ with signature Σ (a Σ-algebra) is a pair $(A, F_\Sigma^{\mathbf{A}})$ such that: (i) the *carrier* A is a set of elements typed over S_Σ; (ii) $F_\Sigma^{\mathbf{A}} = \{f_\Sigma^{\mathbf{A}} : A_{\tau_1} \times \ldots \times A_{\tau_n} \to A_\tau \mid f_\Sigma : \tau_1 \times \ldots \times \tau_n \to \tau \in F_\Sigma\}$ is a family of functions on A typed over S_Σ^*, where $A_\tau = \{a \in A \mid a : \tau\}$, and each $f_\Sigma \in F_\Sigma$ corresponds to a function $f_\Sigma^{\mathbf{A}}$ in $F_\Sigma^{\mathbf{A}}$.

Given two Σ-algebras $\mathbf{A}$ and $\mathbf{B}$, a *(partial) morphism* ϱ is a family of partial functions $\{\varrho_\tau : A_\tau \rightharpoonup B_\tau \mid \tau \in S_\Sigma\}$ typed over S_Σ, such that, for each function symbol $f_\Sigma : \tau_1 \times \ldots \times \tau_n \to \tau \in F_\Sigma$ and list of elements $a_1, \ldots, a_n$, if each function ϱ_{τ_i} is defined for the element a_i of type τ_i, then ϱ_τ is defined for the element $f_\Sigma^{\mathbf{A}}(a_1, \ldots, a_n)$ of type τ and the elements $\varrho_\tau(f_\Sigma^{\mathbf{A}}(a_1, \ldots, a_n))$ and $f_\Sigma^{\mathbf{B}}(\varrho_{\tau_1}(a_1), \ldots, \varrho_{\tau_n}(a_n))$ coincide. A morphism is injective, surjective or bijective if all the ϱ_τ are so, meaning that they are so over its domain of definition.

Example 1. The signature for directed graphs is (S_{Gr}, F_{Gr}). The set S_{Gr} consists of the sorts of nodes τ_N and edges τ_E, while the set F_{Gr} is composed by the function symbols $s : \tau_E \to \tau_N$ and $t : \tau_E \to \tau_N$, which determine the source and the target node of an edge. For example, in Fig. 1 the graph tagged with w_1 is $(N \uplus E, \{s, t\})$, where $N = \{u, v\}$, $E = \{e_1\}$, $s = \{e_1 \mapsto u\}$ and $t = \{e_1 \mapsto v\}$.

A basic ingredient of our logic are open terms. For this purpose we consider signatures Σ_X obtained by extending a many-sorted signature Σ with a denumerable set X of variables typed over S_Σ. We let X_τ denote the τ-typed subset of variables and with x_τ or $x : \tau$ a variable with sort τ. Similarly, we let ϵ_τ or $\epsilon : \tau$ indicate a τ-sorted term. The set $T(\Sigma_X)$ of (possibly open) *terms* obtained from Σ_X is the smallest set such that $X \subseteq T(\Sigma_X)$ and $f(\epsilon_1, \ldots, \epsilon_n) : \tau \in T(\Sigma_X)$ for any $f : \tau_1 \times \ldots \times \tau_n \to \tau \in F_\Sigma$ and $\epsilon_i : \tau_i \in T(\Sigma_X)$.

For ease of presentation, we omit the sort when it is clear from the context or when it is not necessary. Moreover, we fix a generic many-sorted signature Σ.

We are finally ready to introduce counterpart models, which can be seen as a generalization of graph transition systems (see e.g. [2]).

Definition 1 (Counterpart model). *Let $\mathcal{A}$ be the set of Σ-algebras. A counterpart model M is a triple $(W, \rightsquigarrow, d)$ such that W is a set of worlds, $d : W \rightarrow \mathcal{A}$ is a function assigning to each world a Σ-algebra, and $\rightsquigarrow \subseteq W \times (\mathcal{A} \rightharpoonup \mathcal{A}) \times W$ is the* accessibility relation *over W, enriched with (partial) morphisms (*counterpart relations*) between the algebras of the connected worlds.*

In the following we may use $w_1 \overset{cr}{\rightsquigarrow} w_2$ for $(w_1, cr, w_2) \in \rightsquigarrow$. In particular, for each $w_1 \overset{cr}{\rightsquigarrow} w_2$ we have that $cr : d(w_1) \rightarrow d(w_2)$ defines the counterparts of (the algebra of) w_1 in (the algebra of) w_2. Counterpart relations allow hence to avoid *trans-world identity*, the implicit identification of elements of different worlds sharing the same name. Element names thus have a meaning that is local to their world. For this reason, these relations allow for the creation, deletion, and type-respecting renaming and merging of elements. Duplication is forbidden: no cr associates any element of $d(w_1)$ to more than one of $d(w_2)$.

Should Σ be a signature for graphs, a counterpart model is a two-level hierarchical graph: at the higher level the nodes are the worlds $w \in W$, and the edges are the evolution steps labeled with the associated counterpart relation; at the lower level, each world w contains a graph representing its internal structure. In standard

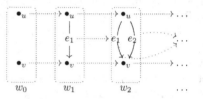

Fig. 1. A counterpart model

terminology, we consider a transition system labeled with algebra morphisms, as an immediate generalization of *graph transition systems* [2].

Example 2. The counterpart model in Fig. 1 is made of a sequence of worlds w_i, where world w_i is essentially associated to a graph $d(w_i)$ with i edges between nodes u and v. The counterpart relations (drawn with dotted lines) reflect the fact that each transition (w_i, cr_i, w_{i+1}) is such that cr_i is the identity for $d(w_i)$.

2.2 A Logic to Reason about Counterpart Models

We now present a logic for counterpart models. The main idea is that the interpretation of a formula in a model M provides sets of pairs (w, σ_w) where w is a world of M and σ_w associates first- and second-order variables to elements and to sets of elements, respectively, of $d(w)$. In what follows we fix a model M with signature Σ, and let X, $\mathcal{X}$ and $\mathcal{Z}$ ranged by x, Y and Z, respectively, denote denumerable sets of first-order, second-order, and fix-point variables, respectively.

Definition 2 (Quantified modal formulae). *The set $\mathcal{F}_{\Sigma X}$ of formulae ψ of our logic is generated by*

$$\psi ::= \ tt \mid \epsilon \in_\tau Y \mid \neg\psi \mid \psi \vee \psi \mid \exists_\tau x.\psi \mid \exists_\tau Y.\psi \mid \Diamond\psi \mid Z \mid \mu Z.\psi$$

where ϵ is a τ-term over Σ_X, $\in_\tau$ is a family of membership predicates typed over S_Σ (stating that a term with sort τ belongs to a second-order variable with the same sort), $\exists_\tau$ quantifies over elements (sets of elements) with sort τ, $\Diamond$ is the "possibility" one-step modality, and μ denotes the least fixed point operator.

The semantics of the logic is given for *formulae-in-context* $\psi[\Gamma; \Delta]$, where $\Gamma \subset X$ and $\Delta \subset \mathcal{X}$ are the *first-* and *second-order contexts* of ψ, containing at least its free variables. However, we may omit types and contexts for the sake of presentation. As usual, we restrict to *monotonic* formulae, where fix-point variables occur under an even number of negations to ensure well-definedness.

The logic is simple, yet reasonably expressive. We can derive useful operators other than boolean connectives $\wedge$, $\rightarrow$, $\leftrightarrow$, and universal quantifiers $\forall_\tau$. For instance "$=_\tau$", the family of equivalence operators for terms in $T(S_\Sigma)$, typed over S_Σ, can be derived as $\epsilon_1 =_\tau \epsilon_2 \equiv \forall_\tau Y. (\epsilon_1 \in_\tau Y \leftrightarrow \epsilon_2 \in_\tau Y)$. The greatest fix-point operator can be derived as $\nu Z.\psi \equiv \neg\mu Z.\neg\psi$, and the "necessarily" one-step modality as $\Box\psi \equiv \neg\Diamond\neg\psi$ (ψ *holds in all the next one-steps*). Moreover, we can derive the standard CTL* temporal operators, as explained in detail in [8].

The semantic domain of our formulae are sets of assignments.

Definition 3 (Assignments). *An assignment (σ_w^1, σ_w^2) for a world $w \in W$ is a pair of partial functions typed over S_Σ with $\sigma_w^1 : X \rightharpoonup d(w)$ and $\sigma_w^2 : \mathcal{X} \rightharpoonup 2^{d(w)}$. We use Ω_M (or just Ω) to denote the set of pairs (w, σ_w), for σ_w an assignment for w. A fix-point variable assignment is a partial function $\rho : \mathcal{Z} \rightharpoonup 2^{\Omega_M}$.*

Given a term ϵ and an assignment $\sigma = (\sigma^1, \sigma^2)$, we denote with $\sigma(\epsilon)$ or $\sigma^1(\epsilon)$ the lifting of σ^1 to $T(\Sigma_X)$. Intuitively, it evaluates ϵ under the assignment σ for its variables. If σ is undefined for any variable in ϵ, then $\sigma(\epsilon)$ is undefined as well. We let $\lambda = (\lambda_1, \lambda_2)$ indicate the empty (or undefined) assignment.

Example 3. Let $r, z, x, y \in X$. In our logic it is easy to define a predicate regarding the presence of an entity with sort τ in a world as $\mathbf{present}_\tau(z) \equiv \exists_\tau r. z = r$. The predicate evaluates in pairs $(w, (\{z \mapsto a\}, \lambda_2))$, with $a : \tau \in d(w)$. Now, the predicate (omitting typings) $\mathbf{p}(x, y) \equiv \mathbf{present}(z) \wedge s(z) = x \wedge t(z) = y$ regards the existence of an edge connecting two node terms. The evaluation of $\mathbf{p}(u, v)$, with u and v nodes, provides assignments of z to edges connecting u to v.

We denote by $\Omega_M^{[\Gamma; \Delta]}$ the set of all pairs $(w, (\sigma_w^1, \sigma_w^2))$ such that the domain of definition of σ_w^1 is contained in Γ, and the one of σ_w^2 is exactly Δ. Note the asymmetry in the definition: σ may be undefined over the elements of Γ, yet not over those of Δ. Intuitively, $\sigma(x)$ may be undefined if the element it was denoting has been deallocated, while we can always assign the empty set to $\sigma(Y)$. We hence use partial first-order assignments to treat item deallocations.

Given models $M = (W, \leadsto, d)$, $M' = (W', \leadsto', d')$, worlds $w \in W$, $w' \in W'$, morphism $\phi : d(w) \rightarrow d'(w')$, and assignment $\sigma_w = (\sigma_w^1, \sigma_w^2)$ for w, we use $\phi \circ \sigma_w$ to denote the assignment $\sigma_{w'}$ (for w') obtained applying ϕ to the components of $\sigma_{w'}$, i.e. $\sigma_{w'}^1 = \phi \circ \sigma_w^1$, and $\sigma_{w'}^2 = 2^\phi \circ \sigma_w^2$, for 2^ϕ the lifting of ϕ to sets.

Assignments can be *restricted to* and *extended by* variables. Given an assignment $\sigma = (\sigma^1, \sigma^2)$ such that $(w, \sigma) \in \Omega^{[\Gamma, x; \Delta]}$, its *restriction* $\sigma \downarrow_x$ wrt. $x \notin \Gamma$ is the assignment $(\sigma^1 \downarrow_x, \sigma^2)$, such that $(w, \sigma \downarrow_x) \in \Omega^{[\Gamma; \Delta]}$, obtained by removing x from the domain of definition of σ^1. Vice versa, the *extension* $\sigma[^a/_x]$ of an assignment $\sigma = (\sigma^1, \sigma^2)$ such that $(w, \sigma) \in \Omega^{[\Gamma; \Delta]}$ wrt. mapping $x \mapsto a$ (for $x \notin \Gamma$ and $a \in d(w)$) is the assignment $(\sigma^1[^a/_x], \sigma^2)$ such that $(w, \sigma[^a/_x]) \in \Omega^{[\Gamma, x; \Delta]}$.

The notation above is analogously and implicitly given also for second-order variables, as well as for their lifting to sets $2^{\downarrow x}$ and $2^{\uparrow x}$. Intuitively, by extending $\Omega^{[\Gamma;\Delta]}$ with respect to a variable $x_\tau \notin \Gamma$, we replace every pair $(w, \sigma_w) \in \Omega^{[\Gamma;\Delta]}$ with the set $\{(w, \sigma_w[^a/_x]) \mid a : \tau \in d(w)\}$. Note that extensions may shrink the set of assignments, should the algebra associated to the world have no element of the correct type. In general terms, the cardinality of $2^{\uparrow x_\tau}(\{(w, \sigma_w)\})$ is the cardinality of $d(w)_\tau$, i.e. the cardinality of the set of elements of type τ in $d(w)$.

Given a transition $w \overset{cr}{\rightsquigarrow} w'$ and $(w, \sigma_w) \in \Omega^{[\Gamma;\Delta]}$, the *counterpart assignment* of σ_w relatively to cr (denoted $\sigma_w \overset{cr}{\rightsquigarrow} \sigma_{w'}$) is the assignment $\sigma_{w'} = cr \circ \sigma_w$. Thus, for $x \in \Gamma$, if $\sigma_w(x)$ is undefined, then $\sigma_{w'}(x)$ is undefined as well, meaning that if $\sigma_w(x)$ refers to an element deallocated in w, then also $\sigma_{w'}(x)$ does in w'; if $\sigma_w(x)$ is defined, but $cr(\sigma_w(x))$ is not, then the considered transition deallocates $\sigma_w(x)$. Whenever both $\sigma_w(x)$ and $cr(\sigma_w(x))$ are defined, then $\sigma_w(x)$ has to evolve in $\sigma_{w'}(x)$ accordingly to cr. As for $Y \in \Delta$, the elements in $\sigma_w(Y)$ preserved by cr are mapped in $\sigma_{w'}(Y)$. If $\sigma_w(Y)$ is defined, then $\sigma_{w'}(Y)$ is also defined, with a cardinality equal or smaller, due to fusion or deletion of elements induced by cr.

We now introduce the evaluation of formulae in a model M, as a mapping from formulae $\psi[\Gamma;\Delta]$ into sets of pairs contained in $\Omega^{[\Gamma;\Delta]}$. Hence, the domain of the assignments in these pairs is, respectively, contained in Γ, and exactly Δ. Intuitively, a pair (w, σ_w) belongs to the semantics of $\psi[\Gamma;\Delta]$ if the formula holds in w under the assignment σ_w for its free variables. We assume that all the bound variables are different among themselves, and from the free ones.

Definition 4 (Semantics). *The evaluation of a formula $\psi[\Gamma;\Delta]$ in M under assignment $\rho : \mathcal{Z} \to 2^{\Omega^{[\Gamma;\Delta]}}$ is given by the function $[\![\cdot]\!]_\rho : \mathcal{F}^{[\Gamma;\Delta]} \to \Omega^{[\Gamma;\Delta]}$*

$$[\![tt[\Gamma;\Delta]]\!]_\rho = \Omega^{[\Gamma;\Delta]}$$
$$[\![(\epsilon \in_\tau Y)[\Gamma;\Delta]]\!]_\rho = \{(w, \sigma_w) \in \Omega^{[\Gamma;\Delta]} \mid \sigma_w(\epsilon) \text{ is defined and } \sigma_w(\epsilon) \in \sigma_w(Y)\}$$
$$[\![\neg\psi[\Gamma;\Delta]]\!]_\rho = \Omega^{[\Gamma;\Delta]} \setminus [\![\psi[\Gamma;\Delta]]\!]_\rho$$
$$[\![\psi_1 \vee \psi_2[\Gamma;\Delta]]\!]_\rho = [\![\psi_1[\Gamma;\Delta]]\!]_\rho \cup [\![\psi_2[\Gamma;\Delta]]\!]_\rho$$
$$[\![\exists_\tau x. \psi[\Gamma;\Delta]]\!]_\rho = 2^{\downarrow x}(\{(w, \sigma_w) \in [\![\psi[\Gamma, x;\Delta]]\!]_{(2^{\uparrow x} \circ \rho)} \mid \sigma_w(x) \text{ is defined}\})$$
$$[\![\exists_\tau Y. \psi[\Gamma;\Delta]]\!]_\rho = 2^{\downarrow Y}([\![\psi[\Gamma;\Delta, Y]]\!]_{(2^{\uparrow Y} \circ \rho)})$$
$$[\![\Diamond\psi[\Gamma;\Delta]]\!]_\rho = \{(w, \sigma_w) \in \Omega^{[\Gamma;\Delta]} \mid \exists w \overset{cr}{\rightsquigarrow} w'. \exists (w', \sigma_{w'}) \in [\![\psi[\Gamma;\Delta]]\!]_\rho . \sigma_w \overset{cr}{\rightsquigarrow} \sigma_{w'}\}$$
$$[\![Z[\Gamma;\Delta]]\!]_\rho = \rho(Z)$$
$$[\![\mu Z.\psi[\Gamma;\Delta]]\!]_\rho = lfp(\lambda Y.[\![\psi[\Gamma;\Delta]]\!]_{\rho[^Y/_Z]})$$

Notice how in order to evaluate $\exists_\tau x.\psi[\Gamma;\Delta]$, we first evaluate ψ extending Γ with x. Then, by dropping the pairs with undefined assignment for x, we obtain the ones whose worlds contain items satisfying ψ if assigned to x. The second-order case is similar, but assignments are defined for all the variables in Δ. Note that ρ is modified accordingly, thus ensuring a proper sorting for $\rho(Z)$.

Another interesting case arises evaluating formulae $\Diamond\psi[\Gamma;\Delta]$, where we search for pairs (w, σ_w) such that there exists a transition $w \overset{cr}{\rightsquigarrow} w'$ and a $\sigma_{w'}$ with $\sigma_w \overset{cr}{\rightsquigarrow} \sigma_{w'}$ and $(w', \sigma_{w'})$ belonging to the evaluation of $\psi[\Gamma;\Delta]$. In words, $\sigma_{w'}$ has to respect the relation induced by cr between the items of the two worlds.

Finally, the evaluation of a closed formula, i.e. with empty context, is a set of pairs (w, λ), for λ the empty assignment, ensuring that our proposal properly extends standard semantics for propositional modal logics.

Example 4 (Evaluation of formulae). Consider the formula of Example 3, the model M of Fig. 1 and the empty assignment $\lambda = (\lambda_1, \lambda_2)$. Evaluating $[\![\mathbf{p}(u, v)]\!]^M$ results in $\{(w_1, (\{z \mapsto e_1\}, \lambda_2)), (w_2, (\{z \mapsto e_1\}, \lambda_2)), (w_2, (\{z \mapsto e_2\}, \lambda_2)), \dots\}$.

3   Behavioural Equivalences for Counterpart Models

In this section we lift classical behavioural preorders and equivalences to counterpart models. For the sake of presentation, for the rest of the paper we fix two models $M = (W, \rightsquigarrow, d)$ and $M' = (W', \rightsquigarrow', d')$. Intuitively, we define relations from M to M' as sets of triples $(w, \phi, w') \in R$ formed by a world $w \in W$, a world $w' \in W'$ and a morphism $\phi : d(w) \to d(w')$ relating their respective structures.

Definition 5 (Simulation). *Let $R \subseteq W \times (\mathcal{A} \to \mathcal{A}) \times W'$ be a set of triples (w, ϕ, w'), with $\phi : d(w) \to d'(w')$ a morphism. R is a simulation from M to M' if for every $(w_1, \phi_1, w_1') \in R$ we have that $w_1 \overset{cr}{\rightsquigarrow} w_2$ implies $w_1' \overset{cr'}{\rightsquigarrow} w_2'$ for some $w_2' \in W'$, with $(w_2, \phi_2, w_2') \in R$ and $\phi_2 \circ cr = cr' \circ \phi_1$. If $R^{-1} = \{(w', \phi^{-1}, w) \mid (w, \phi, w') \in R\}$ is well defined, and it is also a simulation, then R (as well as R^{-1}) is called bisimulation.*

Notice how the ϕ components of bisimulations are precisely injections. We call "iso" a bisimulation whose ϕ components are isomorphisms. We may abbreviate $(w, \phi, w') \in R$ in wRw' if ϕ is irrelevant. As usual, we define (bi)similarity as the greatest (bi)simulation, and say that M *is similar to* M' or that M' *simulates* M, written $M \sqsubseteq_R M'$ (where we may omit R), if there exists a simulation R from M to M' such that, for every $w \in W$, there exists at least a $w' \in W'$ with wRw'.

Example 5. Fig. 2 depicts three models: M (center), $\overline{M}$ (top) and $\underline{M}$ (bottom). The model M, taken from Example 2, is infinite-state and $\overline{M}$ and $\underline{M}$ can be understood as its over- and under-approximations, respectively. Indeed, we have relations $\overline{R}$ and $\underline{R}$ (denoted with double arrows) such that $\underline{M} \sqsubseteq_{\underline{R}} M \sqsubseteq_{\overline{R}} \overline{M}$.

Intuitively, $\underline{M}$ is a truncation of M considering only the first two transitions of M. Every tuple $(\underline{w}, \phi, w)$ in $\underline{R}$ is such that $\phi : \underline{d}(\underline{w}) \to d(w)$ is the identity.

On the other hand, $\overline{M}$ can be seen as "M modulo the fusion of edges". That is, every tuple $(w, \overline{\phi}, \overline{w})$ in $\overline{R}$ is such that $\overline{\phi} : d(w) \to \overline{d}(\overline{w})$ is a bijection for nodes

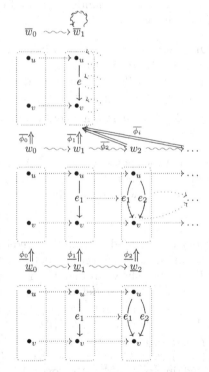

Fig. 2. Approximations

(in particular, the identity restricted to the nodes of $d(w)$) and a surjection on edges mapping every edge e_i into edge e.

Given a set of pairs $\omega \subseteq \Omega_M$ and a simulation R from M to M' we use $R(\omega)$ to denote the set $\{(w', \phi \circ \sigma_w) \mid (w, \sigma_w) \in \omega \wedge (w, \phi, w') \in R\}$.

In the following, with an abuse of notation we use $R \circ \rho$ to indicate the composition of R with the fix-point assignment ρ, defined as $R \circ \rho = \{(Z \mapsto R(\omega)) \mid (Z \mapsto \omega) \in \rho\}$. Note that R^{-1} is not always well-defined since the morphisms in the triples (w, ϕ, w') may not be injective. However, we often use the pre-image $R^{-1}[\cdot]$ of R, defined for a set of pairs $\omega' \subseteq \Omega_{M'}$ as $R^{-1}[\omega'] = \{(w, \sigma_w) \in \Omega_M \mid \exists (w, \phi, w') \in R. (w', \phi \circ \sigma_w) \in \omega'\}$.

4 Preservation and Reflection

As usual, the evaluation of formulae in a model M may be only approximated by a simulation M'. We hence introduce the usual notions of *preserved* formulae, those whose "satisfaction" in M implies their "satisfaction" in M', and *reflected* formulae, those whose "satisfaction" in M' implies their "satisfaction" in M. Of course, since the semantic domain of our logic are assignment pairs, the notion of "satisfaction" corresponds to the existence of such pairs.

Definition 6 (Preserved and reflected formulae). *Let R be a simulation from M to M' (i.e., $M \sqsubseteq_R M'$), $\psi[\Gamma; \Delta]$ a formula, and ρ an assignment. We say that ψ is* preserved *under R (written $\psi :_R \Rightarrow$) if $[\![\psi[\Gamma; \Delta]]\!]^{M'}_{R \circ \rho} \supseteq R([\![\psi[\Gamma; \Delta]]\!]^M_\rho)$;* reflected *under R (written $\psi :_R \Leftarrow$) if $R^{-1}[[\![\psi[\Gamma; \Delta]]\!]^{M'}_{R \circ \rho}] \subseteq [\![\psi[\Gamma; \Delta]]\!]^M_\rho$; and* strongly preserved *under R (written $\psi :_R \Leftrightarrow$) if $\psi :_R \Rightarrow$ and $\psi :_R \Leftarrow$.*

Note that the choice of ρ, Γ, and Δ is irrelevant. In the definition of $\psi :_R \Leftarrow$ we use $R^{-1}[\cdot]$ rather than R^{-1} because the latter is not always defined and, moreover, if $\psi :_R \Leftarrow$, then we additionally have that $R([\![\neg\psi[\Gamma; \Delta]]\!]^M_\rho) \cap [\![\psi[\Gamma; \Delta]]\!]^{M'}_{R \circ \rho} = \emptyset$, i.e. that a pair in $\Omega^{[\Gamma; \Delta]}_M \setminus [\![\psi[\Gamma; \Delta]]\!]^M_\rho$ cannot be similar to any pair in $[\![\psi[\Gamma; \Delta]]\!]^{M'}_{R \circ \rho}$.

Example 6. Consider again the predicate $\mathbf{p}(x, y)$ of Example 3 stating the existence of an edge connecting node x to node y, and the models $\underline{M} \sqsubseteq_{\underline{R}} M \sqsubseteq_{\overline{R}} \overline{M}$ of Example 5 shown in Fig. 2. It is easy to see that $\mathbf{p}(u, v)$ is strongly preserved both under $\underline{R}$ and under $\overline{R}$. Recall that in Example 4 we saw that $[\![\mathbf{p}(u, v)]\!]^M = \{(w_1, (\{z \mapsto e_1\}, \lambda_2)), (w_2, (\{z \mapsto e_1\}, \lambda_2)), (w_2, (\{z \mapsto e_2\}, \lambda_2)), \dots\}$ (for any ρ, thus neglected). Now, $[\![\mathbf{p}(u, v)]\!]^{\underline{M}} = \{(\underline{w_1}, (\{z \mapsto e_1\}, \lambda_2)), (\underline{w_2}, (\{z \mapsto e_1\}, \lambda_2)), (\underline{w_2}, (\{z \mapsto e_2\}, \lambda_2))\}$, and hence $\underline{R}([\![\mathbf{p}(u, v)]\!]^{\underline{M}})$ is $\{(w_1, (\{z \mapsto e_1\}, \lambda_2)), (w_2, (\{z \mapsto e_1\}, \lambda_2)), (w_2, (\{z \mapsto e_2\}, \lambda_2))\}$, which is clearly contained in $[\![\mathbf{p}(u, v)]\!]^M$. Moreover we also have that $\underline{R}^{-1}[[\![\mathbf{p}(u, v)[\Gamma; \Delta]]\!]^M] \subseteq [\![\mathbf{p}(u, v)[\Gamma; \Delta]]\!]^{\underline{M}}$. We hence have that $\mathbf{p}(u, v) :_{\underline{R}} \Leftrightarrow$. Similarly, we have that $[\![\mathbf{p}(u, v)]\!]^{\overline{M}} = \{(\overline{w}_1, (\{z \mapsto e\}, \lambda_2))\}$, and $\overline{R}([\![\mathbf{p}(u, v)]\!]^M) = \{(\overline{w}_1, (\{z \mapsto e\}, \lambda_2))\}$. Both conditions are again satisfied, and hence we have $\mathbf{p}(u, v) :_{\overline{R}} \Leftrightarrow$.

Of course, determining whenever a formula is preserved (or reflected) cannot be done in practice by performing the above check, since that would require to calculate the evaluation of the formula in the (possibly infinite) original model M, which is precisely what we want to avoid. Moreover, note that determining whenever a formula is preserved (and the same occurs for being reflected) is an undecidable problem, since our logic subsumes that of [4].

Nevertheless, we can apply the same approach of [4] and define a type system that approximates the preservation and reflection of formulae. In particular, our type system generalizes the one of [4] in several directions: (i) we consider counterpart models, a generalization of graph transition systems; (ii) our type system is parametric with respect to the simulations R (while the original one is given for graph morphisms that are total and bijective for nodes and total and surjective for edges, we exploit the injectivity, surjectivity and totality of the morphisms of R for each sort τ); (iii) we use the type system to reason on all formulae (while the original proposal restricts to closed ones); and (iv) we propose a technique exploiting over- and under-approximations of a model to obtain more precise approximated formulae evaluations, and we handle part of the untyped formulae.

The type system is parametric with respect to the properties of R. In particular, we consider the properties of the morphisms in R, namely, for each sort τ, if they are τ-total (τ_t), τ-surjective (τ_s) or τ-bijective (τ_b). To ease the presentation, we say "$\tau_{prop}\ R$", with $prop \in \{t, s, b\}$, whenever all $(w, \phi, w') \in R$ are such that ϕ is τ-$prop$. Moreover, we shall consider the case in which R is an iso-bisimulation.

Definition 7 (Type system). *Let R be a simulation from M to M' (i.e., $M \sqsubseteq_R M'$), ψ a formula, and $\mathcal{T} = \{\leftarrow, \rightarrow, \leftrightarrow\}$ a set of types. We say that ψ has type $d \in \mathcal{T}$ if $\psi : d$ can be inferred using the following rules*

$$\frac{}{tt :_R \leftrightarrow} \qquad d=\begin{cases} \rightarrow \ for\ \tau_t\ R \\ \leftarrow\ for\ \tau_b\ R \end{cases} \qquad \frac{\psi :_R \rightarrow \quad \psi :_R \leftarrow}{\psi :_R \leftrightarrow} \qquad \frac{\psi :_R \leftrightarrow}{\psi :_R d}$$

$$\frac{\psi_i :_R d}{\psi_1 \vee \psi_2 :_R d} \qquad \frac{\psi :_R d\ with\ d=\begin{cases} \rightarrow\ for\ \tau_t\ R \\ \leftarrow\ for\ \tau_s\ R \end{cases}}{\exists_\tau x.\psi :_R d\ and\ \exists_\tau Y.\psi :_R d} \qquad \frac{\psi :_R d}{\neg\psi :_R d^{-1}}$$

$$\frac{}{Z :_R \leftrightarrow} \qquad \frac{\psi :_R d\ with\ d=\begin{cases} \rightarrow \quad\quad\quad\quad\ for\ any\ R \\ \leftarrow\ for\ R\ an\ iso\text{-}bisimulation \end{cases}}{\Diamond\psi :_R d} \qquad \frac{\psi :_R d}{\mu Z.\psi :_R d}$$

where it is intended that $\rightarrow^{-1}=\leftarrow$, $\leftarrow^{-1}=\rightarrow$ and $\leftrightarrow^{-1}=\leftrightarrow$.

The type system is not complete, meaning that some formulae cannot be typed: if ψ cannot be typed, we then write $\psi :_R \bot$. However, the next proposition states its soundness.

Proposition 1 (Type system soundness). *Let R be a simulation from M to M' (i.e., $M \sqsubseteq_R M'$) and ψ a formula. Then (i) $\psi :_R\rightarrow$ implies $\psi :_R\Rightarrow$; (ii) $\psi :_R\leftarrow$ implies $\psi :_R\Leftarrow$; and (iii) $\psi :_R\leftrightarrow$ implies $\psi :_R\Leftrightarrow$.*

As we already noted, our type system can be instantiated for graph signatures, in order to obtain the one of [4] as a subsystem. In fact, the authors there consider only simulation relations R that are total on both sorts, as well as being $(\tau_N)_b$ (that is, bijective on nodes) and $(\tau_E)_s$ (surjective on edges).

Another instance is for iso-bisimulations. This is the case of the analysis of graph transition systems up to isomorphism (e.g. as implemented in [13]). In this case the type system is complete and correctly types every formula as $\psi :\leftrightarrow$.

Example 7. Consider the models $\underline{M} \sqsubseteq_R M \sqsubseteq_{\overline{R}} \overline{M}$ of Fig. 2, and the formula $\mathbf{p}(u, v)$ of Example 6, where we saw that $\mathbf{p}(u, v) :_R\leftrightarrow$ and $\mathbf{p}(u, v) :_{\overline{R}}\leftrightarrow$. Our type system provides the types $\mathbf{p}(u, v) :_R\leftrightarrow$ and (since $\overline{R}$ is not injective on edges) $\mathbf{p}(u, v) :_{\overline{R}}\rightarrow$. Note that the type for $\underline{R}$ is exactly inferred, while for $\overline{R}$ it is only approximated as we get *preserved* while it is actually *strongly preserved*.

5 Approximated Semantics

Approximations can be used to estimate the evaluation of formulae. Consider the case of three models $\underline{M}$, M and $\overline{M}$, with $\underline{M} \sqsubseteq_R M \sqsubseteq_{\overline{R}} \overline{M}$, as in Fig. 2, where $\underline{M}$ and $\overline{M}$ are under- and over-approximations of M, respectively. Intuitively, an approximated evaluation of a formula ψ in $\underline{M}$ or $\overline{M}$ may provide us a lower- and upper-bound, defined for either $\underline{M}$ or $\overline{M}$, of the actual evaluation of ψ in M. We call under- and over-approximated evaluations the ones obtained using, respectively, under- (e.g. $\underline{M}$), and over-approximations (e.g. $\overline{M}$).

Exploiting approximated evaluations, we may address the *local* model checking problem: *"does a given assignment pair belong to the evaluation of the formula ψ in M?"*. Given that approximated semantics compute lower- and upper-bounds, we cannot define a complete procedure, i.e. one answering either *true* or *false*. A third value is required for the cases of uncertainty. For this purpose we use a standard three valued logic (namely Kleene's one) whose domain consists of the set of values $\mathbb{K} = \{T, F, ?\}$ (where ? reads "unknown"), and whose operators extend the standard Boolean ones with $T \vee ? = T$, $F \vee ? = ?$, $\neg ? = ?$ (i.e. where disjunction is the join in the complete lattice induced by the *truth* ordering relation $F < ? < T$). Moreover, we also consider a *knowledge addition* (binary, associative, commutative, partial) operation $\oplus : \mathbb{K} \times \mathbb{K} \rightharpoonup \mathbb{K}$ defined as $T \oplus T = T$, $F \oplus F = F$ and $x \oplus ? = x$ for any $x \in \mathbb{K}$. Notice how we intentionally let undefined the case of contradictory addition "$F \oplus T$".

In particular, given a formula, with our approximated semantics we are able to group the pairs of an approximating model in three distinct sets: the ones associated with T, the ones associated with F, and the ones associated with ?. For instance, the *over-approximated* semantics is defined as follows.

Definition 8 (Over-approximated semantics). *Let $\overline{R}$ be a simulation from M to $\overline{M}$ (i.e. $M \sqsubseteq_{\overline{R}} \overline{M}$) and ρ an assignment. The over-approximated semantics of $[\![\cdot]\!]_\rho^M$ in $\overline{M}$ via $\overline{R}$ is given by the function $\{\![\cdot]\!\}_\rho^{\overline{R}} : \mathcal{F}^{[\Gamma;\Delta]} \rightarrow (\Omega_{\overline{M}}^{[\Gamma;\Delta]} \rightarrow \mathbb{K})$, defined as $\{\![\psi[\Gamma;\Delta]]\!\}_\rho^{\overline{R}} = \{(\overline{p}, \overline{k}(\overline{p}, \psi[\Gamma;\Delta], \overline{R})) \mid \overline{p} \in \Omega_{\overline{M}}^{[\Gamma;\Delta]}\}$, where*

	$\{[\mathbf{p}(u,v)]\}$	$\{[\neg\mathbf{p}(u,u)]\}$	$\{[\mathbf{p}(u,v) \vee \neg\mathbf{p}(u,u)]\}$	$^+\{[\mathbf{p}(u,v) \vee \neg\mathbf{p}(u,u)]\}$
$(\overline{w}_0,\lambda), (\overline{w}_1,\lambda)$	F	T	?	T
$(\overline{w}_1,(z\mapsto e,\lambda_2))$	?	T	?	T
$(\underline{w}_0,\lambda),(\underline{w}_1,\lambda),(\underline{w}_2,\lambda)$	F	T	T	T
$(\underline{w}_1,(z\mapsto e_1,\lambda_2))$	T	T	T	T
$(\underline{w}_2,(z\mapsto e_1,\lambda_2))$	T	T	T	T
$(\underline{w}_2,(z\mapsto e_2,\lambda_2))$	T	T	T	T
$(w_2,(z\mapsto e_2,\lambda_2)) \models^{\overline{R}} [\cdot]$	?	T	?	T
$(w_2,(z\mapsto e_2,\lambda_2)) \models_{\underline{R}} [\cdot]$	T	T	T	T
$(w_2,(z\mapsto e_2,\lambda_2)) \models_{\underline{R}}^{\overline{R}} [\cdot]$	T	T	T	T

Fig. 3. Approximated semantics and checks for some formulae

$$\overline{k}(\overline{p}, \psi[\Gamma;\Delta], \overline{R}) = \begin{cases} T & \text{if } \psi :_{\overline{R}}\leftarrow \text{ and } \overline{p} \in [\![\psi[\Gamma;\Delta]]\!]_{\overline{R}\circ\rho}^M \\ F & \text{if } \psi :_{\overline{R}}\rightarrow \text{ and } \overline{p} \notin [\![\psi[\Gamma;\Delta]]\!]_{\overline{R}\circ\rho}^M \\ ? & \text{otherwise} \end{cases}$$

Intuitively, the mapping of the pairs in $\Omega_{\overline{M}}^{[\Gamma;\Delta]}$ depends on the type of ψ. If it is typed as reflected, then all pairs in $[\![\psi[\Gamma;\Delta]]\!]_{\overline{R}\circ\rho}^M$ are mapped to T, since their counterparts in M do certainly belong to the evaluation of ψ. Nothing can be said about the rest of the pairs, which are hence mapped to ?.

Dually, if ψ is typed as preserved, then all those pairs that do not belong to $[\![\psi[\Gamma;\Delta]]\!]_{\overline{R}\circ\rho}^M$ are mapped to F because we know that their counterparts in M do certainly not belong to the evaluation of ψ. Again, nothing can be said about the rest of the pairs, which are hence mapped to ?.

Finally, if ψ cannot be typed, then all pairs are mapped to ?.

Notice how, in practice, we rarely have to explicitly compute $\overline{R} \circ \rho$. In fact, formulae of our logic are thought to be evaluated under an initial empty assignment for fix-point variables, which is manipulated during the evaluation. Clearly $\overline{R} \circ \emptyset = \emptyset$ for any $\overline{R}$, and it can be shown that the rules of the semantics manipulating the fix-point assignment preserve this equivalence.

We can hence use the over-approximated semantics to decide whether an assignment pair belongs to the evaluation of a formula in M as formalized below.

Definition 9 (Over-check). *Let $\overline{R}$ be a simulation from M to $\overline{M}$ (i.e. $M \sqsubseteq_{\overline{R}} \overline{M}$) and ρ an assignment. The over-approximated model check (shortly, overcheck) of $[\![\cdot]\!]_\rho^M$ in $\overline{M}$ via $\overline{R}$ is given by the function $\cdot \models^{\overline{R}} [\![\cdot]\!]_\rho^M : \Omega_M^{[\Gamma;\Delta]} \times \mathcal{F}^{[\Gamma;\Delta]} \to \mathbb{K}$, defined as*

$$p \models^{\overline{R}} [\![\psi[\Gamma;\Delta]]\!]_\rho^M = \begin{cases} ? & \text{if } \overline{R}(p) = \emptyset \\ \bigvee_{\overline{p}\in\overline{R}(p)} \{[\psi[\Gamma;\Delta]]\}_\rho^{\overline{R}}(\overline{p}) & \text{otherwise} \end{cases}$$

Example 8. Consider again the predicate $\mathbf{p}(x,y)$ of Example 3 stating the existence of an edge connecting node x to node y, and the models M and $\overline{M}$ with $M \sqsubseteq_{\overline{R}} \overline{M}$ of Example 5 shown in Fig. 2. In the first group of lines of Fig. 3 we exemplify the over-approximated semantics in $\overline{M}$ via $\overline{R}$ of $[\![\mathbf{p}(u,v)]\!]^M$, $[\![\neg\mathbf{p}(u,u)]\!]^M$, and $[\![\mathbf{p}(u,v) \vee \neg\mathbf{p}(u,u)]\!]^M$, considering the pairs $\Omega_{\overline{M}}^{[z;\emptyset]} = \{(\overline{w}_0,\lambda),(\overline{w}_1,\lambda),(\overline{w}_1,(z \mapsto e,\lambda_2))\}$. We recall from Example 7 that

$\mathbf{p}(u, v) :_{\overline{R}}\rightarrow$, and, hence, $\neg\mathbf{p}(u, u) :_{\overline{R}}\leftarrow$ and $\mathbf{p}(u, v) \vee \neg\mathbf{p}(u, u) :_{\overline{R}} \perp$. Moreover we know that $[\![\mathbf{p}(u, v)]\!]^{\overline{M}} = \{(w_1, (z \mapsto e, \lambda_2))\}$, and $[\![\neg\mathbf{p}(u, u)]\!]^{\overline{M}} = \Omega_{\overline{M}}^{[z;\emptyset]}$. Following Definition 8, we hence have that $(\overline{w}_0, \lambda)$ and $(\overline{w}_1, \lambda)$ are mapped to F for $\mathbf{p}(u, v)$, and to T for $\neg\mathbf{p}(u, u)$, while $(\overline{w}_1, (z \mapsto e, \lambda_2))$ is mapped to ? and to T. Different is the case of $\mathbf{p}(u, v) \vee \neg\mathbf{p}(u, u)$: it cannot be typed and its approximation hence maps the three pairs to ?.

In the third group of lines of Fig. 3 we find the over-check "$\cdot \models^{\overline{R}} [\![\cdot]\!]$" of $(w_2, (z \mapsto e_2, \lambda_2))$ in $\overline{M}$ via $\overline{R}$ for the three formulae. Note that $\overline{R}((w_2, (z \mapsto e_2, \lambda_2))) = (\overline{w}_1, (z \mapsto e, \lambda_2))$, hence the over-checks of $\mathbf{p}(u, v)$ and of $\mathbf{p}(u, v) \vee \neg\mathbf{p}(u, u)$ give ?, because no pair in $\overline{R}((w_2, (z \mapsto e_2, \lambda_2)))$ is mapped to either T or F. Instead, given that $\{[\neg\mathbf{p}(u, u)]\}((\overline{w}_1, (z \mapsto e, \lambda_2))) = T$, then we have $(w_2, (z \mapsto e_2, \lambda_2)) \models^{\overline{R}} [\![\neg\mathbf{p}(u, u)]\!] = T$.

With the next proposition we state that the above described check is sound.

Proposition 2 (Soundness of over-check). *Let $\overline{R}$ be a simulation from M to $\overline{M}$ (i.e. $M \sqsubseteq_{\overline{R}} \overline{M}$), $\psi[\Gamma; \Delta]$ a formula, and ρ an assignment. Then (i) $p \models^{\overline{R}}$ $[\![\psi[\Gamma; \Delta]]\!]_\rho^M = T$ implies $p \in [\![\psi[\Gamma; \Delta]]\!]_\rho^M$; and (ii) $p \models^{\overline{R}} [\![\psi[\Gamma; \Delta]]\!]_\rho^M = F$ implies $p \notin [\![\psi[\Gamma; \Delta]]\!]_\rho^M$.*

Now, we can define the under-approximated semantics in a specular way.

Definition 10 (Under-approximated semantics). *Let $\underline{R}$ be a simulation from $\underline{M}$ to M (i.e. $\underline{M} \sqsubseteq_{\underline{R}} M$) and ρ an assignment. Then, the under-approximated semantics of $[\![\cdot]\!]_\rho^M$ in $\underline{M}$ via $\underline{R}$ is the function $\{\!\![\cdot]\!\!\}_\rho^{\underline{R}} : \mathcal{F}^{[\Gamma;\Delta]} \to$ $(\Omega_{\underline{M}}^{[\Gamma;\Delta]} \to \mathbb{K})$, defined as $\{\!\![\psi[\Gamma; \Delta]]\!\!\}_\rho^{\underline{R}} = \{\underline{p} \mapsto \underline{k}(\underline{p}, \psi[\Gamma; \Delta], \underline{R}) \mid \underline{p} \in \Omega_{\underline{M}}^{[\Gamma;\Delta]}\}$, where*

$$\underline{k}(\underline{p}, \psi[\Gamma; \Delta], \underline{R}) = \begin{cases} T & \text{if } \psi :_{\underline{R}}\rightarrow \text{ and } \underline{p} \in [\![\psi[\Gamma; \Delta]]\!]_{\underline{R}^{-1}[\cdot]\circ\rho}^M \\ F & \text{if } \psi :_{\underline{R}}\leftarrow \text{ and } \underline{p} \notin [\![\psi[\Gamma; \Delta]]\!]_{\underline{R}^{-1}[\cdot]\circ\rho}^M \\ ? & \text{otherwise} \end{cases}$$

We can define an under-approximated model checking procedure as follows.

Definition 11 (Under-check). *Let $\underline{R}$ be a simulation from $\underline{M}$ to M (i.e. $\underline{M} \sqsubseteq_{\underline{R}} M$) and ρ an assignment. The under-approximated model check (shortly, under-check) of $[\![\cdot]\!]_\rho^M$ in $\underline{M}$ via $\underline{R}$ is given by the function $\cdot \models_{\underline{R}} [\![\cdot]\!]_\rho^M : \Omega_M^{[\Gamma;\Delta]} \times$ $\mathcal{F}^{[\Gamma;\Delta]} \to \mathbb{K}$, defined as*

$$p \models_{\underline{R}} [\![\psi[\Gamma; \Delta]]\!]_\rho^M = \begin{cases} ? & \text{if } \underline{R}^{-1}[p] = \emptyset \\ \bigvee_{\underline{p} \in \underline{R}^{-1}[p]} \{\!\![\psi[\Gamma; \Delta]]\!\!\}_\rho^{\underline{R}}(\underline{p}) & \text{otherwise} \end{cases}$$

Next proposition states the soundness of the under-check procedure.

Proposition 3 (Soundness of under-check). *Let $\underline{R}$ be a simulation from $\underline{M}$ to M (i.e., $\underline{M} \sqsubseteq_{\underline{R}} M$), $\psi[\Gamma; \Delta]$ a formula, and ρ an assignment. Then (i) $p \models_{\underline{R}} [\![\psi[\Gamma; \Delta]]\!]_\rho^M = T$ implies $p \in [\![\psi[\Gamma; \Delta]]\!]_\rho^M$; and (ii) $p \models_{\underline{R}} [\![\psi[\Gamma; \Delta]]\!]_\rho^M = F$ implies $p \notin [\![\psi[\Gamma; \Delta]]\!]_\rho^M$.*

We finally show how to combine sets of under- and over-approximations.

Definition 12 (Approximated check). *Let $\{\underline{R_0} \ldots \underline{R_n}\}$ be simulations from $\{\underline{M_0} \ldots \underline{M_n}\}$ to M and $\{\overline{R_0} \ldots \overline{R_m}\}$ simulations from M to $\{\overline{M_0} \ldots \overline{M_m}\}$ (i.e. $\underline{M_i} \sqsubseteq_{\underline{R_i}} M \sqsubseteq_{\overline{R_j}} \overline{M_j}$ for any $i \in \{0 \ldots n\}$ and $j \in \{0 \ldots m\}$). The approximated check of $[\![\cdot]\!]_\rho^M$ in $\{\underline{M_0} \ldots \underline{M_n}\}$ and $\{\overline{M_0} \ldots \overline{M_m}\}$ via $\{\underline{R_0} \ldots \underline{R_n}\}$ and $\{\overline{R_0} \ldots \overline{R_m}\}$ is the function $\cdot \models_{\{\underline{R_0} \ldots \underline{R_n}\}}^{\{\overline{R_0} \ldots \overline{R_m}\}} [\![\cdot]\!]_\rho^M : \Omega_M^{[\Gamma;\Delta]} \times \mathcal{F}^{[\Gamma;\Delta]} \rightharpoonup \mathbb{K}$, defined as*

$$ p \models_{\{\underline{R_0} \ldots \underline{R_n}\}}^{\{\overline{R_0} \ldots \overline{R_m}\}} [\![\psi[\Gamma;\Delta]]\!]_\rho^M = \bigoplus_j (p \models^{\overline{R_j}} [\![\psi[\Gamma;\Delta]]\!]_\rho^M) \bigoplus_i (p \models_{\underline{R_i}} [\![\psi[\Gamma;\Delta]]\!]_\rho^M) $$

Note that, even if $\oplus$ is partial, the approximated check is well-defined since Propositions 2 and 3 ensure that we never have to combine contradictory results (e.g. $T \oplus F$). It is also easy to see that the soundness result of Propositions 2 and 3 allows us to conclude the soundness of the approximated check.

Theorem 1 (Soundness of approximated check). *Let $\{\underline{R_0} \ldots \underline{R_n}\}$ be simulations from $\{\underline{M_0} \ldots \underline{M_n}\}$ to M and $\{\overline{R_0} \ldots \overline{R_m}\}$ from M to $\{\overline{M_0} \ldots \overline{M_m}\}$ (i.e. $\underline{M_i} \sqsubseteq_{\underline{R_i}} M \sqsubseteq_{\overline{R_j}} \overline{M_j}$ for any $i \in \{0 \ldots n\}$ and $j \in \{0 \ldots m\}$). Let $\psi[\Gamma;\Delta]$ a formula, and ρ an assignment. Then (i) $p \models_{\{\underline{R_0} \ldots \underline{R_n}\}}^{\{\overline{R_0} \ldots \overline{R_m}\}} [\![\psi[\Gamma;\Delta]]\!]_\rho^M = T$ implies $p \in [\![\psi[\Gamma;\Delta]]\!]_\rho^M$; and (ii) $p \models_{\{\underline{R_0} \ldots \underline{R_n}\}}^{\{\overline{R_0} \ldots \overline{R_m}\}} [\![\psi[\Gamma;\Delta]]\!]_\rho^M = F$ implies $p \notin [\![\psi[\Gamma;\Delta]]\!]_\rho^M$.*

Approximated semantics provide us with a suitable evaluation of any formula, even though its result may not be meaningful, since we may have empty lower-bounds or unbounded upper-bounds as particular instances, namely when all the pairs are assigned to ?. Indeed, this is the case of formulae that cannot be typed with our type system. In order to obtain a more significant approximation also in those cases, we may try to enrich our approximated semantics by rules exploiting the structure of formulae.

We can thus extend both under- and over-approximated semantics (Definitions 8 and 10). In the following we present the enrichment for over-approximated semantics only, with the under-approximated case treated similarly.

Definition 13 (Enriched over-approximated semantics). *Let $\overline{R}$ be a simulation from M to $\overline{M}$ (i.e. $M \sqsubseteq_{\overline{R}} \overline{M}$) and ρ an assignment, such that $\{\!|\cdot|\!\}_\rho^{\overline{R}}$ is the over-approximated semantics of $[\![\cdot]\!]_\rho^M$ in $\overline{M}$ via $\overline{R}$. The enriched over-approximated semantics of $[\![\cdot]\!]_\rho^M$ in $\overline{M}$ via $\overline{R}$ is given by the function $^+\{\!|\cdot|\!\}_\rho^{\overline{R}} : \mathcal{F}^{[\Gamma;\Delta]} \rightarrow (\Omega_{\overline{M}}^{[\Gamma;\Delta]} \rightarrow \mathbb{K})$ defined as*

$$ ^+\{\!|\psi[\Gamma;\Delta]|\!\}_\rho^{\overline{R}} = \begin{cases} ^+\{\!|\psi_1[\Gamma;\Delta]|\!\}_\rho^{\overline{R}} \vee {}^+\{\!|\psi_2[\Gamma;\Delta]|\!\}_\rho^{\overline{R}} & \text{if } \psi :_{\overline{R}} \bot \text{ and } \psi \equiv \psi_1 \vee \psi_2 \\ \neg^+\{\!|\psi_1[\Gamma;\Delta]|\!\}_\rho^{\overline{R}} & \text{if } \psi :_{\overline{R}} \bot \text{ and } \psi \equiv \neg\psi_1 \\ \{\!|\psi[\Gamma;\Delta]|\!\}_\rho^{\overline{R}} & \text{otherwise} \end{cases} $$

Example 9. Consider again the predicate $\mathbf{p}(x,y)$ of Example 3, and the models M and $\overline{M}$ with $M \sqsubseteq_{\overline{R}} \overline{M}$ of Example 5 shown in Fig. 2. In Example 8 we have seen that the over-approximated semantics of $[\![\mathbf{p}(u,v) \vee \neg\mathbf{p}(u,u)]\!]^M$ in $\overline{M}$ via $\overline{R}$ does not provide us any information. This happens because $\mathbf{p}(u,v) :_{\overline{R}}\to$ and $\neg\mathbf{p}(u,u) :_{\overline{R}}\leftarrow$, and hence $\mathbf{p}(u,v) \vee \neg\mathbf{p}(u,u) :_{\overline{R}} \bot$. In particular, as depicted in the third column of the first group of lines of Fig. 3, all the pairs in $\Omega_{\overline{M}}^{[z;\emptyset]}$ are assigned to "?". The enriched over-approximated semantics is instead more interesting. Following Definition 13, we evaluate separately the (enriched) over-approximated semantics of the two disjuncts (first and second column of the first group of lines of Fig. 3), and then combine them as shown in the last column of Fig. 3. Considering for example the pair $(\overline{w}_0, \lambda)$, we have $^+\{\!\{\mathbf{p}(u,v)[z;\emptyset]\}\!\}_\rho^{\overline{R}}((\overline{w}_0, \lambda)) = F$, and $^+\{\!\{\neg\mathbf{p}(u,u)[z;\emptyset]\}\!\}_\rho^{\overline{R}}((\overline{w}_0, \lambda)) = T$, and hence $^+\{\!\{\mathbf{p}(u,v)) \vee \neg\mathbf{p}(u,u)[z;\emptyset]\}\!\}_\rho^{\overline{R}}((\overline{w}_0, \lambda)) = F \vee T = T$. Considering instead $(\overline{w}_1, (z \mapsto e, \lambda_2))$ we have $^+\{\!\{\mathbf{p}(u,v))\} \vee \neg\mathbf{p}(u,u)[z;\emptyset]\}\!\}_\rho^{\overline{R}}((\overline{w}_1, (z \mapsto e, \lambda_2))) = ? \vee T = T$.

We may enrich the under-approximated semantics exactly in the same way, and thus straightforwardly define an enriched version "$\cdot^+\models_{\{R_i\}}^{\{R_j\}} [\![\cdot]\!]_\rho^M$" of the approximated checking by replacing both approximated semantics with their enriched variants. It is also easy to verify that also this new check is sound.

6 Conclusions and Further Works

In the present work we proposed a general framework for simulation-based approximations, and we exploited them for developing a verification technique based on a suitable type system for formulae of a second-order modal logic with fix-point operators. The logic was previously introduced for the specification of systems with dynamic topology [8, 11], and it is thus now equipped with a powerful abstraction mechanism.

Our approach can be seen as an evolution of the verification technique for graph transformation systems based on temporal graph logics and unfoldings [2, 4], which is generalized to counterpart models, and extended for the kind of simulations under analysis.

Our proposal may provide interesting insights for other approximation techniques, such as *neighbourhood abstraction* [5], where states are *shapes* (i.e. graph algebras extended with an operation for abstraction purposes), and suitable abstraction morphisms (i.e. surjective graph morphisms, similar to the morphisms of our simulations) coalesce nodes and edges of concrete states according to their neighbourhood similarity. The logic adopted is less expressive than ours (as well as of the one used in [2]), but it offers the advantage that all formulae are strongly preserved by the approximation.

We foresee several directions for further research. First, we plan to enrich our prototypal model checker for finite models [11] with the techniques presented here, possibly making use of existing approximation techniques, like the ones previously mentioned. Second, we would like to investigate the enrichment of approximated semantics in order to deal with more untyped formulae. An interesting question in this regard is whether we can use both an under- and an over-approximation *simultaneously*, by translating assignment pairs back and forth via the composition of the corresponding abstraction and concretization functions.

References

1. Baldan, P., Corradini, A., König, B.: A Static Analysis Technique for Graph Transformation Systems. In: Larsen, K.G., Nielsen, M. (eds.) CONCUR 2001. LNCS, vol. 2154, pp. 381–395. Springer, Heidelberg (2001)
2. Baldan, P., Corradini, A., König, B., Lluch Lafuente, A.: A Temporal Graph Logic for Verification of Graph Transformation Systems. In: Fiadeiro, J.L., Schobbens, P.-Y. (eds.) WADT 2006. LNCS, vol. 4409, pp. 1–20. Springer, Heidelberg (2007)
3. Baldan, P., König, B.: Approximating the Behaviour of Graph Transformation Systems. In: Corradini, A., Ehrig, H., Kreowski, H.-J., Rozenberg, G. (eds.) ICGT 2002. LNCS, vol. 2505, pp. 14–29. Springer, Heidelberg (2002)
4. Baldan, P., König, B., König, B.: A Logic for Analyzing Abstractions of Graph Transformation Systems. In: Cousot, R. (ed.) SAS 2003. LNCS, vol. 2694, pp. 255–272. Springer, Heidelberg (2003)
5. Bauer, J., Boneva, I., Kurbán, M.E., Rensink, A.: A Modal-Logic Based Graph Abstraction. In: Ehrig, H., Heckel, R., Rozenberg, G., Taentzer, G. (eds.) ICGT 2008. LNCS, vol. 5214, pp. 321–335. Springer, Heidelberg (2008)
6. Courcelle, B., Engelfriet, J.: Graph Structure and Monadic Second-Order Logic, A Language Theoretic Approach. Cambridge University Press, Cambridge (2012)
7. Dawar, A., Gardner, P., Ghelli, G.: Expressiveness and complexity of graph logic. Information and Computation 205(3), 263–310 (2007)
8. Gadducci, F., Lluch Lafuente, A., Vandin, A.: Counterpart semantics for a second-order mu-calculus. Fundamenta Informaticae 118(1-2) (2012)
9. Ghamarian, A.H., de Mol, M., Rensink, A., Zambon, E., Zimakova, M.: Modelling and Analysis Using GROOVE. Software Tools for Technology Transfer 14(1), 15–40 (2012)
10. König, B., Kozioura, V.: Counterexample-Guided Abstraction Refinement for the Analysis of Graph Transformation Systems. In: Hermanns, H., Palsberg, J. (eds.) TACAS 2006. LNCS, vol. 3920, pp. 197–211. Springer, Heidelberg (2006)
11. Lluch Lafuente, A., Vandin, A.: Towards a Maude Tool for Model Checking Temporal Graph Properties. In: Gadducci, F., Mariani, L. (eds.) GT-VMT. ECEASST, vol. 42, EAAST (2011)
12. Rensink, A.: Towards model checking graph grammars. In: Leuschel, M., Gruner, S., Lo Presti, S. (eds.) AVOCS. DSSE-TR, vol. 2003-2. University of Southampton (2003)
13. Rensink, A.: Isomorphism checking in GROOVE. In: Zündorf, A., Varró, D. (eds.) GraBaTs. ECEASST, vol. 1. EAAST (2006)
14. Rensink, A., Distefano, D.: Abstract graph transformation. In: Mukhopadhyay, S., Roychoudhury, A., Yang, Z. (eds.) SVV. ENTCS, vol. 157(1), pp. 39–59. Elsevier, Amsterdam (2006)
15. Rensink, A., Zambon, E.: Neighbourhood abstraction in GROOVE. In: de Lara, J., Varró, D. (eds.) GraBaTs. ECEASST, vol. 32. EAAST (2010)
16. Zambon, E., Rensink, A.: Using Graph Transformations and Graph Abstractions for Software Verification. In: Corradini, A. (ed.) ICGT - Doctoral Symposium. ECEASST, vol. 38. EASST (2011)

Pattern-Based Graph Abstraction*

Arend Rensink and Eduardo Zambon

Formal Methods and Tools Group,
Computer Science Department,
University of Twente
P.O. Box 217, 7500 AE, Enschede, The Netherlands
{rensink,zambon}@cs.utwente.nl

Abstract. We present a new abstraction technique for the exploration
of graph transformation systems with infinite state spaces. This technique is based on *patterns*, simple graphs describing structures of interest
that should be preserved by the abstraction. Patterns are collected into
pattern graphs, layered graphs that capture the hierarchical composition
of smaller patterns into larger ones. Pattern graphs are then abstracted
to a finite universe of *pattern shapes* by collapsing equivalent patterns.
This paper shows how the application of production rules can be lifted to
pattern shapes, resulting in an over-approximation of the original system
behaviour and thus enabling verification on the abstract level.

1 Introduction

Graph transformation (GT) is a framework that, on one hand, is intuitive and
flexible enough to serve as a basic representation of many kinds of structures,
and, on the other hand, is precise and powerful enough to formally describe system behaviour. Many techniques and tools have been proposed to analyse the
behaviour of GT systems. In particular, systems with infinite state spaces pose a
challenge since they require some form of abstraction. A key aspect when designing such abstractions is the trade-off between preserving the expressive power of
GT and managing the complexity of the abstraction mechanism. This trade-off
has been considered at various points of its scale on the different abstractions
given in the literature [1,3,17].

In this paper we present a new abstraction based on *graph patterns* (simple
edge-labelled graphs), to be used in the analysis of infinite-state GT systems.
The novelty of the approach lies in the flexibility for tuning the abstraction
according to the substructures of interest, represented via a *type graph*. At the
concrete level we work with *pattern graphs*, a layered structure that describes the
composition of patterns. The abstraction of pattern graphs gives rise to *pattern
shapes*, which are bounded structures forming a finite universe.

In this work we define how pattern graphs and pattern shapes are constructed
from simple graphs, and we show how the application of GT rules can be lifted

---

* The work reported herein is being carried out as part of the GRAIL project, funded
by NWO (Grant 612.000.632).

H. Ehrig et al.(Eds.): ICGT 2012, LNCS 7562, pp. 66–80, 2012.

to these new structures. Our major result is a proof that the abstract state space obtained by transforming pattern shapes is an over-approximation of the original system behaviour, thus enabling verification on the abstract level. This text is an abridged version of [16] where all technical details, including proofs, are given.

2   Simple Graphs

In its basic form, a graph is composed of nodes and directed binary edges.

Definition 1 (graph). *A graph is a tuple* $G = \langle N_G, E_G, \mathsf{src}_G, \mathsf{tgt}_G \rangle$ *where*
- N_G *is a finite set of nodes;*
- E_G *is a finite set of edges, disjoint from N_G; and*
- $\mathsf{src}_G : E_G \to N_G$ *and* $\mathsf{tgt}_G : E_G \to N_G$ *are mappings associating each edge to its source and target nodes, respectively.*   ◀

For a node $v \in N_G$, we consider the set of edges *outgoing* from and *incoming* to v, defined as $v \triangleright_G = \{e \in E_G \mid \mathsf{src}_G(e) = v\}$ and $v \triangleleft_G = \{e \in E_G \mid \mathsf{tgt}_G(e) = v\}$, respectively. A *path* in G is a non-empty sequence of edges $\pi = e_1 \cdots e_k$ such that $\mathsf{tgt}_G(e_i) = \mathsf{src}_G(e_{i+1})$ for $1 \leq i < k$. For convenience, we write $\mathsf{src}(\pi) = \mathsf{src}_G(e_1)$ and $\mathsf{tgt}(\pi) = \mathsf{tgt}_G(e_k)$. Paths π_1, π_2 are *parallel* if $\mathsf{src}(\pi_1) = \mathsf{src}(\pi_2)$ and $\mathsf{tgt}(\pi_1) = \mathsf{tgt}(\pi_2)$. Furthermore, v is a *predecessor* of w in G, denoted $v \leq_G w$, if either $v = w$ or there is a path π with $\mathsf{src}(\pi) = v$ and $\mathsf{tgt}(\pi) = w$.

Definition 2 (graph morphism). *A graph morphism between graphs G, H is a function $m : (N_G \cup E_G) \to (N_H \cup E_H)$, such that $m(N_G) \subseteq N_H$, $m(E_G) \subseteq E_H$, $\mathsf{src}_H \circ m = m \circ \mathsf{src}_G$, and $\mathsf{tgt}_H \circ m = m \circ \mathsf{tgt}_G$.*   ◀

If function m is injective (surjective, bijective) then the morphism is called injective (surjective, bijective). A bijective morphism is also called an *isomorphism* and we write $G \simeq H$ to denote that there is an isomorphism between G and H. We use $m : G \to H$ as a short-hand notation for $m : (N_G \cup E_G) \to (N_H \cup E_H)$. If $G \subseteq H$, we use $\mathsf{emb}(G, H)$ to denote the *embedding* of G into H.

Let Lab be a finite set of labels, partitioned into disjoint unary and binary label sets, denoted $\mathsf{Lab}^{\mathsf{U}}$ and $\mathsf{Lab}^{\mathsf{B}}$, respectively.

Definition 3 (simple graph). *A simple graph G is a graph extended with an edge labelling function $\mathsf{lab}_G : E_G \to \mathsf{Lab}$, where*
- *for all $e \in E_G$, if $\mathsf{lab}_G(e) \in \mathsf{Lab}^{\mathsf{U}}$ then $\mathsf{src}_G(e) = \mathsf{tgt}_G(e)$; and*
- *for any $e_1, e_2 \in E_G$, if $\mathsf{src}_G(e_1) = \mathsf{src}_G(e_2)$, $\mathsf{tgt}_G(e_1) = \mathsf{tgt}_G(e_2)$, and $\mathsf{lab}_G(e_1) = \mathsf{lab}_G(e_2)$, then $e_1 = e_2$.*   ◀

The second condition in the definition above prohibits parallel edges with the same label, justifying the choice of the term simple graph. The first condition limits the occurrence of unary labels to self-edges, which are used to encode node labels. We write $\langle v, l, w \rangle$ to represent an edge e with $\mathsf{src}_G(e) = v$, $\mathsf{lab}_G(e) = l$, and $\mathsf{tgt}_G(e) = w$. The universe of simple graphs is denoted by SGraph.

Figure 1(a) shows an example of a simple graph representing a single-linked list composed of five cells and a sentinel node to mark the head and tail elements

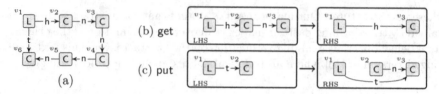

Fig. 1. (a) Simple graph representing a linked list. (b,c) Two transformation rules.

of the list. Unary labels are shown inside their associated node and node identities are displayed at the top left corner of each node.

Definition 4 (simple graph morphism). *A simple graph morphism between simple graphs* $G, H \in$ SGraph *is a graph morphism* $m : G \rightarrow H$ *that preserves edge labels, i.e.,* $\mathsf{lab}_H \circ m = \mathsf{lab}_G$. ◀

We write SMorph to represent the universe of simple graph morphisms. Simple graphs are transformed by simple graph rules.

Definition 5 (simple graph transformation rule). *A simple graph transformation rule* $r = \langle L, R \rangle$ *consists of two simple graphs* $L, R \in$ SGraph, *called left-hand side (LHS) and right-hand side (RHS), respectively.* ◀

The relation between L and R is established by their common elements, via an implicit identity morphism on L. We distinguish the following sets:
- $N^{\mathsf{del}} = N_L \setminus N_R$ and $E^{\mathsf{del}} = E_L \setminus E_R$ are the sets of elements deleted; and
- $N^{\mathsf{new}} = N_R \setminus N_L$ and $E^{\mathsf{new}} = E_R \setminus E_L$ are the elements created by the rule.

We write $U = L \cup R$ and $U^{\mathsf{new}} = N^{\mathsf{new}} \cup E^{\mathsf{new}}$ as short-hand notation, and we use SRule to denote the universe of simple graph transformation rules. Figure 1(b,c) shows rules removing the head element of a list (**get**) and inserting a new element at the tail of the list (**put**).

Definition 6 (simple graph transformation). *Let* G *be a simple graph and* $r = \langle L, R \rangle$ *a simple graph transformation rule such that* G *and* U^{new} *are disjoint. An application of* r *into* G *involves finding a match* m *of* r *into* G, *which is an injective simple graph morphism* $m : L \rightarrow G$. *Extend* m *to* U *by taking* $m \cup \mathsf{id}_{U^{\mathsf{new}}}$. *Given such* m, *rule* r *transforms* G *into a new simple graph* H, *where*
- $N_H = (N_G \setminus m(N^{\mathsf{del}})) \cup N^{\mathsf{new}}$;
- $E_H = (\{e \in E_G \mid \mathsf{src}_G(e), \mathsf{tgt}_G(e) \notin m(N^{\mathsf{del}})\} \setminus m(E^{\mathsf{del}})) \cup E^{\mathsf{new}}$;
- $\mathsf{src}_H = (\mathsf{src}_G \cup (m \circ \mathsf{src}_U))|_{E_H}$, $\mathsf{tgt}_H = (\mathsf{tgt}_G \cup (m \circ \mathsf{tgt}_U))|_{E_H}$; *and*
- $\mathsf{lab}_H = (\mathsf{lab}_G \cup \mathsf{lab}_U)|_{E_H}$. ◀

In the definition above, dangling edges are deleted, following the SPO approach. Since we do not distinguish between isomorphic graphs, the assumption that U^{new} and G are disjoint can be satisfied without loss of generality by taking an isomorphic copy of U where the elements of U^{new} are fresh with respect to G. We write $G \xrightarrow{r} H$ to denote that the application of r to G (under some m) gives rise to the transformed graph H.

Definition 7 (simple graph grammar). *A simple graph grammar is a tuple* $\mathcal{G} = \langle \mathcal{R}, G_0 \rangle$, *with* $\mathcal{R}$ *a set of simple graph rules and* G_0 *an initial simple graph.*◀

Our standard model of behaviour is a simple graph transition system (SGTS).

Definition 8 (SGTS). *A simple graph transition system is a tuple* $SGTS = \langle \mathcal{S}, \rightarrow, \iota \rangle$ *where* $\mathcal{S} \subseteq$ SGraph *is a set of states,* $\rightarrow \subseteq \mathcal{S} \times$ SRule $\times \mathcal{S}$ *a set of rule applications, and* $\iota \in \mathcal{S}$ *is the initial state. Grammar* $\mathcal{G}$ *generates a* $SGTS_{\mathcal{G}}$ *if* $\iota = G_0$ *and* $\mathcal{S}$ *is the minimal set of graphs such that* $G \in \mathcal{S}$ *and* $G \xrightarrow{r} H$ *for* $r \in \mathcal{R}$ *implies that there exists* $H' \in \mathcal{S}$ *where* $H \simeq H'$ *and* $G \xrightarrow{r} H'$ *is a transition.* ◀

3 Pattern Graphs

Pattern graphs are the cornerstone graph representation in this work. We first give a general definition, but in this paper we restrict ourselves to pattern graphs that are *well-formed*, a condition that will be presented shortly.

Definition 9 (pattern graph). *A pattern graph* P *is a graph extended with a labelling function* $\mathsf{lab}_P : N_P \cup E_P \rightarrow$ SGraph $\cup$ SMorph *that maps nodes to simple graphs* $(\mathsf{lab}_P(N_P) \subseteq$ SGraph), *and edges to simple graph morphisms* $(\mathsf{lab}_P(E_P) \subseteq$ SMorph), *such that* $\mathsf{lab}_P(d) : \mathsf{lab}_P(\mathsf{src}_P(d)) \rightarrow \mathsf{lab}_P(\mathsf{tgt}_P(d))$ *is an injective, non-surjective simple graph morphism, for all* $d \in E_P$. ◀

In categorical terms, a pattern graph corresponds to a diagram in the category SGraph. Elements of N_P and E_P are called *pattern nodes* and *pattern edges*, respectively. For $p \in N_P$, $\mathsf{lab}_P(p)$ is the *pattern* of p. As with simple graphs, we may write $\langle p, f, q \rangle$ as a short-hand for a pattern edge d with $\mathsf{src}_P(d) = p$, $\mathsf{lab}_P(d) = f$, and $\mathsf{tgt}_P(d) = q$. Note that the restriction to non-surjective simple graph morphisms means that the pattern of $\mathsf{tgt}_P(d)$ is always strictly larger than that of $\mathsf{src}_P(d)$, which in turn implies that a pattern graph is always a dag. We layer the nodes of P according to the number of simple edges in their patterns (for $i \geq 0$): $N_P^i = \{p \in N_P \mid |E_G| = i, G = \mathsf{lab}_P(p)\}$ and $N_P^{i^+} = \bigcup_{j \geq i} N_P^j$.

Figure 2 shows an example of a pattern graph. Pattern nodes are drawn with dashed lines and the associated patterns are shown inside the node. Pattern edges are depicted as arrows labelled with their corresponding simple graph morphisms, except that embeddings are omitted to avoid clutter. Layers are indicated on the right. Note that there is no distinction between simple edges with unary or binary labels for the purpose of layer assignment. From here on we simplify the figures by showing only the patterns with labelled simple nodes.

Let $d = \langle p, f, q \rangle \in E_P$ be a pattern edge and let $G = \mathsf{lab}_P(p)$. The *image* of d is defined as $\mathsf{img}_P(d) = H$ where $N_H = f(N_G)$, $E_H = f(E_G)$, $\mathsf{src}_H = f \circ \mathsf{src}_G$, $\mathsf{tgt}_H = f \circ \mathsf{tgt}_G$, and $\mathsf{lab}_H \circ f = \mathsf{lab}_G$. It is easy to see that $H \subset \mathsf{lab}_P(q)$. We say that every pattern edge d *covers* the sub-graph $\mathsf{img}_P(d)$. Furthermore, we call a set of pattern edges $E' \subseteq E_P$ *jointly surjective* if $\mathsf{tgt}_P(d_1) = \mathsf{tgt}_P(d_2) = p$ for all $d_1, d_2 \in E'$, and $\bigcup_{d \in E'} \mathsf{img}_P(d) = \mathsf{lab}_P(p)$. As an equivalent term we say that the pattern edges of E' *together cover* $\mathsf{lab}_P(p)$.

A pattern graph P is called *well-formed* if (with $G = \mathsf{lab}_P(p)$)

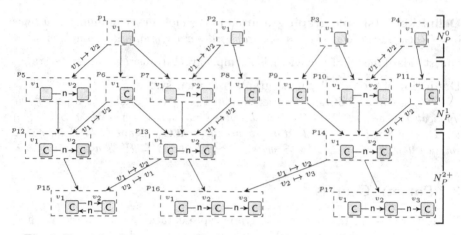

Fig. 2. Example of a pattern graph that is not well-formed and not commuting

- for all $p \in N_P^0$, $|N_G| = 1$;
- for all $p \in N_P^1$ and the unique $e \in E_G$, $N_G = \{\text{src}_G(e), \text{tgt}_G(e)\}$ and $p \triangleleft_P$ together cover N_G; and
- for all $p \in N_P^{2+}$, $p \triangleleft_P$ together cover G.

In words, the patterns of level-0 pattern nodes consist of a single node, the patterns of level-1 pattern nodes consist of a single edge and its end nodes, and the patterns on any other level are determined by the combined images of their predecessors. Another consequence of pattern morphisms being non-surjective is that, on well-formed pattern graphs, there are no pattern edges between nodes of the same layer. The universe of (well-formed) pattern graphs is denoted PGraph. Note that the pattern graph in Figure 2 is not well-formed: pattern nodes p_5, p_{12}, and p_{17} are not sufficiently covered by the incoming morphisms.

Definition 10 (pattern graph morphism). *A pattern graph morphism between pattern graphs* $P, Q \in$ PGraph *is a graph morphism* $m : P \to Q$, *where,*

1. *for all* $p \in N_P$, *there exists an isomorphism* $\varphi_p : \text{lab}_P(p) \to \text{lab}_Q(m(p))$; *and*
2. *for all* $d = \langle p, f, q \rangle \in E_P$, $\varphi_q \circ f = f' \circ \varphi_p$, *with* $f' = \text{lab}_Q(m(d))$.

Moreover, m *is called* closed *if, in addition,*

3. *for all* $p \in N_P$ *and* $d' \in m(p) \triangleleft_Q$, *there exists* $d \in p \triangleleft_P$ *with* $m(d) = d'$; *and*
4. *for all* $N' \subseteq N_P$ *and jointly surjective* $\{d'_k \in m(p) \triangleright_Q \mid p \in N'\}_{k \in K}$ *(where* K *is some index set), there are jointly surjective* $\{d_k \in p \triangleright_P \mid p \in N'\}_{k \in K}$ *with* $m(d_k) = d'_k$, *for all* $k \in K$. ◀

The definition above states that m maps pattern nodes with isomorphic patterns (condition 1) and that m is compatible with the simple graph morphisms of pattern edges modulo isomorphism (condition 2). Closedness indirectly imposes conditions on P: every pattern edge in Q whose target pattern node is in the morphism image should itself also be the image of some pattern edge in P (condition 3), and so should every jointly surjective set of pattern edges in Q whose source pattern nodes are in the morphism image (condition 4).

Definition 11. *Let P be a pattern graph.*

- *P is called* commuting *when for all $q \in N_P^{2+}$ and any distinct $d_1, d_2 \in q \lhd_P$, if $G = \text{img}_P(d_1) \cap \text{img}_P(d_2)$ is not an empty graph, then there exist $p \in N_P$ and parallel paths π_1, π_2 in P such that $\text{lab}_P(p) = G$, $\text{src}(\pi_i) = p$, $\text{tgt}(\pi_i) = q$, and $d_i \in \pi_i$, for $i = 1, 2$.*
- *P is called* concrete *if it satisfies the following properties:*
  1. *for all distinct $p, q \in N_P$, $\text{lab}_P(p) \neq \text{lab}_P(q)$; and*
  2. *for all $\langle p, f, q \rangle \in E_P$, $f = \text{emb}(\text{lab}_P(p), \text{lab}_P(q))$.* ◄

The commutativity condition states that common simple nodes and edges in patterns always stem from a common ancestor. The pattern graph in Figure 2 is not commuting: the pattern associated with pattern node p_{16} cannot be constructed from its predecessors since there is no common ancestor for simple node v_2.

The first concrete pattern graph condition states that all patterns are distinct and the second condition that identities of simple graph elements are preserved along pattern graph edges. For concrete pattern graphs P, we define the *flattening* of P as $\text{flat}(P) = \bigcup_{p \in N_P} \text{lab}_P(p)$. The following states that we can essentially always treat commuting pattern graphs as concrete.

Proposition 12. *Let P be a pattern graph.*
1. *If P is concrete, then P is commuting.*
2. *If P is commuting, there is a concrete pattern graph Q isomorphic to P.* ◄

So far we have not restricted the patterns occurring in a pattern graph, but in practice we will only use *typed* patterns.

Definition 13 (pattern type graph). *A pattern type graph T is a pattern graph such that $\text{lab}_T(p) \not\simeq \text{lab}_T(q)$ for all distinct $p, q \in N_T$. A T-type morphism is a closed pattern graph morphism to T.* ◄

Figure 3 shows an example of a pattern type graph. We call P a *T-pattern graph* if it is typable by T, *i.e.*, has a T-type morphism. It is easy to see that for a given pattern type graph T, any pattern graph P has at most one morphism to T. If this morphism exists but is not a type morphism (*i.e.*, is not closed), it is always possible to extend P to a T-pattern graph Q, namely by adding the elements required by the morphism closure conditions.

Proposition 14. *Let P be a concrete pattern graph and T a pattern type graph. If there exists a morphism $m : P \to T$, then there exists an unique (modulo isomorphism) concrete T-pattern graph $Q \supseteq P$, and $\text{flat}(P) = \text{flat}(Q)$.* ◄

We call Q the *closure* of P with respect to T, and denote it $\text{close}_T(P)$. Given a pattern type graph T we can define the *lifting* operation $P = \text{lift}_T(G)$ from simple graphs G to concrete T-pattern graphs P with typing morphism t:

- For all $H \subseteq G$ such that there exists an isomorphism $\varphi_H : H \to \text{lab}_T(p')$ for some $p' \in N_T$, add a fresh p_H to N_P and let $\text{lab}_P : p_H \mapsto H$, and $t : p_H \mapsto p'$.
- For all $p_H \in N_P$ and for every $d' \in t(p_H) \lhd_T$, let $F \subseteq H$ be defined by $F = \varphi_H^{-1}(\text{img}_T(d'))$. Note that this implies $F \simeq \text{lab}_T(\text{src}_T(d'))$, and therefore

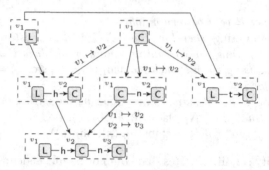

Fig. 3. Example of a pattern type graph

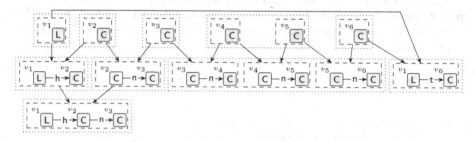

Fig. 4. Concrete pattern graph lifted from the simple graph of Figure 1 according to the pattern type graph of Figure 3. The pattern equivalence relation ≡ (Definition 25) over nodes of the pattern graph is shown with dotted rectangles.

there exists a $p_F \in N_P$. Add a fresh edge d to E_P and let $\mathsf{src}_P \colon d \mapsto p_F$, $\mathsf{tgt}_P \colon d \mapsto p_H$, $\mathsf{lab}_P \colon d \mapsto \mathsf{emb}(F, H)$, and $t \colon d \mapsto d'$.

An example of a lifted concrete T-pattern graph is shown in Figure 4.

Proposition 15. *Let T be a pattern type graph.*
1. *For any simple graph G, $\mathsf{lift}_T(G)$ is a concrete T-pattern graph.*
2. *For any concrete T-pattern graph P, $\mathsf{lift}_T(\mathsf{flat}(P)) \simeq P$.*   ◀

Concrete T-pattern graphs are transformed by lifting simple graph rules to pattern graph equivalents. Usually we only consider simple rules whose left hand sides are patterns in T, i.e., for any simple graph rule $r = \langle L, R \rangle$, $L \simeq \mathsf{lab}_T(p)$ for some $p \in N_T$. In this sense, a set of simple rules $\mathcal{R}$ constrains the choice of type graph T, or, equivalently, T is extracted from set $\mathcal{R}$.

Definition 16 (pattern graph transformation rule). *A pattern graph rule $r = \langle \lambda, \rho \rangle$ consists of two concrete T-pattern graphs λ (the LHS) and ρ (the RHS), where for any $x_1 \in \lambda$ and $x_2 \in \rho$, $x_1 = x_2$ if and only if $\mathsf{lab}_\lambda(x_1) = \mathsf{lab}_\rho(x_2)$.* ◀

The definition above implies that there exists an identity pattern graph morphism on λ ensuring the equality of patterns. Based on this identity, we distinguish the sets of pattern graph elements deleted and created by pattern graph

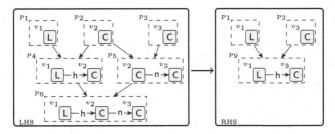

Fig. 5. Pattern graph rule equivalent to the get rule of Figure 1

rules, as done previously for simple graphs. Also, we use $\Upsilon = \lambda \cup \rho$ to denote the pattern graph resulting from the union of LHS and RHS, and Υ^{new} to represent the set of pattern graph elements created by the rule. We use PRule to denote the universe of pattern graph transformation rules, with one shown in Figure 5. A simple graph rule $r = \langle L, R \rangle$ and a pattern graph rule $r' = \langle \lambda, \rho \rangle$ are *equivalent* if $\text{flat}(\lambda) = L$ and $\text{flat}(\rho) = R$.

Essentially, what happens when a rule is applied is that graph elements are removed and others are added. In a concrete pattern graph, each simple graph element x is represented by a pattern node in N_P^0 (if x is a simple node) or N_P^1 (if x is a simple edge), and it also contributes to all successor patterns; so when x is removed, all those pattern nodes disappear. Conversely, adding simple graph elements to a pattern graph means adding new pattern nodes to N_P^0 and N_P^1, and then closing the resulting structure with respect to T.

Definition 17 (pattern graph transformation). *Let P be a concrete pattern graph, $r = \langle L, R \rangle$ a simple graph transformation rule, and $r' = \langle \lambda, \rho \rangle$ an equivalent pattern graph rule such that P and Υ^{new}, and $\text{flat}(P)$ and U^{new} are disjoint. An application of r' into P involves finding a match μ of r' into P, which is an injective pattern graph morphism $\mu : \lambda \to P$. Match μ implicitly induces a simple graph match $m : L \to \text{flat}(P)$. Extend μ to Υ and m to U by taking $\mu \cup \text{id}_{\Upsilon^{\text{new}}}$ and $m \cup \text{id}_{U^{\text{new}}}$, respectively, and let $N' = \{q \in N_P \mid p \in \mu(N^{\text{del}}), p \leq_P q\}$ and $E' = \{d \in E_P \mid \text{src}_P(d) \in N' \text{ or } \text{tgt}_P(d) \in N'\}$. Given such μ, r' transforms P into $\text{close}_T(Q)$, where Q is defined by*

- $N_Q = (N_P \setminus N') \cup N^{\text{new}}$ *and* $E_Q = (E_P \setminus E') \cup E^{\text{new}}$;
- $\text{src}_Q = (\text{src}_P \cup (\mu \circ \text{src}_\Upsilon))|_{E_Q}$ *and* $\text{tgt}_Q = (\text{tgt}_P \cup (\mu \circ \text{tgt}_\Upsilon))|_{E_Q}$;
- *for all* $p \in N_Q$, $\text{lab}_Q : \begin{cases} p \mapsto \text{lab}_P(p) & \text{if } p \notin N^{\text{new}}, \\ p \mapsto m(\text{lab}_\rho(p)) & \text{otherwise; and} \end{cases}$
- *for all* $d \in E_Q$, $\text{lab}_Q : d \mapsto \text{emb}(\text{lab}_Q(\text{src}_Q(d)), \text{lab}_Q(\text{tgt}_Q(d)))$. ◀

As with simple graph transformations, we can satisfy the disjointness assumptions of the definition above by taking isomorphic copies of r and r' and the result of the transformation is the same, modulo isomorphism. It is easy to see that Q is a concrete pattern graph and therefore its closure w.r.t. T is well-defined. Figure 6 shows an example of a pattern graph transformation.

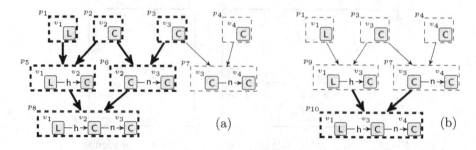

Fig. 6. Example of a pattern graph transformation. (a) Pattern graph to be transformed, with match of rule **get** shown in bold. (b) Resulting pattern graph, with elements added by the closure operation shown in bold.

We come now to the first major result of this paper: simple and pattern graph transformations are equivalent.

Theorem 18 (transformation equivalence). *Let G be a simple graph, $r = \langle L, R \rangle$ a simple graph rule, and $r' = \langle \lambda, \rho \rangle$ an equivalent pattern graph rule.*

1. If $G \xrightarrow{r} H$ is a simple graph transformation then there is a pattern graph transformation $\mathsf{lift}_T(G) \xrightarrow{r'} Q$ with $Q \simeq \mathsf{lift}_T(H)$.

2. If $\mathsf{lift}_T(G) \xrightarrow{r'} Q$ is a pattern graph transformation then there is a simple graph transformation $G \xrightarrow{r} H$ with $Q \simeq \mathsf{lift}_T(H)$. ◀

The equivalence between simple graph and pattern graph transformations is used to show the equivalence between simple graph transition systems (SGTS, Definition 8) and pattern graph transition systems (PGTS, defined below).

Definition 19 (pattern graph grammar). *A pattern graph grammar $\mathcal{P}_T = \langle \mathcal{R}_T, P_0 \rangle$ has a set of pattern graph rules $\mathcal{R}_T$ and an initial T-pattern graph P_0.* ◀

We say that a simple graph grammar $\mathcal{G} = \langle \mathcal{R}, G_0 \rangle$ and a pattern graph grammar $\mathcal{P}_T = \langle \mathcal{R}'_T, P_0 \rangle$ are *equivalent* if for any simple graph rule $r \in \mathcal{R}$ there exists an equivalent pattern graph rule $r' \in \mathcal{R}'_T$, and vice versa; and $P_0 = \mathsf{lift}_T(G_0)$.

Definition 20 (PGTS). *A pattern graph transition system $PGTS = \langle \mathcal{S}, \Rightarrow, \iota \rangle$ consists of a set of states $\mathcal{S} \subseteq \mathsf{PGraph}$, a set of rule applications $\Rightarrow \subseteq \mathcal{S} \times \mathsf{PRule} \times \mathcal{S}$, and an initial state $\iota \in \mathcal{S}$. Grammar $\mathcal{P}_T$ generates a $PGTS_{\mathcal{P}}$ if $\iota = P_0$ and $\mathcal{S}$ is the minimal set of graphs such that $P \in \mathcal{S}$ and $P \xrightarrow{r} Q$ for $r \in \mathcal{R}_T$ implies that there exists $Q' \in \mathcal{S}$ such that $Q \simeq Q'$ and $P \xrightarrow{r} Q'$ is a transition.* ◀

We conclude this section with our second major result, which establishes the relation between a SGTS and a PGTS generated by equivalent grammars.

Theorem 21. *Let T be a pattern type graph, $\mathcal{G}$ a simple graph grammar and $\mathcal{P}_T$ a pattern graph grammar equivalent to $\mathcal{G}$. Transition systems $SGTS_{\mathcal{G}}$ and $PGTS_{\mathcal{P}}$ are isomorphic.* ◀

Isomorphism is a quite interesting result because it implies that satisfaction of μ-calculus formulae (and thus also CTL*, CTL, and LTL formulae) are preserved among the two systems. This in turn means that we can discard the SGTS and perform verification (model-checking) on the pattern graph level. However, a PGTS may still be infinite, effectively preventing its construction.

4   Pattern Shapes

Pattern graphs are abstracted into pattern shapes. As usual with structural abstraction, equivalent structures (patterns) are collapsed into an abstract representative, while keeping an approximate count of the number of concrete elements collapsed. We use ω to denote an upper bound on the set of natural numbers and we write $\mathbb{N}^\omega = \mathbb{N} \cup \{\omega\}$. A *multiplicity* is an element of set $\mathcal{M} = \{\langle i, j \rangle \in (\mathbb{N} \times \mathbb{N}^\omega) \mid i \leq j\}$ that is used to represent an interval of consecutive values taken from $\mathbb{N}^\omega$. Given $\langle i, j \rangle \in \mathcal{M}$, if $i = j$ we write it as $\mathbf{i}$ and if $j = \omega$, we use $\mathbf{i}^+$ as short-hand. Multiplicity $\mathbf{1}$ is called *concrete*. Set $\mathcal{M}$ is infinite, since i and j are taken from infinite sets. To ensure finiteness, we need to define a bound of precision, which limits the possible values of i and j.

Definition 22 (bounded multiplicity). *A bounded multiplicity is an element of set* $\mathcal{M}^\mathbf{b} \subset \mathcal{M}$, *defined, for a given bound* $\mathbf{b} \in \mathbb{N}$, *as* $\mathcal{M}^\mathbf{b} = \{\langle i, j \rangle \in \mathcal{M} \mid i \leq \mathbf{b} + 1, \ j \in \{0, \ldots, \mathbf{b}, \omega\}\}$. ◀

It is straightforward to define arithmetic operations and relations over multiplicities. For this paper, it suffices to consider *(i)* the *subsumption* relation $\sqsubseteq$, defined as $\langle i, j \rangle \sqsubseteq \langle i', j' \rangle$ if $i \geq i'$ and $j \leq j'$, and *(ii)* relation $\leq$, defined as $\langle i, j \rangle \leq \langle i', j' \rangle$ if $j \leq j'$. Also, it is simple to define a function $\beta^\mathbf{b} : \mathcal{M} \to \mathcal{M}^\mathbf{b}$ that approximates multiplicities according to a bound $\mathbf{b}$. Let $M \subset \mathcal{M}$ be a set of multiplicities. We write $\sum^\mathbf{b} M$ to denote the bounded multiplicity sum over elements of M, as a short-hand notation for $\beta^\mathbf{b}(\sum M)$. From here on we assume the existence of two bounds, $\mathbf{n}, \mathbf{e} \in \mathbb{N}$, called *node* and *edge* bounds, respectively.

Definition 23 (pattern shape). *A pattern shape* S *is a pattern graph with additional node and edge multiplicity functions, denoted* $\mathsf{mult}^\mathsf{n}_S : N_S \to \mathcal{M}^\mathbf{n}$ *and* $\mathsf{mult}^\mathsf{e}_S : E_S \to \mathcal{M}^\mathbf{e}$, *respectively.* ◀

Function $\mathsf{mult}^\mathsf{n}_S$ indicates how many concrete patterns were folded into an abstract pattern node, up to bound $\mathbf{n}$. Function $\mathsf{mult}^\mathsf{e}_S$, on the other hand, counts locally, *i.e.*, it indicates how many edges of a certain type *each* of the concrete nodes had, up to bound $\mathbf{e}$. We write PShape to denote the universe of pattern shapes and we consider $\mathsf{mult}_S = \mathsf{mult}^\mathsf{n}_S \cup \mathsf{mult}^\mathsf{e}_S$. Pattern graphs can be *trivially extended* to pattern shapes by associating multiplicity maps according to the kind of pattern graph. For a pattern type graph T we associate the most abstract multiplicity to all elements of T, *i.e.*, $\mathsf{mult}_T(x) \mapsto \mathbf{0}^+$, for all $x \in T$. For any other pattern graph P, its trivial extension is obtained by making $\mathsf{mult}_P(x) \mapsto \mathbf{1}$, for all $x \in P$. From here on, we consider that trivial extensions of pattern graphs are taken when necessary. The distinct choice for multiplicities in pattern type graphs is motivated by the definition below.

Definition 24 ($\preceq$-morphism). *A $\preceq$-morphism between pattern shapes $X, Y \in$ PShape is a pattern graph morphism $m : X \to Y$ that relates multiplicities according to relation $\preceq$, i.e.,*

- *for all $p' \in N_Y$, $\sum_{p \in m^{-1}(p')}^{n} \mathsf{mult}_X^n(p) \preceq \mathsf{mult}_Y^n(p')$; and*
- *for all $d' \in E_Y$ and all $p \in N_X$, $\sum_{d \in C}^{e} \mathsf{mult}_X^e(d) \preceq \mathsf{mult}_Y^e(d')$, where $C = m^{-1}(d') \cap p \rhd_X$.*

A pattern shape morphism *is defined to be a $\leq$-morphism, and a $\sqsubseteq$-morphism is called a* subsumption morphism. ◀

We write $\mathsf{depth}(S)$ to denote the maximum layer of pattern shape S that is not empty, *i.e.*, $\mathsf{depth}(S) = i \in \mathbb{N}$ such that $|N_S^i| \neq 0$ and for all $j > i$, $|N_S^j| = 0$. Let A be a set and $\equiv \subseteq A \times A$ be an equivalence relation over A. For $x \in A$, we write $[x]_\equiv$ to denote the equivalence class of x induced by $\equiv$, *i.e.*, $[x]_\equiv = \{y \in A \mid y \equiv x\}$ and we write $A/\equiv$ to denote the set of equivalence classes in A, *i.e.*, $A/\equiv = \{[x]_\equiv \mid x \in A\}$. For any $C_1, C_2 \subseteq A$, $C_1 \equiv C_2$ if for all $x_1 \in C_1$ there exists $x_2 \in C_2$ such that $x_1 \equiv x_2$, and vice versa.

Definition 25 (pattern equivalence). *Let S be a T-pattern shape and $t : S \to T$ be the typing morphism. The* pattern equivalence $\equiv$ *is defined as the smallest symmetrical relation over $N_S \times N_S$ and $E_S \times E_S$ where*

- *for any $p_1, p_2 \in N_S$, $p_1 \equiv p_2$ if $t(p_1) = t(p_2)$ and for all $C_1 \in (p_1 \rhd_S)/\equiv$, there exists $C_2 \in (p_2 \rhd_S)/\equiv$ such that $C_1 \equiv C_2$ and $\sum_{d_1 \in C_1}^{e} \mathsf{mult}_S^e(d_1) = \sum_{d_2 \in C_2}^{e} \mathsf{mult}_S^e(d_2)$; and*
- *for any $d_1, d_2 \in E_S$, $d_1 \equiv d_2$ if $t(d_1) = t(d_2)$ and $\mathsf{tgt}_S(d_1) \equiv \mathsf{tgt}_S(d_2)$.* ◀

The definition above implies that only nodes of the same layer can be equivalent, and that equivalent nodes have the same number of outgoing edges of each type into the same classes. Also, note that the second condition for node equivalence is vacuously true for nodes in layer $\mathsf{depth}(S)$, which gives a base case for the inductive definition. Given $\equiv$ we can derive a finer relation $\triangleq$ that groups edges per source equivalence classes: $d_1 \triangleq d_2$ if $d_1 \equiv d_2$ and $\mathsf{src}_S(d_1) \equiv \mathsf{src}_S(d_2)$.

Definition 26 (canonical pattern shape). *Let X be a T-pattern shape and let $t : X \to T$ be the typing morphism. The* canonical pattern shape *of X w.r.t. equivalence relation $\equiv$ is the pattern shape Y, where $N_Y = N_X/\equiv$ and $E_Y = E_X/\triangleq$, and for all $[p]_\equiv \in N_Y$, $p \in N_X$, $[d]_\triangleq \in E_Y$ and $d \in E_X$:*

- $\mathsf{src}_Y : [d]_\triangleq \mapsto [\mathsf{src}_X(d)]_\equiv$ *and* $\mathsf{tgt}_Y : [d]_\triangleq \mapsto [\mathsf{tgt}_X(d)]_\equiv$;
- $\mathsf{lab}_Y : [p]_\equiv \mapsto \mathsf{lab}_T(t(p))$ *and* $\mathsf{lab}_Y : [d]_\triangleq \mapsto \mathsf{lab}_T(t(d))$;
- $\mathsf{mult}_Y^n : [p]_\equiv \mapsto \sum_{p' \in [p]_\equiv}^{n} \mathsf{mult}_X^n(p')$; *and*
- $\mathsf{mult}_Y^e : [d]_\triangleq \mapsto \sum_{d' \in C}^{e} \mathsf{mult}_X^e(d')$, *where $C = (\mathsf{src}_X(d) \rhd_X) \cap [d]_\triangleq$.* ◀

In words, nodes and edges of canonical pattern shape Y are the equivalence classes of X, the labelling of Y takes the type associated with each equivalence class of X, and the multiplicities of Y are the bounded sum of the multiplicities of elements in the equivalence classes of X. Let P be a pattern graph and S be a pattern shape. We write $\mathsf{abstract}(P)$ and $\mathsf{normalise}(S)$ to denote the canonical pattern shape of P and S, respectively. Morphism $\alpha : P \to \mathsf{abstract}(P)$ is called

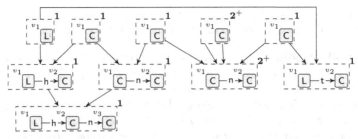

Fig. 7. Canonical pattern shape obtained when considering the pattern graph and equivalence relation of Figure 4. Pattern edge morphisms are not explicitly shown. Node multiplicities are given at the upper right corner of each pattern node. All edge multiplicities are **1** and are not shown.

an *abstraction morphism* and morphism $\Omega : S \rightarrow$ normalise(S) is called a *normalisation morphism*. Both α and Ω are instances of subsumption morphisms.

Figure 7 shows an example of a canonical pattern shape. We use CanPShape$_T^{n,e}$ to denote the universe of canonical shapes typable by T and bounded by n and e. Our third major result follows, establishing finiteness of the abstract state space.

Theorem 27. *Given a pattern type graph T and bounds $n, e \in \mathbb{N}$, universe* CanPShape$_T^{n,e}$ *is finite (under isomorphism).* ◄

We now proceed to define how pattern shapes can be transformed by rules. Given a pattern shape S and a set of pattern nodes $N' \subseteq N_S$, let $N'_{\square} = \{q \in N_S \mid \exists o \in N_S, p \in N' : o \leq_S p, o \leq_S q\}$ and $E'_{\square} = \{d \in E_S \mid \text{src}_S(d) \in N'_{\square}\}$. Pair env$_S(N') = \langle N'_{\square}, E'_{\square} \rangle$ is called the *environment* of N' in S, i.e., the pattern graph elements that can be affected by a pattern graph transformation matched on N'. Environment env$_S(N')$ can be trivially turned into a sub-graph of S.

Definition 28 (rule pre-match/match into pattern shapes). *Let S be a T-pattern shape and $r = \langle \lambda, \rho \rangle$ be a pattern graph rule. A* pre-match *of r into S is a pattern shape morphism $\mu : \lambda \rightarrow S$. We call μ a* match *if* env$_S(\mu(N_\lambda))$ *is a concrete pattern graph and for all $x \in$ env$_S(\mu(N_\lambda))$, mult$_S(x) = 1$.* ◄

Given a match, a *concrete pattern shape transformation* proceeds as a pattern graph transformation on the environment sub-graph.

Definition 29 (concrete pattern shape transformation). *Let X be a T-pattern shape, $r = \langle \lambda, \rho \rangle$ be a pattern graph rule, $\mu : \lambda \rightarrow X$ be a match of r into X, $X_{\square} =$ env$_X(\mu(N_\lambda)) \subseteq X$ be the environment sub-graph of $\mu(N_\lambda)$ in X, and let $X_{\square} \overset{r}{\Rightarrow} Y'$ be a pattern graph transformation. The result of a concrete pattern shape transformation of X is the T-pattern shape Y, where*

- $N_Y = (N_X \setminus N_{X_{\square}}) \cup N_{Y'}$ *and* $E_Y = \{d \in E_X \mid \text{src}_X(d), \text{tgt}_X(d) \in N_Y\} \cup E_{Y'}$;
- src$_Y = (\text{src}_X \cup \text{src}_{Y'})|_{E_Y}$ *and* tgt$_Y = (\text{tgt}_X \cup \text{tgt}_{Y'})|_{E_Y}$;
- lab$_Y = (\text{lab}_X \cup \text{lab}_{Y'})|_{(N_Y \cup E_Y)}$; *and*
- mult$_Y^n = (\text{mult}_X^n \cup \text{mult}_{Y'}^n)|_{N_Y}$ *and* mult$_Y^e = (\text{mult}_X^e \cup \text{mult}_{Y'}^e)|_{E_Y}$. ◄

We write $X \overset{r}{\Rightarrow} Y$ to denote a concrete pattern shape transformation, which we can now use to define transformations for canonical pattern shapes.

Definition 30 (canonical pattern shape transformation). *Given a canonical T-pattern shape X, a rule $r = \langle \lambda, \rho \rangle$, and a pre-match $\mu : \lambda \to X$, let X' be a T-pattern shape such that $\Omega_X : X' \to X$ is a normalisation morphism, $\mu' : \lambda \to X'$ is a match of r into X', and $\mu = \Omega_X \circ \mu'$. The* canonical pattern shape transformation *of X is the canonical T-pattern shape Y, where $X' \overset{r}{\Rightarrow} Y'$ is a concrete pattern shape transformation and $\Omega_Y : Y' \to Y$ is a normalisation morphism.*◄

Pattern shape X' is called a *materialisation* of X according to pre-match μ. An essential property is that a materialisation always exists, for any pre-match. We write $X \Rightarrow^r Y$ to denote a canonical pattern shape transformation. Similarly to what was done with simple graphs and pattern graphs, rule applications on pattern shapes produce a *pattern shape transition system* (PSTS).

Definition 31 (PSTS). *A pattern shape transition system $PSTS = \langle S, \Rightarrow, \iota \rangle$ consists of a set of states $S \subseteq \mathsf{CanPShape}_T^{n,e}$, a set of rule applications $\Rightarrow \subseteq S \times \mathsf{PRule} \times S$, and an initial state $\iota \in S$. A pattern graph grammar $\mathcal{P}_T = \langle \mathcal{R}_T, P_0 \rangle$ generates a $PSTS_\mathcal{P}$ if $\iota = \mathsf{abstract}(P_0)$ and S is the minimal set of canonical pattern shapes such that $X \in S$ and $X \Rightarrow^r Y$ for $r \in \mathcal{R}_T$ implies that there exists $Y' \in S$ such that $Y \simeq Y'$ and $X \Rightarrow^r Y'$ is a transition.*   ◄

Our last result establishes the connection between concrete and abstract spaces.

Theorem 32. *Transition system $PSTS_\mathcal{P}$ simulates $PGTS_\mathcal{P}$.*   ◄

An immediate consequence of PSTS being an over-approximation is that verification can then be carried out on the abstract domain, which is always finite. As usual with over-approximations, if a property holds at the pattern shape level then it is guaranteed to hold at the pattern graph level; however if a property is false for pattern shapes then we cannot be sure that it is also false for pattern graphs since the abstraction can introduce spurious behaviour [1]. The results at the pattern graph level transfer directly to simple graphs due to Theorem 21.

5   Related Work

Perhaps the most well-known method for the verification of infinite-state GT systems is the Petri graph unfolding technique, by König *et al.*, initially presented in [1]. Given a graph grammar this method extracts an *approximated unfolding*: a finite structure (called *Petri graph*) that is composed of a hyper-graph and a Petri net. The Petri graph captures all structure that can occur in the reachable graphs of the system, and dependencies for rule applications are recorded by net transitions. The structure obtained can then be used to check safety properties in the original system. If a spurious counter-example is introduced, the abstraction can be incrementally refined [10]. These techniques are implemented in the tool AUGUR which is now in its second version [11].

The approaches presented in [19,17] use a backwards reachability analysis for hyper-edge replacement grammars, where a search is started from a forbidden graph configuration and traverses backwards trying to reach an initial graph. This technique is applied to ad-hoc network routing and heap analysis, respectively, but is not guaranteed to terminate. In [5], an invariant checking method is developed for determining statically if a given forbidden pattern is an inductive invariant of a given rule. States that may lead to a forbidden graph are described symbolically, yielding a representation that is both complete and finite.

Our own take on GT abstractions was inspired by the seminal work on shape analysis by Sagiv *et al.* [18] and lead to theoretical results on *graph shapes* [13,14]. A similar approach called *partner abstraction* was developed in parallel by Bauer [2,4] and later these two approaches were unified in the framework of *neighbourhood abstraction* [3]. In [15,20] we describe the implementation effort for integrating neighbourhood abstraction into GROOVE [12,8], our GT tool set.

The main advantage of pattern abstraction over neighbourhood abstraction is the flexibility in specifying which structures should be collapsed. Precision of the neighbourhood method can be adjusted via a radius parameter but this radius is always the same when analysing the equivalence of nodes and edges. On the other hand, the pattern-based method is much more fine-grained: patterns of various sizes can be represented in the type graph, and are considered independently by the abstraction. Roughly speaking, a radius i neighbourhood abstraction can be simulated by a pattern abstraction using a type graph T with depth i, where all possible simple graphs of size smaller or equal to i occur as patterns in T.

6   Conclusions and Future Work

This paper presented a new method for a property-driven abstraction of graphs, based on a pre-defined collection of patterns of interest, represented as a pattern type graph. Pattern-based abstraction leads to a finite over-approximation of (infinite-state) graph transformation systems and as such the abstraction can be used for system verification. Furthermore, the technique lends itself nicely to abstraction refinement: one can start with a rather minimal type graph and then add more patterns to it to make the abstraction more precise when necessary.

The theory here presented, while sound, still leaves certain steps under-specified. In particular, the materialisation of canonical pattern shapes is of importance, since it may have a significant influence on the size of abstract state spaces. Based on our previous experience of implementing the theory of neighbourhood abstraction, we defer the definition of a materialisation algorithm to a later implementation phase, when a proper practical analysis of the trade-off between performance and abstraction precision can be made.

An interesting side-effect of developing a theory of pattern graph transformation is that structures used in incremental pattern matching like RETE networks [9] can now also be formalised as pattern graphs. To the best of our knowledge, this is the first time such structures were considered under a more formal focus.

We plan to implement this new theory into GROOVE, so that experiments can be carried out to gauge how suitable the proposed abstraction is in practice.

References

1. Baldan, P., Corradini, A., König, B.: A Static Analysis Technique for Graph Transformation Systems. In: Larsen, K.G., Nielsen, M. (eds.) CONCUR 2001. LNCS, vol. 2154, pp. 381–395. Springer, Heidelberg (2001)
2. Bauer, J.: Analysis of Communication Topologies by Partner Abstraction. PhD thesis, Universität des Saarlandes (2006)
3. Bauer, J., Boneva, I., Kurban, M., Rensink, A.: A modal-logic based graph abstraction. In: [7]
4. Bauer, J., Wilhelm, R.: Static Analysis of Dynamic Communication Systems by Partner Abstraction. In: Riis Nielson, H., Filé, G. (eds.) SAS 2007. LNCS, vol. 4634, pp. 249–264. Springer, Heidelberg (2007)
5. Becker, B., Beyer, D., Giese, H., Klein, F., Schilling, D.: Symbolic invariant verification for systems with dynamic structural adaptation. In: ICSE. ACM (2006)
6. de Lara, J., Varro, D. (eds.): GraBaTs. ECEASST, vol. 32. EASST (2010)
7. Ehrig, H., Heckel, R., Rozenberg, G., Taentzer, G. (eds.): ICGT 2008. LNCS, vol. 5214. Springer, Heidelberg (2008)
8. Ghamarian, A.H., de Mol, M., Rensink, A., Zambon, E., Zimakova, M.: Modelling and analysis using GROOVE. STTT 14(1) (2012)
9. Ghamarian, A.H., Rensink, A., Jalali, A.: Incremental pattern matching in graph-based state space exploration. In: [6]
10. König, B., Kozioura, V.: Counterexample-Guided Abstraction Refinement for the Analysis of Graph Transformation Systems. In: Hermanns, H., Palsberg, J. (eds.) TACAS 2006. LNCS, vol. 3920, pp. 197–211. Springer, Heidelberg (2006)
11. König, B., Kozioura, V.: Augur 2 - a new version of a tool for the analysis of graph transformation systems. ENTCS 211 (2008)
12. Rensink, A.: The GROOVE Simulator: A Tool for State Space Generation. In: Pfaltz, J.L., Nagl, M., Böhlen, B. (eds.) AGTIVE 2003. LNCS, vol. 3062, pp. 479–485. Springer, Heidelberg (2004)
13. Rensink, A.: Canonical Graph Shapes. In: Schmidt, D. (ed.) ESOP 2004. LNCS, vol. 2986, pp. 401–415. Springer, Heidelberg (2004)
14. Rensink, A., Distefano, D.: Abstract graph transformation. In: Workshop on Software Verification and Validation (SVV). ENTCS, vol. 157 (2006)
15. Rensink, A., Zambon, E.: Neighbourhood abstraction in GROOVE. In: [6]
16. Rensink, A., Zambon, E.: Pattern-based graph abstraction (extended version). Technical report, University of Twente, Enschede, The Netherlands (2012)
17. Rieger, S., Noll, T.: Abstracting complex data structures by hyperedge replacement. In: [7]
18. Sagiv, S., Reps, T.W., Wilhelm, R.: Parametric shape analysis via 3-valued logic. ToPLaS 24(3) (2002)
19. Saksena, M., Wibling, O., Jonsson, B.: Graph Grammar Modeling and Verification of Ad Hoc Routing Protocols. In: Ramakrishnan, C.R., Rehof, J. (eds.) TACAS 2008. LNCS, vol. 4963, pp. 18–32. Springer, Heidelberg (2008)
20. Zambon, E., Rensink, A.: Graph subsumption in abstract state space exploration. In: GRAPHITE (Pre-proceedings) (2012)

Well-Structured Graph Transformation Systems with Negative Application Conditions*

Barbara König and Jan Stückrath

Abteilung für Informatik und Angewandte Kognitionswissenschaft,
Universität Duisburg-Essen, Germany
{barbara_koenig,jan.stueckrath}@uni-due.de

Abstract. Given a transition system and a partial order on its states, the coverability problem is the question to decide whether a state can be reached that is larger than some given state. For graphs, a typical such partial order is the minor ordering, which allows to specify "bad graphs" as those graphs having a given graph as a minor. Well-structuredness of the transition system enables a finite representation of upward-closed sets and gives rise to a backward search algorithm for deciding coverability.

It is known that graph tranformation systems without negative application conditions form well-structured transition systems (WSTS) if the minor ordering is used and certain condition on the rules are satisfied. We study graph transformation systems with negative application conditions and show under which conditions they are well-structured and are hence amenable to a backwards search decision procedure for checking coverability.

1 Introduction

Graph transformation systems (GTS) [17] are a Turing-complete model of computation, which means that many properties of interest, especially concerning reachability and coverability ("Is it possible to reach a graph that contains a given graph as a subgraph?") are undecidable. Naturally, one obtains decidability of both problems when restricting to finite-state graph transformation systems, i.e., systems where only finitely many graphs up to isomorphism are reachable from a given start graph. However, similar to the case of Petri nets [5], it is possible to obtain decidability results also for certain (restricted) classes of infinite-state graph transformation systems [2]. This is important for many applications, since systems with infinitely many states arise easily in practice.

A good source of decidability results for the coverability problem are so-called well-structured transition systems [7,1]. They consist of a (usually infinite) set of states, together with a transition relation and a well-quasi-order (see Definition 1), such that the well-quasi-order is a simulation relation for the transition system. Standard place/transition nets can be seen as well-structured transition systems, furthermore systems with some degree of lossiness (such as lossy channel systems, where channels might lose messages) are well-structured.

---

* Research partially supported by DFG project GaReV.

H. Ehrig et al.(Eds.): ICGT 2012, LNCS 7562, pp. 81–95, 2012.

Well-structuredness implies that every upward-closed set of states can be represented by a finite set of minimal states (this is a direct consequence of the properties of a well-quasi-order). Under some additional conditions it is possible to perform a backwards search in order to compute and represent (via minimal states) all predecessors of an upward-closed set. This allows to answer coverability questions algorithmically.

In [11,10] we have shown how (single-pushout) graph transformation systems with edge contraction rules can be seen as well-structured transition systems. As well-quasi-order we used the minor ordering on graphs, which is shown to be a well-quasi-order in the famous Robertson-Seymour theorem [15,16].

However, the theory in [11] does not apply to graph transformation systems with negative application conditions [8,4], which often arise in practice. Such negative application conditions disallow the application of a rule if a certain "forbidden" subgraph is present.

Here we study such graph transformation systems with negative application conditions and show that they are well-structured under certain conditions (for instance in the presence of deletion and contraction rules that arise naturally in lossy systems). While this result is fairly straightforward to prove, it is more difficult to perform a backwards step and hence obtain a decision algorithm. We here give a general procedure for computing the predecessor set and show that it terminates in specific cases, i.e., for certain types of negative application conditions. We illustrate the theory with various examples, especially we study a (faulty) termination detection protocol and apply the decision procedure to the set of rewriting rules describing the protocol.

Proofs are published as a technical report [12].

2   Preliminaries

2.1   Well-Structured Transition Systems

We will now give the definitions concerning well-structured transition systems, following the presentation in [7].

Definition 1 (wqo and upward closure). *A quasi-order[1] $\leq$ (over some set X) is a well-quasi-order (wqo) if for any infinite sequence $x_0, x_1, x_2, \ldots$ of elements of X, there exist indices $i < j$ with $x_i \leq x_j$.*

An upward-closed set *is any set $I \subseteq X$ such that $x \leq y$ and $x \in I$ implies $y \in I$. A* downward-closed set *can be defined analogously.*

For a subset $Y \subseteq I$, we define its upward closure *$\uparrow Y = \{x \mid \exists y \in Y : y \leq x\}$. Then, a* basis *of an upward-closed set I is a set I^b such that $I = \uparrow I^b$.*

The definition of well-quasi-orders gives rise to some properties which are especially important for the backwards algorithm presented later.

Lemma 1. *Let $\leq$ be a well-quasi-order, then the following two statements hold:*

---

[1] Note that a quasi-order is the same as a preorder.

1. *Any upward-closed set I has a finite basis.*
2. *For any infinite, increasing sequence of upward-closed sets $I_0 \subseteq I_1 \subseteq I_2 \subseteq \ldots$ there exists an index $k \in \mathbb{N}$ such that $I_i = I_{i+1}$ for all $i \geq k$.*

Definition 2 (Well-structured transition system). *A well-structured transition system (WSTS) is a transition system $T = (S, \Rightarrow, \leq)$, where S is a set of states and $\Rightarrow \subseteq S \times S$, such that the following conditions hold:*

1. **Well-quasi-ordering:** *$\leq$ is a well-quasi-order on S.*
2. **Compatibility:** *For all $s_1 \leq t_1$ and a transition $s_1 \Rightarrow s_2$, there exists a sequence $t_1 \Rightarrow^* t_2$ of transitions such that $s_2 \leq t_2$.*

$$
\begin{array}{ccc}
t_1 & \xRightarrow{\;*\;} & t_2 \\
\rotatebox{90}{\!\!\text{VI}} & & \rotatebox{90}{\!\!\text{VI}} \\
s_1 & \Longrightarrow & s_2
\end{array}
$$

Given a set $I \subseteq S$ of states we denote by $Pred(I)$ the set of direct predecessors of I, i.e., $Pred(I) = \{s \in S \mid \exists s' \in I : s \Rightarrow s'\}$. Furthermore $Pred^*(I)$ is the set of all predecessors which can reach some state of I with an arbitrary sequence of transitions. Let $(S, \Rightarrow, \leq)$ be a WSTS. Backward reachability analysis involves the computation of $Pred^*(I)$ as the limit of the sequence $I_0 \subseteq I_1 \subseteq I_2 \subseteq \ldots$ where $I_0 = I$ and $I_{n+1} = I_n \cup Pred(I_n)$. However, in general this may not terminate. For WSTS, if I is upward-closed then it can be shown that $Pred^*(I)$ is also upward-closed (compatibility condition) and that termination is guaranteed (see Lemma 1).

Definition 3 (Effective pred-basis). *A WSTS has an effective pred-basis if there exists an algorithm accepting any state $s \in S$ and returning $pb(s)$, a finite basis of $\uparrow Pred(\uparrow\{s\})$.*

Now assume that T is a WSTS with effective pred-basis. Pick a finite basis I^b of I and define a sequence $K_0, K_1, K_2, \ldots$ of sets with $K_0 = I^b$ and $K_{n+1} = K_n \cup pb(K_n)$. Let m be the first index such that $\uparrow K_m = \uparrow K_{m+1}$. Such an m must exist by Lemma 1 and we have $\uparrow K_m = Pred^*(I)$.

The *covering problem* is to decide, given two states s and t, whether starting from a state s it is possible to cover t, i.e. to reach a state t' such that $t' \geq t$. The decidability of the covering problem follows from the previous argument: we define $I = \uparrow\{t\}$ and check whether $s \in Pred^*(I)$, i.e., if there exists a $\overline{s} \in K_m$ such that $\overline{s} \leq s$.

Theorem 1 (Covering problem [7]). *The covering problem is decidable for a WSTS with an effective pred-basis and a decidable wqo $\leq$.*

Thus, if T is a WSTS satisfying the extra conditions of Theorem 1 and the "error states" can be represented as an upward-closed set I, then it is decidable whether any element of I is reachable from the start state.

2.2   Graph Transformation Systems

We will now introduce the necessary preliminaries in order to define single-pushout (SPO) graph rewriting. Note that the minor relation used in the following admits a characterization via partial graph morphisms and hence straightforwardly integrates with SPO, which is defined in the category of partial morphisms. The concatenation of a rule and a minor morphism is again a rule.

Definition 4 (Hypergraph). *Let Λ be a finite set of labels and a function $ar\colon \Lambda \to \mathbb{N}$ that assigns an* arity *to each label. A (Λ-)hypergraph is a tuple (V_G, E_G, c_G, l_G) where V_G is a finite set of nodes, E_G is a finite set of edges, $c_G\colon E_G \to V_G^*$ is a connection function and $l_G\colon E_G \to \Lambda$ is the labelling function for edges. We require that $|c_G(e)| = ar(l_G(e))$ for each edge $e \in E_G$.*

We will simply use *graph* to denote a hypergraph. To simplify the necessary computations we will only use hypergraphs with at most binary edges, i.e. $|ar(\ell))| \leq 2$ for all labels ℓ. However note that large parts of the theory (except Proposition 2) can be extended to the general case.

Definition 5 (Graph morphism). *Let G, G' be (Λ-)graphs. A partial graph morphism (or simply* morphism*) $\varphi\colon G \rightharpoonup G'$ consists of a pair of partial functions $(\varphi_V : V_G \rightharpoonup V_{G'}, \varphi_E : E_G \rightharpoonup E_{G'})$ such that for every $e \in E_G$ it holds that $l_G(e) = l_{G'}(\varphi_E(e))$ and $\varphi_V(c_G(e)) = c_{G'}(\varphi_E(e))$ whenever $\varphi_E(e)$ is defined. Furthermore if a morphism is defined on an edge, it must be defined on all nodes adjacent to it. Total morphisms are denoted by an arrow of the form $\to$.*

We will now introduce the notion of (SPO) graph rewriting ([13]) with negative application conditions.

Definition 6 (Graph rewriting). *A rewriting rule is a partial morphism $r\colon L \rightharpoonup R$ together with a finite set of negative application conditions. A negative application condition (NAC) is a total, injective morphism $n_i\colon L \to N_i$. A match is a total, injective morphism $m\colon L \to G$. We say that a match m satisfies a NAC n_i if there is no total, injective morphism $n_i'\colon N_i \to G$ such that $n_i' \circ n_i = m$. A rule is* applicable *to a graph G if there is a match satisfying all NACs.*

For a pair of a rule r and a match m applicable to G, a rewriting step is obtained by taking the pushout of m and r in the category of partial graph morphisms. Then G is rewritten to the pushout object H (written as $G \overset{r,m}{\Rightarrow} H$ or simply $G \Rightarrow H$).

In [11] so called *conflict-free* matches are used, which may be non-injective. However, these matches cannot be used with our variant of negative application conditions, since the NAC and its match to G are injective and can therefore not commute with any non-injective match m. Note that any injective match is also conflict-free.

In this paper a graph transformation system (GTS) is simply a finite set of rules, not necessarily associated with an initial graph. For verification purposes initial and final graphs can be used as shown in the example below.

Example 1. For the later illustration of a backward step and the backward search algorithm, we introduce the following termination detection protocol as a running example. A similar protocol was used in [2] but without negative application

(a) A possible initial graph   (b) The final graph

Fig. 1. The protocols initial graph and it error configuration

conditions. For lossy systems this protocol is erroneous and we will show how the backwards search will detect the error.

The protocol consists of normal processes which can be active (A) or passive (P) and a detector process which can be active (DA) or passive (DP). The label $(D)A$ thereby stands both for an active detector and an active normal process ($(D)P$ is used analogously). The initial graph is a directed circle with an active detector and an active (normal) process (see Fig. 1a). Additional active processes can be generated (Fig. 2d) and active processes may become passive (Fig. 2b). When the detector becomes passive *start* and *end* flags are created and attached to the corresponding detector (Fig. 2a). The *end* flag can be forwarded along passive processes (Fig. 2e) and when the *end* flag was forwarded around the entire ring to reach the detector again, a *termination* flag is created (Fig. 2f) stating that all processes are passive. Any active process can reactivate a passive process (Fig. 2c) if there is no *start* flag between them. This ensures that all processes (including the detector) between the *start* and the *end* flag are passive. The absence of the *start* flag is thereby ensured by a negative application condition, which is indicated by the dashed edge.

Additionally there are rules simulating the lossiness of the system. Processes can leave the ring (Fig. 2g, Fig. 2h) and flags can be lost (Fig. 2i, Fig. 2j). We will later show that this GTS is well-structured.

The protocol is correct if and only if from the initial graph no graph can be reached which contains the final graph (Fig. 1b), because this would mean that a *termination* flag was generated although there still exists an active process.

2.3   Minors and Minor Morphisms

We will now introduce the notion of graph minor [15,16] and recall some results from [11].

Definition 7 (Minor). *A graph M is a* minor *of a graph G (written $M \leq G$), if M can be obtained from G by (repeatedly) performing the following operations on G:*

1. *Contraction of an edge. In this case we remove the edge, choose an arbitrary partition on the nodes connected to the edge and merge the nodes as specified by the partition. (This includes edge deletion as a special case.)*
2. *Deletion of an isolated node.*

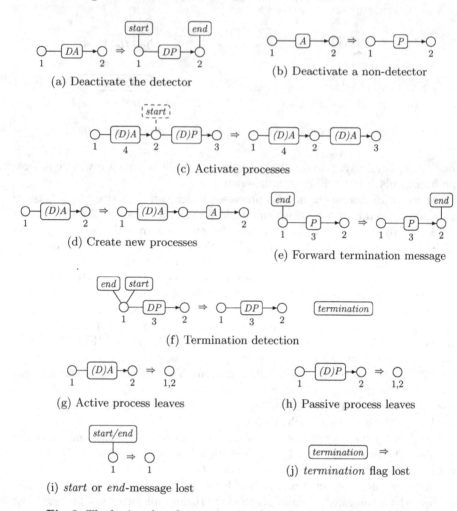

(a) Deactivate the detector

(b) Deactivate a non-detector

(c) Activate processes

(d) Create new processes

(e) Forward termination message

(f) Termination detection

(g) Active process leaves

(h) Passive process leaves

(i) *start* or *end*-message lost

(j) *termination* flag lost

Fig. 2. The basic rules of a termination detection protocol with NACs

Note that since we restrict the arity of any edge to at most two, there is only one possible partition if the arity is zero or one and only two possible partitions if the arity is two (one of which coincides with edge deletion).

The *Robertson-Seymour Theorem* [15] says that the minor order is a well-quasi-order even if the edges and vertices of the graphs are labelled from a well-quasi-ordered set, and also for hypergraphs and directed graphs (see [10,16]). Here we use the minor ordering presented in [11], but since we restrict the arity of edges, the ordering is essentially the same as for directed graphs.

Now any GTS with negative application conditions which satisfies the compatibility condition of Definition 2 (with respect to the minor ordering), can be analysed using the theory of WSTS. But before we characterize such GTS we first need the definition of minor morphisms and their properties. A *minor morphism* is a partial morphism that identifies a minor of a graph.

Definition 8 (Minor morphism). *A partial morphism* $\mu : G \rightharpoonup M$ *is a minor morphism (written* $\mu : G \mapsto M$ *) if*

1. *it is surjective,*
2. *it is injective on edges and*
3. *whenever* $\mu(v) = \mu(w) = z$ *for some* $v, w \in V_G$ *and* $z \in V_M$, *there exists a path between* v *and* w *in* G *where all nodes on the path are mapped to* z *and* μ *is undefined on every edge on the path.*

For a minor morphism μ *we define* $\| \mu \|$ *to be the number of nodes and edges on which* μ *is undefined.*

We call a minor morphism μ *a* one-step *minor morphism if* μ *deletes exactly one node or contracts or deletes exactly one edge, i.e. additionally to the above restrictions it holds that either:*

1. μ *is injective, defined on all edges and defined on all but one node or*
2. μ *is defined on all nodes, defined on all but one edge and is injective on all nodes not attached to the undefined edge.*

In [16] a different way to characterize minors is proposed: a function, going in the opposite direction, mapping nodes of M to subgraphs of G. This however cannot be seen as a morphism in the sense of Definition 5 and we would have problems integrating it properly into the theory of graph rewriting. However, in [10] it is proven, that our minor ordering is a wqo and that the following facts about minor morphisms hold.

Lemma 2 ([11]). *The class of minor morphisms is closed under composition.*

Lemma 3 ([11]). M *is a minor of* G *iff there exists a minor morphism* $\mu : G \mapsto M$.

Lemma 4 ([11]). *Pushouts preserve minor morphisms in the following sense: If* $f : G_0 \mapsto G_1$ *is a minor morphism and* $g : G_0 \to G_2$ *is total, then the morphism* f' *in the pushout diagram below is a minor morphism.*

$$
\begin{array}{ccc}
G_0 & \xmapsto{\ f\ } & G_1 \\
\Big\downarrow{\scriptstyle g} & & \Big\downarrow{\scriptstyle g'} \\
G_2 & \xmapsto{\ f'\ } & G_3
\end{array}
$$

Note that Lemma 4 is also valid if g is not total and we require that f does not contract any edge deleted by g. In the following we will also use this variant of the lemma. Finally we need the following lemma, which is a weaker version of a related lemma in [11].

Lemma 5 ([11]). *Let* $\psi_1 : L \to G$ *be total and injective. If the diagram below on the left is a pushout and* $\mu : H \mapsto S$ *a minor morphism, then there exist minors* M *and* X *of* R *and* G *respectively, such that:*

1. *the diagram below on the right commutes and the outer square is a pushout.*
2. *the morphisms $\mu_G \circ \psi_1 : L \to X$ and $\varphi_1 : M \to S$ are total.*

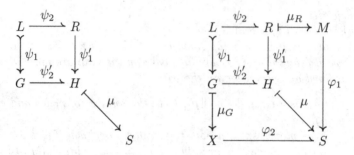

Furthermore we need the following additional properties.

Lemma 6 (Reordering of minor decompositions). *Every minor morphism μ can be decomposed into a finite sequence of one-step minor morphisms μ_i such that $\mu = \mu_1 \circ \ldots \circ \mu_n$ and there are k, ℓ with $1 \le k \le \ell \le n+1$ and*

1. *μ_i with $1 \le i < k$ is undefined on one node (node deletion),*
2. *μ_i with $k \le i < \ell$ is undefined on one edge and not injective on nodes (edge contraction) and*
3. *μ_i with $\ell \le i < n+1$ is undefined on one edge and injective on nodes (edge deletion).*

3   Well-Structuredness and Negative Application Conditions

One condition for the backwards algorithm to compute correct results is well-structuredness. In the following proposition we formulate a rather general condition for a GTS with negative application conditions which ensures that it is well-structured with respect to the minor ordering. We have to ensure that whenever $M \le G$ and a rule $r : L \twoheadrightarrow R$ can be applied to M, the same rule can be applied to G. There are two problems that might disallow the rule application: G might contain a disconnected copy of the left-hand side L (which must be contracted to form a match of L in M) and G may contain more structure such that a valid match of L in M might not satisfy the NACs when extended to G. In both cases we ensure that G can be rewritten to G' in which both problems disappear.

Proposition 1. *A GTS containing rules with negative application conditions is well-structured wrt. the minor ordering if for every rule $r : L \twoheadrightarrow R$ with negative application conditions $n_i : L \to N_i$ and every minor morphism $\mu : G \mapsto M$ the following holds: for every match $m : L \to M$ satisfying all NACs (indicated by the crossed-out arrows), G can be rewritten to some graph G' such that there is a match $m' : L \to G'$ satisfying all NACs and a minor morphism $\mu' : G' \mapsto M$ with $m = \mu' \circ m'$ (see diagram below).*

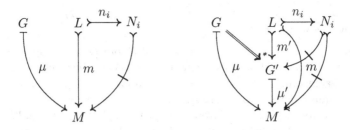

"Lossy systems" (such as the GTS of Example 1) usually already satisfy Proposition 1 and other GTS can be transformed into GTS satisfying the conditions by adding rules for lossiness. In general this is possible by introducing rules that contract and deleted nodes and edges for every label (so called minor rules). The contraction rules enable the simulation of edge contractions performed by μ to generate a graph where the rule matches, while the deletion rules can be used to destroy matches of negative application condition. In the latter case it is sufficient to introduce rules which only delete edges that are associated with a negative application condition, such as rules $\nu_e \colon N \to N \setminus \{e\}$ for all edges e not in the range of $n \colon L \to N$. Note that any introduction of new rules will cause the backward search to compute an overapproximation for the original GTS.

For the computation of the finite basis of $Pred(S)$ we first introduce the notion of edge decontraction and bounding functions.

Definition 9 (Edge decontraction). *Let G be a graph. We define $expand(G)$ to be the set of all graphs $G' \geq G$ such that there is an edge e in G' and the contraction of e results in a graph isomorphic to G.*

Definition 10 (Bounding function). *Let $\mathcal{T}$ be a GTS with rule set $\mathcal{R}$ and let $b_r^{\mathcal{T}} \colon Matches \to \mathbb{N}_0$ be a function, where $(r \colon L \rightharpoonup R, \mathcal{N}) \in \mathcal{R}$ is a rule and Matches is the set of all matches $m \colon L \to G$ of r to some graph G, not necessarily satisfying all NACs $\mathcal{N}$. We call $b_r^{\mathcal{T}}$ a bounding function if every minor morphism $\mu \colon G' \mapsto G$, where there is a match $m' \colon L \to G'$ satisfying all NACs, can be decomposed into minor morphisms $\mu' \colon G' \mapsto M$ and $\mu'' \colon M \mapsto G$ satisfying the following properties:*

1. $\mu = \mu'' \circ \mu'$,
2. $\mu' \circ m'$ is a total, injective match of L in M and satisfies all NACs and
3. $\|\mu''\| \leq b_r^{\mathcal{T}}(m)$.

The bounding function is used to calculate the maximal number of decontractions needed to compute a graph of the predecessor basis. The existence of such a function guarantees termination of a backward step and we will prove its existence for special cases. In the following we will omit the superscript $\mathcal{T}$, since the GTS of interest is fixed.

We will now describe a procedure for computing an effective pred-basis (see Definition 3). In essence, given a graph S, we have to apply a rule $r \colon L \rightharpoonup R$ backwards. However, there are several complications, caused by the fact that S does not simply stand for itself, but represents a whole set of graphs, the upward-closure of S (i.e., all graphs that have S as a minor). Hence we have to consider the following facts:

- S might not contain an entire (co-)match of R, but it might represent graphs that contain such matches. In order to solve this problem we do not simply apply r backwards, but first compute all minors of R and look for matches of those minors (see Step 2 below).
- Whenever, after doing a backwards step, we find that the resulting graph X' contains a non-injective match of L, we do not discard X'. Again, this is because X' directly is not a predecessor, but it represents other graphs which are predecessors of graphs represented by S. Those graphs can be obtained by "forcing" the match to be injective via edge decontractions (Step 3).
- Finally, we have to solve the problem that the backwards step might result in a graph X which may contain an injective match, but does not satisfy the NACs. Similar to the case above, we have to find larger graphs represented by X by edge decontractions such that the resulting graphs do satisfy the NACs (Step 4).

The last item is more complex than the second-last: while we can bound the number of steps needed to "make" a match injective via decontractions by the number of nodes which are merged, this is not so easy in the case of NACs: by decontractions we may destroy a NAC, but create a new one. Since we want to represent upwards-closed sets wrt. a well-quasi-order, we can be sure that there are finitely many representatives. However, when searching for them, we might not know whether we have already found all of them. This is where the bounding function of Definition 2 comes into play in order to terminate the search.

Procedure 1. Let $\mathcal{T}$ be a GTS with the rule set $\mathcal{R}$ satisfying the compatibility condition, as described in Proposition 1. We assume that there is a bounding function for every rule of $\mathcal{T}$.

In the following we give a description of a procedure $pbn(S)$ which generates a finite basis for the set of graphs reaching a graph larger or equal to S in one step. The first two steps are basically identical to the procedure presented in [11].

1. For each rule $(r\colon L \rightharpoonup R,\ \mathcal{N}) \in \mathcal{R}$, where $\mathcal{N}$ is the set of all negative application conditions of r, let $\mathcal{M}_\mathcal{R}$ be the set of all minor morphisms with source R. Furthermore let b_r be the bounding function of r.
2. For each $(\mu\colon R \mapsto M) \in \mathcal{M}_\mathcal{R}$ consider the rule $\mu \circ r\colon L \rightharpoonup M$ and perform the following steps.
3. For each total match $m''\colon M \to S$ compute all minimal pushout complements X' such that $m'\colon L \to X'$ below is total and injective on edges.[2] Then repeatedly apply the *expand*-function in order to split non-injectively matched nodes and to obtain a basis for all graphs X, of which X' is a minor and which contain an injective match of L (see diagram below).[3]

---

[2] The term *minimal* is used with respect to the minor ordering. For more details see [11].

[3] Alternatively the graph can be partially enlarged before applying the rule backwards to create an initial, injective match as explained in [2].

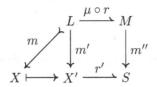

4. For each such X compute every total, injective morphism $n_i^j \colon N_i \to X$ of any NAC $(n_i \colon L \to N_i) \in \mathcal{N}$ that commutes with the match. If there is none, store X and proceed with the next X.

   If there is at least one such morphism, compute $expand(X)$. For each $\overline{X} \in expand(X)$ do the following:
   - Let $\overline{\mu} \colon \overline{X} \mapsto X$ be the minor morphism that exists by Definition 9 and Lemma 3.
   - If there is no $\overline{m} \colon L \to \overline{X}$, such that $m = \overline{\mu} \circ \overline{m}$, i.e. the decontraction destroyed the match, discard X.
   - If there is a match $\overline{m}$ satisfying all NACs, store $\overline{X}$ and proceed with the next $\overline{X}$.

   If none of the previous two conditions hold, repeat Step 4 with the graph $\overline{X}$ and the match $\overline{m}$. Stop the recursion after $b_r(m)$ steps.[4]
5. The set $pbn(S)$ contains all graphs X stored after the last repetition of the previous step.

Example 2. To illustrate the handling of negative application conditions we exemplarily apply rule 2c of our running example backward to the graph G in Figure 3a ($\rightarrow$ indicates a backward step). In Step 3 of Procedure 1 among others the graph H in Figure 3a is generated (μ is the identity), but the rule cannot be applied (in forward direction) to H to reach G since the negative application condition is not satisfied. Hence, in Step 4 the minimal set of graphs larger than H has to be found which can be rewritten to some graph larger than G. All eight resulting graphs are shown in Figure 3b. The "decontracted" edge in the middle can point in both directions and be labeled A, P, DA or DP. All other decontractions either destroy the match or do not destroy the match of the NAC, which in both cases produces graphs where the rule is not applicable.

Additional graphs will be generated if different matches of the left-hand side of rule 2c are used. For instance the graph in Figure 3c is generated (among others) if only one of the A-edges is matched to the right A-edge of G. An additional A-edge is created while generating an injective match. Note that the last graph is immediately deleted since it is larger than G and therefore already represented. A more comprehensive demonstration of backward steps is done in Section 4.

We will now state our main theorems: especially we will show that every graph generated by $pbn(S)$ is in the predecessor set of the upwards-closure of S and vice versa.

---

[4] Note that applicability of the rule $\mu \circ r$ does not depend on μ and hence we do not need a bounding function $b_{\mu \circ r}$.

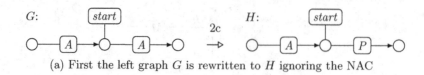

(a) First the left graph G is rewritten to H ignoring the NAC

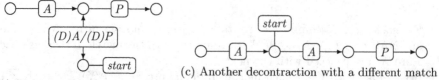

(b) Eight possible decontractions

(c) Another decontraction with a different match

Fig. 3. Backward step of rule 2c containing a negative application condition

Fig. 4. A simple rule with a NAC (the dashed part)

Theorem 2. *The procedure pbn(S) (Procedure 1) computes a finite subset of* $Pred(\uparrow S)$.

Theorem 3. *If there is a bounding function* b_r *for every rule* r, *the set generated by pbn(S) is a finite basis of* $\uparrow Pred(\uparrow S)$.

Theorem 3 depends on the existence of a bounding function. As we will illustrate later, bounding functions are non-trivial to obtain, but we have bounding functions for specific negative application conditions.

Proposition 2. *There is a bounding function for all rules if the corresponding GTS satisfies the following properties:*

1. *$ar(\ell) \leq 2$ for all labels $\ell \in \Lambda$ and*
2. *for every negative application condition $n\colon L \to N$ the graph N has at most one edge and two nodes in addition to $n(L)$.*

Example 3. In the general case the complexity of the bounding function can grow with the complexity of the match and the NAC. In particular there are rather simple NACs where decontractions cannot be discarded if they generate new matches of a NAC. Figure 4 shows a rule together with a NAC (dashed part of the graph). Assume the left graph in Figure 5 was generated by step 3 of the procedure. The two displayed decontractions lead to the right graph, but the middle graph does not satisfy the NAC and there is no single decontraction where the NAC is satisfied and the rightmost graph is represented.

Negative application conditions for arbitrary hypergraphs can therefore require a complex bounding functions.

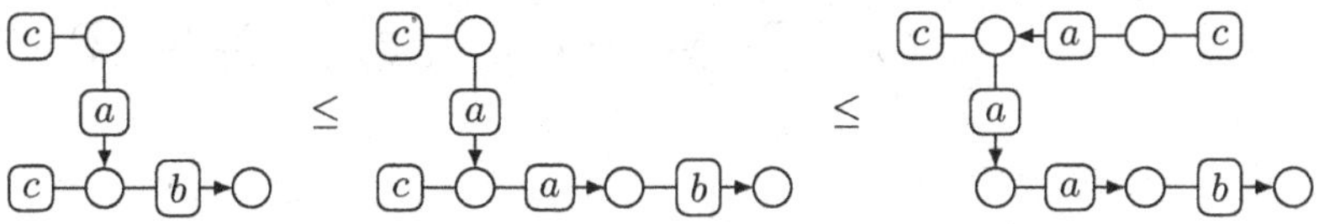

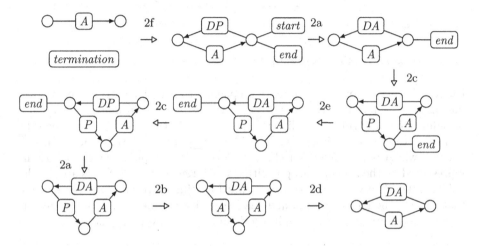

Fig. 5. Graphs generated from left to right via decontractions

Fig. 6. An exemplary sequence of backward steps

4   Verification of a Termination Detection Protocol

As already mentioned the termination detection protocol presented in Example 1 is erroneous for lossy systems. Note also that its NAC satisfies the restrictions of Proposition 2 and can hence be analyzed automatically. One possibility to derive (a minor of) the initial graph from the final graph is shown in Figure 6. This derivation is found by the backward search within eight backward steps each indicated by $\rightarrow$ (together with the used rule). The rules used in the first three steps and the sixth step only partially match the graph, i.e. not the corresponding rule r is applied backwards, but the rule $\mu \circ r$ for some minor morphism μ (see Step 2 of Procedure 1). For instance the minor morphism μ used in the first step is injective and only undefined on the DP-edge, hence this edge is added when the rule is applied backwards. In the second step the minor morphism is undefined for the *end* flag, which is non-existent at the right place in the graph, but it is not added, since the rule creates it when applied forward. In all other steps (except the sixth) r is applied backwards directly, i.e. μ is the identity.

The most interesting step is the third, because rule 2c cannot be applied directly to the graph, since the match would be non-injective, but the rule can be applied to some larger graphs which are represented by the given graph. This backward step is shown in detail in Figure 7, where the diagram of Step 3 of the procedure is shown with the generated graphs. First the rule is applied

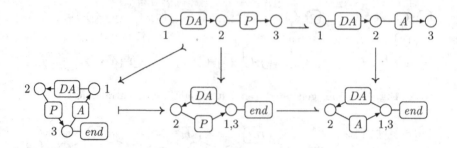

Fig. 7. The third backward step of the sequence in Fig. 6 in detail

backwards with a non-injective match replacing the A-edge with a P-edge, producing another non-injective match. Since this is not a valid match in the sense of Definition 6, the rule is not directly applicable and the node 1,3 has to be split such that the match is injective. This is done by decontraction generating the A-edge (which is just one possibility). The resulting graph is one of the graphs represented by the previous graph, with the difference that the rule is applicable.

The other five steps are straightforward or similar to the first three. Note that since in the last step a graph is reached which is smaller or equal to (in this case isomorphic) the initial graph, the final state is coverable and the protocol erroneous.

Note that, seen in the other direction, this is not quite a sequence of rewriting steps of the GTS. Rather, by applying the rules in the forward direction, we sometimes obtain a graph different from the one in the figure, but represented by it. Still, following the sequence against the direction of the arrows, it is possible to reconstruct why the error occurs in the protocol: after the detector first turns passive, the *start* flag is lost. Then, the *end* flag as well as the activation zone make their tour around the ring. Note that this is possible for the activation zone since the *start* flag is no longer there to block rule applications. Then the detector turns again passive and creates another *start* flag that reacts with the *end* flag of the previous round, leading to an erroneous *termination* message.

5   Conclusion

We have shown how graph transformation systems with negative application conditions can be viewed as well-structured transition systems. This is similar to the case of lossy vector addition systems [3] where lossiness is used in a similar way in order to deal with inhibiting conditions.

Furthermore we have described a generic backwards search decision procedure and proved that it terminates in specific cases. Termination depends on the existence of bounding functions and a question left open in this paper is to prove their existence in the general case.

Once this problem is solved, it could be interesting to study also general nested application conditions [14,9] and to check whether the results of this paper can be generalized. Furthermore we plan to look at other kinds of well-quasi-orders for graphs, possibly for restricted sets of graphs (see [6]).

We also plan to extend an existing implementation to deal with negative application conditions.

References

1. Abdulla, P.A., Čerāns, K., Jonsson, B., Tsay, Y.-K.: General decidability theorems for infinite-state systems. In: Proc. of LICS 1996, pp. 313–321. IEEE (1996)
2. Bertrand, N., Delzanno, G., König, B., Sangnier, A., Stückrath, J.: On the decidability status of reachability and coverability in graph transformation systems. In: Proc. of RTA 2012. LIPIcs. Schloss Dagstuhl – Leibniz Center for Informatics (2012)
3. Bouajjani, A., Mayr, R.: Model Checking Lossy Vector Addition Systems. In: Meinel, C., Tison, S. (eds.) STACS 1999. LNCS, vol. 1563, pp. 323–333. Springer, Heidelberg (1999)
4. Ehrig, H., Ehrig, K., Prange, U., Taentzer, G.: Fundamentals of Algebraic Graph Transformation. Monographs in Theoretical Computer Science. Springer (2006)
5. Esparza, J., Nielsen, M.: Decidability issues for Petri nets. Technical Report RS-94-8, BRICS (May 1994)
6. Fellows, M.R., Hermelin, D., Rosamond, F.A.: Well-Quasi-Orders in Subclasses of Bounded Treewidth Graphs. In: Chen, J., Fomin, F.V. (eds.) IWPEC 2009. LNCS, vol. 5917, pp. 149–160. Springer, Heidelberg (2009)
7. Finkel, A., Schnoebelen, P.: Well-structured transition systems everywhere! Theoretical Computer Science 256(1-2), 63–92 (2001)
8. Habel, A., Heckel, R., Taentzer, G.: Graph grammars with negative application conditions. Fundam. Inf. 26(3-4), 287–313 (1996)
9. Habel, A., Pennemann, K.-H.: Nested Constraints and Application Conditions for High-Level Structures. In: Kreowski, H.-J., Montanari, U., Yu, Y., Rozenberg, G., Taentzer, G. (eds.) Formal Methods in Software and Systems Modeling. LNCS, vol. 3393, pp. 293–308. Springer, Heidelberg (2005)
10. Joshi, S., König, B.: Applying the graph minor theorem to the verification of graph transformation systems. Technical Report 2012-01, Abteilung für Informatik und Angewandte Kognitionswissenschaft, Universität Duisburg-Essen (2012)
11. Joshi, S., König, B.: Applying the Graph Minor Theorem to the Verification of Graph Transformation Systems. In: Gupta, A., Malik, S. (eds.) CAV 2008. LNCS, vol. 5123, pp. 214–226. Springer, Heidelberg (2008)
12. König, B., Stückrath, J.: Well-structured graph transformation systems with negative application conditions. Technical Report 2012-03, Abteilung für Informatik und Angewandte Kognitionswissenschaft, Universität Duisburg-Essen (2012)
13. Löwe, M.: Algebraic approach to single-pushout graph transformation. Theoretical Computer Science 109, 181–224 (1993)
14. Rensink, A.: Representing First-Order Logic Using Graphs. In: Ehrig, H., Engels, G., Parisi-Presicce, F., Rozenberg, G. (eds.) ICGT 2004. LNCS, vol. 3256, pp. 319–335. Springer, Heidelberg (2004)
15. Robertson, N., Seymour, P.: Graph minors XX. Wagner's conjecture. Journal of Combinatorial Theory Series B 92, 325–357 (2004)
16. Robertson, N., Seymour, P.: Graph minors XXIII. Nash-Williams' immersion conjecture. Journal of Combinatorial Theory Series B 100, 181–205 (2010)
17. Rozenberg, G. (ed.): Handbook of Graph Grammars and Computing by Graph Transformation. Foundations, vol. 1. World Scientific (1997)

Parallelism and Concurrency of Stochastic Graph Transformations

Reiko Heckel[1], Hartmut Ehrig[2], Ulrike Golas[3], and Frank Hermann[4]

[1] University of Leicester, UK
reiko@mcs.le.ac.uk
[2] Technische Universität Berlin, Germany
ehrig@cs.tu-berlin.de
[3] Konrad-Zuse-Zentrum für Informationstechnik Berlin, Germany
golas@zib.de
[4] University of Luxembourg, Luxembourg
frank.hermann@uni.lu

Abstract. Graph transformation systems (GTS) have been proposed for high-level stochastic modelling of dynamic systems and networks. The resulting systems can be described as semi-Markov processes with graphs as states and transformations as transitions. The operational semantics of such processes can be explored through stochastic simulation. In this paper, we develop the basic theory of stochastic graph transformation, including generalisations of the Parallelism and Concurrency Theorems and their application to computing the completion time of a concurrent process.

1 Introduction

Stochastic graph transformation systems (SGTS) [4,12] support the integrated modelling of structural reconfiguration and non-functional aspects such as performance and reliability. An SGTS is a graph transformation system (GTS) where each rule name is associated with a probability distribution governing the delay of its application once a match has been found.

SGTSs have been used to model P2P networks, biochemical and cellular systems, as well as business processes. Typical questions to ask of such models include quality-of-service properties, such as if in 90% of cases of submitting a request, an answer will be received within two minutes. Such guarantees combine a statement of a time interval for an event with the probability for the event to happen within that interval. Properties like these can be verified by stochastic model checking or simulation.

As an alternative to the "execution" of models or the generation and explicit representation of their state space, we are interested in dealing with the stochastic aspect symbolically. For example, if two processes are composed sequentially and their individual durations are known, it is possible to compute the overall duration. The theory of graph transformation knows a number of constructions

H. Ehrig et al.(Eds.): ICGT 2012, LNCS 7562, pp. 96–110, 2012.

on rules and transformations, such as embedded rules, specialised by additional context, concurrent or parallel rules representing the execution of two or more (independent) steps without observing intermediate states, etc. In each case, we would like to give the distribution associated to the composed rule in terms of the distributions of the components. Thereby, we have to make sure that an application of the composed rule enjoys the same timing and probability as the corresponding applications of its component rules.

In this paper, we start developing the corresponding theory of stochastic graph transformation, investigating parallel and sequential compositions which allow us to construct certain classes of concurrent processes. Graph transformations are a natural choice for the modelling of concurrent processes where actions have local causes and effects. Such a process is represented by a partial order or equivalence class of sequences of transformations.

If we are interested in the completion time of a process, the symbolic approach can be used to compute the distribution of the overall duration of the process from the distributions of individual transformations. The problem turns out to be that of determining the *makespan* for a stochastic scheduling problem given by a set of tasks with stochastic durations and a partial order describing their precedence [10]. Such problems can be solved efficiently for a class of processes that are series-parallel, i.e., can be obtained by parallel and sequential composition. We solve the problem of approximating the delay of a graph transformation process by representing it as a series-parallel composition of rules.

This is of theoretical interest because it provides us with the stochastic equivalent of the basic notion of concurrent process of DPO graph transformation. From a practical point of view, they provide a model of business processes with concurrent actions, but deterministic outcomes. Being able to predict the likely runtime of such processes, we can select the fastest on average or the one most likely to finish by a certain deadline.

Example 1 (Business Process). As running example, we use a business process that is a small and slightly simplified part of a loan granting process model used in [1,5] based on a case study at Credit Suisse. The business process model depicted in Fig. 1 is specified as an event driven process chain (EPC) and contains a sequence of 3 business functions (B1,B2,B3). Function B1 specifies the last step of the customer interview, where the relationship manager (RM) identifies the customer demand and stores this information in a data base (DB1). In the second step, the rating application is used to calculate the rating and credit worthiness of the customer (business function B2). For this purpose, the application uses customer data (address and customer value) that were determined at the beginning of the customer interview (not contained in Fig. 1). Finally, the third step specifies that the credit advisor (CA) uses the price engine (PE) to create an optimized product, which is again stored in a data base (DB1). This last step requires that the customer demand (CD) as well as the credit worthiness of the customer were determined in previous steps.

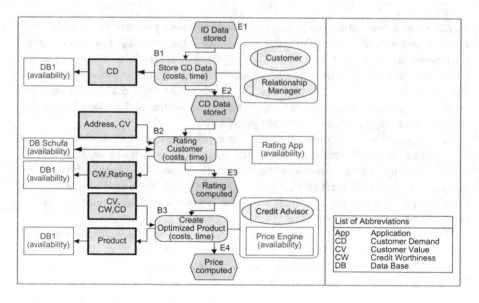

Fig. 1. Business process (extended EPC diagram)

2   Preliminaries

In this section, we provide the necessary background on probability theory [3]. The delay of a rule application, between the time it is enabled by a match and its application, is treated as a real-valued random variable v. This means that v's values are not known in advance, but they are known to fall into particular real-valued intervals with certain probabilities described by a probability distribution $f : \mathbb{R}_+ \to [0,1]$. From such a distribution we can compute a cumulative distribution function $F : \mathbb{R}_+ \to [0,1]$ by $F(x) = \int_0^x f(t)dt$, whose value at each real x is the probability that the value of the random variable is smaller than or equal to x. Values for v can be chosen randomly based on the probability distribution f. We denote such choice of a *random number* by $v = RN_F$.

All graphs and computations in our example were computed and generated with Wolfram Mathematica [8].

Example 2 (Distributions). To each business function (B1-B3) of our business process model in Fig. 1, we assign a fictive, but realistic distribution as shown in Fig. 2. All distributions $f(v_1) - f(v_3)$ are given by log-normal distribution functions with parameters (μ_i, σ_i) according to the table depicted in the figure. Log-normal distributions are a typical model for response times of both humans fulfilling a task [7] as well as servers or transaction systems [9]. Such a distribution is characterized by a location parameter μ, representing the mean, and a scale parameter σ, representing the standard deviation, on a logarithmized scale.

By inspecting the distribution functions in the diagram, we can observe the following properties. Function $f(v_1)$ (customer interview) has a peak at app. 22

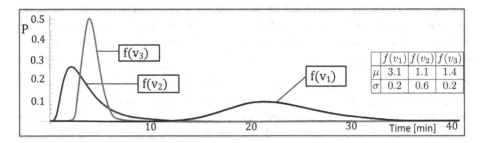

Fig. 2. Distribution functions for completion times of variables v_1, v_2 and v_3 (business functions B1, B2, B3)

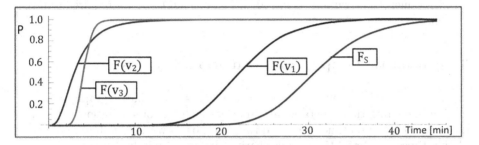

Fig. 3. Cumulative distribution functions for completion times of variables v_1, v_2 and v_3 (business functions B1, B2, B3)

minutes, function $f(v_2)$ (rating of the customer) peaks at 2 minutes and function $f(v_3)$ (product optimization) at 4 minutes. Moreover, the variance of function $f(v_2)$ is higher than the one of f_3, because an external data base (DBSchufa) is involved and may lead to delays. Naturally, the variance of $f(v_1)$ is much higher than the other two, because it concerns the customer interview, i.e., communication between humans. The cumulative distributions $F(v_1) - F(v_3)$ of our example are depicted in Fig. 3 and are computed from the distribution functions $f(v_1) - f(v_3)$.

We will make use of the sum and the maximum of independent random variables in order to derive the overall completion time of a process consisting of the sequential or concurrent composition of transformation steps. The corresponding operations at the level of distributions are the convolution and product of distribution functions. Given distribution functions f_1 and f_2 associated with independent real-valued random variables v_1 and v_2, $f_1 * f_2(u) = \int_{-\infty}^{\infty} f_1(x) f_2(u-x) dx$ defines their convolution, i.e., the distribution function associated with the sum $v_1 + v_2$. By definition, we obtain the cumulative distribution function $F_1 * F_2(x) = \int_{0}^{x} f_1 * f_2(t) dt$. The operation is commutative and associative, which allows us to write the convolution of a set of distributions $\{F_1, \ldots, F_n\}$

as $F_1 * \cdots * F_n$ or $*_{i \in \{1,\ldots,n\}} F_i$. The product $F_1 \cdot F_2$ of cumulative distribution functions, defined by $F_1 \cdot F_2(x) = F_1(x) \cdot F_2(x)$, represents the cumulative distribution of the maximum $max(v_1, v_2)$ of the independent random variables v_1 and v_2.

Example 3 (Unaccomplished Guarantee). Considering the example introduced in Ex. 2, we can compute the completion time of a business process executing the function $B1, B2, B3$ by the convolution $F_S = F(v_1) * F(v_2) * F(v_3)$ as the sum of the single processing times and derive the cumulative distribution function F_S shown in Fig. 3. In the present scenario, the required guarantee concerning the completion time is given by $F(40) \geq 95\%$, i.e., in less than 5% of all cases, the completion time may exceed the limit of 40 minutes. Note that the result for F_S (sequential execution) is $F_S(40) = 93.90\%$ and thus, the required criteria are not guaranteed.

3   Stochastic Graph Transformation Systems

In this section we first introduce the basic notions of stochastic graph transformation, including their operational semantics in terms of simulation. Then we define the completion time of a timed run and show how the distribution of completion times of a set of timed runs sharing the same underlying transformation sequence is derived from the delay distributions of the rules in the sequence.

A typed graph transformation system $\mathcal{G} = (TG, P, \pi)$ consists of a type graph TG, a set of rule names P and a function π associating with each name $p \in P$ a span of injective TG-typed graph morphisms $\pi(p) = (L \xleftarrow{l} K \xrightarrow{r} R)$. The application $G \stackrel{p,m}{\Longrightarrow} H$ of rule p at a match $m : L \to G$ subject to the usual gluing conditions is defined by a double-pushout construction. A *stochastic graph transformation system* $\mathcal{SG} = (TG, P, \pi, F)$ consists of a graph transformation system $\mathcal{G} = (TG, P, \pi)$ and a function $F : P \to \mathbb{R}_+ \to [0,1]$ which associates with all rule names in P cumulative probability distribution functions $F(p) : \mathbb{R}_+ \to [0,1]$.

Given a start graph G_0, the behaviour of $\mathcal{SG}$ can be explored by simulation. To this effect the simulation tool GrabS [12] has been developed, using the graph transformation tool VIATRA [13] as its underlying engine to represent graphs, find matches, and perform transformations. The simulation works as follows.

For a graph G, *events* in G are defined by $E(G) = \{(p, m) \mid p \in P \wedge \pi(p) = L \leftarrow K \to R \wedge \exists m : L \to G$ satisfying p's gluing conditions$\}$, i.e., each event is an enabled rule match. States (G, t, s) of the simulation are given by the current graph G, the simulation time t, and the schedule $s : E(G) \to \mathbb{R}_+$ mapping events to their scheduled times.

1. Initially, the current graph is the start graph $G = G_0$, the time is set to $t = 0$ and the scheduled time $s(p, m) = RN_{F(p)}$ for each enabled event (p, m) is selected randomly based on p's probability distribution.
2. For each simulation step

(a) the first event $e = (p, m)$ is identified, i.e., such that for all events e', $s(e) \leq s(e')$. Rule p applied at match m to the current graph G produces the new current graph H, i.e., $G \overset{p,m}{\Longrightarrow} H$
(b) the simulation time is advanced to $t = s(e)$
(c) the new schedule s', based on the updated set of enabled events $E(H)$, is defined for all $(p', n') \in E(H)$ with $\pi(p') = L' \leftarrow K' \rightarrow R'$ in the diagram below as
   - $s'(p', n') = s(p', n)$, if $(p', n) \in E(G)$ and there exists a morphism $k : L' \rightarrow D$ with $l^* \circ k = n$, such that $n' = r^* \circ k$;
   - $s'(p', n') = t + RN_{F(p')}$, otherwise.

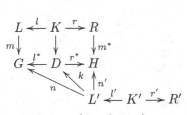

The result is a *(simulation) run* $r = (G_0 \overset{p_1,m_1,t_1}{\Longrightarrow} \cdots \overset{p_n,m_n,t_n}{\Longrightarrow} G_n)$, i.e., a transformation sequence where steps are labelled by time stamps $t_1, \ldots, t_n \in \mathbb{R}_+$ with $t_i < t_{i+1}$ for all $i \in \{1, \ldots, n-1\}$. We call $ct(r) = t_n$ the *completion time* of the run.

Our simulation algorithm represents an optimal strategy for executing runs in $\mathcal{SG}$, in the sense that every step is performed as early as possible: The stochastic delay begins as soon as the rule is enabled by the match, and it is applied precisely when that time has passed. This is described more abstractly as follows.

- A step $G_{i-1} \overset{p_i,m_i}{\Longrightarrow} G_i$ has an enabling time $et(i)$ and an application time $at(i)$, but no duration.
- The enabling time is the point at which match m_i comes to exist, either at the start of the run $(et(i) = 0)$, or by the application of an earlier step $G_{j-1} \overset{p_j,m_j}{\Longrightarrow} G_j$ with $j < i$, in which case $et(i) = at(j)$. In the latter case, that step j will be the last step that step i depends on in the sequence.
- The delay $delay(i) = at(i) - et(i)$ of step i is a random variable governed by the distribution function $F(p_i)$.

Given a start graph G_0, we say that run r *follows* sequence s if they both start in G_0 and contain the same steps (rules and matches) in the same order. Formally, a run following a sequence of n steps extends s by an ordered vector $\bar{r} \in (\mathbb{R}_+)^n$ of time stamps.

We can imagine the entire behaviour of the underlying GTS $\mathcal{G}$ with start graph G_0 to be given by a transition system with states as graphs and transitions given by transformation steps labelled by rule-match pairs. A transformation sequence s starting in G_0 is a path in this transition system. The probability for a run in $\mathcal{SG}$ starting in G_0 to follow a particular path s is thus determined by the probability of choosing the "right" outgoing transitions from each state on the path. For this to happen, the transformation corresponding to the desired transition needs

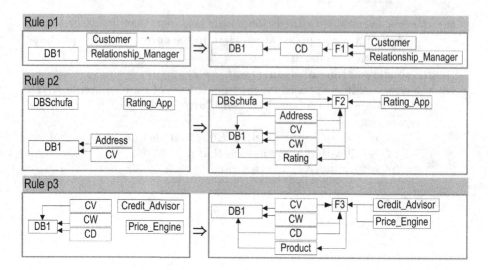

Fig. 4. Derived operational rules for dependency analysis (for business functions B1, B2, B3)

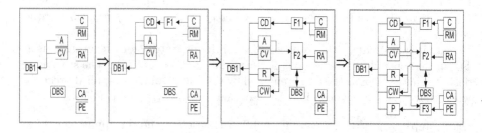

Fig. 5. Transformation sequence according to the business process model

to be the next on the schedule. The probability for that depends on the delay distribution of the respective rules. We denote the probability for a run in $\mathcal{SG}$ starting in G_0 to follow a particular sequence s by $Prob(r$ follows $s)$.

Example 4 (Generated Transformation System). In order to analyse and improve the business process in Fig. 1, we generate a corresponding transformation system according to [5]. Each business function corresponds to a graph transformation rule. In our example, we derive the three rules in Fig. 4 specifying the three business functions B1-B3. The left hand sides of the rules contain the data elements that are read by the business function from their corresponding resources and they contain the involved actors. Each rule (p1-p3) creates a node for its corresponding business function (B1-B3), the data elements that are created by it and edges from the actors who execute the business function.

The transformation sequence in Fig. 5 shows the three transformation steps via rules p1-p3 in the same order as in the given business process model in Fig. 1. Matches are given by naming.

Fixing SG and G_0, the completion time distribution of all runs following sequence s (briefly completion time distribution of s) is defined as the function $ctd(s)$: $\mathbf{R}^+ \to [0,1]$ assigning each positive real number x the conditional probability for a run to complete within time x if the run follows s.

$$ctd(s)(x) = Prob\{ct(r) \le x \mid \text{run } r \text{ follows } s\}$$

For a transformation sequence $s = (G_0 \overset{r_1,m_1}{\Longrightarrow} \cdots \overset{r_n,m_n}{\Longrightarrow} G_n)$ we find its completion time distribution by first identifying the critical steps of the sequence, i.e., the chain of latest steps required before the next step is enabled, and then adding up the individual delays of these steps.

Definition 1 (enabling steps). *Assume a sequence $s = (G_0 \overset{p_1,m_1}{\Longrightarrow} \cdots \overset{p_n,m_n}{\Longrightarrow} G_n)$ in SG. We write $j \rightsquigarrow k$ if step $G_{j-1} \overset{p_j,m_j}{\Longrightarrow} G_j$ enables $G_{k-1} \overset{p_k,m_k}{\Longrightarrow} G_k$, that is, in the diagram below there exists n_k with $h \circ n_k = m_k$ such that $g \circ n_k$ satisfies p_k's gluing condition, and there is no n_j with $r_j^* \circ n_j = g \circ n_k$ such that $l_j^* \circ n_j$ satisfied p_k's gluing condition.*

The span $G_j \overset{g}{\longleftarrow} D \overset{h}{\longrightarrow} G_{k-1}$ represents the derived span, i.e. the composition via pullbacks of the sequence of transformation spans of steps $G_j \overset{p_{j+1},m_{j+1}}{\Longrightarrow} \cdots \overset{p_{k-1},m_{k-1}}{\Longrightarrow} G_{k-1}$.

That means, step p_k is applicable to G_j, but not to G_{j-1}, at a match compatible with m_k. G_j is therefore the first graph enabling an event (namely $(p_k, g \circ n_k)$) evolving into (p_k, m_k). Therefore, step k is scheduled when step j is performed, and the delay of step k spans the interval $[at(j) = et(k), at(k)]$. The following proposition decomposes a run into consecutive intervals, so that the sum of their delays makes up the overall completion time.

Proposition 1 (completion time distribution). *Assume a non-empty sequence $s = (G_0 \overset{p_1,m_1}{\Longrightarrow} \cdots \overset{p_n,m_n}{\Longrightarrow} G_n)$ in SG. The set of critical steps $CS(s) \subseteq \{1,\dots,n\}$ of s is the smallest subset of step indices such that*

- $n \in CS(s)$ *and*
- *for all $k \in CS(s)$ and $j \le k$ with $j \rightsquigarrow k$, also $j \in CS(s)$.*

The completion time distribution of runs following s is given by

$$ctd(s) = *_{i \in CS(s)} F(p_i)$$

Proof. By induction on the cardinality of $CS(s)$: If the final step n is enabled in G_0, $CS(s) = \{n\}$. In this case, the completion time $ct(r)$ of a run following s equals the delay $at(n) - et(n)$ of the last step, while the (shorter) delays of steps $1 \ldots n-1$ are irrelevant because they are contained in the delay of step n.

If $s = (G_0 \overset{p_1,m_1}{\Longrightarrow} \ldots \overset{p_k,m_k}{\Longrightarrow} G_k \overset{p_{k+1},m_{k+1}}{\Longrightarrow} \ldots \overset{p_n,m_n}{\Longrightarrow} G_n)$ such that $k \rightsquigarrow n$, $CS(s) = \{n\} \cup CS(s_{1k})$ with $s_{1k} = G_0 \overset{p_1,m_1}{\Longrightarrow} \ldots \overset{p_k,m_k}{\Longrightarrow} G_k$. In this case, $ct(r) = ct(r_{1k}) + ct(r_{kn})$ where $r_{1k}; r_{kn}$ is the decomposition of r into runs following s_{1k} and the remainder sequence s_{kn}.

Given a run r following s, the overall time is thus obtained by decomposing the sequence into intervals $[et(i), at(i)]_{i \in CS(s)}$. The intervals are consecutive, i.e., if j is the smallest index in $CS(s)$ larger than i, $at(i) = et(j)$. Due to the construction of $CS(s)$ the smallest index has enabling time 0 while the largest has application time $at(n)$. Thus, the completion time of a run r following s is given by

$$ct(r) = \sum_{i \in CS(s)} at(i) - et(i)$$

By definition, $ct(r)$ is a random variable distributed according to $ctd(s)$. Since the random variables representing the delays are independent, the convolution of their distributions $F(p_i)$ provides the distribution of the sum of all intervals, making up the completion time distribution.

Example 5 (Completion time distribution of a run). The dependencies between the three steps in the graph transformation sequence $s = (G_0 \overset{p_1,m_1}{\Longrightarrow} G_1 \overset{p_2,m_2}{\Longrightarrow} G_2 \overset{p_3,m_3}{\Longrightarrow} G_3)$ in Fig. 5 are $1 \rightsquigarrow 3$ and $2 \rightsquigarrow 3$, because both steps 1 and 2 are necessary for the enabling of step 3. These dependencies can be computed as presented in [1] based on Mathematica, but also using the dependency analysis of the tool AGG [11] using the extended type graph according to [1].

Since steps 1 and 2 are independent, the simulation will tend to execute the rule with the smaller delay first, depending on the concrete values $RN_{F(v_1)}$ and $RN_{F(v_2)}$ for a certain run. This means that a run r follows the sequence $s = (G_0 \overset{p_1,m_1}{\Longrightarrow} G_1 \overset{p_2,m_2}{\Longrightarrow} G_2 \overset{p_3,m_3}{\Longrightarrow} G_3)$ if $RN_{F(v_1)} \leq RN_{F(v_2)}$. In this case we derive the set $CS(s) = \{2,3\}$ of critical steps and $ct(r) = F(v_2) * F(v_3)$. Otherwise, if $RN_{F(v_1)} \geq RN_{F(v_2)}$, the run r will follow the sequence $s' = (G_0 \overset{p_2,m_2'}{\Longrightarrow} G_1' \overset{p_1,m_1'}{\Longrightarrow} G_2 \overset{p_3,m_3}{\Longrightarrow} G_3)$ and the set $CS(s') = \{1,3\}$ of critical steps leads to $ct(r) = F(v_1) * F(v_3)$. That means, completion times can be different for the two runs, even if they are shift equivalent.

In our model, steps have delays but no duration. A sequence can show apparent concurrency when delays overlap, even if the actual steps are strictly sequential. In the next section we consider parallelism and concurrency of steps in the traditional sense.

4   Parallelism and Concurrency

Much of the basic theory of algebraic graph transformation follows a common pattern. In order to describe a structured notion of transformation, a composition operation op on rules is introduced such that applications of composed rules represent op-structured transformations. For example, with $op = +$, parallel rules are defined, leading to parallel transformations. A corresponding theoretical result defines a relation $ser : (op\ \mathcal{G})^* \to \mathcal{P}(\mathcal{G}^*)$ where $\mathcal{G}^*$ denotes the set of transformation sequences of $\mathcal{G}$ and $(op\ \mathcal{G})^*$ is the set of sequences of op-structured rules. For example, each parallel transformation sequence can be serialised into several sequential sequences. Often, it is possible to characterise the codomain of this serialisation, giving rise to a precise correspondence between op-structured and suitable sequential derivations. For example, parallel steps are in correspondence with sequentially independent two-step sequences.

We will exploit this idea to extend the definition of completion time from sequential to structured (i.e., parallel and concurrent transformations), and then ask for a characterisation of completion time distributions of complex steps in terms of those of their components. The delay distribution of composed rules will be defined accordingly. The approach is illustrated in the diagram below. The completion time distribution for an op-structured transformation sequence d is defined by $ctd(d) =_{def} ctd(ser(d))$ where $ser(d)$ is the set of sequential transformations related to d and ctd is extended to sets of sequences S by

$$ctd(S)(x) = Prob\{ct(r) \le x \mid \text{run } r \text{ follows a sequence } s \in S\}$$

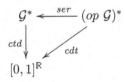

Below, we consider this situation for parallel rules and transformations. Let the parallel rule for rules p_1 and p_2 be given by $p_1 + p_2$, with $\pi(p_1 + p_2) = \pi(p_1) + \pi(p_2)$ the parallel span obtained by componentwise coproduct, and $s = (G \overset{p_1 + p_2, m}{\Longrightarrow} H)$ denote the corresponding parallel step. A *parallel sequence* is a sequence of sequential and parallel transformation steps. Extending the sequential case, we say that a run r *follows a parallel sequence* d if there exists a sequence s obtained by replacing parallel steps in d by their corresponding serialisations, such that r follows s. Accordingly, the notion of completion time distribution carries over to (runs following) parallel sequences.

Proposition 2 (Parallelism). *The completion time of a parallel step* $d = (G \overset{p_1 + p_2, m}{\Longrightarrow} H)$ *is distributed according to the product* $ctd(d) = F(p_1) \cdot F(p_2)$ *of the distribution functions for the delays of* p_1 *and* p_2. *Therefore, the cumulative distribution function for the delay of the parallel rule is defined by* $F(p_1 + p_2) = F(p_1) \cdot F(p_2)$.

Proof. If r is a run following (either of the two serialisations s_1 and s_2 of) d, then its completion time $ct(r)$ is given by the greater of the delays d_1 and d_2 scheduled for p_1 and p_2, i.e., $ct(r) = max(d_1, d_2)$. These delays are distributed according to $F(p_1)$ and $F(p_2)$, respectively. The product $F(p_1) \cdot F(p_2)$ of the distribution functions represents the distribution of the maximum of these delays.

For rules p_1 and p_2 with a specified overlap of the right-hand side R_1 with the left-hand side L_2 we build their concurrent rule [2], intuitively, as sequential composition of p_1 and p_2.

Definition 2 (Concurrent rule). *Assume a dependency relation $R_1 \xrightarrow{e_1} E \xleftarrow{e_2} L_2$ between rules $p_1 : L_1 \leftarrow K_1 \rightarrow R_1$ and $p_2 : L_2 \leftarrow K_2 \rightarrow R_2$ as shown in the diagram below, such that e_1 satisfies the gluing condition of the inverse rule $p_1^{-1} : R_1 \leftarrow K_1 \rightarrow L_1$ and e_2 that of p_2. The concurrent rule $p_1 \star_{e_1,e_2} p_2 : L \leftarrow K \rightarrow R$ is constructed as follows. First the reduction of graph E to the image E' of R_1 and L_2 in E is defined via the coproduct $R_1 + L_2$ and the epi-mono factorisation $i \circ e$ of the unique morphism into E that commutes with the coproduct injections i_1, i_2 and morphisms e_1 and e_2. Second, applying the inverse rule $p_1^{-1} : R_1 \leftarrow K_1 \rightarrow L_1$ to $m_1^* = e \circ i_1$ and rule p_2 to $m_2 = e \circ i_2$, we construct the double pushout diagrams representing transformations $L \xRightarrow{p_1,m_1} E' \xRightarrow{p_2,m_2} R$. Extracting K, r^*, l^* as pullback of r_1^* and l_2^*, the concurrent rule is given by*
$$p_1 \star_{e_1,e_2} p_2 : L \xleftarrow{l_1^* \circ r^*} K \xrightarrow{r_2^* \circ l^*} R.$$

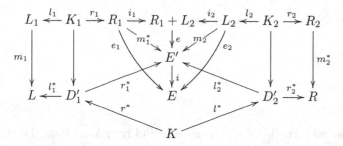

Rule $p_1 \star_{e_1,e_2} p_2$ is called sequentially independent *if the two DPO diagrams representing its construction form a sequentially independent transformation sequence $L \xRightarrow{p_1,m_1} E' \xRightarrow{p_2,m_2} R$, that is, there exist morphisms $k_1 : R_1 \rightarrow D_2'$ such that $l_2^* \circ k_1 = m_1^*$ and $k_2 : L_2 \rightarrow D_1'$ such that $r_1^* \circ k_2 = m_2$. If $p_1 \star_{e_1,e_2} p_2$ is not sequentially independent, we write $p_1;_{e_1,e_2} p_2$.*

The rule is disjoint *if $e_1(R_1) \cap e_2(L_2) = \emptyset$, in which case we write $p_1 + p_2$.*

The definition is slightly non-standard in the sense that it allows graph E to be larger than necessary for the dependency relation to be expressed. Therefore, a reduction step is needed to cut out the unnecessary context in E, leaving only the union of the images of R_1 and L_2. After that the construction proceeds as usual. The variation is required due to Def. 4 in the following section where graph E will be fixed while, in a composition of rules p_1 and p_2, p_1 may be replaced by a subrule p_1'. While composing p_1 and p_2, E may represent a union of p_1's right- and p_2's left-hand side, but this won't be true of the composition of p_1' and p_2. In such a case we manipulate the embeddings explicitly, e.g., replacing in $p_1 \star_{e_1,e_2} p_2$, rule p_1 by $p_1' : L_1' \leftarrow K_1' \rightarrow R_1'$ with $j_1 : R_1' \rightarrow R_1$, we write $p_1' \star_{e_1 \circ j_1, e_2} p_2$.

If $p_1 \star_{e_1,e_2} p_2$ is independent it is possible to replace its applications by applications of $p_1 + p_2$ at a match identifying the parts where the rules overlap in (e_1, e_2). Thus, concurrent rules and transformations subsume parallel ones, and we can limit ourselves to considering two cases: non-independent concurrent rules representing sequential composition and disjoint concurrent rules representing parallel composition.

The Concurrency Theorem [2] states that steps $G \overset{p_1 \star_{e_1,e_2} p_2, m}{\Longrightarrow} H$ using concurrent rules are in one-to-one correspondence with so-called e_1, e_2-related transformation sequences $s_1 = (G \overset{p_1, m_1}{\Longrightarrow} G_1 \overset{p_2, m_2}{\Longrightarrow} H)$, where the overlap between p_1 and p_2 in G_1 conforms to the dependency relation e_1, e_2.

A *concurrent sequence* is a sequence of (possibly concurrent) transformation steps. We say that a run r *follows a concurrent sequence* d if there exists a sequence s obtained by substituting concurrent steps in d by their corresponding sequences, such that r follows s. The notion of completion time distribution carries over to (runs following) concurrent sequences.

Proposition 3 (Concurrency). *Let* $d = (G \overset{p_1 \star_{e_1,e_2} p_2, m}{\Longrightarrow} H)$ *be a transformation using the concurrent rule and* $s_1 = (G \overset{p_1, m_1}{\Longrightarrow} G_1 \overset{p_2, m_2}{\Longrightarrow} H)$ *be the corresponding* (e_1, e_2)*-related transformation sequence according to the Concurrency Theorem. If* s_1 *is not sequentially independent,* $ctd(d) = F(p_1) * F(p_2)$, *otherwise* $ctd(d) = F(p_1) \cdot F(p_2)$.

Therefore, if $p_1 \star_{e_1,e_2} p_2$ *is sequentially independent, the cumulative distribution function for the concurrent rule is defined by* $F(p_1 \star_{e_1,e_2} p_2) = F(p_1) \cdot F(p_2)$, *otherwise it is defined by* $F(p_1 \star_{e_1,e_2} p_2) = F(p_1) * F(p_2)$.

Proof. First, observe that (e_1, e_2)-related transformation sequence s_1 is sequentially independent if and only if $p_1 \star_{e_1,e_2} p_2$ is sequentially independent. Therefore, if the concurrent rule is independent, s_1 is an independent sequence and by Prop. 2 this implies that $ctd(d) = F(p_1) \cdot F(p_2)$. Otherwise, it follows from Prop. 1 that $ctd(d) = F(p_1) * F(p_2)$.

5   Series-Parallel Transformations

Using the results of the previous section, we should be able to determine the completion time of derivations formed by sequential and/or parallel composition of basic steps. We first introduce a notion of series-parallel rule and then study the conditions under which the completion time distributions calculated via their structure match the actual distributions of corresponding derivations.

Definition 3 (series-parallel rules and derivations). *Assuming* $S\mathcal{G} = (TG, P, \pi, F)$ *and a* TG*-typed graph* E, *series-parallel (sp) rule expressions over* E *are defined by*

$$c ::= p \mid c_{;e_1,e_2} c \mid c + c$$

where $p \in P$ and c represent basic and composed rules, respectively, and e_1, e_2 are morphisms into E. The mappings π and F are extended inductively from P.

1. *For $\pi(c_i) = L_i \leftarrow K_i \rightarrow R_i$, $e_1 : R_1 \rightarrow E$ and $e_2 : L_2 \rightarrow E$, such that $c_{1;e_1,e_2} c_2$ is not sequentially independent, we define $\pi(c_{1;e_1,e_2} c_2) = \pi(c_1) \star_{e_1,e_2} \pi(c_2)$ and $F(c_{1;e_1,e_2} c_2) = F(c_1) * F(c_2)$.*
2. *$\pi(c_1 + c_2) = \pi(c_1) + \pi(c_2)$ and $F(c_1 + c_2) = F(c_1) \cdot F(c_2)$*

A *series-parallel (sp) rule* is a rule expression with rule span $c : \pi(c)$. A *series-parallel (sp) derivation* is a sequence of steps using sp rules.

Let $d = G \overset{c,m}{\Longrightarrow} H$ be an sp transformation. If $c = c_{1;e_1,e_2} c_2$ or $c = c_1 + c_2$, $ser(d) = \{s_1 s_2 \mid s_i \in ser(d_i)$ where $d_1 d_2$ is a serialisation of $d.\}$

Note that, due to clause 1 above, π and F are only partially defined, i.e., there are rule expressions which, in the context of a given stochastic GTS, do not give rise to rules. If $\pi(c)$ is defined, we call c a *well-formed* rule expression.

Serialisation of sp transformations is defined in terms of the serialisation of parallel and (non-independent) concurrent transformations. It is well-defined because serialisations have the same start and end graphs of the original sequence. Note that in the parallel case, d has two serialisations $d_1 d_2$, up to isomorphism, whereas in the concurrent case there is just one e_1, e_2-dependent derivation $d_1 d_2$.

A run r *follows* an sp sequence d if it follows any sequence s obtained by serialising d. The notion of completion time distribution carries over to (runs following) sp sequences.

The distribution of an sp rule implies an execution strategy where, whenever a sequential composition $c_{1;e_1,e_2} c_2$ is encountered, all components of c_1 have to finish before any components of c_2 are scheduled. This is consistent with the behaviour of serialisations if each parallel component following a sequential composition $;_{e_1,e_2}$ depends (directly or indirectly) on all parallel components before $;_{e_1,e_2}$. If this is the case, $c_{1;e_1,e_2} c_2$ allows an exact characterisation of completion time distribution by $F(c_1) * F(c_2)$.

Definition 4 (strongly dependent series parallel rules). *We distinguish the class of strongly dependent sp rules by defining it inductively, following the structure of their rule expressions. When referring to a rule expression, we assume that it is well-formed, otherwise the corresponding clause is not applicable.*

1. *basic rule names p and their sequential compositions $p_{1;e_1,e_2} p_2$ are strongly dependent;*
2. *$c_1 + c_2$ is strongly dependent if both c_1 and c_2 are;*
3. *$c_{;e_1,e_2} (c_1 + c_2)$ is strongly dependent if $c_{;e_1,e_2 \circ j_1} c_1$ and $c_{;e_1,e_2 \circ j_2} c_2$ are, where $j_i : L_i \rightarrow L_1 + L_2$ are the coproduct injections;*
4. *$(c_1 + c_2)_{;e_1,e_2} c$ is strongly dependent if $c_{1;e_1 \circ j_1,e_2} c$ and $c_{2;e_1 \circ j_2,e_2} c$ are, where $j_i : R_i \rightarrow R_1 + R_2$ are the coproduct injections;*
5. *$(c_{1;e_1,e_2} c_2)_{;e'_2,e'_3} c_3$ is strongly dependent if $c_{1;e_1,e_2} c_2$ and $c_{2;e'_2 \circ m^*_2,e'_3} c_3$ are, where m^*_2 is the morphism mapping the right-hand side of c_2 into that of $c_{1;e_1,e_2} c_2$ (cf. Def. 2);*

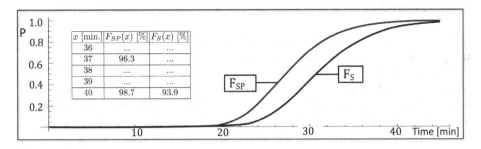

Fig. 6. Cumulative distribution functions for completion times of transformation sequence $s = (G_0 \xrightarrow{p_1, m_1} G_1 \xrightarrow{p_2, m_2} G_2 \xrightarrow{p_3, m_3} G_3)$

6. $c_{1;e_1,e_2} (c_{2;e_2',e_3'} c_3)$ *is strongly dependent if* $c_{1;e_1,e_2 \circ m_1} c_2$ *and* $c_{2;e_2',e_3'} c_3$ *are, where* m_1 *is the morphism mapping the left-hand side of* c_2 *into that of* $c_{2;e_2',e_3'} c_3$ *(cf. Def. 2);*

Proposition 4 (series-parallel composition). *If transformation* $d = (G \xRightarrow{c,m} H)$ *is using an sp rule* c, $ctd(d) \leq F(c)$. *If* c *is strongly dependent,* $ctd(d) = F(c)$.

Proof. By induction on the structure of rule expressions c we prove the stronger statement for the strongly dependent case first. If $c = p$ is basic, $ctd(d) = F(p)$ by Prop. 1. For the induction step, if $c = c_{1;e_1,e_2} c_2$, then $F(c) = F(c_{1;e_1,e_2} c_2) = F(c_1) * F(c_2)$. Assuming $ctd(d_i) = F(c_i)$ we have to show that $ctd(d) = ctd(d_1) * ctd(d_2)$ for e_1, e_2-related transformations $d_1 d_2$ corresponding to d. Serialisations of d are concatenations of serialisations $s_1 s_2$ of $d_1 d_2$ and the completion time distribution of an sp sequence is defined via the completion time distributions of its serialisations. For these we have $ctd(s_1 s_2) = ctd(s_1) * ctd(s_2)$ because strong dependency implies that no step of s_2 is enabled before the last step of s_1. If $c = c_1 + c_2$, then $F(c) = F(c_1 + c_2) = F(c_1) \cdot F(c_2) = ctd(d)$, by the same reasoning as in Prop. 3.

If c is not strongly dependent, a step of s_2 could be enabled before the last step of s_1, leading for the completions times of runs of s_1 and s_2 to overlap, thus leading to $ctd(s_1 s_2) \leq ctd(s_1) * ctd(s_2)$ and therefore $ctd(d) \leq ctd(d_1) * ctd(d_2) = F(c_1) * F(c_2) = F(c)$.

Example 6 (Completion time distributions of a run with series parallel rule). According to Prop. 4 we derive an sp rule c from the transformation sequence $s = (G_0 \xrightarrow{p_1, m_1} G_1 \xrightarrow{p_2, m_2} G_2 \xrightarrow{p_3, m_3} G_3)$, where steps 1 and 2 are combined via parallel composition, because the steps are independent. The completion time distribution F_{SP} for the corresponding run $d = (G_0 \xRightarrow{c,m} G_3)$ is shown in Fig. 6. Concerning the given deadline of 40 minutes, we derive $F_{SP}(40) = 98.7\%$ and thus, the given requirement is met. In particular, we have that $F_{SP}(37) = 96.3\%$, i.e., there is even a buffer of 3 minutes. This means that we have not only found these business functions that can be executed in parallel, but have also proven that in this case the time constraints are guaranteed.

6 Conclusion

We extended the parallel and sequential composition of rules to stochastic graph transformations and used them to characterise the overall completion time of a series-parallel process. We would like to explore in more detail the relation of our approach with stochastic scheduling theory [10], using the causal order of a process to derive a series-parallel over-approximation by adding new dependencies. This would provide an upper bound to the CTD analogous to Prop. 4.

References

1. Brandt, C., Hermann, F., Groote, J.F.: Generation and Evaluation of Business Continuity Processes; Using Algebraic Graph Transformation and the mCRL2 Process Algebra. Journal of Research and Practice in Information Technology 43(1), 65–85 (2011)
2. Ehrig, H., Ehrig, K., Prange, U., Taentzer, G.: Fundamentals of Algebraic Graph Transformation. Monographs in Theoretical Computer Science. An EATCS Series. Springer-Verlag New York, Inc., Secaucus (2006)
3. Grinstead, C., Snell, J.: Introduction to Probability. Dartmouth Chance Project (2005), http://www.dartmouth.edu/~chance/teaching_aids/books_articles/probability_book/book.html
4. Heckel, R.: Stochastic Analysis of Graph Transformation Systems: A Case Study in P2P Networks. In: Van Hung, D., Wirsing, M. (eds.) ICTAC 2005. LNCS, vol. 3722, pp. 53–69. Springer, Heidelberg (2005)
5. Hermann, F.: Analysis and Optimization of Visual Enterprise Models Based on Graph and Model Transformation. Ph.D. thesis, TU Berlin (2011), http://opus.kobv.de/tuberlin/volltexte/2011/3008/
6. Lack, S., Sobociński, P.: Adhesive Categories. In: Walukiewicz, I. (ed.) FOSSACS 2004. LNCS, vol. 2987, pp. 273–288. Springer, Heidelberg (2004)
7. van der Linden, W.J.: A lognormal model for response times on test items. Journal of Educational and Behavioral Statistics 31(2), 181–204 (2006)
8. Wolfram Research: Mathematica 8.0 (2012), http://www.wolfram.com/
9. Mielke, A.: Elements for response-time statistics in ERP transaction systems. Perform. Eval. 63(7), 635–653 (2006)
10. Möhring, R.H.: Scheduling under Uncertainty: Bounding the Makespan Distribution. In: Alt, H. (ed.) Computational Discrete Mathematics. LNCS, vol. 2122, pp. 79–97. Springer, Heidelberg (2001)
11. TFS, TU Berlin: AGG: The Attributed Graph Grammar System, http://tfs.cs.tu-berlin.de/agg/
12. Torrini, P., Heckel, R., Ráth, I.: Stochastic Simulation of Graph Transformation Systems. In: Rosenblum, D.S., Taentzer, G. (eds.) FASE 2010. LNCS, vol. 6013, pp. 154–157. Springer, Heidelberg (2010)
13. Varró, D., Balogh, A.: The model transformation language of the VIATRA2 framework. Sci. Comput. Program. 68(3), 214–234 (2007)

Refined Graph Rewriting in Span-Categories

A Framework for Algebraic Graph Transformation

Michael Löwe

FHDW-Hannover
Freundallee 15, 30173 Hannover, Germany
`michael.loewe@fhdw.de`

Abstract. There are three major algebraic approaches to graph transformation, namely the double-pushout (DPO), single-pushout (SPO), and sesqui-pushout approach (SqPO). In this paper, we present a framework that generalises all three approaches. The central issue is a gluing construction, which is a generalisation of the construction introduced in [14]. It has pushout-like properties wrt. composition *and* decomposition, which allow to reestablish major parts of the theory for the algebraic approaches on a general level. We investigate parallel independence here.

1 Introduction and Preliminaries

There are three algebraic frameworks for graph transformation, namely the double-pushout (DPO) [6], single-pushout (SPO) [12,13,1], and sesqui-pushout approach (SqPO) [3]. In all approaches, a transformation rule t is represented by a span $t = L \xleftarrow{l} K \xrightarrow{r} R$ of morphisms in a basic category. A match m for a rule $t = L \xleftarrow{l} K \xrightarrow{r} R$ in a host graph G is also similar in all three approaches, namely a morphism $L \xrightarrow{m} G$. The notion of direct derivation, however, is different.

In this paper, we show that all three concepts of direct derivation can be understood as special cases of a single gluing construction of pairs of spans. As a basis, we use the gluing construction introduced in [14], which was already able to model DPO, SPO at conflict-free matches, and both variants of SqPO presented in [3]. The refined version of this gluing of spans is presented in section 2. In section 3, we show that SPO is a special case of the new framework. Section 4 demonstrates that the refined gluing possesses pushout-like composition and decomposition properties[1]. These properties can serve as a basis for a general theory for a broad variety of new rewriting mechanisms. Section 5 presents one example, namely a refined version of contextual graph rewriting, compare [14]. Section 5 also demonstrates the conceptual power of the new framework by a proof of a general local Church-Rosser theorem.

Due to space limitations the paper does not contain the proofs for all propositions and theorems. The missing proofs are contained in [15].

In the following we assume for the underlying category $\mathcal{G}$:

---

[1] Composition is already known, see [14]. Decomposition is new.

H. Ehrig et al.(Eds.): ICGT 2012, LNCS 7562, pp. 111–125, 2012.
© Springer-Verlag Berlin Heidelberg 2012

($\mathcal{G}_1$) $\mathcal{G}$ has all limits of small diagrams[2].

($\mathcal{G}_2$) $\mathcal{G}$ has all finite colimits.

($\mathcal{G}_3$) All pullback functors in $\mathcal{G}$ possess a right adjoint.

Given a morphism $m : A \rightarrow B$ in the category $\mathcal{G}$, $m^* : \mathcal{G} \downarrow B \rightarrow \mathcal{G} \downarrow A$ denotes the pullback functor and $m_* : \mathcal{G} \downarrow A \rightarrow \mathcal{G} \downarrow B$ its right-adjoint. The co-unit of m_* for $t \in \mathcal{G} \downarrow A$ is denoted by $\varepsilon_t^m : m^*(m_*(t)) \rightarrow t$. Note the following facts about this adjunction:

Proposition 1. (Properties of the pullback functor and its right-adjoint)

(a) *If an object* $t \in \mathcal{G} \downarrow A$ *is monic,* $m_*(t)$, $m^*(m_*(t))$, *and* ε_t^m *are monic.*

(b) *If* t *is isomorphism in* $\mathcal{G} \downarrow A$, $m_*(t)$, $m^*(m_*(t))$, ε_t^m *are isomorphisms.*

(c) *If* m *is monic, the co-unit* ε_t^m *is monic for all* $t \in \mathcal{G} \downarrow A$.

(d) *If* f *is monomorphism in* $\mathcal{G} \downarrow A$, $m_*(f)$ *and* $m^*(m_*(f))$ *are monic.*

A concrete span in $\mathcal{G}$ is a pair of $\mathcal{G}$-morphisms (p, q) such that $\mathrm{domain}(p) = \mathrm{domain}(q)$. A concrete span (p, q) is a *relation* if p and q are jointly monic.[3] Two spans (p_1, q_1) and (p_2, q_2) are equivalent and denote the same abstract span if there is an isomorphism i such that $p_1 \circ i = p_2$ and $q_1 \circ i = q_2$; in this case we write $(p_1, q_1) \equiv (p_2, q_2)$ and $[(p, q)]_\equiv$ for the whole class of spans that are equivalent to (p, q). The *category of abstract spans* $S(\mathcal{G})$ over $\mathcal{G}$ has the same objects as $\mathcal{G}$ and equivalence classes of spans wrt. $\equiv$ as arrows. The identities are defined by $\mathrm{id}_A^{S(\mathcal{G})} = [(\mathrm{id}_A, \mathrm{id}_A)]_\equiv$ and composition of two spans $[(p, q)]_\equiv$ and $[(r, s)]_\equiv$ such that $\mathrm{codomain}(q) = \mathrm{codomain}(r)$ is given by $[(r, s)]_\equiv \circ^{S(\mathcal{G})} [(p, q)]_\equiv = [(p \circ r', s \circ q')]_\equiv$ where (r', q') is a pullback of the pair (q, r).

2   The Refined Gluing Construction

Definition 2. (Gluing construction) *The central construction in this paper is depicted in Figure 1. Let* (l, r) *and* (p, q) *be two spans such that the codomain of* l *and* p *coincide:*

1. *Construct* (l_0, p_0) *as the pullback of* (l, p).

2. *Let* $p_1' = r_*(p_0)$, (p_1, r_1) *be the pullback of* (r, p_1'), *i. e.* $p_1 = r^*(r_*(p_0))$, *and* $u_0 = \varepsilon_{p_0}^r$ *satisfying* $p_0 \circ u_0 = p_1$ *the corresponding co-unit. The co-unit* u_0 *will also be called* $\bar{v}_0$ *in the following.*

3. *Symmetrically, let* $l_1' = q_*(l_0)$, (l_1, q_1) *be the pullback of* (q, l_1'), *i. e.* $l_1 = q^*(q_*(l_0))$, *and* $v_0 = \varepsilon_{l_0}^q$ *satisfying* $l_0 \circ v_0 = l_1$ *the corresponding co-unit. The co-unit* v_0 *will also be called* $\bar{u}_0$ *in the following.*

4. *Iterate the following process for* $i \geq 1$:

---

[2] A diagram is small, if its collection of objects and its collection of morphisms are both (possibly infinite) sets.

[3] We do not need the category of relations in the following. Therefore, we do not define composition on relations. In this paper, being a relation is just a property of a span.

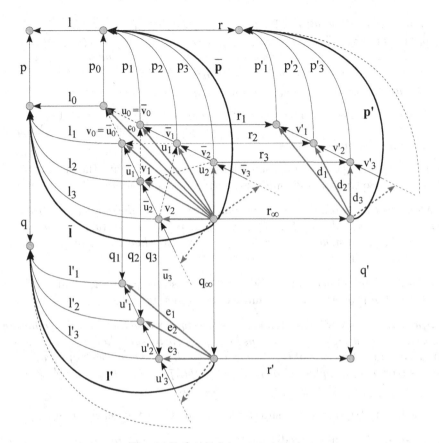

Fig. 1. Refined gluing construction

(a) *Let $v_i' = r_*(\overline{u}_{i-1})$ and $\overline{v}_i = r^*(v_i')$.*
[This implies that (i) $p_{i+1}' := r_*(p_0 \circ \overline{u}_0 \ldots \circ \overline{u}_{i-1})$, (ii) (p_{i+1}, r_{i+1}) is the pullback of (r, p_{i+1}'), i. e. $p_{i+1} = r^*(r_*(p_0 \circ \overline{u}_0 \ldots \circ \overline{u}_{i-1}))$, (iii) $r_i \circ \overline{v}_i = v_i' \circ r_{i+1}$ and $p_i \circ \overline{v}_i = p_{i+1}$, (iv) there is the co-unit $u_i = \varepsilon_{p_0 \circ \ldots \circ \overline{u}_{i-1}}^r$ satisfying $(p_0 \circ \overline{u}_0 \ldots \circ \overline{u}_{i-1}) \circ u_i = p_{i+1}$, and (v) $\overline{u}_{i-1} \circ u_i = u_{i-1} \circ \overline{v}_i$, since $\overline{v}_i = r^*(r_*(\overline{u}_{i-1}))$.]
(b) *Symmetrically, let $u_i' = q_*(\overline{v}_{i-1})$ and $\overline{u}_i = q^*(u_i')$.*
[This implies that (vi) $l_{i+1}' := q_*(l_0 \circ \overline{v}_0 \ldots \circ \overline{v}_{i-1})$, (vii) (l_{i+1}, q_{i+1}) is the pullback of (r, q_{i+1}'), i. e. $l_{i+1} = r^*(r_*(l_0 \circ \overline{v}_0 \ldots \circ \overline{v}_{i-1}))$, (viii) $q_i \circ \overline{u}_i = u_i' \circ q_{i+1}$ and $l_i \circ \overline{u}_i = l_{i+1}$ (ix) there is the co-unit $v_i = \varepsilon_{l_0 \circ \ldots \circ \overline{v}_{i-1}}^r$ satisfying $(l_0 \circ \overline{v}_0 \ldots \circ \overline{v}_{i-1}) \circ v_i = l_{i+1}$, and (x) $\overline{v}_{i-1} \circ v_i = v_{i-1} \circ \overline{u}_i$, since $\overline{u}_i = q^*(q_*(\overline{v}_{i-1}))$.]

5. *Construct $(\mathcal{C}, (c_i^r)_{i \in \mathbb{N}_0}, (c_i^q)_{i \in \mathbb{N}_0})$ as the limit of the double chain $(\overline{v}_i, \overline{u}_i)_{i \in \mathbb{N}_0}$, such that $c_0^r = c_0^q$ and for all $i \in \mathbb{N}$ $c_{i-1}^r = \overline{v}_{i-1} \circ c_i^r$ and $c_{i-1}^q = \overline{u}_{i-1} \circ c_i^q$. In the following $c_0^r = c_0^q$ is also called c_0. Construct $(\mathcal{D}, (d_i)_{i \in \mathbb{N}})$ as the limit of the chain $(v_i')_{i \in \mathbb{N}}$ with $(v_i' \circ d_{i+1} = d_i)_{i \in \mathbb{N}}$, and $(\mathcal{E}, (e_i)_{i \in \mathbb{N}})$ as the limit of the chain $(u_i')_{i \in \mathbb{N}}$ with $(u_i' \circ e_{i+1} = e_i)_{i \in \mathbb{N}}$.*

6. *Since we have for all $i \in \mathbb{N}$: $r_i \circ c_i^r = r_i \circ \bar{v}_i \circ c_{i+1}^r = v_i' \circ r_{i+1} \circ c_{i+1}^r$ and $q_i \circ c_i^q = q_i \circ \bar{u}_i \circ c_{i+1}^q = u_i' \circ q_{i+1} \circ c_{i+1}^q$, $(r_i \circ c_i^r)_{i \in \mathbb{N}}$ and $(q_i \circ c_i^q)_{i \in \mathbb{N}}$ are cones for $(v_i')_{i \in \mathbb{N}}$ resp. $(u_i')_{i \in \mathbb{N}}$. Thus, we obtain $r_\infty : \mathcal{C} \to \mathcal{D}$ and $q_\infty : \mathcal{C} \to \mathcal{E}$ with $(d_i \circ r_\infty = r_i \circ c_i^r)_{i \in \mathbb{N}}$ resp. $(e_i \circ q_\infty = q_i \circ c_i^q)_{i \in \mathbb{N}}$.*

7. *Construct (q', r') as the pushout of (r_∞, q_∞).*

8. *Set $\bar{p} := p_0 \circ c_0$, $\bar{l} := l_0 \circ c_0$, $p' := p_1' \circ d_1$, and $l' := l_1' \circ e_1$.*

The pair of spans $((l', r'), (p', q'))$ is the gluing of (l, r) and (p, q). A gluing is bounded if c_0 is monomorphisms. □

Note that the refined gluing construction above coincides with the one in [14] in the special case that u_0 and v_0 are isomorphisms, compare also [3].

Proposition 3. (Commutative diagram in $S(\mathcal{G})$) *In the gluing construction above, $(\bar{p}, r_\infty)$ is pullback of (r, p') and $(\bar{l}, q_\infty)$ is pullback of (q, l'). This means that the gluing construction provides a commutative diagram in the category of spans, i. e. $(p', q') \circ (l, r) = (l', r') \circ (p, q)$.*

The result of the refined gluing construction possesses properties that are comparable to the properties of the gluing construction in [14].

Proposition 4. (Mediating triple) *The gluing construction has the following property: Given a tuple of morphisms (x', x, k, y', y, h) such that (i) $l \circ x' = p \circ y'$, (ii) (x', k) is the pullback of (r, x), and (iii) (y', h) is the pullback of (q, y), then there is a triple of morphisms (g, y^q, x^r) such that $\bar{p} \circ g = x'$, $\bar{l} \circ g = y'$, $p' \circ x^r = x$, $x^r \circ k = r_\infty \circ g$, $l' \circ y^q = y$, and $y^q \circ h = q_\infty \circ g$.*

The mediating triple of Proposition 4 is not unique in the general case.

Proposition 5. (Unique mediating triple) *The triple in Proposition 4 is unique if the gluing is bounded.*

This means that $(\bar{p}, p', r_\infty, \bar{l}, l', q_\infty)$ is final among all comparable 6-tuples in the bounded case. Thus, the gluing produces a "final triple diagram" in the sense of [17].

Proposition 6. (Bounded) *Let $((l', r'), (p', q'))$ be the gluing of $((l, r), (p, q))$:*

(a) *If l and p are monic, then $\bar{l}$, $\bar{p}$, l', and p' are monic and the gluing is bounded.*

(b) *If r and q are monic, the gluing is bounded.*

(c) *If p and q are monic, the gluing is bounded and $\bar{p}$ and p' are monic.*

3   Gluing of Partial and Co-partial Morphisms

An abstract span $[(p, q)]_\equiv$ in $\mathcal{G}$ is a *partial morphism* if p is a monomorphism. Note that the identity spans are partial morphisms and that the span composition of partial morphisms yields a partial morphism again, since pullbacks preserve monic arrows. We denote the subcategory of $S(\mathcal{G})$ which consists of all abstract partial morphisms by $P(\mathcal{G})$.

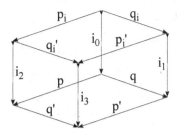

Fig. 2. Pushout/pullback cube

Definition 7. (Hereditary pushout) *A pushout (q', p') of (p, q) in $\mathcal{G}$ is hereditary if for each commutative cube as in Figure 2, which has pullback squares (p_i, i_0) and (q_i, i_0) of (i_2, p) resp. (i_1, q) as back faces such that i_1 and i_2 are monomorphisms, the top square (q_i', p_i') is pushout of (p_i, q_i) if and only if the front faces (p_i', i_1) and (q_i', i_2) are pullbacks of (i_3, p') resp. (i_3, q') and morphism i_3 is monic.*[4]

Theorem 8. (Pushout of partial morphisms) *Given a category $\mathcal{G}$ with hereditary pushouts, the gluing $((l', r'), (p', q'))$ of partial morphisms (l, r) and (p, q) is the pushout of (l, r) and (p, q) in $P(\mathcal{G})$.*

Proof. Commutativity of the resulting diagram is guaranteed by Proposition 3. That (l', r') and (p', q') are partial morphisms and the fact that the gluing is bounded is provided by Proposition 6(a).

Now suppose $(p^*, q^*) \circ (l, r) = (l^*, r^*) \circ (p, q)$. Then we obtain the situation depicted in Figure 3, where (p'', r'') is pullback of (r, p^*), (l'', q'') is pullback of (q, l^*), $l \circ p'' = p \circ l''$, and $q^* \circ r'' = r^* \circ q''$. Due to Proposition 5, there are unique morphisms x_1, x_2, and x_3 such that the resulting diagram commutes. These morphisms are monic since p'' resp. l'', p^*, and l^* are. Now construct $(\bar{q}, \bar{r})$ as the pushout of (r'', q''). Since $q' \circ x_2 \circ r'' = q' \circ r_\infty \circ x_1 = r' \circ q_\infty \circ x_1 = r' \circ x_3 \circ q''$, there is x_4 with $x_4 \circ \bar{q} = q' \circ x_2$ and $x_4 \circ \bar{r} = r' \circ x_3$. Since $q^* \circ r'' = r^* \circ q''$, there is x_5 with $x_5 \circ \bar{q} = q^*$ and $x_5 \circ \bar{r} = r^*$. Since the pushout (q', r') is hereditary, x_4 is a monomorphism, $(x_2, \bar{q})$ is pullback of (q', x_4), and $(x_3, \bar{r})$ is pullback of (r', x_4). Thus, (x_4, x_5) is a mediating partial morphism with $(x_4, x_5) \circ (p', q') = (p^*, q^*)$ and $(x_4, x_5) \circ (l', r') = (l^*, r^*)$.

It remains to show that the mediating partial morphism is unique. Suppose there are two mediating partial morphisms (x_4, x_5) and (x_4', x_5'), such that $(x_4, x_5) \circ (p', q') = (p^*, q^*) = (x_4', x_5') \circ (p', q')$ and $(x_4, x_5) \circ (l', r') = (l^*, r^*) = (x_4', x_5') \circ (l', r')$. Then we obtain the situation depicted in Figure 4: $(x_2, \bar{q}_1)$, $(x_2', \bar{q}_2)$, $(x_3, \bar{r}_1)$, and $(x_3', \bar{r}_2)$ are pullbacks and we have the isomorphisms i and j with (i) $p' \circ x_2 \circ i = p' \circ x_2'$, (ii) $x_5 \circ \bar{q}_1 \circ i = x_5' \circ \bar{q}_2$, (iii) $l' \circ x_3 \circ j = l' \circ x_3'$, and (iv) $x_5 \circ \bar{r}_1 \circ j = x_5' \circ \bar{r}_2$. Since p' and l' are monomorphisms, we can conclude (i') $x_2 \circ i = x_2'$ and (iii') $x_3 \circ j = x_3'$.

Since $(\bar{p} \circ x_1, r_1'')$ is pullback of $(r, p' \circ x_2)$ and $(\bar{p} \circ x_1', r_2'')$ is pullback of $(r, p' \circ x_2')$, we get an isomorphism i' with (v) $r_1'' \circ i' = i \circ r_2''$ and (vi) $\bar{p} \circ x_1 \circ i' = \bar{p} \circ x_1'$.

---

[4] For details on hereditary pushouts see [9].

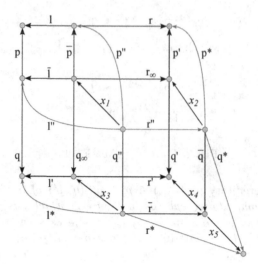

Fig. 3. Mediating span

Similar arguments provide an isomorphism j' with (vii) $q_1'' \circ j' = j \circ q_2''$ and (viii) $\bar{l} \circ x_1 \circ j' = \bar{l} \circ x_1'$. From (vi) and (viii) we get $l \circ \bar{p} \circ x_1 \circ i' = l \circ \bar{p} \circ x_1' = p \circ \bar{l} \circ x_1' = p \circ \bar{l} \circ x_1 \circ j' = l \circ \bar{p} \circ x_1 \circ j'$. Since l, and $\bar{p} \circ x_1$ are monomorphisms, (ix) $i' = j'$. Since pushouts are hereditary, we know that (x) $(\bar{q}_1, \bar{r}_1)$ and (xi) $(\bar{q}_2, \bar{r}_2)$ are pushouts of (r_1'', q_1'') resp. (r_2'', q_2''). Now we have $\bar{q}_1 \circ i \circ r_2'' \overset{(v)}{=} \bar{q}_1 \circ r_1'' \circ i' \overset{(x)}{=} \bar{r}_1 \circ q_1'' \circ i' \overset{(ix)}{=} \bar{r}_1 \circ q_1'' \circ j' \overset{(vii)}{=} \bar{r}_1 \circ j \circ q_2''$. Thus, there is k with (xii) $k \circ \bar{q}_2 = \bar{q}_1 \circ i$ and (xiii) $k \circ \bar{r}_2 = \bar{r}_1 \circ j$. A symmetric argument provides the inverse for k, which, thereby, turns out to be an isomorphism. That $x_5 \circ k = x_5'$, follows from $x_5 \circ k \circ \bar{r}_2 \overset{(xiii)}{=} x_5 \circ \bar{r}_1 \circ j \overset{(iv)}{=} x_5' \circ \bar{r}_2$ and $x_5 \circ k \circ \bar{q}_2 \overset{(xii)}{=} x_5 \circ \bar{q}_1 \circ i \overset{(ii)}{=} x_5' \circ \bar{q}_2$; and $x_4 \circ k = x_4'$ is implied by $x_4 \circ k \circ \bar{q}_2 \overset{(xii)}{=} x_4 \circ \bar{q}_1 \circ i = q' \circ x_2 \circ i \overset{(i')}{=} q' \circ x_2' = x_4' \circ \bar{q}_2$ and $x_4 \circ k \circ \bar{r}_2 \overset{(xiii)}{=} x_4 \circ \bar{r}_1 \circ j = r' \circ x_3 \circ i \overset{(iii')}{=} r' \circ x_3' = x_4' \circ \bar{r}_2$.   □

Note that the refined gluing construction of partial morphisms is an exact model of single-pushout rewriting with a rule (l, r) at a conflict-free or a conflicting match (p, q), compare [13]. At the same time, it is also a generalisation of sesqui-pushout rewriting with monic left-hand sides in rules at conflict-free matches [3]. In both cases p is not only a monomorphism but an isomorphism. Conflict-freeness can be reformulated here by the requirement that v_0 is a isomorphism in Figure 1. Theorem 8 shows that the direct derivation in this variant of the sesqui-pushout approach is a pushout of spans.

There are other restrictions of the general span category in which the gluing construction turns out to be pushout: An abstract span $[(p, q)]_\equiv$ in $\mathcal{G}$ is a *co-partial morphism* if q is a monomorphism. The identity spans are co-partial morphisms and composition is closed within co-partial morphisms. $C(\mathcal{G})$ denotes the subcategory of $S(\mathcal{G})$ which contains all abstract co-partial morphisms.

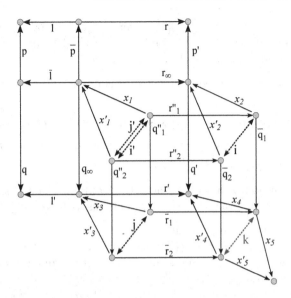

Fig. 4. Unique mediating span

Definition 9. (Van-Kampen pushout) *A pushout (q', p') of (p, q) in $\mathcal{G}$ is a van-Kampen diagram if for each commutative cube as in Figure 2, which has pullback squares (p_i, i_0) and (q_i, i_0) of (i_2, p) resp. (i_1, q) as back faces, the top square (q'_i, p'_i) is pushout of (p_i, q_i) if and only if the front faces (p'_i, i_1) and (q'_i, i_2) are pullbacks of (i_3, p') resp. (i_3, q'). A pushout is a semi-van-Kampen diagram if it satisfies only the if-part of the property above.*

Note that, due to requirement $\mathcal{G}_3$, each pushout in the underlying category possesses the semi-van-Kampen property, compare "stability" in [16].

Theorem 10. (Pushout of co-partial morphisms) *Given a category $\mathcal{G}$ in which (i) pushouts preserve monomorphisms and (ii) pushouts of monic morphisms are van-Kampen, the gluing $((l', r'), (p', q'))$ of co-partial morphisms (l, r) and (p, q) is the pushout of (l, r) and (p, q) in $C(\mathcal{G})$.*

4   Gluing Beyond Partial Morphisms

The constructions within the proof of Theorem 8, that the gluing construction of definition 2, or a suitable specialisation, can be characterised by universal properties in a more general set-up. We investigate these properties here: The construction of the mediating morphism is always possible if the pushout of the gluing is van-Kampen. Uniqueness requires a *bounded* gluing and two additional ingredients, compare Figure 4:

1. The implication $(*)$ $x_5 \circ \bar{q}_1 \circ r''_1 \circ i' = x_5 \circ \bar{q}_1 \circ r''_1 \circ j' \implies \bar{q}_1 \circ r''_1 \circ i' = \bar{q}_1 \circ r''_1 \circ j'$, provides an isomorphism k with $x_5 \circ k = x'_5$. $(*)$ is guaranteed if x_5 is monic.

2. If $x_1 \circ i' = x_1'$ and $x_1 \circ j' = x_1'$ is given, we obtain by the properties of bounded gluings that $x_2 \circ i = x_2'$ and $x_3 \circ j = x_3'$. This provides the second half of the required property of k, namely $x_4 \circ k = x_4'$.

The formalisation of the first requirement is the contents of the rest of the section.

Definition 11. (Semi-pushout) *A co-span* $B \xrightarrow{h} D \xleftarrow{k} C$ *is a semi-pushout of the span* $B \xleftarrow{f} A \xrightarrow{g} C$ *if* (i) $h \circ f = k \circ g$ *and* (ii) *the mediating morphism* $u : P \to D$ *from the pushout* $B \xrightarrow{g'} P \xleftarrow{f'} C$ *of* (f, g) *to* D *is a monomorphism.*

Note that every pushout is a semi-pushout by definition. Semi-pushouts are the necessary tool for an abstract definition of a gluing with "good" properties:

Definition 12. (Span semi-pushout and perfect gluing) *A span semi-pushout of* $((l, r), (p, q))$ *is a pair of spans* $((l', r'), (p', q'))$ *such that* (i) $(l', r') \circ (p, q) = (p', q') \circ (l, r)$, *and* (ii') (r', q') *is semi-pushout. A perfect gluing for* $((l, r), (p, q))$ *is a span semi-pushout* $((l', r'), (p', q'))$, *such that* (ii) (r', q') *is pushout and* (iii) *for any span semi-pushout* $((p^*, q^*), (l^*, r^*))$ *of* $((l, r), (p, q))$, *there is a unique span* (x, y) *with* $(x, y) \circ (l', r') = (l^*, r^*)$ *and* $(x, y) \circ (p', q') = (p^*, q^*)$.

Corollary 13. *Perfect gluings are unique up to isomorphism, if they exist.*

In the rest of this section, we show that perfect gluings can be composed and decomposed like pushouts. As a prerequisite for this, we need that semi-pushouts themselves can be composed and decomposed like pushouts.

Proposition 14. (Composition and decomposition of semi-pushouts) *In a category where pushouts preserve monomorphisms the following holds:* (i) *If* (c, d) *is semi-pushout of* (a, b) *and* (g, f) *is semi-pushout of* (d, e), *then* $(f \circ c, g)$ *is semi-pushout of* $(a, e \circ b)$, *compare left part of Figure 5.* (ii) *If* (c, d) *is pushout of* (a, b) *and* $(f \circ c, g)$ *is semi-pushout of* $(a, e \circ b)$, *then* (g, f) *is semi-pushout of* (d, e), *compare right part of Figure 5.*

Proposition 15. *Let a category be given in which pushouts preserve monomorphisms: If* (i) $(l', r') \circ (n, m) = (n', m') \circ (l, r)$, (ii) $(p', q') \circ (n', m') = (n'', m'') \circ (p, q)$, (iii) $(l \circ x, q \circ y) = (p, q) \circ (l, r)$, (iv) $(l' \circ h, q' \circ k) = (p', q') \circ (l', r')$, *and* (v) (r', m') *is pushout, then* (m'', q') *is pushout or semi-pushout, if and only if* $(q' \circ k, m'')$ *is pushout resp. semi-pushout, compare Figure 6.*

Proof. In Figure 6, let $(r_\infty, \overline{n})$ and $(m_\infty, \overline{l})$ be the two pullbacks for (i) such that $n \circ \overline{l} = l \circ \overline{n}$ and (r', m') is pushout of (r_∞, m_∞). Let $(q_\infty, \overline{n'})$ and $(m'_\infty, \overline{p})$ be the two pullbacks for (ii) such that $n' \circ \overline{p} = p \circ \overline{n'}$ and $m'' \circ q_\infty = q' \circ m'_\infty$. Let (f, g) be the pullback of $(r_\infty, \overline{p})$ and v and w the mediating morphisms that make the whole diagram commute. Then (f, w) is pullback of (m_∞, h), since (h, k) and $(f, m'_\infty \circ g)$ are pullbacks. Symmetrically, (v, g) is pullback of $(y, \overline{n'})$. Thus $(v, q_\infty \circ g)$ and $(w, \overline{l} \circ f)$ are the two pullbacks representing $(n'', m'') \circ ((p, q) \circ (l, r)) = ((p', q') \circ (l', r')) \circ (n, m)$. The semi-van-Kampen property guarantees that (k, m'_∞) is pushout of (w, g).

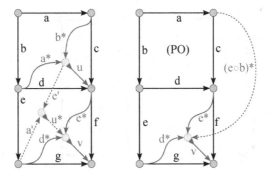

Fig. 5. Composition of semi-pushouts

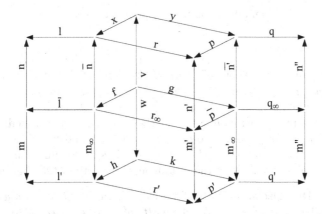

Fig. 6. Composition and decomposition of perfect gluings (Diagram)

"$\Longleftarrow$": (q', m'') is pushout or semi-pushout of (m'_∞, q_∞). The composition property of pushouts resp. semi-pushouts (compare Proposition 14) provides that $(q' \circ k, m'')$ is pushout resp. semi-pushout.

"$\Longrightarrow$": $(q' \circ k, m'')$ is pushout or semi-pushout and (k, m'_∞) is pushout. The decomposition property of pushouts/semi-pushouts (Proposition 14) guarantees that (q', m'') is pushout/semi-pushout of (m'_∞, q_∞). □

The definition of span semi-pushouts implies:

Corollary 16. (Inheritance of span semi-pushouts) *If a pair of spans (u, v) is a span semi-pushout for the span pair $(f, h \circ g)$, i. e. $v \circ f = u \circ (h \circ g)$, then $(u \circ h, v)$ is a span semi-pushout for (f, g).*

Proposition 17. (Composition of perfect gluings) *If (f', g') is a perfect gluing of (f, g) (subdiagram (1a) in Figure 7) and (f'', h') is a perfect gluing of (f', h) (subdiagram (2a) in Figure 7), then $(f'', h' \circ g')$ is a perfect gluing of $(f, h \circ g)$ ((1a)+(2a) in Figure 7).*

Proof. We have to proof the properties (i) - (iii) of definition 12. Commutativity is obvious. Property (ii) follows from Proposition 15. It remains to show the

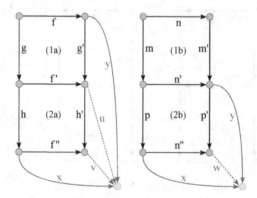

Fig. 7. Composition and decomposition of perfect gluings (universal property)

universal property (iii). Let (x, y) be a span semi-pushout for $(f, h \circ g)$. Then $(x \circ h, y)$ is a span semi-pushout of (f, g) by Corollary 16. Thus, there is a unique span u with $u \circ g' = y$ and $u \circ f' = x \circ h$. Since (x, y) is span semi-pushout and (1a) is perfect gluing, (x, u) is span semi-pushout for (h, f') by Proposition 15. Thus, we get unique v with $v \circ h' = u$ and $v \circ f'' = x$. This v is a mediating span from $(f'', h' \circ g')$ to (x, y). Suppose v' with $v' \circ f'' = x$ and $v' \circ h' \circ g' = y$. Then $v' \circ h' \circ f' \circ g = v' \circ f'' \circ h \circ g = x \circ h \circ g = y \circ f$. Thus by the uniqueness property of (1a), we get $v' \circ h' = u$. Now the uniqueness property of (2a) implies $v = v'$.

Proposition 18. (Decomposition of perfect gluings) *Let a commutative diagram* (1b)+(2b) *as in Figure 7 be given: If* (n', m') *is perfect gluing of* (n, m) *((1b) in Figure 7) and* $(n'', p' \circ m')$ *is perfect gluing of* $(n, p \circ m)$ *((1b)+(2b) in Figure 7) then* (n'', p') *is perfect gluing of* (n', p) *((2b) in Figure 7).*

Proof. Commutativity is given. Proposition 15 guarantees the pushout property. If (x, y) is a span semi-pushout for (n', p), then $(x, y \circ m')$ is a span semi-pushout for $(n, p \circ m)$ by Proposition 15. Therefore, there is unique w with $w \circ n'' = x$ and $w \circ (p' \circ m') = y \circ m'$. Since $(x, y \circ m')$ is a span semi-pushout for $(n, p \circ m)$, $(x \circ p, y \circ m')$ is a span semi-pushout for (n, m) by Corollary 16 providing $w \circ p' = y$. Any other w' with $w' \circ n'' = x$ and $w' \circ p' = y$ has also the property $w' \circ p' \circ m' = y \circ m'$. Since (1b)+(2b) is perfect gluing $w = w'$. □

Here is the first instance of pairs of spans that possess a perfect gluing.

Theorem 19. (Gluing of a relation with a monic span) *Given a category where* (i) *pushouts preserve monomorphisms and* (ii) *all pushouts along monomorphisms are van-Kampen, the gluing (definition 2) of a relation* (l, r) *and a pair of monomorphisms* (p, q) *is a perfect gluing.*

Note that the gluing of a relation and a monic span is a model for sesqui-pushout rewriting with a rule (l, r) at a monic match (p, q) [3], namely for the special case, where p and v_0 (compare Figure 1) are isomorphisms. Theorem 19 shows that direct derivations in this type of SqPO-rewriting have pushout-like properties.

5   Gluing of Graphs

The gluing construction and its theory developed above can be used as a general framework for algebraic graph transformation in the category of graphs:

Definition 20. *(Category of graphs) A graph $G = (V; E; s, t : E \to V)$ consists of a set of vertices V, a set of edges E, and two total mappings s and t assigning the source and target vertex to every edge. A graph morphism $f : G \to H = (f_V : V_G \to V_H, f_E : E_G :\to E_H)$ is a pair of total mappings such that $f_V \circ s^G = s^H \circ f_E$ and $f_V \circ t^G = t^H \circ f_E$. All graphs and graph morphisms with component-wise composition and identities constitute the category $\mathcal{GRAPH}$.*

The category $\mathcal{GRAPH}$ will be used in this section. Note its properties:
1. It has all small limits and colimits
2. Pullback functors have right-adjoints [16].
3. Pushouts preserve monomorphisms [16].
4. Pushouts are hereditary [13].
5. Pushouts along monomorphisms are van-Kampen [6].

Definition 21. *(Algebraic graph rewriting system) An algebraic graph rewriting system $GRS = (T, M = (M_t)_{t \in T})$ consists of a class of abstract spans T, called* transformation rules, *and a class of* matching *abstract spans M_t for each $t \in T$, such that $m = (p, q) \in M_{t=(l,r)}$ implies codomain$(l) = $ codomain(p). The gluing $((l', r'), (p', q'))$ of $(t = (l, r), m = (p, q) \in P_t)$ is the direct derivation with t at m. A direct derivation of a Graph G to H with a rule (l, r) at a match (p, q) is written $(l', r') : G \rightsquigarrow H$.*

All algebraic graph transformation approaches fit into this framework, namely:
1. Double-pushout approach [14].
2. Single-pushout approach (Theorem 8).
3. Single-pushout approach with co-partial morphisms (Theorem 10).
4. Sesqui-pushout approach at monic matches [3] (Theorem 19).
5. Sesqui-pushout approach at monic partial matches (Theorem 19).
6. Sesqui-pushout approach with monic left-hand sides in rules [3] (Theorem 8).
7. Contextual graph rewriting in [14].

Note that direct derivations in DPO, SqPO, and contextual graph rewriting [14] are perfect gluings. The theory of SPO is based on composition and decomposition properties of pushouts. If we want to obtain a similar theory for contextual graph rewriting [14] based on Propositions 17 and 18, we need additional properties such that the underlying gluing construction becomes a perfect gluing.

Definition 22. *(Straight co-span) A co-span $\xrightarrow{p} \xleftarrow{l}$ is straight for a span $\xleftarrow{l'} \xrightarrow{p'}$ in $\mathcal{GRAPH}$ if $p \circ l' = l \circ p'$ and for all vertices $v_1^p, v_2^p, v_1^l, v_2^l$ with preimages wrt. l' resp. p': $p(v_1^p) = p(v_2^p) = l(v_1^l) = l(v_2^l) \implies v_1^p = v_2^p \vee v_1^l = v_2^l$ and for all edges $e_1^p, e_2^p, e_1^l, e_2^l$ with preimages wrt. l' resp. p', such that $s(e_1^p) = s(e_2^p) \wedge t(e_1^p) = t(e_2^p)$ and $s(e_1^l) = s(e_2^l) \wedge t(e_1^l) = t(e_2^l)$, we get: $p(e_1^p) = p(e_2^p) = l(e_1^l) = l(e_2^l) \implies e_1^p = e_2^p \vee e_1^l = e_2^l$.*

Now we are able to formulate a general rewriting system whose direct derivations are perfect. It can provide the theoretical basis for concrete and practical rewriting systems. For example contextual rewriting introduced in [14] is an instance.

Definition 23. (Abstract span rewriting at co-partial matches) *The algebraic graph rewriting system $GRS_{PSR} = (T_{PSR}, M^{PSR})$ has (i) all relations as transformation rules and $m = (p, q) \in M^{PSR}_{t=(l,r)}$, if (ii) q is a monomorphism, (iii) the gluing of t and m is bounded, and (iv) (p, l) is straight for $(\bar{l}, \bar{p})$.*

Theorem 24. (Perfect rewriting) *Direct derivations in GRS_{PSR} are perfect.*

Proof. The assumptions (ii) and (iii) in definition 23 imply that a mediating span always exists. It remains to show uniqueness. Consider Figure 4. Note: since the gluing is bounded, $(\bar{l}, \bar{p})$ is a sub-object of the pullback of (l, p) and the pair $(\bar{l}, \bar{p})$ is jointly monic $(*)$.

For a given vertex v, consider the three vertices $v_1 = x'_1(v)$, $v_2 = x_1(i'(v))$, and $v_3 = x_1(j'(v))$. Since x_5 and $\bar{q}_1$ are monic, $r''_1 \circ i' = r''_1 \circ j'$. Thus, (a) $r_\infty(v_2) = r_\infty(x_1(i'(v))) = x_2(r''_1(i'(v))) = x_2(r''_1(j'(v))) = r_\infty(x_1(j'(v))) = r_\infty(v_3)$. We know (b) $\bar{p}(v_1) = \bar{p}(v_2)$ and (c) $\bar{l}(v_1) = \bar{l}(v_3)$. Since (l, p) is straight, we obtain either (d1) $\bar{p}(v_1) = \bar{p}(v_2) = \bar{p}(v_3)$ or (d2) $\bar{l}(v_1) = \bar{l}(v_2) = \bar{l}(v_3)$. In case of (d1), it follows $v_1 = v_3$ due to $(*)$ (with (c) and (d1)). Properties (a) and (d1) imply $v_2 = v_3$, since $(\bar{p}, r_\infty)$ is pullback. All together, we get $v_1 = v_2 = v_3$, which means $x'_1(v) = x_1(i'(v)) = x_1(j'(v))$. In case of (d2), it follows $v_1 = v_2$ due to $(*)$ (with (b) and (d2)). It remains to show $v_2 = v_3$: (d2) provides (e) $l(\bar{p}(v_2)) = p(\bar{l}(v_2)) = p(\bar{l}(v_3)) = l(\bar{p}(v_3))$. By (a), we obtain (f) $r(\bar{p}(v_2)) = p'(r_\infty(v_2)) = p'(r_\infty(v_3)) = r(\bar{p}(v_3))$. (e) and (f) imply (g) $\bar{p}(v_2) = \bar{p}(v_3)$, since (l, r) is a relation. Finally, (d2) and (g) guarantee $v_2 = v_3$.

The proof for edges is analog, since for every edge e we know, by the proof above, that the source and target nodes for the edges $x'_1(e)$, $x_1 \circ i'(e)$, and $x_1 \circ j'(e)$ coincide.   □

The properties (i) – (iv) of the system GRS_{PSR} (definition 23) have the following operational interpretation: Rules can *delete* and *add* items, if l resp. r is not surjective. Rules can *copy* and *identify* items, if l resp. r is not injective. It is forbidden to take a copy of an item and to identify copy and original afterwards (rule must be a relation; def. 23(i)). This is reasonable since such an action has no effect from the operational point of view.

The *injective part of the match*, namely q, identifies the part of the host graph that gets affected by rule application (def. 23(ii)). A non-injective p allows to match many items in the host graph with a single item in the rule. *Universal quantification* can be formulated this way, compare for example [14].

Straightness of (p, l) forbids copy-copy conflicts (def. 23(iv)): Items that get copied by rule application (non-injective l) must be matched with a single object in the host graph. The *boundedness* of the gluing (def. 23(iii)) forbids identify-copy conflicts, since bounded gluings in $\mathcal{GRAPH}$ always result in straight pairs (r, p') for $(\bar{p}, r_\infty)$, compare Figure 3. Items that get identified by rule application

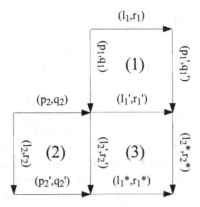

Fig. 8. Parallel independence

(non-injective r) must be matched with a single object in the host graph. If one item of a group of objects that get identified is not matched at all, the identification results in a deletion.[5] If all items are matched with at least one object and one of them is matched with more than one object in the host graph, the gluing result is not bounded. In this case the rule is not applicable at (q, p).[6]

Some theorems can be formulated once and for all (perfect-gluing-based) algebraic approaches if the refined gluing construction of definition 2 and the Propositions 17 and 18 (see above) are used as a common basis.

Definition 25. (Parallel independence) *Two transformations* $(l_1', r_1') : G \rightsquigarrow H$, *and* $(l_2', r_2') : G \rightsquigarrow K$ *with rule* $t_1 = (l_1, r_1)$ *at match* (p_1, q_1) *and with* $t_2 = (l_2, r_2)$ *at* (p_2, q_2) *are* parallel independent *in a graph rewriting system, if* $(l_1', r_1') \circ (p_2, q_2) \in P_{t_2}$ *and* $(l_2', r_2') \circ (p_1, q_1) \in P_{t_1}$.[7]

Theorem 26. (Local Church-Rosser) *Let an algebraic graph transformation system be given in which all direct derivations are perfect gluings: If two transformations* $G \rightsquigarrow H$ *with rule* t_1 *and* $G \rightsquigarrow K$ *with rule* t_2 *are parallel independent, then there are transformations* $H \rightsquigarrow X$ *with rule* t_2 *and* $K \rightsquigarrow X$ *with rule* t_1.

Proof. Parallel independence guarantees that t_2 is applicable after t_1 and vice versa at the induced matches. Consider Figure 8. Construct the gluing of (l_1, r_1) and $(l_2', r_2') \circ (p_1, q_1)$. Then we get (l_2^*, r_2^*) with $(l_2^*, r_2^*) \circ (l_1', r_1') = (l_1^*, r_1^*) \circ (l_2', r_2')$, compare subdiagram (3) in Figure 8. By Proposition 18, (3) is a perfect gluing. Due to Proposition 17, (2)+(3) is the perfect gluing that represents the transformation with t_2 after t_1. Thus, t_1 after t_2 produces the same graph as t_2 after t_1. Even the two traces $(l_2^*, r_2^*) \circ (l_1', r_1')$ and $(l_1^*, r_1^*) \circ (l_2', r_2')$ coincide.   □

---

[5] Single-pushout behaviour at conflicting matches [13].
[6] The result will be infinite even when rule and match comprise finite graphs only.
[7] Note the parallel independence is purely syntactical compared to [14].

6   Related Work and Conclusion

Besides SqPO, there are some approaches to graph transformation that provide mechanisms for copying and identification in the rules and universal quantification in the matches.[8]

Graph rewriting with polarised cloning [5] is a refinement of the SqPO-approach and provides more control on the cloning process of edges in the context of copied vertices. These effects can be simulated by the gluing construction introduced above, if the context for each copied vertex, namely a prototype incoming and a prototype outgoing edge, is added in the rule with the requirement that all edges of this type in the host graph must be mapped to these prototype edges.

The relation-algebraic approach [10,11] uses a completely different set-up in collagories. The effects of direct derivations in this approach are similar to the gluing construction introduced here. But due to the completely different base category, there is no easy correlation of the framework in [10,11] to the algebraic approaches to graph transformation.

In the double-pushout approach, universal quantification for matches is added by amalgamation [8]. The gluing construction introduced here is able to provide a simpler notion of direct derivation for many situations where amalgamation is needed in the double-pushout approach.

There is a good chance that the gluing construction of definition 2 can serve as a common framework for many mechanisms in SPO-, SqPO-, and DPO-rewriting and as a basis for higher-level operational effects like controlled cloning [4] and matches that require that some edges in the rule's left-hand side are universally quantified. It is up to future research to investigate how far the known theory for DPO or SPO can be reestablished. First results on parallel independence have been achieved. Theorem 26 improves the corresponding result in [14], since independence is purely syntactical if all gluings involved are perfect. And contextual rewriting [14] with relations as rules turned out to be perfect, compare Theorem 24. It can be expected that other fragments of the theory can be lifted to a more abstract level.

References

1. Bauderon, M., Jacquet, H.: Pullback as a generic graph rewriting mechanism. Applied Categorical Structures 9(1), 65–82 (2001)
2. Corradini, A., Ehrig, H., Montanari, U., Ribeiro, L., Rozenberg, G. (eds.): ICGT 2006. LNCS, vol. 4178. Springer, Heidelberg (2006)
3. Corradini, A., Heindel, T., Hermann, F., König, B.: Sesqui-pushout rewriting. In: Corradini, et al. (eds.) [2], pp. 30–45
4. Drewes, F., Hoffmann, B., Janssens, D., Minas, M., Van Eetvelde, N.: Adaptive star grammars. In: Corradini, et al. (eds.) [2], pp. 77–91
5. Duval, D., Echahed, R., Prost, F.: Graph rewriting with polarized cloning. CoRR, abs/0911.3786 (2009)

---

[8] Note that DPO and SPO do not provide these features per se.

6. Ehrig, H., Ehrig, K., Prange, U., Taentzer, G.: Fundamentals of Algebraic Graph Transformation. Springer (2006)
7. Ehrig, H., Rensink, A., Rozenberg, G., Schürr, A. (eds.): ICGT 2010. LNCS, vol. 6372. Springer, Heidelberg (2010)
8. Golas, U., Ehrig, H., Habel, A.: Multi-amalgamation in adhesive categories. In: Ehrig, et al. (eds.) [7], pp. 346–361
9. Heindel, T.: Hereditary pushouts reconsidered. In: Ehrig, et al. (eds.) [7], pp. 250–265
10. Kahl, W.: A relation-algebraic approach to graph structure transformation. Habil. Thesis 2002-03, Fakultät für Informatik, Univ. der Bundeswehr München (2001)
11. Kahl, W.: Amalgamating pushout and pullback graph transformation in collagories. In: Ehrig, et al. (eds.) [7], pp. 362–378
12. Kennaway, R.: Graph Rewriting in Some Categories of Partial Morphisms. In: Ehrig, H., Kreowski, H.-J., Rozenberg, G. (eds.) Graph Grammars 1990. LNCS, vol. 532, pp. 490–504. Springer, Heidelberg (1991)
13. Löwe, M.: Algebraic approach to single-pushout graph transformation. Theor. Comput. Sci. 109(1&2), 181–224 (1993)
14. Löwe, M.: Graph rewriting in span-categories. In: Ehrig, et al. (eds.) [7], pp. 218–233
15. Löwe, M.: A unifying framework for algebraic graph transformation. Technical Report 2012/03, FHDW-Hannover (2012)
16. McLarty, C.: Elementary Categories, Elementary Toposes. Oxford Science Publications, Clarendon Press, Oxford (1992)
17. Monserrat, M., Rossello, F., Torrens, J., Valiente, G.: Single pushout rewriting in categories of spans i: The general setting. Technical Report LSI-97-23-R, Department de Llenguatges i Sistemes Informtics, Universitat Politcnica de Catalunya (1997)

Borrowed Contexts for Attributed Graphs

Fernando Orejas[1,*], Artur Boronat[1,2,**], and Nikos Mylonakis[1]

[1] Universitat Politècnica de Catalunya, Spain
{orejas,nicos}@lsi.upc.edu
[2] University of Leicester, UK
aboronat@mcs.le.ac.uk

Abstract. Borrowed context graph transformation is a simple and powerful technique developed by Ehrig and König that allows us to derive labeled transitions and bisimulation congruences for graph transformation systems or, in general, for process calculi that can be defined in terms of graph transformation systems. Moreover, the same authors have also shown how to use this technique for the verification of bisimilarity. In principle, the main results about borrowed context transformation do not apply only to plain graphs, but they are generic in the sense that they apply to all categories that satisfy certain properties related to the notion of adhesivity. In particular, this is the case of attributed graphs. However, as we show in the paper, the techniques used for checking bisimilarity are not equally generic and, in particular they fail, if we want to apply them to attributed graphs. To solve this problem, in this paper, we define a special notion of symbolic graph bisimulation and show how it can be used to check bisimilarity of attributed graphs.

Keywords: Attributed graph transformation, symbolic graph transformation, borrowed contexts, bisimilarity.

1 Introduction

Bisimilarity is possibly the most adequate behavioural equivalence relation. In [5] Ehrig and König show how a notion of bisimilarity could be defined for graph transformation systems. In particular, they introduced borrowed context graph transformation as a simple and powerful technique that allows us to derive labelled transitions and bisimulation congruences for graph transformation systems or, in general, for process calculi that can be defined in terms of graph transformation systems (e.g. [1]). These results are quite general since they apply to any kind of transformation system on a category that satisfies some properties related to adhesivity [7,3], as the category of attributed graphs. Moreover, in [12], Rangel, König and Ehrig showed how to use these techniques for the verification of bisimilarity. Unfortunately, the approach to verification introduced informally

---

* This work has been partially supported by the CICYT project (ref. TIN2007-66523) and by the AGAUR grant to the research group ALBCOM (ref. 00516).
** Supported by a Study Leave from University of Leicester.

H. Ehrig et al.(Eds.): ICGT 2012, LNCS 7562, pp. 126–140, 2012.

in [5] and studied in detail in [12] does not work when dealing with attributed graphs. This is especially unfortunate, because one of the motivations for [12] is being able to show that model transformations (and, in particular, refactorings) preserve behavioural equivalence, and in these cases we work with attributed graphs. The problem is related with the fact that, when applying an attributed graph transformation rule, we must match all the variables in the left-hand side of the rule to some given values. This implies, in the case of borrowed context transformation, that the partial application of a single rule may give rise to an infinite number of different borrowed context transformation, where all of them have different labels. The reason is that, if the borrowed context includes some variable, each substitution of that variable by any data value defines a different borrowed context transformation.

The nature of the problem immediately suggests using borrowed context symbolic graph transformation to check bisimilarity of attributed graphs, since in symbolic graph transformation variables do not need to be immediately substituted by values [8,10]. Actually, since symbolic graphs form an adhesive HLR category, all the results in [5] apply to this class of graphs. Unfortunately, as a counter-example in Sect. 4 shows, bisimilarity for symbolic graphs, as defined in [5], does not coincide with bisimilarity for attributed graphs. Hence, in this work we define a new notion of symbolic bisimilarity, which is also a congruence, and we show that it coincides with attributed graph bisimilarity. Moreover, using a variation of the case study presented in [5], we show how this new notion can be used for checking bisimilarity for attributed graphs. In particular, the example not only shows that the problem with the substitution of the variables is solved, but it also shows how symbolic graphs allow us to decouple a bisimilarity proof in two parts. The first one, that has to do with graph structure, is based on borrowed context transformation, while the second one, which is related to data in the graphs, has to do with reasoning in the given data algebra.

The paper is organized as follows. In Sect. 2 we introduce the main results and constructions presented in [5] and we introduce our case study, describing the problems when attributed graph transformation is considered. In the following section we introduce symbolic graph transformation. Sect. 4 is the core of the paper, where we present our main constructions and results. In Sect 5, we extend some techniques used in [5] for the verification of bisimilarity and we apply them to our case study. Finally, in Sect. 6 we present some related work and draw some conclusions. Due to lack of space, the paper includes no proofs. Anyhow they can be found in the long version of the paper.

2   Graph Transformation with Borrowed Contexts

Given a set of transformation rules, graph transformation with borrowed contexts is a technique that allows us to describe and analyze how a graph could evolve when embedded in different contexts. This technique is based on several ideas. The first one is that we have to specify explicitly what is the *open* (or visible) part of the given graph G, i.e. what part of G can be extended by a context. This part is called the *interface* of the graph and, in general, it may be any arbitrary subgraph of G. A consequence of this is that a context should be a graph with two interfaces. The reason is that, when we connect a context to G, by matching the interface of the graph with a corresponding interface of the context, the result is also a graph G' with an interface, so that it can also be embedded into a new context. More precisely, the resulting graph is obtained gluing together, by means of a pushout, G and the context.

Definition 1. *A graph with interface J is an injective morphism (usually an inclusion) $J \to G$, and a context is a cospan of injective morphisms $J \to C \leftarrow J'$. The result of embedding $J \to G$ into the context $J \to C \leftarrow J'$ is the graph with interface $J' \to G'$, defined by the pushout diagram:*

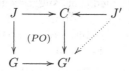

the resulting graph G' will also be denoted as $C[G]$.

The second idea underlying this technique is to allow for a *partial match* between the left-hand side of a rule L and the graph G. Then, the associated transformation would start adding to G the missing part of L and, afterwards, applying a standard graph transformation. That is, we add to G a minimal context, so that the given rule can be applied. As this context is the part of L that has not been matched with G, we say that G *borrows this context* from the rule. The third idea is to consider that the interface of the resulting graph is the old interface plus the borrowed context, minus the parts deleted by the rule. Finally, the last idea is to label the borrowed context transformations with the context used in the transformation step.

Definition 2. *Given a graph with interface $J \to G$ and a graph transformation rule $p : L \leftarrow K \to R$, we say that $J \to G$ reduces to $I \to H$ with label $J \to F \leftarrow I$, denoted $(J \to G) \xrightarrow{J \to F \leftarrow I} (I \to H)$ if there are graphs C, G^+, D and additional morphisms such that all the squares in the diagram below are pushouts (PO) or pullbacks (PB) and all the morphisms are injective:*

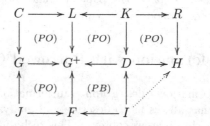

The intuition here is that C is the subgraph of L that completely matches G; $J \to F \leftarrow I$ is the context borrowed to extend G; and G^+ is the graph G enriched with the borrowed context. In particular, F, defined as the pushout complement (if it exists) of the left lower square, extends J with all the elements in G^+ which are not in G. In Fig. 3 we can see an example of attributed borrowed context transformation that is explained in Example 1. But not all borrowed context transformations are useful for characterizing the behaviour of a graph when embedded in any arbitrary context. This is the case of transformations where the partial match is included in the part of the interface that remains invariant after the transformation, since the same transformation could be applied to any graph with the same interface. These transformations are called *independent*.

In labelled transition systems, a bisimulation is a symmetric relation between states that is compatible with their observational behaviour. This means that if two states s_1 and s_2 are related then for every transition from s_1 labelled with l there should be a transition from s_2 with the same label and the resulting states should again be related. Then, bisimilarity is the largest bisimulation relation.

Definition 3. *Given a set $\mathcal{T}$ of transformation rules, a relation $\mathcal{R}$ on graphs with interface is a bisimulation if it is symmetric and moreover if $(J \to G_1)\mathcal{R}(J \to G_2)$, for every transformation $(J \to G_1) \xrightarrow{J \to F \leftarrow I} (I \to H_1)$ there exists a transformation $(J \to G_2) \xrightarrow{J \to F \leftarrow I} (I \to H_2)$ such that $(I \to H_1)\mathcal{R}(I \to H_2)$.*

Bisimilarity, denoted $\sim$ is the largest bisimulation relation (or, equivalently, the union of all bisimulation relations).

In [5] it is proved that the condition to prove that a relation is a bisimulation can be restricted to dependent transformations. Moreover, the main result in that paper is that bisimilarity is a congruence.

Example 1. The running example that we use in this paper is an adaptation of the example used in [5], but now including some attributes. The rules in Fig. 1 describe communication in a network. For simplicity we have omitted the interface part of the rules, which is assumed to be the common part between their left and right-hand sides. Round nodes represent locations in the network and edges between these nodes represent communication links. There are two kinds of links, simplex communication links, represented by thin arrows, and duplex communication links, represented by thick arrows. The rules specify that messages can be sent in one direction via simplex links and in two directions via duplex links. The difference with respect to the example in [5], is that we use some attributes to describe a simple form of encrypted communication. We assume that the data algebra includes two sorts of values, *messages* and *keys*, and two operations, e and d, for encrypting and decrypting messages, respectively. Both operations have two parameters, a key and a message, and return a message. In this sense, the rules also include square and elliptic nodes that include some data values (i.e. they are attributed). Square nodes represent messages and include a value of sort *message*, and elliptic nodes represent keys and include a value of sort *key*. Message nodes may be connected to a round node meaning that the message is at that location. Elliptic nodes may also be connected to round nodes indicating that this key is shared by the connected locations.

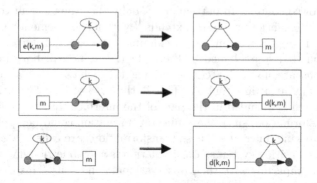

Fig. 1. Transformation rules

s The three rules describe how (encrypted) messages are sent through the network. In particular, the first rule describes how a message is sent from the source to the target node of a simplex communication link, while the second and third rules show how a message can be sent in both directions through a duplex communication link. Moreover, the three rules describe in a different (and rather artificial) way that the messages are assumed to be encrypted before they are sent, and they are decrypted when they are received. In the first rule, the message to be sent is explicitly assumed to be encrypted, since it is the result of the term $e(k, m)$, where k is the key shared by the two locations. After the transformation, the decrypted message m is associated to the target node. In the other two rules, it is not explicitly stated if the message m attached to the sending node is encrypted or not. However, the message received is explicitly decrypted, since the message received is the result of $d(k, m)$.

Fig. 2. Bisimilar graphs

The example used in [5] is similar but graphs have no attributes: there are no keys or encryption/decryption of messages, and messages are not assumed to be values of any kind, but just nodes. Then, using the techniques described in the paper, it is shown that the two graphs $J \to G$ and $J \to G'$, depicted in Fig. 2, are bisimilar. In particular, they analyze what are all the borrowed context transformations that can be applied to the two graphs and check that for every transformation applied to one of them, there is another transformation that can be applied to the other one, with the same label, such that the resulting graphs $I \to H$ and $I \to H'$ are extensions, with the same context, of $J \to G$ and $J \to G$, respectively. And this means that $J \to G$ and $J \to G'$ are bisimilar.

However, in our example, that kind of proof is not possible. The main problem is that, when applying an attributed graph transformation rule to an attributed

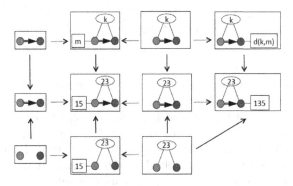

Fig. 3. Attributed borrowed context transformation

graph, all the variables in the rule must be matched to some given data values and, accordingly, the terms included must be evaluated. This applies also to borrowed context transformation, but variables in the context would not be matched but must be substituted by arbitrary values. For instance, in Fig. 3 we show a possible borrowed context transformation by applying the second rule in Fig. 1 to the graph $J \to G'$ on the left in Fig. 2, assuming that messages and keys are integers, that the variables m and k are substituted by 23 and 15, respectively, and that the value of $d(23, 15) = 135$. and other substitutions of m and k would lead to a different borrowed context transformation. This causes that, to prove the bisimilarity of the above two graphs we would need to prove the bisimilarity of an infinite number of graphs. A second problem is that the data (the attributes) in the rules do not exactly coincide. Actually, as we will see in Sect. 5, the two graphs are bisimilar if the encrypting and decrypting operations are the inverse of each other. Then it may be unclear how to prove this fact in the context of attributed graph transformation.

3   Symbolic Graphs and Symbolic Graph Transformation

A symbolic graph is a graph that specifies a class of attributed graphs. More precisely, a symbolic graph $SG = \langle G, \Phi \rangle$ over a given data algebra $\mathcal{A}$ is a kind of labelled graph G (technically, an E-graph [3]), whose nodes and edges may be decorated with labels from a given set of variables X, together with a set of formulas Φ over these variables and over the values in $\mathcal{A}$. The intuition is that each substitution $\sigma : X \to \mathcal{A}$ of the variables in X by values of $\mathcal{A}$ such that $\mathcal{A} \models \sigma(\Phi)$, defines an attributed graph $\sigma(G)$ in the semantics of G, obtained replacing each variable x in G by the corresponding data value $\sigma(x)$. That is, the semantics of SG is defined:

$$Sem(SG) = \{\sigma(G) \mid \mathcal{A} \models \sigma(\Phi)\}$$

For instance, the graph below on the left specifies a class of attributed graphs, including distances in the edges, that satisfy the well-known triangle inequality, and the graph in the center would belong to its semantics

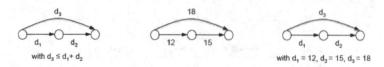

It may be noticed that every attributed graph may be seen as a symbolic graph by just replacing all its values by variables, and by including an equation $x_v = v$, into the corresponding set of formulas, for each value v, where x_v is the variable that has replaced the value v. We call these kind of symbolic graphs *grounded symbolic graphs*. In particular, $GSG(G)$ denotes the grounded symbolic graph defined by G. For instance, the graph above on the right, can be seen as the symbolic representation of the attributed graph in the center.

A morphism $h : \langle G_1, \Phi_1 \rangle \to \langle G_2, \Phi_2 \rangle$ is a graph morphism $h : G_1 \to G_2$ such that $\mathcal{A} \models \Phi_2 \Rightarrow h(\Phi_1)$, where $h(\Phi_1)$ is the set of formulas obtained when replacing in Φ_1 every variable x_1 in the set of labels of G_1 by $h(x_1)$. Symbolic graphs and morphisms over a given data algebra $\mathcal{A}$ form the category **SymbGraph**$_{\mathcal{A}}$.

To write $h(\Phi_1)$ is, actually, an abuse of notation, since h is assumed to be a graph morphism and, therefore, it is not defined on formulas. To be rigorous, and less readable, we should have written $h_X^{\#}(\Phi_1)$, where h_X is the restriction of h to the set of labels of G_1 (i.e. h_X maps the variables in G_1 to the variables in G_2) and $h_X^{\#}$ is the (unique) extension of h_X to terms over the given signature.

For (technical) simplicity, we assume that in our graphs no variable is bound to two different elements of the graph. We call these graphs in *normal form*. It should be clear that this is not a limitation since every symbolic graph SG is equivalent to a symbolic graph SG' in normal form, in the sense that $Sem(SG) = Sem(SG')$. It is enough to replace each repeated occurrence of a variable x by a fresh variable y, and to include the equality $x = y$ in $\Phi_{G'}$.

In [8], we showed that **SymbGraph**$_{\mathcal{A}}$ is an adhesive HLR category [7,4] taking as M-morphisms all injective graph morphisms where the formulas constraining the source and target graphs are equivalent. In particular, the proposition below shows how pushouts of symbolic graphs are defined:

Proposition 1. *[8] Diagram* (1) *below is a pushout if and only if diagram* (2) *is also a pushout and* $\mathcal{A} \models \Phi_3 \equiv (g_1(\Phi_1) \cup g_2(\Phi_2))$.

$$
\begin{array}{ccc}
\langle G_0, \Phi_0 \rangle & \xrightarrow{\ h_1\ } & \langle G_1, \Phi_1 \rangle \\
{\scriptstyle h_2}\downarrow & (1) & \downarrow{\scriptstyle g_1} \\
\langle G_2, \Phi_2 \rangle & \xrightarrow[\ g_2\ ]{} & \langle G_3, \Phi_3 \rangle
\end{array}
\qquad
\begin{array}{ccc}
G_0 & \xrightarrow{\ h_1\ } & G_1 \\
{\scriptstyle h_2}\downarrow & (2) & \downarrow{\scriptstyle g_1} \\
G_2 & \xrightarrow[\ g_2\ ]{} & G_3
\end{array}
$$

In this paper, a *symbolic graph transformation rule* is a pair $\langle L \hookleftarrow K \hookrightarrow R, \Phi \rangle$, where L, K and R are graphs over the sets of variables X_L, X_K and X_R, respectively, and Φ is a set of formulas over X_R and over the values in $\mathcal{A}$. We consider that a rule is a span of symbolic graph inclusions $\langle L, \emptyset \rangle \hookleftarrow \langle K, \emptyset \rangle \hookrightarrow \langle R, \Phi \rangle$, Intuitively, Φ relates the attributes in the left and right-hand side of the rule.

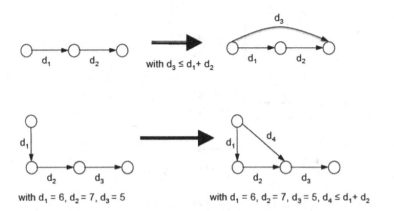

Fig. 4. Symbolic graph transformation

This means that we implicitly assume that $X_L = X_K \subseteq X_R$. In [10] we allow for more general rules.

As usual, we can define the application of a graph transformation rule $\langle L \hookleftarrow K \hookrightarrow R, \Phi \rangle$ by a double pushout in the category of symbolic graphs [10]).

Definition 4. *Given a transformation rule* $r = \langle L \hookleftarrow K \hookrightarrow R, \Phi \rangle$ *over a data algebra* $\mathcal{A}$ *and a morphism* $m : L \to G$, $\langle G, \Phi' \rangle \Longrightarrow_{r,m} \langle H, \Phi' \cup m'(\Phi) \rangle$ *if and only if the diagram below is a double pushout and* $\Phi' \cup m'(\Phi)$ *is satisfiable in the given data algebra.*

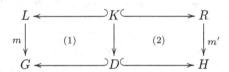

For instance in the upper part of figure 4, we can see a symbolic graph transformation rule stating that if a given graph has two consecutive edges, e_1 and e_2, with some given distances d_1 and d_2, respectively, then we can add a new edge from the source of e_1 to the target of e_2 whose distance must be smaller or equal than $d_1 + d_2$. Moreover, in the bottom part of the figure we may see the result of applying that rule to the graph on the left.

We may notice that, in general, $\Phi' \cup m'(\Phi)$ may be unsatisfiable. This would mean that the resulting graph $\langle H, \Phi' \cup m'(\Phi) \rangle$ would have an empty semantics, i.e. it would be inconsistent. This is avoided by requiring explicitly that $\Phi' \cup m'(\Phi)$ must be satisfiable. It is not difficult to show that the above construction defines a double pushout in the category of symbolic graphs [10].

A symbolic graph transformation rule can be seen as a specification of a class of attributed graph transformation rules. More precisely, we may consider that the rule $r = \langle L \hookleftarrow K \hookrightarrow R, \Phi \rangle$ denotes the class of all rules $\sigma(L) \hookleftarrow \sigma(K) \hookrightarrow \sigma(R)$, where σ is a substitution such that $\mathcal{A} \models \sigma(\Phi)$, i.e.:

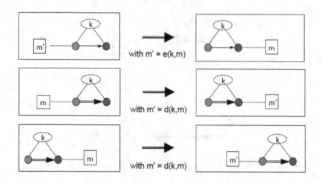

Fig. 5. Symbolic transformation rules

$$Sem(r) = \{\sigma(L) \hookleftarrow \sigma(K) \hookrightarrow \sigma(R) \mid \mathcal{A} \models \sigma(\Phi)\}$$

It is not difficult to see that given a rule r and a symbolic graph SG, $SG \Longrightarrow_r SG'$ if and only if for every graph $G \in Sem(SG)$, $G \Longrightarrow_r G'$, with $G' \in Sem(SG')$ and moreover, for every graph $G' \in Sem(SG')$, $G \Longrightarrow_r G'$, for some $G \in Sem(SG)$ [8].

We will require that transformation rules must be *strict* meaning that the result of a rule application to a grounded model must also be grounded[1]:

Definition 5. *A transformation rule* $\langle L \hookleftarrow K \hookrightarrow R, \Phi \rangle$ *is* strict *if for every transformation* $\langle G, \Phi' \rangle \Longrightarrow_{p,m} \langle H, \Phi' \cup m'(\Phi) \rangle$, *whenever* $\langle G, \Phi' \rangle$ *is grounded* $\langle H, \Phi' \cup m'(\Phi) \rangle$ *is also grounded.*

For instance, the rule in figure 4 is not strict. However, assuming that all rules must be strict is not really a restriction since every non-strict rule may be made strict including in X_L all the variables in X_R. This forces to match every variable in X_R to some given value causing that the result of the transformation $\langle H, \Phi' \cup m'(\Phi) \rangle$ must be grounded. Moreover, it is easy to see that the semantics of a non-strict rule (as defined above) and its associated strict one coincide.

As shown in [8], symbolic graph transformation is more powerful than attributed graph transformation. In particular, any attributed graph transformation rule r can be represented by a symbolic graph transformation rule $SR(r)$ such that an attributed graph G can be transformed by r into a graph H if and only if the grounded graph associated to G, SG, can be transformed by $SR(r)$ into the grounded graph associated to H. However, the converse is not true.

For example, in Fig 5 we show the symbolic rules associated to the attributed rules depicted in Fig 1.

---

[1] This does not mean that the result of a borrowed context transformation of a grounded graph using a strict rule must also be grounded. In particular, the rules in Example 1 are strict but, as shown in Example 3, the results of their borrowed context application to two grounded graphs are not grounded.

4   Symbolic Bisimilarity

In this section we study a notion of bisimilarity for symbolic graphs to be used for proving the bisimilarity of attributed graphs. First we show that a definition of symbolic bisimulation just applying the concepts introduced in Section 2 to **SymbGraph**$_A$ is not adequate for checking the bisimilarity of attributed graphs. Then, we define a notion of symbolic bisimulation and show that it coincides with attributed bisimulation, when restricted to grounded graphs. Finally, we show that this symbolic bisimilarity is also a congruence on **SymbGraph**$_A$.

Symbolic graphs form an M-adhesive category [8]. Hence, all the definitions and results in [5] concerning borrowed context transformation and bisimilarity apply to this category. For instance, we have notions of symbolic graph with interface, of context, and of borrowed context transformation exactly as in Section 2, but within the category of symbolic graphs. To be precise, we consider that graph interfaces are not arbitrary symbolic graphs, but graphs with an empty set of conditions, since we consider that the interface must only specify the open part of a graph. Then, we may think that a direct application of these results may be a solution for the problem of checking bisimilarity for attributed graphs, since we may directly work with terms and variables, without having to compute all its possible substitutions, as described in Example 1. This means that for deciding if two attributed graphs are bisimilar we could check if their associated grounded graphs are bisimilar in the category of symbolic graphs. Unfortunately, as we can see in the counter-example below, two attributed graphs may be bisimilar as attributed graphs, while their associated grounded symbolic graphs are not bisimilar as symbolic graphs.

Example 2. Let us consider the attributed graph transformation system, consisting of three rules, depicted below.

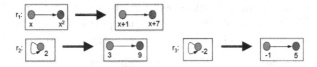

The first rule includes a variable x used in different expressions including x^2. The remaining two rules are simple attributed rules with integer values. Now, let us consider the two attributed graphs, including as interface just the source node of the edges in the graphs, $I \to G_1$ and $I \to G_2$ that are depicted below.

We can see that these graphs are bisimilar. The only dependent borrowed context transformations that we can apply on both graphs are depicted below and, in both cases, the resulting graphs are equal.

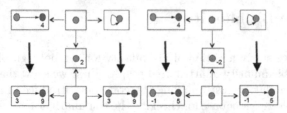

Notice that, in the figure, we do not depict the data values that are not bound to any node or edge. However, if we consider the symbolic versions of the above rules we can show that $GSG(I \to G_1)$ and $GSG(I \to G_2)$ are not bisimilar. In particular, below we can find the symbolic borrowed context transformation of the two grounded graphs using the symbolic versions of r_3 and r_1, and we may see that no direct transformation can be applied to the resulting graph on the left, $GSG(I \to G_4)$. However, we may also see that the conditions associated to the resulting graph on the right are equivalent to $(z = 4 \wedge x = 2 \wedge t = 3 \wedge u = 9) \vee (z = 4 \wedge x = -2 \wedge t = -1 \wedge u = 5)$, which means that we can transform this graph using the symbolic version of r_1 (matching x to 3).

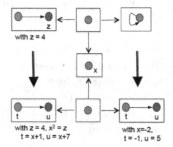

The problem in the above counter-example is that, when considering attributed graph transformation, each instance of the rule r_1 (i.e. when $x = 2$ or when $x = -2$) is simulated by r_2 and r_3, respectively, and vice versa. However, when considering symbolic transformation, we need to say that r_1 is simulated by r_2 and r_3 *together*, and vice versa. This is not possible if we define symbolic bisimulation as in [5]. Instead, we define a new notion that solves this problem:

Definition 6. *A relation $\mathcal{R}$ on symbolic graphs with interface is a symbolic bisimulation with respect to a set of transformation rules, if it is symmetric and moreover if $(J \to SG_1)\mathcal{R}(J \to SG_2)$, for every transformation $(J \to SG_1) \xrightarrow{J \to F \leftarrow I} (I \to SG_1')$, with $SG_1' = \langle G_1', \Phi_1' \rangle$ there exist a family of conditions $\{\Psi_i\}_{i \in \mathcal{I}}$ and a family of transformations $\{(J \to SG_2) \xrightarrow{J \to F \leftarrow I} (I \to SH_i)\}_{i \in \mathcal{I}}$, with $SH_i = \langle H_i, \Pi_i \rangle$ such that:*

- *For every substitution σ_1' such that $\mathcal{A} \models \sigma_1'(\Phi_1')$, there is an index i and a substitution σ_i such that $\mathcal{A} \models \sigma_i(\Psi_i \cup \Pi_i)$ and $\sigma_1' \upharpoonright_I = \sigma_i \upharpoonright_I$, where $\sigma \upharpoonright_I$ denotes the restriction of σ to the variables in I.*
- *For every i, $(I \to \langle G_1', \Phi_1' \cup \Psi_i \rangle) \mathcal{R} (I \to \langle H_i, \Pi_i \cup \Psi_i \rangle)$.*

Symbolic bisimilarity, denoted $\sim_S$, is the largest symbolic bisimulation relation.

Intuitively, each condition Ψ_i specifies which instances of the transformation $(J \rightarrow SG_1) \xrightarrow{J \rightarrow F \leftarrow I} (I \rightarrow SG_1')$ are simulated by instances of the transformation $(J \rightarrow SG_2) \xrightarrow{J \rightarrow F \leftarrow I} (I \rightarrow SH_i)$. For instance, in the above counterexample, to show $GSG(I \rightarrow G_1) \sim_S GSG(I \rightarrow G_2)$, the symbolic transformation of $GSG(I \rightarrow G_2)$ via r_3 depicted above on the left, would be simulated by the transformation of $GSG(I \rightarrow G_1)$ via rule r_1, together with the condition $x = -2$. Similarly, the symbolic transformation of $GSG(I \rightarrow G_1)$ via r_1 depicted above on the right, would be simulated by the transformations of $GSG(I \rightarrow G_2)$ via rule r_2, together with the condition $x = 2$ and of $GSG(I \rightarrow G_2)$ via rule r_3, together with the condition $x = -2$.

Using symbolic bisimilarity we can prove the bisimilarity of attributed graphs:

Theorem 1. *Given transformation rules $\mathcal{T}$, $(J \rightarrow G_1) \sim (J \rightarrow G_2)$ with respect to $Sem(\mathcal{T})$ if and only if $GSG(J \rightarrow G_1) \sim_S GSG(J \rightarrow G_2)$ with respect to $\mathcal{T}$.*

To prove this theorem we use two lemmas:

Lemma 1. *Let $\mathcal{R}$ be the following relation defined on symbolic graphs. $(J \rightarrow SG_1)\mathcal{R}(J \rightarrow SG_2)$ if:*

- *For every attributed graph $\sigma_1(J \rightarrow SG_1) \in Sem(J \rightarrow SG_1)$ there is an attributed graph $\sigma_2(J \rightarrow SG_2) \in Sem(J \rightarrow SG_2)$ such that $\sigma_1(J \rightarrow SG_1) \sim \sigma_2(J \rightarrow SG_2)$.*
- *For every attributed graph $\sigma_2(J \rightarrow SG_2) \in Sem(J \rightarrow SG_2)$ there is an attributed graph $\sigma_1(J \rightarrow SG_1) \in Sem(J \rightarrow SG_1)$ such that $\sigma_1(J \rightarrow SG_1) \sim \sigma_2(J \rightarrow SG_2)$.*

Then, $\mathcal{R}$ is a bisimulation.

Lemma 2. *The relation on attributed graphs $(J \rightarrow G_1)\mathcal{R}(J \rightarrow G_2)$ if $GSG(J \rightarrow G_1) \sim_S GSG(J \rightarrow G_2)$ is a bisimulation.*

To prove this theorem we have that, if $(J \rightarrow G_1) \sim (J \rightarrow G_2)$ then $GSG(J \rightarrow G_1)$ and $GSG(J \rightarrow G_2)$ satisfy the conditions of Lemma 1. So they must be bisimilar. Conversely, if $GSG(J \rightarrow G_1) \sim_S GSG(J \rightarrow G_2)$, lemma 2 directly implies that $(J \rightarrow G_1)$ and $(J \rightarrow G_2)$ are bisimilar.

The last theorem shows that symbolic bisimilarity is a congruence:

Theorem 2. *If $(J \rightarrow SG_1) \sim_S (J \rightarrow SG_2)$ then, for every context $J \rightarrow C \leftarrow I$, $(I \rightarrow C[SG_1]) \sim_S (I \rightarrow C[SG_2])$.*

The proof uses a property shown in [5] that if a graph $J \rightarrow G$ is embedded in $J' \rightarrow G'$ with a given context and if $J' \rightarrow F' \leftarrow I'$ is a possible label for transforming $J' \rightarrow G'$ then there is a context $I \rightarrow C' \leftarrow I'$ and a label $J \rightarrow F \leftarrow I$, such that any transformation of $J' \rightarrow G'$ with label $J' \rightarrow F' \leftarrow I'$ can be obtained by first transforming $J \rightarrow G$ with a label $J \rightarrow F \leftarrow I$ and, then, embedding the result in the context $I \rightarrow C' \leftarrow I'$.

5   Checking Bisimilarity

In this section we show the basic ideas of how we can use the previous results to show that two attributed graphs $(J \to G_1)$ and $(J \to G_2)$ are bisimilar. In principle, we would need to consider all the possible borrowed context transformations with the same label of their associated grounded graphs and show that we can group them in pairs, so that under suitable conditions Ψ_i, the resulting graphs are bisimilar. The obvious problem is that this clearly leads to a non-terminating process. To avoid this non-termination (if possible) Sangiorgi [13] defined the notion of bisimulation up to context that is adapted to the case of graph transformation in [5]. In our case this notion would be defined as follows:

Definition 7. *A relation $\mathcal{R}$ is a symbolic bisimulation up to context if whenever $(J \to SG_1)\mathcal{R}(J \to SG_2)$, then for every transformation $(J \to SG_1) \xrightarrow{J \to F \leftarrow I} (I \to SG_1')$, with $SG_1' = \langle G_1', \Phi_1' \rangle$ there exist a family of conditions $\{\Psi_i\}_{i \in \mathcal{I}}$ and a family of transformations $\{(J \to SG_2) \xrightarrow{J \to F \leftarrow I} (I \to SH_i)\}_{i \in \mathcal{I}}$, with $SH_i = \langle H_i, \Pi_i \rangle$ such that:*

- *For every substitution σ_1' such that $\mathcal{A} \models \sigma_1'(\Phi_1')$, there is an index i and a substitution σ_i such that $\mathcal{A} \models \sigma_i(\Psi_i \cup \Pi_i)$ and $\sigma_1' \lceil_F = \sigma_i \lceil_F$.*
- *For every i, $(I \to \langle G_1', \Phi_1' \cup \Psi_i \rangle)$ and $(I \to \langle H_i, \Pi_i \cup \Psi_i \rangle)$ are the result of embedding $(J \to SG_1)$ and $(J \to SG_2)$ in the same context.*

Proposition 2. *[5] If $\mathcal{R}$ is a symbolic bisimulation up to context then $\mathcal{R} \subseteq \sim_S$.*

This means that if, when trying to check if $(J \to SG_1)$ and $(J \to SG_2)$, we have to prove that $(I \to SG_1')$ and $(I \to SG_2')$ are also bisimilar and if the latter graphs can be obtained by embedding the former graphs into the same context, then we can consider that $(I \to SG_1')$ and $(I \to SG_2')$ are bisimilar. So, if this happens for all the pairs of graphs that we have to prove bisimilar, then we can conclude that $(J \to SG_1)$ and $(J \to SG_2)$ are indeed bisimilar.

Moreover, as in [5], we can restrict ourselves to checking only *dependent* transformations. This is important since, if the number of transformation rules is finite, and the given graph, $J \to G$, is also finite, then there is a finite number of possible borrowed context transformations that can be applied to $J \to G$. Let us now show how these ideas would be applied to the example in Section 2.

Example 3. We want to check if the graphs $J \to G$ and $J \to G'$ depicted in Fig. 2 are bisimilar. In this case, since G and G' include no explicit attributes, their associated grounded graphs would look similar. Now, there are only two dependent borrowed context transformations that can be applied to each of the two graphs. In Fig. 6 we depict the transformations that can be applied to $J \to G$ and $J \to G'$, using the first and the second rule in Fig. 5, respectively, when we add to the two graphs a context consisting of a common key k and a message m attached to the leftmost node. In particular, on the left and the right of the figure we depict the transformations of G and G' and, in the middle, we depict the label of the transformation. Fig. 7 is similar, and describes the transformations over $J \to G$ and $J \to G'$, using the first and the third rules, respectively.

To prove $J \to G \sim_S J \to G'$, according to the definition of symbolic bisimulation, we have to show, on the one hand, that the conditions associated to the transformations, $\{m = e(k, m')\}$ and $\{m' = d(k, m)\}$ are equivalent; and, on the other hand, that the resulting graphs are bisimilar, $(I \to SH_1) \sim_S (I \to SH_1')$, with $SH_1 = \langle H1, \{m = e(k, m')\}\rangle$ and $SH_1' = \langle H1', \{m' = d(k, m)\}\rangle$, and $(I \to SH_2) \sim_S (I \to SH_2')$, with $SH_2 = \langle H2, \{m = e(k, m')\}\rangle$ and $SH_2' = \langle H2', \{m' = d(k, m)\}\rangle$. But if the conditions are equivalent, SH_i and SH_i' $(i = 1, 2)$ are just the original graphs extended by the same context. Therefore, the bisimilarity of the graphs $J \to G$ and $J \to G'$ depends on the equivalence of the above conditions. More precisely, if the given data algebra $\mathcal{A}$ satisfies:

$$d(k, e(k, m)) = m$$

$$e(k, d(k, m)) = m$$

i.e. if encryption and decryption are the inverse of each other.

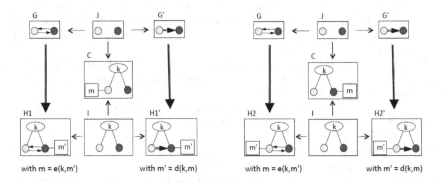

Fig. 6. Transformation 1   **Fig. 7.** Transformation 2

6   Conclusion and Related Work

Bisimilarity was introduced by Park in [11] and, since then, it has been studied by many authors in relation to many different formalisms. In [5], Ehrig and König not only introduced a notion of bisimilarity for graph transformation systems, but they provided a simple and general technique to derive labelled transitions and bisimulation congruences from unlabelled ones. This paper was followed by [12] where they showed how these techniques could be used for the verification of bisimilarity. Moreover, in [6], the borrowed context technique is generalized to transformation rules including application conditions. Unfortunately, as we have seen in this paper, these techniques do not work in the case of attributed graphs. Borrowed context transformations have been used for the definition of other behavioral equivalence relations [2]. On the other hand, symbolic graphs were introduced in [9] in order to define constraints on attributed graphs. Then symbolic graph transformation was studied in detail in [8,10].

In this paper we have presented a new notion of bisimulation for symbolic graph transformation that has been shown to be useful for checking the bisimilarity of attributed graphs. The key issue is that in symbolic graph transformation we do not need to replace all the variables by values. Moreover, the neat separation in symbolic graphs between the graph structure and the algebra of data also helps for this purpose. Currently we are working in devising a specific proof method to implement these ideas.

References

1. Bonchi, F., Gadducci, F., König, B.: Synthesising CCS bisimulation using graph rewriting. Inf. Comput. 207(1), 14–40 (2009)
2. Bonchi, F., Gadducci, F., Monreale, G.V., Montanari, U.: Saturated LTSs for Adhesive Rewriting Systems. In: Ehrig, H., Rensink, A., Rozenberg, G., Schürr, A. (eds.) ICGT 2010. LNCS, vol. 6372, pp. 123–138. Springer, Heidelberg (2010)
3. Ehrig, H., Ehrig, K., Prange, U., Taentzer, G.: Fundamentals of Algebraic Graph Transformation. EATCS Monographs of Theoretical Comp. Sc. Springer (2006)
4. Ehrig, H., Padberg, J., Prange, U., Habel, A.: Adhesive high-level replacement systems: A new categorical framework for graph transformation. Fundamenta Informaticae 74(1), 1–29 (2006)
5. Ehrig, H., König, B.: Deriving bisimulation congruences in the DPO approach to graph rewriting with borrowed contexts. Mathematical Structures in Computer Science 16(6), 1133–1163 (2006)
6. Hülsbusch, M., König, B.: Deriving Bisimulation Congruences for Conditional Reactive Systems. In: Birkedal, L. (ed.) FOSSACS 2012. LNCS, vol. 7213, pp. 361–375. Springer, Heidelberg (2012)
7. Lack, S., Sobocinski, P.: Adhesive and quasiadhesive categories. Theoretical Informatics and Applications 39, 511–545 (2005)
8. Orejas, F., Lambers, L.: Symbolic attributed graphs for attributed graph transformation. In: Int. Coll. on Graph and Model Transformation On the Occasion of the 65th Birthday of Hartmut Ehrig (2010)
9. Orejas, F.: Symbolic graphs for attributed graph constraints. J. Symb. Comput. 46(3), 294–315 (2011)
10. Orejas, F., Lambers, L.: Lazy Graph Transformation. Fundamenta Informaticae 118(1-2), 65–96 (2012)
11. Park, D.: Concurrency and Automata on Infinite Sequences. In: Deussen, P. (ed.) GI-TCS 1981. LNCS, vol. 104, pp. 167–183. Springer, Heidelberg (1981)
12. Rangel, G., König, B., Ehrig, H.: Bisimulation verification for the DPO approach with borrowed contexts. ECEASST 6 (2007)
13. Sangiorgi, D.: On the Proof Method for Bisimulation (Extended Abstract). In: Hájek, P., Wiedermann, J. (eds.) MFCS 1995. LNCS, vol. 969, pp. 479–488. Springer, Heidelberg (1995)

Toward Bridging the Gap between Formal Foundations and Current Practice for Triple Graph Grammars

Flexible Relations between Source and Target Elements

Ulrike Golas[1], Leen Lambers[2], Hartmut Ehrig[3], and Holger Giese[2]

[1] Konrad-Zuse-Zentrum für Informationstechnik Berlin, Germany
golas@zib.de
[2] * Hasso-Plattner-Institut, Universität Potsdam, Germany
{leen.lambers,holger.giese}@hpi.uni-potsdam.de
[3] Technische Universität Berlin
ehrig@cs.tu-berlin.de

Abstract. Triple graph grammars (TGGs) are a common formalism to specify model transformations in a relational way, creating source and target models together with their correspondences. The classical theoretical model of triple graphs is based on a morphism span from the correspondence component to the source and target components. In practice, this formalization often can not be used as for certain applications no proper morphisms between the correspondence and source or target components can be found. In this paper, we introduce TGGs as plain graph grammars with special typing which avoids an extra flattening step and is more directly suitable for implementation and formal analysis due to the more flexible and homogeneous formalization. The typing expresses that each graph can be partitioned into a source, correspondence, and target component allowing arbitrary relationships between the components. We further show that the main decomposition and composition result, which is the formal basis for correctness, completeness, consistency, and functional behavior, holds analogous to the classical approach and demonstrate that classical triple graph transformation is actually a special case – after flattening – of the more flexible one.

1 Introduction

Model transformations can be specified in a *relational (declarative) way* by *triple graph grammars* (TGGs) [1], creating source and target models together with their correspondences. In order to operationalize TGGs, forward, backward, and correspondence rules can be derived. They can be applied to a source model, target model, or pair of source and target model to obtain a forward, backward,

---

* The work of both authors was developed in the course of the DFG-funded project Correct Model Transformations
http://www.hpi.uni-potsdam.de/giese/projekte/kormoran.html?L=1.

H. Ehrig et al.(Eds.): ICGT 2012, LNCS 7562, pp. 141–155, 2012.

or correspondence transformation, respectively. In [2], the basic concepts of classical TGGs were formalized in a set-theoretical way, which was generalized and extended in [3] to typed, attributed graphs. We restrict our considerations in this paper to typed TGGs.

The *classical theoretical model* of triple graphs is based on a morphism span from the correspondence component to the source and target components, while nearly all implementations in contrast map the triple graphs on flat graphs to simplify their handling and be more flexible. However, besides that for certain applications no proper morphisms between the correspondence and source or target components can be found, the classical theoretical model due to its heterogeneous nature also makes formal analysis unnecessarily complex.

In this paper we therefore present a new formalization[1] for TGGs as *plain graph grammars with special typing*. This typing expresses that each graph can be partitioned into a source, correspondence, and target component. Thereby, the correspondence component holds correspondence nodes that can be connected via special edges to source or target nodes. We show that this new formal approach fits very well to most TGG *implementations*, where triple graph transformation is performed as a specially typed plain graph transformation. We show that the *main decomposition and composition result* (cf. Fig. 4) can be carried over in an elegant way to the new approach. This is the formal basis to transfer also the results on correctness, completeness, consistency, and functional behavior of model transformations [5,3,6,4]. Finally, we show a *formal comparison with the classical model*, demonstrating the *flexibility* of the new model with respect to specifying correspondences. Other means to enhance expressiveness are, e.g., application conditions [6,7] or relaxing the bind-only-once semantics of TGGs [8].

In Section 2, we reintroduce the classical triple graph model and motivate our new more flexible and flattened model presented in Section 3. It is demonstrated that triple graph transformation can indeed be performed as specially typed plain graph transformation. We show in Section 4 that for our new model the main decomposition and composition result holds analogous to the classical model [1,3]. In Section 5, we demonstrate that the classical model is actually a special case – after flattening – of the more flexible one. A conclusion and outlook to future work closes the paper.

2   Classical Triple Graphs and Motivation for Flexibility

Classical triple graphs consist of a *morphism span* from the correspondence component to the source and target component. *Classical triple graph morphisms* consist of *three graph morphisms* mapping source, correspondence, and target component to each other in a compatible way. A *classical TGG* consists of a triple graph as start graph and a set of non-deleting triple rules [1].

---

[1] This new model with its corresponding category was introduced in less detail in [4], where it is shown how to bridge the gap between the TGG formal semantics and an efficient corresponding implementation with a specific bookkeeping mechanism.

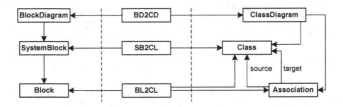

Fig. 1. Type triple graph

Definition 1 (Classical triple graph). *A classical triple graph $G = (G_S \overset{s_G}{\leftarrow} G_C \overset{t_G}{\rightarrow} G_T)$ consists of graphs G_S, G_C, and G_T, called source, correspondence, and target component, and two graph morphisms s_G and t_G mapping the correspondence to the source and target components. A classical triple graph morphism $f : G_1 \rightarrow G_2$ is a tuple $f = (f_S : G_{1S} \rightarrow G_{2S}, f_C : G_{1C} \rightarrow G_{2C}, f_T : G_{1T} \rightarrow G_{2T})$ such that $s_{G_2} \circ f_C = f_S \circ s_{G_1}$ and $t_{G_2} \circ f_C = f_T \circ t_{G_1}$. Classical triple graphs form the category* **CTripleGraphs**.

Definition 2 (Typed classical triple graph). *The typing of a classical triple graph is done in the same way as for standard graphs via a type graph* **TG** *- in this case a classical type triple graph - and a typing morphism* $type_G$ *from the classical triple graph G into this type graph leading to the* typed classical triple graph $(G, type_G)$. *Typed classical triple graphs and morphisms form the category* **CTripleGraphs**$_{TG}$.

While the classical formalization of triple graphs works well in theory and for certain examples, there are other examples showing that it is too restrictive.

Example 1 (Running example). As a running example not adhering to the classical model, we use a model transformation[2] from SDL block diagrams to UML class diagrams. The simplified *type graphs* for these languages are shown on the left and right of Fig. 1, respectively. A *BlockDiagram* may contain *SystemBlocks* which in turn contain *Blocks*. A *ClassDiagram* may contain *Classes* and *Associations* that can connect *Classes* to each other. The middle part of the type graph shown in Fig. 1 expresses potential relations between elements of the source and target languages. A *TGG* is shown for the example in Fig. 2[3]. The axiom relates a *BlockDiagram* with a *ClassDiagram* via the axiom correspondence node. Rule 1 creates a *SystemBlock* and a corresponding *Class*. The *BlockDiagram* and *ClassDiagram* must already exist. Rule 2 creates a *Block* and a corresponding *Class* with an *Association*, connecting thereby the *Block* to the already existing *SystemBlock* and the *Class* with *Association* to the *SystemBlock*'s *Class*. Note that Rule 2 creates triple graphs with *correspondence nodes connected to two*

---

[2] This model transformation is a simplified version of a transformation used in the industrial case study on flexible production control systems [9].

[3] Note, that the types defined in Fig. 1 are abbreviated in Fig. 2. Elements that belong to the left- and right-hand side are drawn in black; elements that belong only to the right-hand side (i.e. which are created by the rule) are green and marked with '++'.

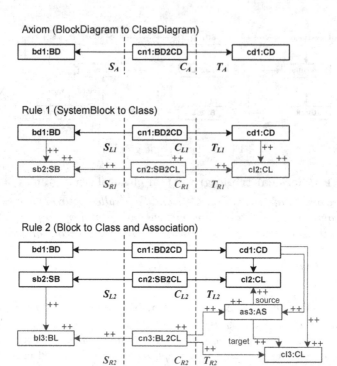

Fig. 2. Triple graph grammar

different target elements, expressing that a *Block* corresponds to two different elements *Class* and *Association* in the target domain. In particular, this does not correspond to the classical formalization, since we cannot define a proper graph morphism from the correspondence to the target component.

In other triple graph papers, such as [10,8,11], more TGG examples and case studies can be found not adhering to the classical triple graph formalization, showing that we need a triple graph formalization with *arbitrary relations between correspondence elements and source or target elements*. A correspondence node may be related to more than one source or target node, or even to none. Both cases cannot be adequately modeled with the current formalization.

Note that formalizing triple graphs as plain graph grammar with special typing instead of as morphism spans has also several practical benefits as witnessed by the choice of nearly all implementations:[4] TGG *implementations* that perform triple graph transformation as specially typed plain graph transformations are directly covered. Furthermore, it simplifies formal analysis [17,18] as concepts

---

[4] Fujaba's TGG Engine [12], MOFLON [13], the TGG Interpreter [8], the TGG Engine [4] and ATOM3 [14] all flatten the triple graphs to obtain a single typed graph containing all three components [15]. The only TGG implementation that we are aware of implementing the classical model directly [16] is based on Mathematica.

for the specification or analysis of typed graph transformations can be employed much easier. E.g., graph constraints [19] can be defined across triple graphs or TGGs can be directly equipped with OCL constraints defined over the same special meta-model [20] .

Summarizing, we propose a *new formalization* using a special type concept to define a formal model of triple graphs which has, as we will demonstrate, no disadvantages as the existing *theoretical results* for triple graph grammars and for plain graph transformation can be carried over but the following advantages: (i) It is *more flexible* than the classical model when it comes to expressing relations between correspondence and source or target elements as needed for various examples. (ii) The new model can be used *without* the need for *flattening as theoretical foundation* for implementations. (iii) The new model is *more suitable* for formal *verification* or *validation* of TGGs. Note, that we use the term "classical triple graph" when referring to the classical definition of triple graphs, while "triple graph" in the following is one w.r.t. our new definition.

3   Flexible Triple Graph Category

The main idea of our more flexible variant of triple graphs is to use a distinguished, fixed triple graph TRIPLE which all triple graphs are typed over. It defines three node types s, c, and t representing the source, correspondence, and target nodes, and corresponding edge types l_s and l_t for source and target graph edges. Moreover, for the connections from correspondence to source or target nodes the edge types e_{cs} and e_{ct} are available.

Definition 3 (Type graph TRIPLE). *The type graph* TRIPLE *is given by* $\text{TRIPLE}_N = \{s, c, t\}$, $\text{TRIPLE}_E = \{l_s, e_{cs}, e_{ct}, l_t\}$ *with source and target functions according to the following signature:*
$$\text{TRIPLE} \quad \overset{l_s}{\underset{}{\circlearrowright}} s \xleftarrow{\quad e_{cs} \quad} c \xrightarrow{\quad e_{ct} \quad} t \overset{l_t}{\circlearrowright}$$

We say that TRIPLE_S, TRIPLE_C, and TRIPLE_T, as shown below,
$$\text{TRIPLE}_S \quad \overset{l_s}{\circlearrowright} s \qquad \text{TRIPLE}_C \quad s \xleftarrow{\quad e_{cs} \quad} c \xrightarrow{\quad e_{ct} \quad} t \qquad \text{TRIPLE}_T \quad t \overset{l_t}{\circlearrowright}$$
are the *source, correspondence,* and *target component* of TRIPLE, respectively. Analogously to the aforementioned case, the projection of a graph G typed over TRIPLE to TRIPLE_S, TRIPLE_C, or TRIPLE_T selects the corresponding component.

We denote a triple graph as a combination of three indexed capitals, as for example $G = S_G C_G T_G$, where S_G denotes the *source* and T_G denotes the *target component* of G, while C_G denotes the *correspondence component*, being the smallest subgraph of G such that all c-nodes as well as all e_{cs}- and e_{ct}-edges are included in C_G. Note that C_G has to be a proper graph, i.e. all target nodes of e_{cs} and e_{ct}-edges have to be included.

Definition 4 (Triple graph). *A triple graph* $(G, triple_G)$ *is a graph* G *with source and target mappings* src *and* tgt, *resp., equipped with a morphism* $triple_G :$ $G \rightarrow \text{TRIPLE}$. *We denote a triple graph* $(G, triple_G)$ *as* $S_G C_G T_G$, *where* $S_G = G_{|\text{TRIPLE}_S}$, $C_G = G_{|\text{TRIPLE}_C} \setminus \{n \mid triple_G(n) \in \{s, t\}, \neg\exists e \in G_{|\text{TRIPLE}_C} : tgt(e) =$

$n\}$, and $T_G = G_{|\text{TRIPLE}_T}$. Consider triple graphs $S_G C_G T_G = (G, triple_G)$ and $S_H C_H T_H = (H, triple_H)$, a triple graph morphism $f : S_G C_G T_G \to S_H C_H T_H$ is a graph morphism $f : G \to H$ such that $triple_G = triple_H \circ f$. The category of triple graphs and triple graph morphisms is called **TripleGraphs**.

Using this notation we can define the restriction of a triple graph and a triple graph morphism to its source and target component.

Definition 5 (Restriction). *The restriction of a triple graph $S_H C_H T_H$ to a triple graph of its source (or target) component is defined by the pullback object $S_H \varnothing \varnothing$ ($\varnothing \varnothing T_H$) of $triple_H$ and $\text{TRIPLE}_S \to \text{TRIPLE}$ (TRIPLE$_T \to$ TRIPLE) as in diagram (1). Analogously, a triple morphism $f : S_G C_G T_G \to S_H C_H T_H$ can be restricted to $f^S : S_G \varnothing \varnothing \to S_H \varnothing \varnothing$ ($f^T : \varnothing \varnothing T_G \to \varnothing \varnothing T_H$) as induced morphism as shown in the diagram.*

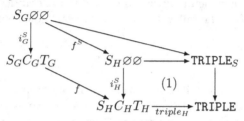

Fact 1. *Given the induced morphisms $i_G^S : S_G \varnothing \varnothing \to S_G C_G T_G$ and $i_H^S : S_H \varnothing \varnothing \to S_H C_H T_H$, for any triple morphism $f : S_G C_G T_G \to S_H C_H T_H$ we have that $f \circ i_G^S = i_H^S \circ f^S$. This holds analogously for the target component.*

Proof. This follows directly from the construction of the restriction as pullback.

Analogously to typed graphs, typed triple graphs are triple graphs typed over a distinguished triple graph, called type triple graph.

Definition 6 (Typed triple graph). *A type triple graph $S_{TT} C_{TT} T_{TT}$ is a distinguished triple graph. A typed triple graph $(S_G C_G T_G, type)$ is a triple graph $S_G C_G T_G$ equipped with a triple graph morphism type : $S_G C_G T_G \to S_{TT} C_{TT} T_{TT}$. Consider typed triple graphs $(S_G C_G T_G, type_G)$ and $(S_H C_H T_H, type_H)$, a typed triple graph morphism $f : (S_G C_G T_G, type_G) \to (S_H C_H T_H, type_H)$ is a triple graph morphism $f : S_G C_G T_G \to S_H C_H T_H$ such that $type_H \circ f = type_G$. The category of typed triple graphs and morphisms is called **TripleGraphs$_{TT}$**.*

In the remainder of this paper, we assume every triple graph $S_G C_G T_G$ and triple graph morphism f to be typed over $S_{TT} C_{TT} T_{TT}$, even if not explicitly mentioned.

Example 2 (Typed triple graph). The type triple graph $S_{TT} C_{TT} T_{TT}$ of our running example is depicted in Fig. 1. In particular, it is a triple graph, since it is typed over TRIPLE. This is visualized by the two dashed vertical lines, distinguishing at the left-hand side the source component S_{TT}, at the right-hand side the target component T_{TT}, and the correspondence edges crossing the dashed lines with incident source, correspondence, and target nodes as C_{TT}. All triple graphs typed over $S_{TT}C_{TT}T_{TT}$ via a triple morphism *type* are instances of $S_{TT}C_{TT}T_{TT}$ respecting the source, correspondence, and target components. As mentioned already in Example 1 and visible in the type triple graph in Fig. 1, a *Block* in the source domain may be connected via an incoming correspondence edge from a single correspondence node and two outgoing edges to two different elements *Class* and *Association* in the target domain. In particular, this does not correspond to the classical formalization, since it would not be possible to define a proper graph morphism mapping a correspondence node of type *BL2CL* to two different node types *Class* and *Association* in the target domain.

Definition 7 (Triple graph rule). *A triple graph rule* $p : S_L C_L T_L \overset{r}{\hookrightarrow} S_R C_R T_R$ *consists of a triple graph morphism* r, *which is an inclusion. The triple graphs* $S_L C_L T_L$ *and* $S_R C_R T_R$ *are called the left-hand side (LHS) and the right-hand side (RHS) of* p, *respectively.*

Definition 8 (Triple graph transformation). *Given a triple graph rule* $p : S_L C_L T_L \overset{r}{\hookrightarrow} S_R C_R T_R$ *and a triple graph* $S_G C_G T_G$, p *can be applied to* $S_G C_G T_G$ *if there is an occurrence of* $S_L C_L T_L$ *in* $S_G C_G T_G$ *i.e. a triple graph morphism* $m : S_L C_L T_L \to S_G C_G T_G$, *called* match. *In this case, a direct triple graph transformation* $S_G C_G T_G \overset{p,m}{\Longrightarrow} S_H C_H T_H$ *from* $S_G C_G T_G$ *to* $S_H C_H T_H$ *via* p *and* m *consists of the pushout (PO) in* **TripleGraphs$_{TT}$** *[21].*

$$
\begin{array}{ccc}
S_L C_L T_L & \overset{r}{\longrightarrow} & S_R C_R T_R \\
{\scriptstyle m}\downarrow & (PO) & \downarrow{\scriptstyle n} \\
S_G C_G T_G & \underset{h}{\longrightarrow} & S_H C_H T_H
\end{array}
$$

Since pushouts along inclusion morphisms in **TripleGraphs$_{TT}$** *always exist, (PO) can always be constructed. In particular,* h *can be chosen to be an inclusion analogous to* r. *A triple graph transformation, denoted as* $S_{G_0} C_{G_0} T_{G_0} \overset{*}{\Rightarrow} S_{G_n} C_{G_n} T_{G_n}$, *is a sequence* $S_{G_0} C_{G_0} T_{G_0} \Rightarrow S_{G_1} C_{G_1} T_{G_1} \Rightarrow \cdots \Rightarrow S_{G_n} C_{G_n} T_{G_n}$ *of direct triple graph transformations.*

Remark 1. Note that pushouts in **TripleGraphs$_{TT}$** are constructed componentwise in the node and edge components, since the involved objects are normal graphs. The *triple* morphism of the pushout object is uniquely induced by the pushout property.

Definition 9 (Triple graph grammar). *A triple graph grammar (TGG)* $GR = (\mathcal{R}, S_A C_A T_A, S_{TT} C_{TT} T_{TT})$ *consists of a set of triple graph rules* $\mathcal{R}$ *and a triple start graph* $S_A C_A T_A$, *called axiom, typed over the type triple graph* $S_{TT} C_{TT} T_{TT}$.

Example 3 (TGG). In Example 1, we already explained our running example TGG (see Fig. 2). The TGG rules are indeed typed over the type triple graph depicted in Fig. 1.

Remark 2 (Formal foundations for implementation). Note that typing a graph over a type triple graph already defines the triples, i.e. any typed graph $(G, type_G)$ and any typed graph morphism $f : (G, type_G) \rightarrow (H, type_H)$ typed over the type triple graph $S_{TT}C_{TT}T_{TT}$ correspond uniquely to a typed triple graph $(S_GC_GT_G, type_G)$ and a typed triple graph morphism $f : (S_GC_GT_G, type_G) \rightarrow (S_HC_HT_H, type_H)$, respectively. Accordingly, if all morphisms in a pushout (PO) in **Graphs** are typed triple graph morphisms, then (PO) is also a pushout in **TripleGraphs** and **TripleGraphs**$_{TT}$. Concluding, regular graph transformation tools can be used to perform (typed) triple graph transformations, since a regular graph transformation can be interpreted as a (typed) triple graph transformation if the rule, match, and graph to be transformed are proper (typed) triple graphs and morphisms.

Remark 3 (Carry over HLR results). By construction, **TripleGraphs**$_{TT}$ is the slice category **TripleGraphs**/TT of the slice category **Graphs**/TRIPLE. Consequently, it follows from [19,22] that the category **TripleGraphs**$_{TT}$ together with the class $\mathcal{M}$ of all monomorphisms forms an $\mathcal{M}$-adhesive category. $\mathcal{M}$-adhesive categories [22] and their variants like adhesive [23] and weak adhesive HLR [19] categories form a framework for graph transformations and their analysis. They are based on a distinguished morphism class $\mathcal{M}$, which is used for the rule morphisms and has to satisfy certain properties, and a special van Kampen property describing the compatibility of gluings and restrictions, i.e. pushouts and pullbacks in categorical terms. Based on this van Kampen property, many results can be proven that hold for different kinds of graphs. In particular, as we will see in Section 4, we will use the Local Church–Rosser and Concurrency Theorems to prove the Composition and Decomposition result for TGGs with flexible correspondences.

4   Classical TGG Results in Flexible Triple Graph Category

For classical triple graphs, *operational transformation rules* from relational TGG rules can be derived as introduced in [1]. Analogously, we define them for our triple graphs. For the forward transformation, all elements belonging to the source domain that were previously created are added to the LHS of the rule. Reversely, a backward rule translates target elements to source elements. A correspondence rule specifies how source and target elements can be connected to each other according to the TGG.

Definition 10 (Operational rules). *Given a triple graph rule $p : S_LC_LT_L \xrightarrow{r} S_RC_RT_R$, construct the pushout (1) over $i_L^S : S_L\varnothing\varnothing \rightarrow S_LC_LT_L$ and $r^S : S_L\varnothing\varnothing \rightarrow S_R\varnothing\varnothing$ with pushout object $S_RC_LT_L$. Since $r \circ i_L^S = i_R^S \circ r^S$ (Fact 1), we obtain from the pushout property the morphism r_{CT}.*

From this construction, we define the source rule $p^S : S_L\varnothing\varnothing \xrightarrow{r^S} S_R\varnothing\varnothing$ and the forward rule $p^F : S_RC_LT_L \xrightarrow{r_{CT}} S_RC_RT_R$.

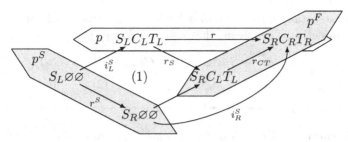

Vice versa, we can construct the target rule $p^T : \varnothing\varnothing T_L \to \varnothing\varnothing T_R$ *and the* backward rule $p^B : S_L C_L T_R \overset{r_{SC}}{\Rightarrow} S_R C_R T_R$. *The* correspondence rule $p^C : S_R C_L T_R \overset{r_C}{\Rightarrow}$ $S_R C_R T_R$ *is obtained by applying this construction twice.*

Example 4. We can derive the forward rule $S_{R2} C_{L2} T_{L2} \to S_{R2} C_{R2} T_{R2}$ of Rule 2 in our running example as depicted in Fig. 3.

Triple graph rules can be understood as a concurrent application of source rules and forward rules, or of target rules and backward rules, or of the parallel application of source and target rules with correspondence rules. A concurrent rule $p_1 *_E p_2$ consists of a sequence of two rules via an overlapping triple graph E of the RHS of the first rule and LHS of the second rule. For the definition of concurrent, parallel and sequentially independent rule applications we refer to [19] where these notions are defined in the context of $\mathcal{M}$-adhesive categories (see also Remark 3). This leads to the main result for triple graph grammars that each triple graph transformation sequence can be decomposed into a source transformation generating the source component followed by a forward transformation translating the source component into its target component, and the other way round.

Fact 2. *Given a triple graph rule* $p : S_L C_L T_L \to S_R C_R T_R$, *then*
(i) $p = p^S *_{S_R C_L T_L} p^F$, *(ii)* $p = p^T *_{S_L C_L T_R} p^B$, *and (iii)* $p = (p^S + p^T) *_{S_R C_L T_R} p^C$.

Proof. Consider the following proof for the source and forward rule, the other decompositions follow analogously. In the following diagram, (1) is the pushout

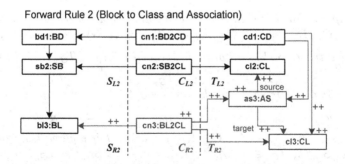

Fig. 3. Type triple graph

from the construction of the operational rules, (2) is also a pushout, and (e, id) are jointly surjective. This means that $p = p^S *_{S_R C_L T_L} p^F$.

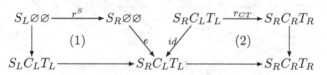

Fact 3. *Whenever a forward transformation is followed by a source transformation, these transformations are sequentially independent.*

Proof. Consider the following diagram, where first a forward transformation $S_G C_G T_G \overset{p_1^F}{\Longrightarrow} S_{G'} C_{G'} T_{G'}$ and then a source transformation $S_{G'} C_{G'} T_{G'} \overset{p_2^S}{\Longrightarrow} S_H C_H T_H$ is applied. Since the forward transformation does not change the source component of G we have $S_G = S_{G'}$ and define $i(x) = m_2(x)$ for all $x \in S_{L_2}$. Since f is an inclusion, the triangle commutes, which shows sequential independence.

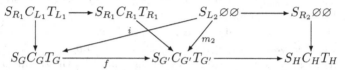

Definition 11 (Source and Match Consistency). *Consider a sequence $(p_i)_{i=1,...,n}$ of triple graph rules leading to corresponding sequences $(p_i^S)_{i=1,...,n}$ and $(p_i^F)_{i=1,...,n}$ of source and forward rules. A triple graph transformation sequence $S_{G_0} C_{G_0} T_{G_0} \xrightarrow{(p_i^S)_{i=1,...,n}} S_{G_n} C_{G_0} T_{G_0} \xrightarrow{(p_i^F)_{i=1,...,n}} S_{G_n} C_{G_n} T_{G_n}$ via first $p_1^S, \ldots, p_n^S$ and then $p_1^F, \ldots, p_n^F$ with matches m_i^S and m_i^F and co-matches n_i^S and n_i^F, respectively, is* match consistent *if the source component of the match m_i^F is uniquely defined by the co-match n_i^S.*

A triple graph transformation $S_{G_n} C_{G_0} T_{G_0} \xrightarrow{(p_i^F)_{i=1,...,n}} S_{G_n} C_{G_n} T_{G_n}$ is called source consistent *if there exists a match consistent sequence as above.*

Theorem 4 (Decomposition and Composition Result). *For triple graph transformation sequences the following holds for triple rules $p_1, \ldots, p_n$ with corresponding source rules $p_1^S, \ldots, p_n^S$ and forward rules $p_1^F, \ldots, p_n^F$:*

1. **Decomposition:** *For each triple graph transformation sequence $S_{G_0} C_{G_0} T_{G_0} \overset{p_1}{\Longrightarrow} S_{G_1} C_{G_1} T_{G_1} \Rightarrow \ldots \overset{p_n}{\Longrightarrow} S_{G_n} C_{G_n} T_{G_n}$ there is a corresponding match consistent triple graph transformation sequence $S_{G_0} C_{G_0} T_{G_0} \overset{p_1^S}{\Longrightarrow} S_{G_1} C_{G_0} T_{G_0}$ $\Rightarrow \ldots \overset{p_n^S}{\Longrightarrow} S_{G_n} C_{G_0} T_{G_0} \overset{p_1^F}{\Longrightarrow} S_{G_n} C_{G_1} T_{G_1} \Rightarrow \ldots \overset{p_n^F}{\Longrightarrow} S_{G_n} C_{G_n} T_{G_n}$.*

2. **Composition:** *For each match consistent triple graph transformation sequence $S_{G_0} C_{G_0} T_{G_0} \overset{p_1^S}{\Longrightarrow} S_{G_1} C_{G_0} T_{G_0} \Rightarrow \ldots \overset{p_n^S}{\Longrightarrow} S_{G_n} C_{G_0} T_{G_0} \overset{p_1^F}{\Longrightarrow}$ $S_{G_n} C_{G_1} T_{G_1} \Rightarrow \ldots \overset{p_n^F}{\Longrightarrow} S_{G_n} C_{G_n} T_{G_n}$ there is a triple graph transformation sequence $S_{G_0} C_{G_0} T_{G_0} \overset{p_1}{\Longrightarrow} S_{G_1} C_{G_1} T_{G_1} \Rightarrow \ldots \overset{p_n}{\Longrightarrow} S_{G_n} C_{G_n} T_{G_n}$.*

3. **Bijective Correspondence:** *Composition and Decomposition are inverse to each other.*

Proof. The proof is similar to that in [3], by substituting classical triple graphs by our new ones. The basic idea is to use the Concurrency Theorem [19] and Fact 2 to split the triple rules and the Local Church–Rosser Theorem [19] to commute sequentially independent transformation steps.

Decomposition: We prove this by induction on the number n of triple graph rule applications. For $n = 1$, as a consequence of Fact 2 and the Concurrency Theorem, a direct triple graph transformation $S_{G_0} C_{G_0} T_{G_0} \xRightarrow{p_1} S_{G_1} C_{G_1} T_{G_1}$ can be uniquely split up into $S_{G_0} C_{G_0} T_{G_0} \xRightarrow{p_1^S} S_{G_1} C_{G_0} T_{G_0} \xRightarrow{p_1^F} S_{G_1} C_{G_1} T_{G_1}$, which is obviously match consistent.

For arbitrary n and the triple graph transformation $S_{G_0} C_{G_0} T_{G_0} \xRightarrow{(p_i)_{i=1,\dots,n}} S_{G_n} C_{G_n} T_{G_n}$, suppose we have a corresponding match consistent transformation sequence $S_{G_1} C_{G_1} T_{G_1} \xRightarrow{p_2^S} S_{G_2} C_{G_1} T_{G_1} \Rightarrow \dots \xRightarrow{p_n^S} S_{G_n} C_{G_0} T_{G_0} \xRightarrow{p_1^F} S_{G_n} C_{G_1} T_{G_1} \Rightarrow \dots \xRightarrow{p_n^F} S_{G_n} C_{G_n} T_{G_n}$. Using the same argument as above, the triple graph transformation $S_{G_0} C_{G_0} T_{G_0} \xRightarrow{p_1} S_{G_1} C_{G_1} T_{G_1}$ can be split up into a match consistent sequence $S_{G_0} C_{G_0} T_{G_0} \xRightarrow{p_1^S} S_{G_1} C_{G_0} T_{G_0} \xRightarrow{p_1^F} S_{G_1} C_{G_1} T_{G_1}$. Now the applications of p_1^F and p_2^S are sequentially independent (Fact 3). This means that the two transformations can be switched leading to square (1) in Fig. 4. Iterating this construction leads to the desired triple graph transformation $S_{G_0} C_{G_0} T_{G_0} \xRightarrow{p_1^S} S_{G_1} C_{G_0} T_{G_0} \Rightarrow \dots \xRightarrow{p_n^S} S_{G_n} C_{G_0} T_{G_0} \xRightarrow{p_1^F} S_{G_n} C_{G_1} T_{G_1} \Rightarrow \dots \xRightarrow{p_n^F} S_{G_n} C_{G_n} T_{G_n}$ which can be shown to be match consistent.

Composition: Vice versa, each match consistent transformation sequence implies that the corresponding subsequences are sequentially independent, such that they can be shifted using again the Local Church–Rosser Theorem. Moreover, the corresponding source and forward rule applications are E-related such that the application of the Concurrency Theorem leads to the required transformation.

Bijective Correspondence: The bijective correspondence of composition and decomposition is a direct consequence of the bijective correspondence in the Local Church–Rosser and the Concurrency Theorem.   □

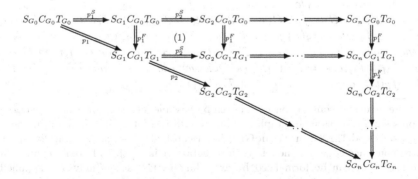

Fig. 4. Decomposition and Composition

5   Comparison of Classical and Flexible Triple Graphs

In this section, we show that the classical theoretical model for triple graphs is a special case of the new model after flattening. In particular, we limit the formal comparison to the full subcategory **CTripleGraphs$_{Dis}$** of **CTripleGraphs**, where triple graphs have a discrete correspondence component, consisting only of nodes and not of edges. The flattening construction for classical triple graphs [15] is limited to the case without edges in the correspondence component as well. Thus, we show that our new model is more flexible and more directly suitable for implementation than the flattened model obtained as described in [15].[5]

We define a flattening functor from **CTripleGraphs$_{Dis}$** to **TripleGraphs$_{Sub}$**, being the full subcategory of **TripleGraphs** where triple graphs have correspondence nodes with exactly one outgoing edge to some source and target node, respectively. Note that **TripleGraphs$_{Sub}$** is a *real* subcategory of **TripleGraphs** – as illustrated by our running example (see Ex. 1) – because in **TripleGraphs** triple graphs with multiple or zero outgoing edges from correspondence nodes are allowed, which is not allowed in **TripleGraphs$_{Sub}$**.

The *flattening functor* translates the source and target component from classical triple graphs directly to source and target components in the new model. The correspondence component together with the morphism span into source and target component of the classical model is translated into a correspondence component in the new model, where the morphism span is flattened into special correspondence edges connecting source nodes and target nodes via correspondence nodes as prescribed by the morphism span.

Definition 12 (Flattening functor). *The* flattening functor $\mathcal{F} = (\mathcal{F}_{Ob}, \mathcal{F}_{Mor})$: **CTripleGraphs$_{Dis}$** → **TripleGraphs$_{Sub}$** *is defined as follows:*

Given a classical triple graph $G = (G_S \xleftarrow{s_G} G_C \xrightarrow{t_G} G_T)$, *where w.l.o.g.* G_S, G_C, *and* G_T *are disjoint, then* $\mathcal{F}_{Ob}(G) = SCT$, *where* $S = G_S$, $T = G_T$, $C_V = G_C \cup s_G(G_C) \cup t_G(G_C)$ *the set of nodes in* C, $C_E = \{(c,s) | c \in G_C \wedge s \in G_S \wedge s_G(c) = s\} \cup \{(c,t) | c \in G_C \wedge t \in G_T \wedge t_G(c) = t\}$ *the set of edges in* C *and source and target mappings for* C *induced by the edge names,* $src_C(c,s) = c$ *and* $tgt_C(c,s) = s$, $src_C(c,t) = c$ *and* $tgt_C(c,t) = t$, *respectively.*

Given a classical triple graph morphism $f : G_1 \to G_2$ *with* $f = (f_S : G_{1S} \to G_{2S}, f_C : G_{1C} \to G_{2C}, f_T : G_{1T} \to G_{2T})$, *then* $\mathcal{F}_{Mor}(f) : \mathcal{F}_{Ob}(G_1) \to \mathcal{F}_{Ob}(G_2)$ *with* $(\mathcal{F}_{Mor}(f))(n) = f_S(n)$ *for each node* n *in* $S_1 = G_{1S}$, $(\mathcal{F}_{Mor}(f))(n) = f_T(n)$ *for each node* n *in* $T_1 = G_{1T}$, $(\mathcal{F}_{Mor}(f))(n) = f_C(n)$ *for each* n *in* $G_{1C} \subseteq C_{1,V}$, $(\mathcal{F}_{Mor}(f))(c,s) = (f_C(c), f_S(s))$ *for each edge* (c,s), *and* $(\mathcal{F}_{Mor}(f))(c,t) = (f_C(c), f_T(t))$ *for each edge* (c,t) *in* $\mathcal{F}_{Ob}(G_1)$.

---

[5] A more extensive comparison would even be possible following the same principle of expressing TGGs as plain graph grammars with special typing. It would be necessary to extend TRIPLE in **TripleGraphs**, accordingly. For example, in [24] special correspondence edges for model synchronization or in [4] special bookkeeping edges are used that can be formalized by extending TRIPLE accordingly. For simplicity reasons, we restrict to the most common variant with discrete correspondence component here.

Using this flattening functor, we can show the equivalence of the categories **TripleGraphs$_{\text{Sub}}$** and **CTripleGraphs$_{\text{Dis}}$**.

Theorem 5 (Equivalence of categories). *The categories* **TripleGraphs$_{\text{Sub}}$** *and* **CTripleGraphs$_{\text{Dis}}$** *are equivalent.*

Proof idea. There exists an inverse functor $\mathcal{G} = (\mathcal{G}_{Ob}, \mathcal{G}_{Mor}) :$ **TripleGraphs$_{\text{Sub}}$** $\to$ **CTripleGraphs$_{\text{Dis}}$**. Given a triple graph $S_G C_G T_G$, then $\mathcal{G}(S_G C_G T_G) = G$ with $G = (G_S \overset{s_G}{\leftarrow} G_C \overset{t_G}{\to} G_T)$, where $G_S = S_G$, $G_C = \{v \in C_G \mid triple_G(v) = c\}$, and $G_T = T_G$. Moreover, $s_G(v) = w$ and $t_G(v) = w'$ for the unique edges that exist from v to the source node w and the target node w' as a prerequisite of $S_G C_G T_G$ being in the subcategory. For a triple graph morphism $f : S_G C_G T_G \to S_H C_H T_H$, $\mathcal{G}(f) = (f_S, f_C, f_T)$ with f_S, f_C, f_T being the restrictions of f to the domain.

Basically, $\mathcal{G}$ translates source and target components and all nodes of the correspondence component of the new model to source, correspondence, and target components of the classical model in a straightforward way. Moreover, correspondence edges from correspondence to source or target nodes are translated into a morphism span in the classical model. Thus, we find natural isomorphisms from both $\mathcal{G} \circ \mathcal{F}$ and $\mathcal{F} \circ \mathcal{G}$ to the identity functor.

This means that **CTripleGraphs$_{\text{Dis}}$** and **TripleGraphs$_{\text{Sub}}$** are *equivalent categories*, demonstrating that the classical theoretical model for triple graphs is a special case of the new model after flattening.

From this result, it also follows that each triple graph transformation in the classical model can be flattened to a corresponding triple graph transformation in our new model.

Corollary 1 (Transformation preservation). *Given a classical triple graph transformation $G \overset{p,m}{\Longrightarrow} H$ consisting of a pushout in* **CTripleGraphs$_{\text{Dis}}$***, there is a corresponding flexible triple graph transformation $\mathcal{F}(G) \overset{\mathcal{F}(p),\mathcal{F}(m)}{\Longrightarrow} \mathcal{F}(H)$ consisting of the pushout flattened via $\mathcal{F}$ in* **TripleGraphs***.*

Proof. This follows directly from the equivalence of the categories **CTripleGraphs$_{\text{Dis}}$** and **TripleGraphs$_{\text{Sub}}$** (Theorem 5), with **TripleGraphs$_{\text{Sub}}$** $\subseteq$ **TripleGraphs**.

Remark 4. Obviously, the reverse direction of Corollary 1 only holds for triple graph transformations in **TripleGraphs$_{\text{Sub}}$**, but not for general ones. A counterexample are triple graph transfomations using the triple rules from Example 1, which do not have a correspondence in classical triple graphs.

6 Conclusion and Future Work

We have presented a new formalization for TGGs as *plain graph grammars with special typing* more suitable for implementation than the classical formalization for TGGs and at the same time also more flexible with respect to specifying relationships between source and target elements. We have proven that the main decomposition and composition result can be carried over from the classical to the new formal TGG model. Moreover, we have presented a formal comparison with the classical model demonstrating that it is actually a special case of the new one, allowing for cross-fertilization between both approaches and the further development of new TGG theory nearer to implementation based on the new formal, flattened and more flexible model.

There exist different approaches to attributed graph transformation [19,25]. It is part of future work to equip our new TGG model with attribution and derive the corresponding important TGG results for triple graphs with attribution as described in one of these approaches and started in [26]. Another future goal is to generalize more recent and specific TGG results like correctness, completeness, consistency, and functional behavior, as presented, for example, in [5,3,6,4], to the new triple graph model in order to bridge the gap between formal semantics and corresponding implementations with different bookkeeping mechanisms. Moreover, the new model and corresponding theory should be generalized to TGGs with application conditions as presented already, for example, in [7,6].

References

1. Schürr, A.: Specification of Graph Translators with Triple Graph Grammars. In: Mayr, E.W., Schmidt, G., Tinhofer, G. (eds.) WG 1994. LNCS, vol. 903, pp. 151–163. Springer, Heidelberg (1995)
2. König, A., Schürr, A.: Tool Integration with Triple Graph Grammars - A Survey. ENTCS 148(1), 113–150 (2006)
3. Ehrig, H., Ehrig, K., Ermel, C., Hermann, F., Taentzer, G.: Information Preserving Bidirectional Model Transformations. In: Dwyer, M.B., Lopes, A. (eds.) FASE 2007. LNCS, vol. 4422, pp. 72–86. Springer, Heidelberg (2007)
4. Giese, H., Hildebrandt, S., Lambers, L.: Bridging the gap between formal semantics and implementation of triple graph grammars - ensuring conformance of relational model transformation specifications and implementations. Software and Systems Modeling (2012)
5. Ehrig, H., Ermel, C., Hermann, F., Prange, U.: On-the-Fly Construction, Correctness and Completeness of Model Transformations Based on Triple Graph Grammars. In: Schürr, A., Selic, B. (eds.) MODELS 2009. LNCS, vol. 5795, pp. 241–255. Springer, Heidelberg (2009)
6. Klar, F., Lauder, M., Königs, A., Schürr, A.: Extended Triple Graph Grammars with Efficient and Compatible Graph Translators. In: Engels, G., Lewerentz, C., Schäfer, W., Schürr, A., Westfechtel, B. (eds.) Nagl Festschrift. LNCS, vol. 5765, pp. 141–174. Springer, Heidelberg (2010)
7. Golas, U., Ehrig, H., Hermann, F.: Formal Specification of Model Transformations by Triple Graph Grammars with Application Conditions. ECEASST 39 (2011)
8. Greenyer, J., Kindler, E.: Comparing relational model transformation technologies: implementing Query/View/Transformation with Triple Graph Grammars. Software and Systems Modeling 9(1), 21–46 (2010)
9. Schäfer, W., Wagner, R., Gausemeier, J., Eckes, R.: An Engineer's Workstation to Support Integrated Development of Flexible Production Control Systems.

In: Ehrig, H., Damm, W., Desel, J., Große-Rhode, M., Reif, W., Schnieder, E., Westkämper, E. (eds.) INT 2004. LNCS, vol. 3147, pp. 48–68. Springer, Heidelberg (2004)

10. Kindler, E., Rubin, V., Wagner, R.: An Adaptable TGG Interpreter for In-Memory Model Transformation. In: Proceedings of Fujaba Days 2004, pp. 35–38 (2004)

11. Guerra, E., de Lara, J.: Model View Management with Triple Graph Transformation Systems. In: Corradini, A., Ehrig, H., Montanari, U., Ribeiro, L., Rozenberg, G. (eds.) ICGT 2006. LNCS, vol. 4178, pp. 351–366. Springer, Heidelberg (2006)

12. Burmester, S., Giese, H., Niere, J., Tichy, M., Wadsack, J.P., Wagner, R., Wendehals, L., Zündorf, A.: Tool Integration at the Meta-Model Level within the FUJABA Tool Suite. Software Tools for Technology Transfer 6(3), 203–218 (2004)

13. Amelunxen, C., Klar, F., Königs, A., Rötschke, T., Schürr, A.: Metamodel-Based Tool Integration with MOFLON. In: Proceedings of ICSE 2008, pp. 807–810. ACM (2008)

14. de Lara, J., Vangheluwe, H.: Using atom^3 as a meta-case tool. In: Proceedings of ICEIS 2002, pp. 642–649 (2002)

15. Ehrig, H., Ermel, C., Hermann, F.: On the Relationship of Model Transformations based on Triple and Plain Graph Grammars. In: Proceedings of GRaMoT 2008, pp. 9–16. ACM (2008)

16. Brandt, C., Hermann, F., Engel, T.: Security and Consistency of IT and Business Models at Credit Suisse Realized by Graph Constraints, Transformation and Integration Using Algebraic Graph Theory. In: Halpin, T., Krogstie, J., Nurcan, S., Proper, E., Schmidt, R., Soffer, P., Ukor, R. (eds.) BPMDS 2009 and EMMSAD 2009. LNBIP, vol. 29, pp. 339–352. Springer, Heidelberg (2009)

17. Giese, H., Lambers, L.: Towards Automatic Verification of Behavior Preservation for Model Transformation via Invariant Checking. In: Ehrig, H., Engels, G., Kreowski, H.-J., Rozenberg, G. (eds.) ICGT 2012. LNCS, vol. 7562, pp. 249–263. Springer, Heidelberg (2012)

18. Giese, H., Hildebrandt, S., Lambers, L.: Toward Bridging the Gap Between Formal Semantics and Implementation of Triple Graph Grammars. In: Proceedings of MoDeVVa 2010, pp. 19–24. IEEE (2010)

19. Ehrig, H., Ehrig, K., Prange, U., Taentzer, G.: Fundamentals of Algebraic Graph Transformation. EATCS Monographs. Springer (2006)

20. Cabot, J., Clarisó, R., Guerra, E., Lara, J.: Verification and validation of declarative model-to-model transformations through invariants. J. Syst. Softw. 83(2), 283–302 (2010)

21. Corradini, A., Montanari, U., Rossi, F., Ehrig, H., Heckel, R., Löwe, M.: Algebraic Approaches to Graph Transformation I: Basic Concepts and Double Pushout Approach. In: Rozenberg, G. (ed.) Handbook of Graph Grammars and Computing by Graph Transformation. Foundations, vol. 1, pp. 163–245. World Scientific (1997)

22. Ehrig, H., Golas, U., Hermann, F.: Categorical Frameworks for Graph Transformation and HLR Systems based on the DPO Approach. BEATCS 102, 111–121 (2010)

23. Lack, S., Sobociński, P.: Adhesive Categories. In: Walukiewicz, I. (ed.) FOSSACS 2004. LNCS, vol. 2987, pp. 273–288. Springer, Heidelberg (2004)

24. Giese, H., Wagner, R.: From model transformation to incremental bidirectional model synchronization. Software and Systems Modeling 8(1), 21–43 (2009)

25. Orejas, F., Lambers, L.: Delaying Constraint Solving in Symbolic Graph Transformation. In: Ehrig, H., Rensink, A., Rozenberg, G., Schürr, A. (eds.) ICGT 2010. LNCS, vol. 6372, pp. 43–58. Springer, Heidelberg (2010)

26. Lambers, L., Hildebrandt, S., Giese, H., Orejas, F.: Attribute Handling for Bidirectional Model Transformations: The Triple Graph Grammar Case. ECEASST (to appear, 2012)

Graph Transformation
with Focus on Incident Edges*

Dominique Duval, Rachid Echahed, and Frédéric Prost

University of Grenoble
B. P. 53, F-38041 Grenoble, France
{Dominique.Duval,Rachid.Echahed,Frederic.Prost}@imag.fr

Abstract. We tackle the problem of graph transformation with particular focus on node cloning. We propose a new approach to graph rewriting, called *polarized node cloning*, where a node may be cloned together with either all its incident edges or with only its outgoing edges or with only its incoming edges or with none of its incident edges. We thus subsume previous works such as the sesqui-pushout, the heterogeneous pushout and the adaptive star grammars approaches. We first define polarized node cloning algorithmically, then we propose an algebraic definition. We use polarization annotations to declare how a node must be cloned. For this purpose, we introduce the notion of polarized graphs as graphs endowed with some annotations on nodes and we define graph transformations with polarized node cloning by means of sesqui-pushouts in the category of polarized graphs.

1   Introduction

Graph transformation [22,11,13] extends string rewriting [3] and term rewriting [1] in several respects. In the literature, there are many ways to define graphs and graph rewriting. The proposed approaches can be gathered in two main streams: (i) the algorithmic approaches, which define a graph rewrite step by means of the algorithms involved in the implementation of graph transformation (see e.g. [2,10]); (ii) the second stream consists of the algebraic approaches, first proposed in the seminal paper [14], and which use categorical constructs to define graph transformation in an abstract way. The most popular algebraic approaches are the double pushout (DPO) [14,5] and the single pushout (SPO) [21,16,17,12].

In this paper we are interested in graph transformation with particular focus on *node cloning*. Indeed, making copies of values is a very useful feature shared by most popular programming languages (see for instance the so-called shallow cloning [15] or deep cloning [20] operations). Informally, by cloning a node n, we mean making zero, one or more copies of n with "some" of its incident edges. The classical DPO and SPO approaches of graph transformation are clearly not well suited to perform cloning of nodes. As far as we are aware of, there are two algebraic attempts to deal with node cloning : the sesqui-pushout approach

---

* This work has been funded by the project CLIMT of the French *Agence Nationale de la Recherche* (ANR-11-BS02-016).

H. Ehrig et al.(Eds.): ICGT 2012, LNCS 7562, pp. 156–171, 2012.

(SqPO) [4] and the heterogeneous pushout approach (HPO) [7]. The sesqui-pushout approach has the ability to clone nodes with all their incident edges whereas the HPO clones a node only with its outgoing edges. Our aim in this paper is to investigate a new flexible way to perfom node cloning, so that every copy of a node n can be made either with all the incident edges (denoted hereafter $n^\pm$), with only its outgoing edges (n^+), with only its incoming edges (n^-), or without any of its incident edges (denoted simply as n). We call this kind of graph transformation *polarized node cloning*. To achieve this task, we introduce the notion of *polarized graphs*. Informally, we define a polarized graph $\mathbb{X}$ as a graph X where each node n is annotated as $n^\pm$, n^+, n^- or just n. The rules in our approach are made of a polarized graph $\mathbb{K}$, consisting of a graph K with annotated nodes, and a span of graphs $L \xleftarrow{l} K \xrightarrow{r} R$. The annotations of $\mathbb{K}$ indicate the cloning strategy of incident edges. We prove that the polarized node cloning can be described as a SqPO rewriting of polarized graphs, preceded by the polarization of every node n in the left hand side as $n^\pm$ and followed by forgetting all polarizations in the right hand side. This is called the *polarized sesqui-pushout* rewriting (PSqPO for short).

The paper is organized as follows. The notion of polarized node cloning of graphs is defined in an elementary algorithmic way in Section 2. In Section 3 we define polarized graphs and the corresponding sesqui-pushout rewriting, from which we get the polarized sesqui-pushout rewriting for the polarized node cloning of graphs. Our approach is adapted to labeled graphs and illustrated through some examples in Section 4. A comparison with related work is made in Section 5 and concluding remarks are given in Section 6. An Appendix is added in order to ease the verification of the accuracy of our results. Detailed proofs can be found in [8]. We use categorical notions which may be found for instance in [19].

2   Polarized Node Cloning of Graphs

In this section we introduce some notations involving graphs and define the notion of *polarized node cloning*.

2.1   Graphs

Definition 1. *A graph X is made of a set of nodes $|X|$, a set of edges $X_\rightarrow$ and two functions* source *and* target *from $X_\rightarrow$ to $|X|$. An edge e with source n and target p is denoted $n \xrightarrow{e} p$. The set of edges from n to p in X is denoted $X_{n \rightarrow p}$. A morphism of graphs $f : X \rightarrow Y$ is made of two functions (both denoted f) $f : |X| \rightarrow |Y|$ and $f : X_\rightarrow \rightarrow Y_\rightarrow$, such that $f(n) \xrightarrow{f(e)} f(p)$ for each edge $n \xrightarrow{e} p$. This provides the category $\mathbf{Gr}$ of graphs.*

In order to build large graphs from smaller ones, we will use the *sum* of graphs and the *edge-sum* for adding edges to a graph, as defined below using the symbol $+$ for the coproduct in the category of sets, i.e., the disjoint union of sets.

Definition 2. *Given two graphs X_1 and X_2, the* sum *X_1+X_2 is the coproduct of X_1 and X_2 in the category of graphs, which means that $|X_1+X_2| = |X_1|+|X_2|$ and $(X_1 + X_2)_\rightarrow = X_{1\rightarrow}+X_{2\rightarrow}$ and the source and target functions for X_1+X_2 are induced by the source and target functions for X_1 and for X_2. Given two graphs X and E such that $|E| \subseteq |X|$, the* edge-sum *$X +_e E$ is the pushout, in the category of graphs, of X and E over their common subgraph made of the nodes of E and no edge. This means that $|X +_e E| = |X|$ and $(X +_e E)_\rightarrow = X_\rightarrow + E_\rightarrow$ and the source and target functions for $X +_e E$ are induced by the source and target functions for X and for E.*

Clearly, the precise set of nodes of E does not matter in the construction of $X +_e E$, as long as it contains the source and target of every edge of E and is contained in $|X|$. This notation is extended to morphisms: let $f_1 : X_1 \rightarrow Y_1$ and $f_2 : X_2 \rightarrow Y_2$, then $f_1 + f_2 : X_1+X_2 \rightarrow Y_1 + Y_2$ is defined piecewise from f_1 and f_2. Similarly, let $f : X \rightarrow Y$ and $g : E \rightarrow F$ with $|E| \subseteq |X|$ and $|F| \subseteq |Y|$, then $f +_e g : X +_e E \rightarrow Y +_e F$ is defined as f on the nodes and piecewise from f and g on the edges.

Remark 1. Let X be a subgraph of a graph Y. Let $\overline{X}$ denote the subgraph of Y induced by the nodes outside $|X|$ and $\widetilde{X}$ the subgraph of Y induced by the edges which are neither in X nor in $\overline{X}$, that is, the edges that are incident to a node (at least) in X but do not belong to X. For all nodes n, p in Y let $\widetilde{X}_{n\rightarrow p}$ denote the subgraph of Y induced by the edges from n to p in $\widetilde{X}$ (so that $\widetilde{X}_{n\rightarrow p}$ is empty whenever both n and p are in $\overline{X}$). Then Y can be expressed as $Y = (X + \overline{X}) +_e \widetilde{X}$ with $\widetilde{X}_\rightarrow = \sum_{n\in|Y|,p\in|Y|} \widetilde{X}_{n\rightarrow p}$ which can also be written as $|Y| = |X| + |\overline{X}|$ and $Y_\rightarrow = X_\rightarrow + \overline{X}_\rightarrow + \sum_{n\in|Y|,p\in|Y|} \widetilde{X}_{n\rightarrow p}$

Definition 3. *A* matching *of graphs is a monomorphism of graphs. Given a matching $m : L \rightarrow G$, the nodes and edges in $m(L)$ are called the* matching nodes *and the* matching edges, *respectively.*

Thus, a morphism of graphs is a matching if and only if it is *injective*, in the sense that both underlying functions (on nodes and on edges) are injections. So, up to isomorphism, every matching of graphs is an inclusion. For simplicity of notations, we now assume that all matchings of graphs are inclusions.

2.2 Polarized Node Cloning, Algorithmically

The *polarized node cloning of graphs* is a graph transformation which allows one to perform flexible cloning of nodes and their incident edges. Given a rewriting rule with a left-hand side L and a right-hand side R and a matching m of L in a graph G, the transformation of the nodes and the matching edges of G is provided by the rule, while the transformation of the non-matching edges (i.e., edges of G not in the image of L) is rather flexible: a node n can be cloned either with all its non-matching edges, or with all its outgoing non-matching edges, or with all its incoming non-matching edges, or with none of its non-matching edges. Definition 4 below provides an algorithmic definition of polarized node cloning (AlgoPC for short). An algebraic approach, more abstract, is presented in Section 3.

Definition 4. *An* AlgoPC rewrite rule *consists of a tuple* $\mu = (L, R, C^+, C^-)$, *where L and R are graphs and $C^+, C^- : |L| \times |R| \to \mathbb{N}$ are mappings. Then L and R are called the* left-hand side *and the* right-hand side, *respectively, and C^+, C^- are called the* cloning multiplicities *of μ. Let $\mu = (L, R, C^+, C^-)$ be an AlgoPC rewrite rule, G a graph and $m : L \to G$ a matching. Thus $|G| = |L| + |\overline{L}|$ and $G_\to = L_\to + \overline{L}_\to + \widetilde{L}_\to$. The AlgoPC rewrite step applying the rule μ to the matching m builds the graph H and the matching $h : R \to H$ such that h is the inclusion and $|H| = |R| + |\overline{L}|$ and $H_\to = R_\to + \overline{L}_\to + \sum_{n \in |H|, p \in |H|} E_{n,p}$ where:*

1. *if $n \in |R|$ and $p \in |R|$ then there is an edge $n \overset{(e,i)}{\to} p$ in $E_{n,p}$ for each edge $n_L \overset{e}{\to} p_L$ in $\widetilde{L}_\to$ and each $i \in \{1, \ldots, C^+(n_L, n) \times C^-(p_L, p)\}$;*

2. *if $n \in |R|$ and $p \in |\overline{L}|$ then there is an edge $n \overset{(e,i)}{\to} p$ in $E_{n,p}$ for each edge $n_L \overset{e}{\to} p$ in $\widetilde{L}_\to$ and each $i \in \{1, \ldots, C^+(n_L, n)\}$;*

3. *if $n \in |\overline{L}|$ and $p \in |R|$ then there is an edge $n \overset{(e,i)}{\to} p$ in $E_{n,p}$ for each edge $n \overset{e}{\to} p_L$ in $\widetilde{L}_\to$ and each $i \in \{1, \ldots, C^-(p_L, p)\}$;*

4. *if $n \in |\overline{L}|$ and $p \in |\overline{L}|$ then $E_{n,p}$ is empty.*

So, when an AlgoPC rule $\mu = (L, R, C^+, C^-)$ is applied to a matching of L in G, the image of L in G is erased and replaced by R, the subgraph $\overline{L}$ remains unchanged, and the edges in $\widetilde{L}$ are handled according to the cloning multiplicities. The subtleties in building clones lie in the treatment of the edges in $\widetilde{L}$.

Example 1. Let us consider the following rule $\mu = (L, R, C^+, C^-)$ where

$C^+(a, c) = 2$, $C^+(a, e) = 1$, $C^-(f, g) = 2$, and every other cloning multiplicity is 0. Now let us consider the graphs G and H:

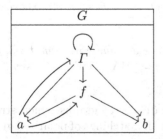   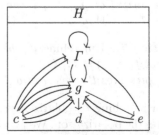

Then G rewrites into H using the rule μ and the matching $L \to G$ defined by the inclusion. Indeed, as specified by the cloning multiplicities, the edge going out of node a towards Γ is cloned three times, two times by the edges going out from c towards Γ $(C^+(a, c) = 2)$ and a third time by the edge going out from e $(C^+(a, e) = 1)$, the node b is erased as well as all its incident edges, and the incoming edges of f are duplicated $(C^-(f, g) = 2)$ and redirected towards g. The edge from a towards f is copied four times $(C^+(a, c) \times C^-(f, g) = 4)$ from c to g and two times $(C^+(a, e) \times C^-(f, g) = 2)$ from e to g.

3   Polarized Sesqui-Pushout of Graphs

In this section, in order to provide an algebraic version of the polarized node cloning of graphs defined in Section 2.2, we introduce polarized graphs, we study their sesqui-pushout rewriting and we use it for defining the notion of *polarized sesqui-pushout* of graphs.

3.1   Polarized Graphs

A polarized graph is a graph where every node may be polarized in the sense that it may be marked either with a "+", with a "−", with both "±" or with no mark. The polarizations will be used as cloning instructions.

Definition 5. *A polarization $X^\pm$ of a graph X is a pair $X^\pm = (|X|^+, |X|^-)$ of subsets of $|X|$. A node n may be denoted n^+ if it is in $|X|^+$, n^- if it is in $|X|^-$ and $n^\pm$ if it is in $|X|^+ \cap |X|^-$. A polarized graph $\mathbb{X} = (X, X^\pm)$ is a graph X together with a polarization $X^\pm$ of X such that the source of each edge e of $X_\to$ is in $|X|^+$ and the target of e is in $|X|^-$. A morphism of polarized graphs $f : \mathbb{X} \to \mathbb{Y}$, where $\mathbb{X} = (X, X^\pm)$ and $\mathbb{Y} = (Y, Y^\pm)$, is a morphism of graphs $f : X \to Y$ such that $f(|X|^+) \subseteq |Y|^+$ and $f(|X|^-) \subseteq |Y|^-$. This provides the category $\mathbf{Gr}^\pm$ of polarized graphs. The notations in defintion 1 are extended to polarized graphs: when $\mathbb{X} = (X, X^\pm)$ then $|\mathbb{X}| = |X|$ and $\mathbb{X}_\to = X_\to$.*

Definition 6. *Given two polarized graphs $\mathbb{X}_1$ and $\mathbb{X}_2$, their sum is the polarized graph $\mathbb{X}_1 + \mathbb{X}_2$ made of the graph $X_1 + X_2$ with the polarization $|X_1 + X_2|^+ = |X_1|^+ + |X_2|^+$ and $|X_1 + X_2|^- = |X_1|^- + |X_2|^-$. Given two polarized graphs $\mathbb{X}$ and $\mathbb{E}$ such that $|E| \subseteq |X|$, $|E|^+ \subseteq |X|^+$ and $|E|^- \subseteq |X|^-$, their edge-sum is the polarized graph $\mathbb{X} +_e \mathbb{E}$ made of the graph $X +_e E$ with the polarization $|X +_e E|^+ = |X|^+$ and $|X +_e E|^- = |X|^-$.*

Definition 7. *A matching of polarized graphs is a monomorphism $f : \mathbb{X} \to \mathbb{Y}$ such that $f(|X|^+) = f(|X|) \cap |Y|^+$ and $f(|X|^-) = f(|X|) \cap |Y|^-$ (we say that f strictly preserves the polarization).*

Thus, a matching of polarized graphs is a matching of graphs which strictly preserves the polarization. We now assume that all matchings of polarized graphs are inclusions, which is the case up to isomorphism.

Remark 2. Let $f : \mathbb{X} \to \mathbb{Y}$ be a matching of polarized graphs. Analogously to Remark 1, using the fact that f strictly preserves the polarization, we can express $\mathbb{Y}$ as $\mathbb{Y} = (\mathbb{X} + \overline{\mathbb{X}}) +_e \widetilde{\mathbb{X}}$ with $\widetilde{\mathbb{X}}_{\to} = \sum_{n \in |\mathbb{Y}|, p \in |\mathbb{Y}|} \widetilde{\mathbb{X}}_{n \to p}$, where $\widetilde{\mathbb{X}}_{n \to p}$ denotes the polarized graph made of the graph $\widetilde{X}_{n \to p}$ as in Remark 1 with its nodes polarized as in $\mathbb{Y}$.

Example 2. Here is a morphism of polarized graphs which is an inclusion although it is not a matching (the condition $f(|X|^+) = f(|X|) \cap |Y|^+$ is not fulfilled):

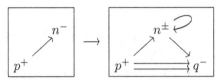

Definition 8. *The* underlying *graph of a polarized graph* $\mathbb{X} = (X, X^\pm)$ *is* X. *This defines a functor* Depol $: \mathbf{Gr}^\pm \to \mathbf{Gr}$. *The polarized graph* $\mathbb{X}$ *induced by a graph* X *is* $\mathbb{X} = (X, X^\pm)$ *where* $|X|^+ = |X|^- = |X|$. *This defines a functor* Pol $: \mathbf{Gr} \to \mathbf{Gr}^\pm$, *which is a right adjoint to* Depol *(this is denoted* Depol $\dashv$ Pol*). Moreover, the functor* Depol $\circ$ Pol *is the identity of* $\mathbf{Gr}$.

3.2   Sesqui-Pushout Rewriting of Polarized Graphs

In this section we describe the sesqui-pushout of polarized graphs. The sesqui-pushout rewriting [4] relies on the well-known categorical notions of pushout (PO) and pullback (PB): a sesqui-pushout rewriting step is made of a final pullback complement followed by a pushout. Pushouts and final pullback complements of polarized graphs are described in Propositions 1 and 2, respectively.

Proposition 1. *Let* $r : \mathbb{K} \to \mathbb{R}$ *be a morphism of polarized graphs and* $d : \mathbb{K} \to \mathbb{D}$ *a matching of polarized graphs. The following square, where* h *is the inclusion, is a pushout of* d *and* r *in* $\mathbf{Gr}^\pm$.

$$
\begin{array}{ccc}
\mathbb{K} & \xrightarrow{\quad r \quad} & \mathbb{R} \\
\Big\downarrow d & & \Big\downarrow h \\
\mathbb{D} = (\mathbb{K} + \overline{\mathbb{K}}) +_e \widetilde{\mathbb{K}} & \xrightarrow[r_1 = (r + \mathrm{id}_{\overline{\mathbb{K}}}) +_e \widetilde{r}]{} & \mathbb{H} = (\mathbb{R} + \overline{\mathbb{K}}) +_e \widetilde{\mathbb{R}}
\end{array}
$$

where $\widetilde{\mathbb{R}}_{n \to p} = \sum_{n_D \in r_1^{-1}(n), p_D \in r_1^{-1}(p)} \widetilde{\mathbb{K}}_{n_D \to p_D}$ *for all* $n, p \in |\mathbb{H}|$, $n_D \in |\mathbb{D}|^+$, $p_D \in |\mathbb{D}|^-$, *and where* $\widetilde{r} : \widetilde{\mathbb{K}} \to \widetilde{\mathbb{R}}$ *maps* $n_D \xrightarrow{e} p_D$ *to* $r_1(n_D) \xrightarrow{e} r_1(p_D)$.

Remark 3. Pushouts are preserved by Depol, because Depol is left adjoint to Pol. With the notations as in Proposition 1, this implies that Depol(h) can also be obtained by computing a pushout of Depol(d) and Depol(r) in $\mathbf{Gr}$.

Let us now define final pullback complements in the naive way, this definition coincides with the one in [9,4] when both exist.

Definition 9. *In a category $\mathcal{M}$, let $a : X \to Y$ and $g : Y \to Y_1$ be consecutive morphisms. A pullback complement (PBC) of a and g is an object X_1 with a pair of morphisms $f : X \to X_1$, $a_1 : X_1 \to Y_1$ such that there is a pullback:*

$$
\begin{array}{ccc}
Y & \xleftarrow{\;a\;} & X \\
{\scriptstyle g}\big\downarrow & & \big\downarrow{\scriptstyle f} \\
Y_1 & \xleftarrow[a_1]{} & X_1
\end{array}
$$

A morphism $k : (X_1, f, a_1) \to (X_1', f', a_1')$ of pullback complements of a and g is a morphism $k : X_1 \to X_1'$ in $\mathcal{M}$ such that $k \circ f = f'$ and $a_1' \circ k = a_1$. This yields the category of pullback complements of a and g, and the final pullback complement (FPBC) of a and g is defined as the final object in this category, if it does exist.

Proposition 2. *Let $l : \mathbb{K} \to \mathbb{L}$ be a morphism and $m : \mathbb{L} \to \mathbb{G}$ a matching of polarized graphs. The following square, where d is the inclusion, is a FPBC of l and m in $\mathbf{Gr}^{\pm}$:*

$$
\begin{array}{ccc}
\mathbb{L} & \xleftarrow{\qquad l \qquad} & \mathbb{K} \\
{\scriptstyle m}\big\downarrow & & \big\downarrow{\scriptstyle d} \\
\mathbb{G} = (\mathbb{L} + \overline{\mathbb{L}}) +_e \widetilde{\mathbb{L}} & \xleftarrow[l_1=(l+\mathrm{id}_{\overline{\mathbb{L}}})+_e\widetilde{l}]{} & \mathbb{D} = (\mathbb{K} + \overline{\mathbb{L}}) +_e \widetilde{\mathbb{K}}
\end{array}
$$

where $\widetilde{\mathbb{K}}_{n_D \to p_D} = \widetilde{\mathbb{L}}_{l_1(n_D) \to l_1(p_D)}$ for all $n_D \in |\mathbb{D}|^+, p_D \in |\mathbb{D}|^-$ (otherwise $\widetilde{\mathbb{K}}_{n_D \to p_D} = \emptyset$) and where $\widetilde{l} : \widetilde{\mathbb{K}} \to \widetilde{\mathbb{L}}$ maps $n_D \xrightarrow{e} p_D$ to $l_1(n_D) \xrightarrow{e} l_1(p_D)$.

The next definition is the usual definition of SqPO rewriting [4], applied to the category of polarized graphs.

Definition 10. *A SqPO rewrite rule of polarized graphs is a span of polarized graphs. Let $\rho = \mathbb{L} \xleftarrow{l} \mathbb{K} \xrightarrow{r} \mathbb{R}$ be a SqPO rewrite rule of polarized graphs and $m : \mathbb{L} \to \mathbb{G}$ a matching of polarized graphs. The SqPO rewrite step applying the rule ρ to the matching m builds the polarized graph $\mathbb{H}$ and the matching of polarized graphs $h : \mathbb{R} \to \mathbb{H}$ such that h is the inclusion, in two steps. First a FPBC of m and l is built as in Proposition 2, which gives rise to a polarized graph $\mathbb{D}$, a morphism $l_1 : \mathbb{D} \to \mathbb{G}$ and a matching $d : \mathbb{K} \to \mathbb{D}$ in $\mathbf{Gr}^{\pm}$. Then a pushout of $d : \mathbb{K} \to \mathbb{D}$ and $r : \mathbb{K} \to \mathbb{R}$ is constructed as in Proposition 1, which gives rise to a graph $\mathbb{H}$, a morphism $r_1 : \mathbb{D} \to \mathbb{H}$ and a matching $h : \mathbb{R} \to \mathbb{H}$ in $\mathbf{Gr}^{\pm}$.*

We represent a SqPO rewrite step of polarized graphs by the following diagram:

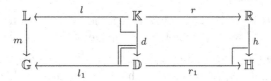

Merging Propositions 1 and 2 yields the following result, which provides an explicit description of a SqPO rewrite step of polarized graphs.

Theorem 1. *In the category of polarized graphs, let* $\rho = (\mathbb{L} \xleftarrow{l} \mathbb{K} \xrightarrow{r} \mathbb{R})$ *be a SqPO rewrite rule and* $m : \mathbb{L} \to \mathbb{G}$ *a matching, so that* $\mathbb{G} = (\mathbb{L} + \overline{\mathbb{L}}) +_e \widetilde{\mathbb{L}}$. *The SqPO rewrite step applying* ρ *to* m *builds the matching* $h : \mathbb{R} \to \mathbb{H}$ *where* h *is the inclusion and* $\mathbb{H} = (\mathbb{R} + \overline{\mathbb{L}}) +_e \widetilde{\mathbb{R}}$ *where, for all nodes* n, p *in* $|\mathbb{H}|$:

$$
\widetilde{\mathbb{R}}_{n \to p} = \begin{cases} \sum_{n_K^+ \in r^{-1}(n), \, p_K^- \in r^{-1}(p)} \widetilde{\mathbb{L}}_{l(n_K) \to l(p_K)} & \text{when } n, p \in |\mathbb{R}| \\ \sum_{n_K^+ \in r^{-1}(n)} \widetilde{\mathbb{L}}_{l(n_K) \to p} & \text{when } n \in |\mathbb{R}|, \, p \in |\overline{\mathbb{L}}| \\ \sum_{p_K^- \in r^{-1}(p)} \widetilde{\mathbb{L}}_{n \to l(p_K)} & \text{when } n \in |\overline{\mathbb{L}}|, \, p \in |\mathbb{R}| \\ \emptyset & \text{when } n, p \in |\overline{\mathbb{L}}| \end{cases}
$$

3.3   Polarized Node Cloning of Graphs, Algebraically

In this section we show that the polarized node cloning of graphs can easily be performed using the sesqui-pushout rewriting of polarized graph. This is called the *polarized sesqui-pushout* rewriting system (PSqPO). In a PSqPO rewriting step, the given matching $m : L \to G$ and the resulting matching $h : R \to H$ are matchings of ordinary graphs, while the interface matching $d : \mathbb{K} \to \mathbb{D}$ is a matching of polarized graphs where the polarization of a node indicates how the rewriting step acts on the non-matching edges incident to this node. The adjoint functors Pol : **Gr** $\to$ **Gr**$^\pm$ (right adjoint) and Depol : **Gr**$^\pm$ $\to$ **Gr** (left adjoint) from Definition 8 are used for moving between categories **Gr** and **Gr**$^\pm$. It should be reminded that Depol $\circ$ Pol is the identity of **Gr**.

Definition 11. *A PSqPO rewrite rule of graphs is a span of graphs* $L \xleftarrow{l} K \xrightarrow{r} R$ *together with a polarized graph* $\mathbb{K}$ *such that* $K = \text{Depol}(\mathbb{K})$. *This is denoted by* $L \xleftarrow{l} \mathbb{K} \xrightarrow{r} R$. *Thanks to the adjunction* Depol $\dashv$ Pol, *each PSqPO rewrite rule* $L \xleftarrow{l} \mathbb{K} \xrightarrow{r} R$ *gives rise to a SqPO rewrite rule* $\mathbb{L} \xleftarrow{l} \mathbb{K} \xrightarrow{r} \mathbb{R}$ *in* **Gr**$^\pm$ *where* $\mathbb{L} = \text{Pol}(L)$ *and* $\mathbb{R} = \text{Pol}(R)$. *The PSqPO rewrite step applying a PSqPO rewrite rule* $\rho = (L \xleftarrow{l} \mathbb{K} \xrightarrow{r} R)$ *to a matching of graphs* $m : L \to G$ *is the following construction of a matching of graphs* $h : R \to H$.

  (i) *Let* $m' = \text{Pol}(m) : \mathbb{L} \to \mathbb{G}$, *so that* m' *is a matching of polarized graphs.*
 (ii) *Let* $h' : \mathbb{R} \to \mathbb{H}$ *be the matching of polarized graphs obtained by applying the SqPO rewriting rule* ρ' *to the matching* m' *in* **Gr**$^\pm$; *note that* Depol($\mathbb{R}$) = Depol(Pol(R)) = R.

(iii) Let $H = \text{Depol}(\mathbb{H})$ and $h = \text{Depol}(h') : R \to H$, this is the required matching of graphs.

This means that H is made of a copy of R together with the non-matching nodes of G (i.e., nodes of G which are not in the image of the matching) and with an edge $n \xrightarrow{(n_D, p_D, e)} p$ for each n_D in $|K|^+ + |\overline{L}|$ such that $r_1(n_D) = n$, each p_D in $|K|^- + |\overline{L}|$ such that $r_1(p_D) = p$ and each $n_G \xrightarrow{e} p_G$ in $G_\rightarrow$ where $n_G = l_1(n_D)$ and $p_G = l_1(p_D)$. Since l_1 and r_1 are the identity on $|\overline{L}|$, whenever both n and p are in $|\overline{L}|$ then $H_{n \to p} = G_{n \to p}$.

A PSqPO rewrite step of graphs can be represented by the following diagram:

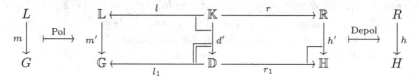

Remark 4. According to Definitions 11 and 10, applying a PSqPO rewrite rule ρ' to a matching m' can be decomposed in four steps: (i) m is mapped to $m' = \text{Pol}(m)$, (ii-a) the FPBC of m and l in $\mathbf{Gr}^\pm$ provides d', (ii-b) the PO of d' and r in $\mathbf{Gr}^\pm$ yields h', (iii) h' is mapped to $h = \text{Depol}(h')$. Thanks to remark 3, steps (ii-b) and (iii) can be "permuted", in the following way: first d' is mapped to $d = \text{Depol}(d')$, then the PO of d and r in $\mathbf{Gr}$ yields h. Thus, a PSqPO rewrite step of graphs can also be represented by the following diagram:

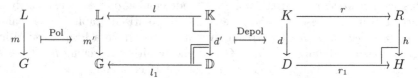

The next result shows that the PSqPO rewriting of graphs does provide an algebraic version of the polarized node cloning of graphs. A proof of a more precise result is provided in [8].

Proposition 3. *Let $\mu = (L, R, C^+, C^-)$ be an AlgoPC rewrite rule. Let $\mathbb{K}$ be the polarized graph without edges and with, for each $\star \in \{+, -\}$, a node $(n_L, n_R)_{i,\star}^\star$ for each pair of nodes $(n_L, n_R) \in |L| \times |R|$ and each $i \in \{1, \ldots, C^\star(n_L, n_R)\}$. Let $\rho = L \xleftarrow{l} \mathbb{K} \xrightarrow{r} R$ be the PSqPO rewrite rule where $l((n_L, n_R)_{i,\star}) = n_L$ and $r((n_L, n_R)_{i,\star}) = n_R$. Then the rules μ and ρ are equivalent, in the sense that for each matching of graphs $m : L \to G$, the AlgoPC rewrite step applying μ to m and the PSqPO rewrite step applying ρ to m yield the same matching $h : R \to H$.*

Example 3. The rewrite rule μ of Example 1 can be translated to the following PSqPO rule:

L		$\mathbb{K}$			R		
f	$\xleftarrow{l}$	f_1^- f_2^-		$\xrightarrow{r}$	g		
a $\searrow$ b		a_1^+ a_2^+	a_3^+		c	d	e

where $l(f_1) = l(f_2) = f$, $l(a_1) = l(a_2) = l(a_3) = a$, $r(f_1) = r(f_2) = g$, $r(a_1) =$
$r(a_2) = c$ and $r(a_3) = e$. As in Example 1, the matching is the inclusion of L in
G (below) and the PSqPO rewrite step builds:

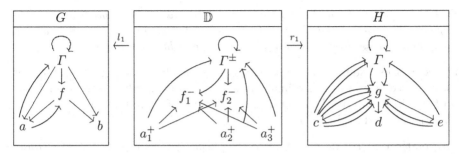

The resulting graph H and matching $h : R \to H$ are the same as in Example 1.

4   An Extension to Labeled Polarized Graphs

For several modeling purposes, it is useful to add labels to nodes and edges.
In this section we discuss an extension of our proposal in order to perform
polarized sesqui-pushout graph transformation on labeled graphs. We provide
syntactic conditions which ensure the existence of the constructions involved
in the rewriting process. Hereafter, two sets $\mathcal{L}_N$ and $\mathcal{L}_E$ are given, they are
called the set of *labels* for nodes and for edges, respectively. Moreover, all the
constructions are considered up to isomorphism.

Definition 12. *A labeled graph (X, lab) is a graph X together with two partial
functions* $\mathrm{lab} : |X| \rightharpoonup \mathcal{L}_N$ *for the labeling of nodes and* $\mathrm{lab} : X_\to \rightharpoonup \mathcal{L}_E$ *for the
labeling of edges. A morphism of labeled graphs $f : (X, \mathrm{lab}_X) \to (Y, \mathrm{lab}_Y)$ is a
morphism of graphs $f : X \to Y$ which preserves the labels, in the sense that if
a node or an edge x in X is labeled with a then $f(x)$ in Y is labeled with a (if
x is unlabeled there is no restriction on the labeling of $f(x)$). This provides the
category* **LGr** *of labeled graphs (with labels in $\mathcal{L}_N$ and $\mathcal{L}_E$).*

A labeled graph (X, lab) is often simply denoted X. A node x is denoted $x : a$ if
it is labeled with a and $x : \circ$ if it is unlabeled. An edge $x \to y$ is denoted $x \xrightarrow{a} y$
if it is labeled with a and simply $x \to y$ if it is unlabeled. A *matching* of labeled
graphs is a matching of graphs which preserves the labels. Since polarizations
and labelings do not interfere, these definitions and results are easily combined
with the definitions and results in Section 3.1. This provides the category **LGr**$^\pm$
of *labeled polarized graphs*, and Proposition 1 and Proposition 2 are generalized
to labeled polarized graphs as follows.

Proposition 4. *Let $r : \mathbb{K} \to \mathbb{R}$ be a morphism of labeled polarized graphs and
$d : \mathbb{K} \to \mathbb{D}$ a matching of labeled polarized graphs. Let us assume that:*

- *For each node or edge x in $\mathbb{K}$, if $r(x) : a$ and $d(x) : b$, then $a = b$.*
- *For each distinct nodes or edges x, y in $\mathbb{K}$, if $r(x) = r(y)$, $d(x) : a$ and $d(y) : b$, then $a = b$.*

Then the pushout of d and r in $\mathbf{LGr}^{\pm}$ exists, its underlying diagram of polarized graphs is the pushout of d and r in $\mathbf{Gr}^{\pm}$ and each node or edge x in $\mathbb{H}$ is labeled if and only if it is the image of a labeled node or edge in $\mathbb{R}$ or in $\mathbb{D}$.

Thanks to the assumptions, no conflict may arise when labeling the graph $\mathbb{H}$: if a node or edge x in $\mathbb{H}$ is the image of several nodes or edges in $\mathbb{R}$ or in $\mathbb{D}$ (at most one in $\mathbb{R}$ and maybe several in $\mathbb{D}$), then all of them have the same label, which becomes the label of x.

Proposition 5. *Let $l : \mathbb{K} \to \mathbb{L}$ be a morphism of labeled polarized graphs and $m : \mathbb{L} \to \mathbb{G}$ a matching of labeled polarized graphs. Then the FPBC of l and m exists, its underlying diagram of polarized graphs is the FPBC in $\mathbf{Gr}^{\pm}$ and each node or edge x_D in the graph $\mathbb{D}$ is labeled as follows: if x_D is not in the image of $\mathbb{K}$ then x_D is labeled in $\mathbb{D}$ like $l_1(x_D)$ in $\mathbb{G}$, otherwise $x_D = d(x_K)$ for a unique x_K in $\mathbb{K}$ and the label of x_D in $\mathbb{D}$ is determined by the labels of x_K in $\mathbb{K}$, $x_L = l(x_K)$ in $\mathbb{L}$ and $x_G = m(x_L)$ in $\mathbb{G}$ according to the following patterns:*

$$
\begin{array}{cccc}
x_L : a \longleftarrow x_K : a & x_L : \circ \longleftarrow x_K : \circ & x_L : \circ \longleftarrow x_K : \circ & x_L : a \longleftarrow x_K : \circ \\
\Big\downarrow \qquad \Big\downarrow & \Big\downarrow \qquad \Big\downarrow & \Big\downarrow \qquad \Big\downarrow & \Big\downarrow \qquad \Big\downarrow \\
x_G : a \longleftarrow x_D : a & x_G : a \longleftarrow x_D : a & x_G : \circ \longleftarrow x_D : \circ & x_G : a \longleftarrow x_D : \circ
\end{array}
$$

The labeled PSqPO rewrite rules cannot be defined simply as PSqPO rewrite rules where the graphs are labeled and the morphisms preserve the labels: indeed, in order to avoid conflicts in labeling the pushout, the assumptions in Proposition 4 must be satisfied after the construction of the polarized FPBC (Proposition 5). This leads to the following definition.

Definition 13. *A labeled PSqPO rewrite rule is a PSqPO rewrite rule $L \xleftarrow{l} \mathbb{K} \xrightarrow{r} R$ (Definition 11) where the graphs are labeled and the morphisms preserve the labels, such that the following conditions are fulfilled (where $K = \mathrm{Depol}(\mathbb{K})$) : (i) for each unlabeled node or edge x in K, if $l(x)$ is unlabeled in L then $r(x)$ is unlabeled in R and (ii) for each distinct unlabeled nodes or edges x, y in K, if $l(x) \neq l(y)$ and $l(x)$ or $l(y)$ is unlabeled in L then $r(x) \neq r(y)$ in R.*

Example 4. The behavior of the "if b then...else..." operator in imperative languages can be modelled thanks to two polarized PSqPO rewrite rules, one when b is **true** and another one when b is **false**. Here is a possible choice when b is **true** (morphisms are represented via node name sharing, for instance $r(m) = r(p) = p, m$ and $l(m) = m$):

These rules for modeling "if...then...else..." are destructive, in the sense that nodes n and q disappear during the rewrite step. Non-destructive rules can also be chosen, here is such a rule for true.

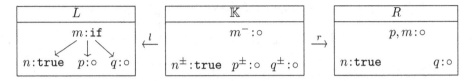

Example 5. The problem of copying objects in object-oriented languages has been thoroughly examined. Two generic ways of copying objects are usually considered: shallow cloning, which is the basic cloning of Java (see the reference of method clone in class Object [15]) and deep cloning, which is implemented by deep_copy in Eiffel [20]. These two ways of cloning can be modelled by our approach. We restrict ourselves here to the cloning of a particular data, say *linked lists of constants* (a constant is implemented as a node without any outgoing edge) and consider two cloning routines: sc (shallow cloning) and dc (deep cloning). Intuitively, we would like to implement rules that transform, for instance, the following graph, where ▼ denotes the end of the list:

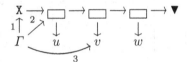

into the following ones, depending whether X is replaced by sc or by dc:

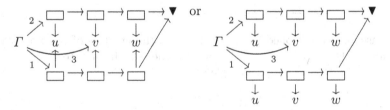

Notice that in the case of deep cloning, the edge from Γ to v is not cloned to point to the "new" occurrence of v (the graph on the right). Indeed, the deep cloning primitives do not modify the environment.

Let us consider labeled graphs. Label c is used to represent the usual cons constructor of lists. The parameters of c are identified by edge labels, the next cell of the list is pointed by an edge labeled n and the element of the cell is pointed by an edge labeled e.

For the shallow cloning (sc) the recursive case is implemented by the following rule where morphisms l, r are represented by node name sharing. All edges that point to m before the execution of this rule will point to the new c node (the node m : c in R). The function sc is recursively called by the node s in R.

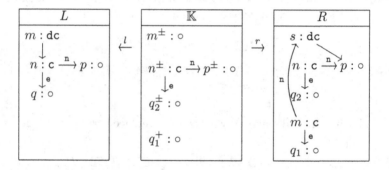

The base (halt) case is implemented by the following rule where the node m, n is the image of both m and n by r:

For deep cloning (dc), the recursive case is implemented by the following rule:

In this case node q is cloned twice in $\mathbb{K}$: the incoming edges of q are not cloned as incoming edges of q_1^+ (as it is the case for the edge from Γ to v).

The base (halt) case for dc is implemented by substituting sc with dc in the corresponding rule.

5   Related Work

Polarized sesqui-pushout graph rewriting (PSqPO) is a new way to perfom graph transformations which offers different possibilities to clone nodes and their incident edges, in addition to classical graph transformations (addition and deletion of nodes and edges). In this section the PSqPO approach is compared with other approaches for graph transformations.

In [7] an algebraic approach of termgraph transformation, based on heterogeneous pushouts (HPO), has been proposed. With respect to cloning abilities, the HPO approach offers the possibility to make one or more copies of a node together with its outgoing edges. Therefore, this way of cloning nodes is limited to the outgoing edges only and contrasts with the flexible possibilities of cloning edges proposed in the present paper. In fact, whenever a graph G rewrites into H according to the HPO approach using a rule (L, R, τ, σ) [7, Definition 5], the graph G can also be rewritten into H according to a rule $L \xleftarrow{l} \mathbb{K} \xrightarrow{r} R$ where morphisms l and r encode the functions τ and σ.

Cloning is also one of the features of the sesqui-pushout approach (SqPO) to graph transformation [4]. The SqPO and PSqPO approaches mainly differ in the way of handling cloning. In [4], the cloning of a node is performed by copying all its incident edges. This is a particular case of PSqPO. The use of polarized graphs helped us to specify for every clone, the way incident edges can be copied. Therefore, a SqPO rewrite step can be simulated by a PSqPO rewrite step by polarizing every node n in the interface graph $\mathbb{K}$ as $n^{\pm}$, but the converse does not hold in general. For instance, in both rules defining the shallow cloning operator sc (see Example 5), all nodes are cloned with polarities $\pm$, thus these rules can be implemented using the SqPO approach in the category of graphs. However, in the recursive case implementing the deep cloning operator, dc, one node, q_1^{+}, in $\mathbb{K}$ is polarized as $+$ only; it follows that this rule cannot be modelled with a standard SqPO transformation of graphs. Furthermore, in [4], the sesqui-pushout approach is compared to the classical DPO and SPO approaches. Therefore, for comparing our approach with the DPO and SPO, we may rely on [4, Propositions 12 and 14].

Cloning is also subject of interest in [6]. The authors consider rewrite rules of the form $S := R$ where S is a *star*, i.e., S is a (nonterminal) node surrounded by its adjacent nodes together with the edges that connect them. Rewrite rules which perform the cloning of a node are given in [6, Definition 6]. These rules show how a star can be removed, kept identical to itself or copied (cloned) more than once. Here again, unlike our approach, each node is cloned together with all its incoming and outgoing edges.

6 Conclusion

We have investigated a new way to perform node cloning in graph transformation with some flexibility in copying incident edges. To obtain this result, we have used an auxiliary category of polarized graphs which allows one to declare how incident edges are cloned. The algebraic definition of a graph rewriting step is based on a sesqui-pushout transformation in the auxiliary category. In [8], the reader may find more results such as the equivalence of the algorithmic and the algebraic definitions of PSqPO as well as the vertical composition of transformations.

In [18], Löwe proposes a general framework of graph rewriting in span-categories. He shows how classical algebraic graph transformation approaches can be seen as instances of his framework. Our approach, which is close to the sesqui-pushout rewriting, could be presented also as an instance of Löwe's framework up to some particular considerations due to the use of two kinds of graphs in our spans, namely polarized and not polarized graphs. Details of the instance, including the complete definitions of abstract spans and matching of abstract spans are matter of further investigation.

References

1. Baader, F., Nipkow, T.: Term rewriting and all that. Cambridge University Press (1998)
2. Barendregt, H.P., van Eekelen, M.C.J.D., Glauert, J.R.W., Kenneway, J.R., Plasmeijer, M.J., Sleep, M.R.: Term Graph Rewriting. In: de Bakker, J.W., Nijman, A.J., Treleaven, P.C. (eds.) PARLE 1987. LNCS, vol. 259, pp. 141–158. Springer, Heidelberg (1987)
3. Book, R.V., Otto, F.: String-rewriting systems. Springer (1993)
4. Corradini, A., Heindel, T., Hermann, F., König, B.: Sesqui-Pushout Rewriting. In: Corradini, A., Ehrig, H., Montanari, U., Ribeiro, L., Rozenberg, G. (eds.) ICGT 2006. LNCS, vol. 4178, pp. 30–45. Springer, Heidelberg (2006)
5. Corradini, A., Montanari, U., Rossi, F., Ehrig, H., Heckel, R., Löwe, M.: Algebraic approaches to graph transformation - part I: Basic concepts and double pushout approach. In: Handbook of Graph Grammars, pp. 163–246 (1997)
6. Drewes, F., Hoffmann, B., Janssens, D., Minas, M., Van Eetvelde, N.: Adaptive Star Grammars. In: Corradini, A., Ehrig, H., Montanari, U., Ribeiro, L., Rozenberg, G. (eds.) ICGT 2006. LNCS, vol. 4178, pp. 77–91. Springer, Heidelberg (2006)
7. Duval, D., Echahed, R., Prost, F.: A Heterogeneous Pushout Approach to Term-Graph Transformation. In: Treinen, R. (ed.) RTA 2009. LNCS, vol. 5595, pp. 194–208. Springer, Heidelberg (2009)
8. Duval, D., Echahed, R., Prost, F.: Graph rewriting with polarized cloning. arXiv:0911.3786,V3 (2012)
9. Dyckhoff, R., Tholen, W.: Exponentiable morphisms, partial products and pullback complements. Journal of Pure and Applied Algebra (1987)
10. Echahed, R.: Inductively Sequential Term-Graph Rewrite Systems. In: Ehrig, H., Heckel, R., Rozenberg, G., Taentzer, G. (eds.) ICGT 2008. LNCS, vol. 5214, pp. 84–98. Springer, Heidelberg (2008)
11. Ehrig, H., Engels, G., Kreowski, H.-J., Rozenberg, G. (eds.): Handbook of Graph Grammars and Computing by Graph Transformations. Applications, Languages and Tools, vol. 2. World Scientific (1999)
12. Ehrig, H., Heckel, R., Korff, M., Löwe, M., Ribeiro, L., Wagner, A., Corradini, A.: Algebraic approaches to graph transformation - part II: Single pushout approach and comparison with double pushout approach. In: Handbook of Graph Grammars, pp. 247–312 (1997)
13. Ehrig, H., Kreowski, H.-J., Montanari, U., Rozenberg, G. (eds.): Handbook of Graph Grammars and Computing by Graph Transformations. Concurrency, Parallelism and Distribution, vol. 3. World Scientific (1999)
14. Ehrig, H., Pfender, M., Schneider, H.J.: Graph-grammars: An algebraic approach. In: 14th Annual Symposium on Foundations of Computer Science (FOCS), The University of Iowa, USA, October 15-17, pp. 167–180. IEEE (1973)
15. Gosling, J., Joy, B., Steele, G., Bracha, G., Buckley, A.: The Java(TM) Language Specification, Java SE 7 edn. Oracle documentation (2005)

16. Kennaway, R.: On "on graph rewritings". Theoretical Computer Science 52, 37–58 (1987)
17. Löwe, M.: Algebraic approach to single-pushout graph transformation. Theoretical Computer Science 109(1&2), 181–224 (1993)
18. Löwe, M.: Graph Rewriting in Span-Categories. In: Ehrig, H., Rensink, A., Rozenberg, G., Schürr, A. (eds.) ICGT 2010. LNCS, vol. 6372, pp. 218–233. Springer, Heidelberg (2010)
19. Mac Lane, S.: Categories for the Working Mathematician, 2nd edn., vol. 5. Springer (1998)
20. Miller, F., Vandome, A., McBrewster, J.: Eiffel (programming language). Alphascript Publishing (2010)
21. Raoult, J.C.: On graph rewriting. Theoretical Computer Science 32, 1–24 (1984)
22. Rozenberg, G. (ed.): Handbook of Graph Grammars and Computing by Graph Transformations. Foundations, vol. 1. World Scientific (1997)

Rational Term Rewriting Revisited: Decidability and Confluence[*]

Takahito Aoto[1] and Jeroen Ketema[2]

[1] RIEC, Tohoku University
2-1-1 Katahira, Aoba-ku, Sendai 980-8577, Japan
aoto@nie.riec.tohoku.ac.jp
[2] Department of Computing, Imperial College London
180 Queen's Gate, London SW7 2BZ, United Kingdom
j.ketema@imperial.ac.uk

Abstract. We consider a variant of rational term rewriting as first introduced by Corradini et al., i.e., we consider rewriting of (infinite) terms with a finite number of different subterms. Motivated by computability theory, we show a number of decidability results related to the rewrite relation and prove an effective version of a confluence theorem for orthogonal systems.

1 Introduction

Cyclic term graph rewriting [1,3,16] can be given an operational semantics based on non-cyclic terms by considering both finite and infinite terms (or trees) and defining a notion of rewriting for these. Two forms of so-called infinitary rewriting exist: (1) Corradini considers finite reductions where in each step infinitely many redexes may be rewritten simultaneously [4]. (2) Kennaway et al. allow for reductions of any countable ordinal length, but in each step only one redex may be rewritten [13,12].

The operational interpretation of cyclic term graph rewriting now takes the form of a so-called adequacy result. For Corradini's notion of infinitary rewriting, this result occurs in [5]. With regard to the notion of infinitary rewriting defined by Kennaway et al., the result can be found in [11].

Infinitary rewriting is strictly more general than any finite form of rewriting. There are, e.g., continuum many infinite terms over any signature consisting of at least two unary function symbols. Unfortunately, this means that infinitary rewriting only exists as a mathematical abstraction, not as something we can actually compute with.

To rectify the above situation, the usual approach is to consider infinite terms that can be finitely represented. In [14] this is done with respect to the work of Kennaway et al. using Turing machines and taking as a starting point methods from computable analysis [18]. In the current work, we consider finite representations within the setting of Corradini's work. More precisely, we consider

---

[*] Takahito Aoto is partially supported by JSPS grant No. 23500002.

H. Ehrig et al.(Eds.): ICGT 2012, LNCS 7562, pp. 172–186, 2012.
© Springer-Verlag Berlin Heidelberg 2012

rewriting of terms that are rational, i.e., rewriting of terms which have only finitely many different subterms. It is well-known that rational terms can be represented finitely [7].

As unraveling a cyclic term graph into an infinite term always yields a term that is rational [11], one may wonder whether we are not simply considering term graph rewriting in disguised form. However, our notion of rewriting is strictly more powerful. From the perspective of term graph rewriting, our rewrite steps combine two operations: The first operation is to transform the considered term graph into *any* other term graph, where the only restriction is that the new graph has the same unraveling. The second operation is the actual rewrite step. The first operation is atypical of graph rewriting, where transformations are usually limited to ones that reduce sharing, e.g., by means of a copying operation [1], or ones that only introduce vertical sharing [15]. Hence, it is not clear whether our notion of rewriting is still decidable.

Related Work. The notion of rational rewriting already occurs in two places in the literature. In [10], rewriting of so-called μ-terms (i.e., term graphs with only vertical sharing) is dubbed "rational rewriting". This work shares with ours the two-phase approach to rewriting as outlined above. However, a redex may only be rewritten when no back edges point to nodes on the path from the root of the graph to the redex.

Similar to us, [6,5] define rational rewriting as a restriction to rational terms of Corradini's notion of infinitary rewriting [4]. However, decidability issues related to the rewrite relation are not considered. Furthermore, although [5] mentions a confluence result for orthogonal systems, no actual proof is provided and, hence, it is unclear to us whether an effective version of confluence is intended.

Outline. We both extend and restrict the notion of rational rewriting from [6,5]. We extend it in the sense that we allow for rewriting in systems with arbitrary rule sets, not just orthogonal ones, as in [6,5]. We restrict the notion of rewriting in the sense that only one rewrite rule may be employed in each rewrite step instead of arbitrary combinations of rules, as again in [6,5]. Currently, this restriction is necessary, as we do not know whether our decidability results hold without it. We hypothesize, however, that this restriction may be lifted.

The remainder of the paper is divided into two parts. In the first, we show decidability of several properties related to rational rewrite steps: Given a regular set Δ of positions, a rewrite rule $l \to r$, and rational terms s and t, we show that $s \to_{l \to r}^{\Delta} t$ is decidable, where $\to_{l \to r}^{\Delta}$ denotes the simultaneous rewrite of the $l \to r$-redexes at positions Δ. More generally, given a rational term s, a rewrite system $\mathcal{R}$, and regular set Δ, we show that the set of all terms t with $s \to_{\mathcal{R}}^{\Delta} t$ can be constructed effectively. Finally, we show that it is decidable whether a rational s is in normal from with respect to a set of rewrite rules $\mathcal{R}$.

In the second part of the paper, we prove for orthogonal systems that, given reductions $s \to^* t_1$ and $s \to^* t_2$, we can effectively construct a rational term u and reductions $t_1 \to^* u$ and $t_2 \to^* u$, i.e., an algorithm exists constructing

u and the two reductions. We also show that this result cannot be extended to so-called weakly orthogonal systems.

Remark that, as we are interested in computability results, we will be explicit regarding the finite representation we use for rational terms (instead of leaving this implicit). We will work with so-called regular systems of equations (see Definition 3.1), which is one of many such finite representations [7]. Note that the same representation has been used in term graph rewriting [1].

2 Preliminaries

We assume basic familiarity with term rewriting [2,17]. However, to fix notation, we recall some basic definitions.

Let Σ be a set of function symbols and $\mathcal{V}$ be a countable, infinite set of variables. Each function symbol $f \in \Sigma$ is equipped with a natural number called the *arity* of f. The set of function symbols of arity n is denoted by Σ_n.

Let $\mathbb{N}^*$ be the set of finite strings over the positive numbers, with ϵ the empty string, $\cdot$ the operator for concatenating strings, and $<$ the prefix order on strings. We can define the set of (finite and infinite) terms as follows [9]:

Definition 2.1. *The set of* (finite and infinite) *terms* $\mathcal{T}(\Sigma, \mathcal{V})$ *is the set of partial functions* $t : \mathbb{N}^* \to \Sigma \cup \mathcal{V}$ *such that* $t(\epsilon)$ *is defined and for all* $p \in \mathbb{N}^*$: *(1) if* $t(p)$ *is defined, then* $t(q)$ *is defined for all* $q < p$; *(2) if* $t(p) \in \Sigma_n$, *then* $t(p \cdot i)$ *is defined for all* $1 \le i \le n$ *and undefined otherwise; (3) if* $t(p) \in \mathcal{V}$, *then* $t(q)$ *is undefined for all* $q > p$.

The domain over which t *is defined is denoted by* $\mathcal{P}os(t)$ *and the elements of the domain are called* positions. *A term* t *is* finite *if* $\mathcal{P}os(t)$ *is finite; the set of finite terms is denoted by* $\mathcal{T}_{fin}(\Sigma, \mathcal{V})$. *Moreover, define* $\mathcal{P}os_\Sigma(t) = \{p \in \mathcal{P}os(t) \mid t(p) \in \Sigma\}$ *and* $\mathcal{P}os_\mathcal{V}(t) = \{p \in \mathcal{P}os(t) \mid t(p) \in \mathcal{V}\}$. *Finally, let* $\mathcal{V}(t)$ *denote the set of variables in* t.

For any $f \in \Sigma_n$ and terms $t_1, \ldots, t_n \in \mathcal{T}(\Sigma, \mathcal{V})$, the term $t = f(t_1, \ldots, t_n)$ is defined by $t(\epsilon) = f$ and $t(i \cdot p) = t_i(p)$ for all $1 \le i \le n$ and $p \in \mathbb{N}^*$. A function $\sigma : \mathcal{V} \to \mathcal{T}(\Sigma, \mathcal{V})$ is called a *substitution* and is extended to a function $\sigma : \mathcal{T}(\Sigma, \mathcal{V}) \to \mathcal{T}(\Sigma, \mathcal{V})$ as usual.

Let the *subterm at position* $p \in \mathcal{P}os(t)$ *in* $t \in \mathcal{T}(\Sigma, \mathcal{V})$, denoted $t|_p$, be defined by $t|_p(q) = t(p \cdot q)$. We can now define the following:

Definition 2.2. *The set of* regular trees *or* rational terms $\mathcal{T}_{reg}(\Sigma, \mathcal{V}) \subseteq \mathcal{T}(\Sigma, \mathcal{V})$ *is the set of terms with a finite number of different subterms, i.e., for each* t *the set* $\{t|_p \mid p \in \mathcal{P}os(t)\}$ *is finite.*

Note that for any rational term t, we have that the set of positions $\mathcal{P}os(t)$ is regular (i.e., there exists a finite automaton that recognizes $\mathcal{P}os(t)$ as a language). Moreover, every finite term is regular. Finally, note that matching is decidable for rational terms [7].

3   Representing Rational Terms

Write $\Sigma_\perp$ for the set of function symbols Σ extended with a fresh constant $\perp$. Recall that rational terms are solutions of regular systems of equations and vice versa [7]:

Definition 3.1. *A* regular system of equations *is a finite set of equations* $E = \{x_1 = t_1, \ldots, x_n = t_n\}$ *with mutually distinct* $x_1, \ldots, x_n \in \mathcal{V}$ *and* $t_i \in \mathcal{T}_{fin}(\Sigma, \mathcal{V})$ *for all* $1 \leq i \leq n$. *In* E, *a variable* x_i *is called* (non-)looping *if there (do not) exist* $1 \leq i_1, \ldots, i_k \leq n$ *with* $x_i = t_{i_1}$ *and* $t_{i_j} = x_{i_{(j \bmod k)+1}}$ *for all* $1 \leq j \leq k$.
    The domain *of* E, *denoted* $\mathcal{D}om(E)$, *is the set of variables* $\{x_1, \ldots, x_n\}$; *the* range *of* E, *denoted* $\mathcal{R}an(E)$, *is the set of terms* $\{t_1, \ldots, t_n\}$. *The solution of* E *for* $x_i \in \mathcal{D}om(E)$ *is the term in* $E^\star(x_i) \in \mathcal{T}(\Sigma_\perp, \mathcal{V})$ *the defined by:*

$$E^\star(x_i)(p) = \begin{cases} t_i(p) & \text{if } p \in \mathcal{P}os(t_i) \text{ and } t_i(p) \notin \mathcal{D}om(E) \\ \perp & \text{if } t_i(p) = x_j \in \mathcal{D}om(E) \text{ looping} \\ E^\star(x_j)(q) & \text{if } t_i(p') = x_j \in \mathcal{D}om(E) \text{ non-looping with } p = p' \cdot q \\ \text{undefined} & \text{otherwise} \end{cases}$$

The pair $\langle E, x_i \rangle$, *or* E_{x_i} *for short, is said to* represent *a term* t *(or is a* regular representation *of* t) *if* $E^\star(x_i) = t$.

Below, we often write $E(x_i)$ for t_i, given a regular system of equations $E = \{x_1 = t_1, \ldots, x_n = t_n\}$.

The above definition extends the usual definition of a regular system of equations by not restricting all variables in the domain to be non-looping. Remark that a non-looping regular system (over $\Sigma_\perp$) can be obtained by replacing every equation $x = t$ with x looping by $x = \perp$.

As in [1], the extension with $\perp$ allows for the treatment of collapsing rules, i.e., rules with a variable on their right-hand side, as all other rules. No special measures need to be taken to handle the looping that may be introduced by collapsing rules (see also Example 4.2).

Remark that each $E^\star$ can be regarded as a substitution and, hence, can be extended to $E^\star : \mathcal{T}_{reg}(\Sigma_\perp, \mathcal{V}) \to \mathcal{T}_{reg}(\Sigma_\perp, \mathcal{V})$ in the usual way. By definition, we have $E^\star(x) = E^\star(t)$ for every $x = t \in E$.

From here onwards, we assume that Σ always includes $\perp$. We now have the following.

Lemma 3.2. *Let* $E = \{x_1 = t_1, \ldots, x_n = t_n\}$ *be a regular system and let* s_i *be a finite term with* $s_i \notin \mathcal{D}om(E)$ *and* $E^\star(x_i) = E^\star(s_i)$. *If* F *is identical to* E *but with* $x_i = t_i$ *replaced by* $x_i = s_i$, *then* $E^\star = F^\star$.

Lemma 3.3. *Let* E *and* F *be regular systems and suppose there exists a surjection* $\delta : \mathcal{D}om(E) \to \mathcal{D}om(F)$ *such that* $\delta(y) = \delta(s) \in F$ *for every* $y = s \in E$, *where* δ *is extended to* $\delta : \mathcal{T}_{fin}(\Sigma, \mathcal{V}) \to \mathcal{T}_{fin}(\Sigma, \mathcal{V})$ *in the usual way. Then,* $E^\star(y) = F^\star(\delta(y))$ *for every* $y \in \mathcal{D}om(E)$.

Surjectivity is required in the above lemma: Consider $E = \{x = f(y)\}$ and $F = \{x = f(y), y = a\}$. If δ is the (non-surjective) identity function, then $E^\star(x) = f(y)$, while $F^\star(x) = f(a)$.

Subterm Positions. For each $x \in \mathcal{D}om(E)$, let $\mathcal{U}_E(x)$ be the smallest set satisfying (1) $x \in \mathcal{U}_E(x)$ and (2) if $y \in \mathcal{U}_E(x)$ and $y = t \in E$, then $\mathcal{V}(t) \cap \mathcal{D}om(E) \subseteq \mathcal{U}_E(x)$. We write $y \sqsubseteq_E x$ if $y \in \mathcal{U}_E(x)$. It is readily checked that $\sqsubseteq_E$ is transitive. The subscript E is omitted if it is obvious from the context. We write $W \sqsubseteq x$ if $y \sqsubseteq x$ for all $x \in W$. For $x \in \mathcal{D}om(E)$, let $E{\restriction}x = \{y = t \in E \mid y \sqsubseteq x\}$. Obviously, $E{\restriction}x$ contains all equations necessary to define $E^\star(x)$. Moreover, we have $E^\star(y) = (E{\restriction}x)^\star(y)$ for each $y \sqsubseteq x$.

For every $y \sqsubseteq x$, there is a set of positions in the rational term $E^\star(x)$ corresponding to occurrences of $E^\star(y)$. Such a set of positions, called a set of subterm positions, is defined next. The notion will be heavily used in our proofs.

Definition 3.4 (Subterm Positions). *Let E be a regular system. For each $x, y \in \mathcal{D}om(E)$ such that $y \sqsubseteq x$, the set $\mathcal{SP}_{E_x}(y)$ of subterm positions is the smallest set satisfying: (1) $\epsilon \in \mathcal{SP}_{E_x}(x)$ and (2) $p \cdot q \in \mathcal{SP}_{E_x}(y)$ if $p \in \mathcal{SP}_{E_x}(z)$ and there exists an equation $z = t \in E$ such that $t|_q = y$.*

For a set $W \subseteq \mathcal{D}om(E)$, we put $\mathcal{SP}_{E_x}(W) = \bigcup_{y \in W} \mathcal{SP}_{E_x}(y)$.

Example 3.5 (Subterm Positions). Let $E = \{x = \mathsf{f}(y), \ y = \mathsf{g}(x)\}$. We have $\mathcal{SP}_{E_x}(x) = \{\epsilon\} \cup \{p \cdot 1 \mid p \in \mathcal{SP}_{E_x}(y)\}$ and $\mathcal{SP}_{E_x}(y) = \{p \cdot 1 \mid p \in \mathcal{SP}_{E_x}(x)\}$. Hence, $\mathcal{SP}_{E_x}(x) = \{1^{2n} \mid n \geq 0\}$ and $\mathcal{SP}_{E_x}(y) = \{1^{2n+1} \mid n \geq 0\}$.

The following is proved in a straightforward way.

Proposition 3.6. *Let E be a regular system and $x \in \mathcal{D}om(E)$.*

1. *If $W \sqsubseteq x$, then $\mathcal{SP}_{E_x}(W)$ is regular.*
2. *If $\mathcal{R}an(E) \cap \mathcal{D}om(E) = \emptyset$ and $p < q$ such that $p \in \mathcal{SP}_{E_x}(y)$ and $q \in \mathcal{SP}_{E_x}(z)$ with $y \neq z$, then $q \notin \{p \cdot p' \mid p' \in \mathcal{P}os_\Sigma(E(y))\}$.*

Canonical Systems. A regular system $E = \{x_1 = t_1, \ldots, x_n = t_n\}$ is said to be *canonical* if for every $1 \leq i \leq n$, either (1) $t_i \in \mathcal{V} \setminus \mathcal{D}om(E)$ or (2) $t_i = \mathsf{f}(y_1, \ldots, y_m)$ for some $\mathsf{f} \in \Sigma$ and $y_1, \ldots, y_m \in \mathcal{D}om(E)$ [7]. A regular representation E_x is said to be canonical if E is so. Canonical regular systems have the following properties.

Proposition 3.7. *Let $E = \{x_1 = t_1, \ldots, x_n = t_n\}$ be a canonical regular system.*

1. *If $E^\star(x) = s$, then for every $p \in \mathcal{P}os(s)$ there exists a variable $y \sqsubseteq x$ such that $s|_p = E^\star(y)$.*
2. *For all $1 \leq i \leq n$, if s_i is a finite term such that $E^\star(x_i) = s_i \sigma$ for some substitution σ, then a variable substitution ρ exists such that $E^\star(x_i) = E^\star(s_i \rho)$.*

The next proposition can be proved in a straightforward way.

Proposition 3.8 (Canonization). *Given a regular system E, a canonical regular system F can be constructed effectively such that (1) $\mathcal{D}om(E) \subseteq \mathcal{D}om(F)$, (2) $E^\star(x) = F^\star(x)$ for every $x \in \mathcal{D}om(E)$, and (3) $\mathcal{SP}_{E_x}(y) = \mathcal{SP}_{F_x}(y)$ for every $x, y \in \mathcal{D}om(E)$ with $y \sqsubseteq x$.*

A canonical regular system $E = \{x_1 = t_1, \ldots, x_n = t_n\}$ is said to be *minimal* if $E^\star(x_i) \neq E^\star(x_j)$ for every $1 \leq i < j \leq n$. A canonical regular representation E_x is minimal if so is E. We have the following.

Proposition 3.9. *Every rational term has a minimal canonical regular representation.*

The next proposition is a consequence of regularity.

Proposition 3.10. *Let Δ be a regular set of positions of a rational term t. A canonical regular representation E_x of t and a set $W \subseteq \mathcal{D}om(E)$ can be constructed effectively such that $\Delta = \mathcal{SP}_{E_x}(W)$.*

4   Rational Rewriting

A *rewrite rule* is a pair (l, r) terms over Σ, invariably written $l \to r$ such that (1) l and r are finite, (2) $l \in \mathcal{T}(\Sigma \setminus \{\bot\}, \mathcal{V})$ and $l \notin \mathcal{V}$, and (3) $\mathcal{V}(r) \subseteq \mathcal{V}(l)$. A *term rewriting system (TRS)* is a finite set of rewrite rules. We define rational rewrite steps.

Definition 4.1 (Rational Rewrite Steps). *Let s and t be rational terms. Define $s \to_{l \to r} t$ if there exist regular representations E_x and F_x of s and t, resp., such that $\mathcal{D}om(E) = \mathcal{D}om(F)$ and $W \subseteq \mathcal{D}om(E)$ with (1) $E(y) = F(y)$ for all $y \in \mathcal{D}om(E) \setminus W$ and (2) for each $y \in W$, there exists a variable substitution ρ such that $E(y) = l\rho$ and $F(y) = r\rho$.*

We write $s \to_{l \to r}^{\Delta} t$ for $\Delta = \mathcal{SP}_{E_x}(W)$. Moreover, if $\mathcal{R}$ is a TRS, we write $s \to_{\mathcal{R}} t$ if there exists $l \to r \in \mathcal{R}$ such that $s \to_{l \to r} t$.

Example 4.2. Let $\mathcal{R} = \{f(x) \to x\}$ be a TRS and $s = f(f(\cdots f(\cdots) \cdots))$.

1. Since $E = \{x = f(x)\}$ and $F = \{x = x\}$ are regular representations of s and $\bot$, resp., we have $s \to_{\mathcal{R}}^{\Delta} \bot$, where $\Delta = \mathcal{SP}_{E_x}(x) = \{1^n \mid n \geq 0\}$.
2. Since $E = \{x = f(y), y = f(x)\}$ and $F = \{x = f(y), y = x\}$ are regular representations of s, we have $s \to_{\mathcal{R}}^{\Delta} s$ with $\Delta = \mathcal{SP}_{E_x}(y) = \{1^{2n+1} \mid n \geq 0\}$.
3. Since $E = \{x = f(y), y = f(y)\}$ and $F = \{x = f(y), y = y\}$ are regular representations of s and $f(\bot)$, resp., we have $s \to_{\mathcal{R}}^{\Delta} f(\bot)$ with $\Delta = \mathcal{SP}_{E_x}(y) = \{1 \cdot 1^n \mid n \geq 0\}$.

Remark 4.3. Contrary to our definition, the definition of rational rewriting as presented in [6] does not require one to explicitly specify the employed rewrite rule. In fact, several rewrite rules may be used in a *single* rewrite step, as long as the set of positions at which the contracted redexes occur is regular. In [6], specifying a regular set of positions and no rewrite rules suffices, as orthogonality is assumed there throughout.

As we do not restrict ourselves to orthogonal systems, specifying just a rational set of positions does not suffice in our case. Consider, e.g., $\mathcal{R} = \{\alpha : f(x) \to g(x), \beta : f(x) \to h(x)\}$, $s = f(f(\cdots f(\cdots) \cdots))$, and $\Delta = \{1^n \mid n \geq 0\}$. We can now choose $\Psi = \{(\epsilon, \alpha), (1, \beta), (1^2, \beta), (1^3, \alpha), (1^4, \beta), (1^5, \beta), (1^6, \beta), \ldots\}$. Hence, s rewrites to $g(h^2(g(h^3(\cdots g(h^n(\cdots)) \cdots))))$, which is clearly not rational.

Remark 4.4. Another difference between our work and [6] surfaces when we consider rules that are not left-linear. Consider, e.g., $\mathcal{R} = \{f(x,x) \rightarrow g(x)\}$, $E = \{x = f(y,y), y = g(x)\}$, $s = E^*(x)$ and $\Delta = \{1^{2n} \mid n \geq 0\}$. Rewriting any redex at a position in Δ is problematic, as it causes any redex at a prefix position to cease to exist.

Employing the definition from [6], the above problematic rewrites would be allowed — albeit falling outside the framework, as $\mathcal{R}$ is not orthogonal. On the other hand, we do not have $s \rightarrow^{\Delta}_{l \rightarrow r} t$ for any t: Suppose a regular representation E and a set $W \subseteq \mathcal{D}om(E)$ exist such that $\mathcal{SP}_{E_x}(W) = \Delta$ and such that for all $y \in W$ we have $E(y) = f(x', x')\rho_y$ with ρ_y a variable substitution. Consider any $y \in W$ and suppose $\rho_y(x') = z$. By definition of Δ, there is a $v \in W$ such that $\mathcal{SP}_{E_z}(v)$ is non-empty. However, since the same variable occurs twice in $f(x', x')$, $\mathcal{SP}_{E_z}(v)$ contains both positions of the form $1 \cdot p$ and of the form $2 \cdot p$, contradicting the assumption that $\mathcal{SP}_{E_x}(W)$ contains only positions with 1s.

Rational Patterns. As witnessed by Example 4.2, there are in general many ways to rewrite a rational term. The notion of a rational pattern is helpful to characterize each rewrite step uniquely. The definition of this notion, uses a labeling of rational terms.

Let Σ be a signature. We define a *marked signature* as $\Sigma^{\bullet} = \Sigma \cup \{f^{\bullet} \mid f \in \Sigma\}$, where for every $f \in \Sigma$ the arity of $f^{\bullet}$ is equal to the arity of f. The notion of a marked signature originates from [11].

Definition 4.5 (Labeling). *Let* $t \in \mathcal{T}(\Sigma, \mathcal{V})$ *and* $\Delta \subseteq \mathcal{Pos}_{\Sigma}(t)$. *The labeled term* $lab(\Delta, t)$ *over the signature* $\Sigma^{\bullet}$ *is defined by*

$$lab(\Delta, t)(p) = \begin{cases} t(p)^{\bullet} & \text{if } p \in \Delta \\ t(p) & \text{if } p \notin \Delta \end{cases}$$

Lemma 4.6. *If* t *is a rational term and* $\Delta \subseteq \mathcal{Pos}_{\Sigma}(t)$ *a regular set of positions, then* $lab(\Delta, t)$ *is rational and a minimal canonical representation* E_x *of* $lab(\Delta, t)$ *and a set* $W \subseteq \mathcal{D}om(E)$ *can be constructed effectively such that* $\Delta = \mathcal{SP}_{E_x}(W)$.

Denoting $lab(\{\epsilon\}, l)$ by $l^{\bullet}$, we define the following.

Definition 4.7 (Rational Pattern). *Let* t *be a rational term. A pair* $\langle l, \Delta \rangle$ *with* l *the left-hand side of a rewrite rule and* $\Delta \subseteq \mathcal{Pos}_{\Sigma}(t)$ *is a* rational pattern *in* t *if (1)* Δ *is a regular set of positions in* t, *and (2) for any* $p \in \Delta$, $lab(\Delta, t)|_p = l^{\bullet}\sigma$ *for some substitution* σ.

Example 4.8. If $E = \{x = f(y,y), y = g(y)\}$ and $s = E^*(x)$, then $\langle f(x, g(x)), \{\epsilon\}\rangle$ is a rational pattern in s. Moreover, if $\Delta = \{1 \cdot 1^n \mid n \geq 0\}$, then $\langle g(x), \Delta\rangle$ is a rational pattern in s, while $\langle g(g(x)), \Delta\rangle$ is not.

If $F = \{x = f(y, z), y = g(y), z = g(h(z))\}$ and $t = F^*(x)$, then $\langle g(x), \Lambda\rangle$ with $\Lambda = \{1 \cdot 1^n \mid n \geq 0\} \cup \{2 \cdot 1^{2n} \mid n \geq 0\}$ is a rational pattern in t. However, if $G = \{x = f(y,y), y = g(x)\}$, $u = G^*(x)$ and $\Gamma = \{1^{2n} \mid n \geq 0\}$, then $\langle f(x,x), \Gamma\rangle$ is not a rational pattern in u (see also Remark 4.4).

To check the second condition in the definition of a rational pattern, the following necessary and sufficient condition is useful, where $\Delta|_p = \{q \mid p \cdot q \in \Delta\}$.

Lemma 4.9. *If t is a rational term, $\Delta \subseteq Pos_\Sigma(t)$, and l is the left-hand side of a rewrite rule, then $lab(\Delta, t) = l^\bullet \sigma$ for a substitution σ iff (1) $t = l\sigma'$ for a substitution σ', (2) $\Delta \cap Pos_\Sigma(l) = \{\epsilon\}$, and (3) $\Delta|_p = \Delta|_q$ for any $p, q \in Pos_V(l)$ with $l(p) = l(q)$.*

The main result of this section is the correspondence between rational rewrite steps and rational patterns. To prove, we introduce two notions: propagation and independence.

Definition 4.10 (Propagation). *Let $E = \{x_1 = t_1, \ldots, x_n = t_n\}$ be a regular system and s a finite term.*

1. *E propagates to a regular system F for variables x_i and x_k, denoted $E \gg^i_k F$, if $t_i = C[x_k]$ in E and $F = \{x_j = t_j \in E \mid j \neq i\} \cup \{x_i = C[t_k]\}$.*
2. *s propagates to a finite term t under E, denoted $s >_E t$, if $s = C[x_i]$ and $t = C[t_i]$ for $x_i = t_i \in E$. The reflexive, transitive closure of $>_E$ is $\gg_E$.*

Observe that $s \gg_E t$ implies $E^\star(s) = E^\star(t)$.

Example 4.11. Let $E = \{x = f(y, z), y = g(z), z = h(y)\}$. We have

$$E \gg^1_2 \{x = f(g(z), z), y = g(z), z = h(y)\}$$
$$\gg^1_3 \{x = f(g(z), h(y)), y = g(z), z = h(y)\}.$$

Moreover,

$$f(y, z) >_E f(g(z), z) >_E f(g(z), h(y)) >_E f(g(h(y)), h(y)).$$

Hence, $f(y, z) \gg_E f(g(h(y)), h(y))$.

Lemma 4.12. *Let $E = \{x_1 = t_1, \ldots, x_n = t_n\}$ be a minimal canonical regular system. If s is a finite term and $E^\star(s) = E^\star(x_i)$ with $s \neq x_i$, then $t_i \gg_E s$.*

Definition 4.13 (Independence). *Let Δ be a set of positions. If l is the left-hand side of a rewrite rule, then Δ is l-independent if $p \cdot q \notin \Delta$ for any $p \in \Delta$ and $q \in Pos_\Sigma(l) \setminus \{\epsilon\}$.*

Lemma 4.14. *Let l be the left-hand side of a rewrite rule, s a rational term, $\Delta \subseteq Pos_\Sigma(s)$, and $p \in \Delta$. If there exists a substitution σ such that $lab(\Delta, s)|_p = l^\bullet \sigma$, then Δ is l-independent.*

Lemma 4.15. *Let $E = \{x_1 = t_1, \ldots, x_n = t_n\}$ be a canonical regular system and $x \in Dom(E)$. Suppose l is the left-hand side of a rewrite rule and $SP_{E_x}(W)$ is l-independent for $W \sqsubseteq x$. Moreover, suppose that for each $x_i \in W$ there exists a variable substitution ρ_i with $t_i \gg_E l\rho_i$. If $F = \{x_i = t_i \in E \mid x_i \notin W\} \cup \{x_i = l\rho_i \mid x_i \in W\}$, then (1) $E \gg^{i_1}_{k_1} \cdots \gg^{i_m}_{k_m} F$ for some $x_{i_1}, \ldots, x_{i_m} \in W$ and $x_{k_1}, \ldots, x_{k_m} \notin W$, (2) $E^\star = F^\star$, and (3) $SP_{E_x}(x_i) = SP_{F_x}(x_i)$ for any $x_i \in W$.*

Rational patterns and rational rewrite steps are related as follows.

Lemma 4.16 (Correspondence). *If s is a rational term, $l \to r$ a rewrite rule, and $\Delta \subseteq \mathcal{P}os(s)$, then $s \to_{l \to r}^{\Delta} t$ for some t iff $\langle l, \Delta \rangle$ is a rational pattern in s.*

Proof. ($\Rightarrow$) By definition, there exist regular representations E_x and F_x of s and t, resp., such that $\mathcal{D}om(E) = \mathcal{D}om(F)$ and $W \subseteq \mathcal{D}om(E)$ with (1) $\Delta = \mathcal{SP}_{E_x}(W)$, (2) $E(y) = F(y)$ for all $y \in \mathcal{D}om(E) \setminus W$, and (3) for each $y \in W$, $E(y) = l\rho$ and $F(y) = r\rho$ for some variable substitution ρ.

Without loss of generality we may assume $\mathcal{D}om(E) \cap \mathcal{R}an(E) = \emptyset$. We show that $\langle l, \Delta \rangle$ is a rational pattern in s. First, Δ is regular by Proposition 3.6(1). Moreover, as l is the left-hand side of a rewrite rule, we have $\Delta \subseteq \mathcal{P}os_\Sigma(t)$. Let $p \in \Delta$, then $lab(\Delta, t)|_p = lab(\Delta|_p, t|_p)$. To show that $lab(\Delta|_p, t|_p) = l^\bullet \sigma$ for some σ, we use Lemma 4.9.

1. As $p \in \Delta$, there is $y \in W$ such that $p \in \mathcal{SP}_{E_x}(y)$ and $E(y) = l\rho$ for a variable substitution ρ. If $\sigma' = E^\star \circ \rho$, then $s|_p = E^\star(y) = E^\star(l\rho) = (E^\star \circ \rho)(l) = l\sigma'$.
2. Obviously, $\epsilon \in \Delta|_p$, as $p \in \Delta$. Suppose there exists a $q \in \mathcal{P}os_\Sigma(l) \setminus \{\epsilon\}$ such that $p \cdot q \in \Delta$. Then, there are $y, z \in W$ with $y \neq z$ such that $p \in \mathcal{SP}_{E_x}(y)$ and $p \cdot q \in \mathcal{SP}_{E_x}(z)$. Since $q \neq \epsilon$, we have $p < q$ and by Proposition 3.6(2), $p \cdot q \notin \{p \cdot p' \mid' \in \mathcal{P}os_\Sigma(l\rho)\}$, where $l\rho = E(z)$, contradicting $q \in \mathcal{P}os_\Sigma(l)$. Hence, $\Delta|_p \cap \mathcal{P}os_\Sigma(l) = \{\epsilon\}$.
3. We have $p \in \mathcal{SP}_{E_x}(y)$ for some $y \in W$, as $p \in \Delta$. Suppose $p_1, p_2 \in \mathcal{P}os_V(l)$ and $l(p_1) = l(p_2) = z$. As $y = l\rho \in E$ for some variable substitution ρ, we have $l\rho(p_1) = l\rho(p_2) = \rho(z)$ If $\rho(z) \notin \mathcal{D}om(E)$, then $\Delta|_{p \cdot p_1} = \emptyset = \Delta|_{p \cdot p_2}$. If $\rho(z) \in \mathcal{D}om(E)$ then $\rho(z) \sqsubseteq y$ and, hence, $\Delta|_{p \cdot p_1} = \mathcal{SP}_{E_{\rho(z)}}(W) = \Delta|_{p \cdot p_2}$.

($\Leftarrow$) By assumption, $\Delta \subseteq \mathcal{P}os_\Sigma(s)$ is regular. Hence, by Lemma 4.6, there exists a minimal canonical representation E_x of $lab(\Delta, s)$ and a set $W \subseteq \mathcal{D}om(E)$ such that $\Delta = \mathcal{SP}_{E_x}(W)$. Let $E = \{x_1 = t_1, \dots, x_n = t_n\}$. By assumption, for every $p \in \Delta$, there exists a substitution σ such that $lab(\Delta, s)|_p = l^\bullet \sigma$. Hence, by Lemma 4.14, Δ is l-independent. Furthermore, for each $x \in \mathcal{V}(l)$, $\sigma(x)$ is a subterm of $lab(\Delta, s)$. Since E is minimal and canonical, we have for each $\sigma(x)$ with $x \in \mathcal{V}(l)$ that there exists unique $x_j \in \mathcal{D}om(E)$ such that $\sigma(x) = E^\star(x_j)$. Thus, for each $x_i \in W$, there exists a variable substitution ρ_i such that $E^\star(x_i) = lab(\Delta, s)|_p = l^\bullet \sigma = E^\star(l^\bullet \rho_i)$. Then, by Lemma 4.12, we obtain $t_i \gg_E l^\bullet \rho_i$. Let $\hat{E} = \{x_i = t_i \in E \mid x_i \notin W\} \cup \{x_i = l^\bullet \rho_i \mid x_i \in W\}$. By Lemma 4.15, $E^\star(x) = \hat{E}^\star(x)$ and $\Delta = \mathcal{SP}_{E_x}(W) = \mathcal{SP}_{\hat{E}_x}(W)$. Let $\tilde{E}$ be the regular system obtained from $\hat{E}$ by removing all marks. Then, $s = \tilde{E}^\star(x)$ and $s \to_{l \to r}^{\Delta} F^\star(x)$ with $F = \{x_i = t_i \in \tilde{E} \mid x_i \notin W\} \cup \{x_i = r\rho_i \mid x_i \in W\}$. $\square$

Our characterization of rewrite steps by rational patterns leads us to obtain following decidability result for rewrite steps.

Lemma 4.17 (Constructiveness). *Let s be a rational term, Δ a regular set of positions, and $l \to r$ a rewrite rule.*

1. *It is decidable whether $\langle l, \Delta \rangle$ is a rational pattern in s.*
2. *If $\langle l, \Delta \rangle$ is a rational pattern in s, then a regular representation of a rational term t can be constructed effectively such that $s \to_{l \to r}^{\Delta} t$.*

Proof. (1) Let s be a rational term and Δ a regular set of positions (possibly with $\Delta \nsubseteq Pos(s)$). We can check whether $\Delta \subseteq Pos(s)$, as both Δ and $Pos(s)$ are regular. If $\Delta \subseteq Pos(s)$ does not hold, then return *no*. By Lemma 4.6, $lab(\Delta, s)$ is regular and its regular representation E_x such that $\Delta = SP_{E_x}(W)$ for some $W \subseteq Dom(E)$ can be constructed effectively. For each $y \in W$ check whether $E^\star(y) = l^\bullet \sigma_y$ for some substitution σ_y. This check is decidable, as the matching problem for rational terms is decidable [7]. (2) As above, construct the set W and the substitution σ_y such that $E^\star(y) = l^\bullet \sigma_y$ for each $y \in W$. Next, define $F = \{y = s \in E \mid y \notin W\} \cup \{y = r\sigma_y \in E \mid y \in W\}$. Then, F_x is a regular representation of t such that $s \rightarrow^{\Delta}_{l \rightarrow r} t$. $\square$

5   Decidability and Constructive Confluence

The proofs in this section heavily depend on the notion of a product:

Definition 5.1. *Let $s \in \mathcal{T}(\Sigma, \mathcal{V})$ and $t \in \mathcal{T}(\Gamma, \mathcal{V})$ be terms. The terms are similar, denoted $s \sim t$, if $Pos_{\mathcal{V}}(s) = Pos_{\mathcal{V}}(t)$ and $Pos_{\Sigma}(s) = Pos_{\Gamma}(t)$.*

Let s and t be similar. The product $s \times t \in \mathcal{T}(\Sigma \times \Gamma, \mathcal{V})$ *of s and t is $(s \times t)(p) = \langle s(p), t(p) \rangle$ for all p, where tuples $\langle x, y \rangle$ are considered to be variables. Moreover, let E and F be canonical regular systems over, resp., Σ and Γ. The* product $E \times F$ *over $\Sigma \times \Gamma$ is*

$$E \times F = \{\langle x, y \rangle = s \times t \mid x = s \in E,\ y = t \in F,\ s \sim t\}.$$

Let $t \in \mathcal{T}(\Sigma \times \Gamma, \mathcal{V})$. The projections π_1 *and* π_2 *to Σ and Γ, resp., are defined by*

$$\pi_i(t)(p) = \begin{cases} f_i & \text{if } t(p) = \langle f_1, f_2 \rangle \in \Sigma \times \Gamma \\ x & \text{if } t(p) = x \in \mathcal{V} \end{cases}$$

where $i \in \{1, 2\}$. Moreover, let E be a canonical regular system over $\Sigma \times \Gamma$. The projections π_1 and π_2 to Σ and Γ, resp., are defined by

$$\pi_i(E) = \{\pi_i(x) = \pi_i(t) \mid x = t \in E\},$$

where $i \in \{1, 2\}$.

The following properties of products are proved in a straightforward way.

Proposition 5.2. *Let s and t be similar rational terms and let E_x and F_y be canonical regular representations of, resp., s and t. If $z = \langle x, y \rangle$ and $G = E \times F$, then for products:*

1. $G^\star(\langle x', y' \rangle) = E^\star(x') \times F^\star(y')$ for $\langle x', y' \rangle \sqsubseteq z$;
2. $SP_{G_z}(\langle x', y' \rangle) = SP_{E_x}(x') \cap SP_{F_y}(y')$.

Moreover, for projections:

3. $(\pi_1(G))^\star(\langle x', y' \rangle) = E^\star(x')$ and $(\pi_2(G))^\star(\langle x', y' \rangle) = F^\star(y')$ for $\langle x', y' \rangle \sqsubseteq z$;
4. $SP_{E_x}(x') = SP_{G_z}(\{x'\} \times Dom(F))$ and $SP_{F_y}(y') = SP_{G_z}(Dom(E) \times \{y'\})$.

The next lemma, which will be used below, follows from Proposition 5.2(2).

Lemma 5.3. *Let E_x and F_y be canonical regular representations of a term s and let $V \sqsubseteq_E x$ and $W \sqsubseteq_F y$ with $\mathcal{SP}_{E_x}(V) = \mathcal{SP}_{F_y}(W) = \Delta$.*

1. *If $U = \{\langle x', y' \rangle \sqsubseteq_{E \times F} \langle x, y \rangle \mid x' \in V, y' \in W \}$, then $\mathcal{SP}_{(E \times F)_{\langle x, y \rangle}}(U) = \Delta$.*
2. *$\{\langle x', y' \rangle \sqsubseteq_{E \times F} \langle x, y \rangle \mid x' \in V, y' \notin W \} = \emptyset$ and $\{\langle x', y' \rangle \sqsubseteq_{E \times F} \langle x, y \rangle \mid x' \notin V, y' \in W \} = \emptyset$.*

The next lemma shows that a rewrite step $s \to^{\Delta}_{l \to r} t$ is uniquely defined by a term s, a regular set Δ of positions in s, and a rewrite rule $l \to r$.

Lemma 5.4. *If $s \to^{\Delta}_{l \to r} t_1$ and $s \to^{\Delta}_{l \to r} t_2$, then $t_1 = t_2$.*

Proof. By assumption, there are representations E_x and F_y of s and representations E'_x and F'_x of t_1 and t_2, resp., with $V \sqsubseteq x$ and $W \sqsubseteq y$ such that $\Delta = \mathcal{SP}_{E_x}(V) = \mathcal{SP}_{F_y}(W)$ and $x_i = l\rho_i$ for all $x_i \in V$ and $y_j = l\delta_j$ for all $y_j \in W$ with ρ_i and δ_j variable substitutions. Without loss of generality, we may assume that E and E', resp. F and F', are canonical with exception of the equations with a variable from V, resp. W, on the left-hand side. Apply Proposition 3.8 to obtain canonical representations $\hat{E}_x$ and $\hat{F}_y$. Since $E^\star(x) = F^\star(y) = s$, we have that $(\hat{E} \times \hat{F})_{\langle x, y \rangle}$ is well-defined and a regular representation of $s \times s$.

Let $U = \{\langle x_i, y_j \rangle \sqsubseteq \langle x, y \rangle \mid x_i \in V, y_j \in W \}$ and $\hat{G} = \pi_1(\hat{E} \times \hat{F}) = \pi_2(\hat{E} \times \hat{F})$. By Lemma 5.3(1), $\mathcal{SP}_{(\hat{E} \times \hat{F})_{\langle x, y \rangle}}(U) = \Delta$ and, by Proposition 5.2(3), $\hat{G}^\star(\langle x_i, y_j \rangle) = \hat{E}^\star(x_i) = \hat{F}^\star(y_j)$, for any $\langle x_i, y_j \rangle \sqsubseteq \langle x, y \rangle$. Furthermore, for every $\langle x_i, y_j \rangle \in U$, we have $E^\star(x_i) = E^\star(l\rho_i)$ and $F^\star(y_j) = F^\star(l\delta_j)$. Hence, by Proposition 5.2(1), for every $\langle x_i, y_j \rangle \in U$, we have $(E \times F)^\star(\langle x_i, y_j \rangle) = E^\star(x_i) \times F^\star(y_j) = E^\star(l\rho_i) \times F^\star(l\delta_j) = (E \times F)^\star((l \times l)\xi_{i,j})$ where $\xi_{i,j} = \{\langle z, z \rangle := \langle \rho_i(z), \delta_j(z) \rangle \mid z \in \mathcal{V}(l) \}$. Hence, $\hat{G}^\star(\langle x_i, y_j \rangle) = \pi_1((E \times F)^\star(\langle x_i, y_j \rangle)) = \pi_1((E \times F)^\star((l \times l)\xi_{i,j})) = \hat{G}^\star(l\xi_{i,j})$. And, by Lemma 3.2, $\hat{G}^\star = G^\star$ where $G = \{\langle x_i, y_j \rangle = w \in \hat{G} \mid \langle x_i, y_j \rangle \notin U \} \cup \{\langle x_i, y_j \rangle = l\xi_{i,j} \mid \langle x_i, y_j \rangle \in U \}$.

Let $G' = \{\langle x_i, y_j \rangle = w \in G \mid \langle x, y \rangle \mid \langle x_i, y_j \rangle \notin U \} \cup \{\langle x_i, y_j \rangle = r\xi_{i,j} \mid \langle x_i, y_j \rangle \in U, \langle x_i, y_j \rangle = l\xi_{i,j} \in G \}$. Define $\delta : \mathcal{D}om(G') \to \mathcal{D}om(E')$ as $\delta(\langle x_i, y_j \rangle) = x_i$ and observe that $\{\langle x_i, y_j \rangle \sqsubseteq \langle x, y \rangle \mid x_i \in V, y_j \notin W \}$ is empty by Lemma 5.3(2). Thus, δ is surjective and for any $\langle x_i, y_j \rangle = w \in G'$ we have $\delta(\langle x_i, y_j \rangle) = \delta(w) \in E'$. By Lemma 3.3, we now have $G'^\star(\langle x_i, y_j \rangle) = E'^\star(x_i)$. Similarly, we have $G'^\star(\langle x_i, y_j \rangle) = F'^\star(y_j)$. Thus, $t_1 = E'^\star(x) = G'^\star(\langle x, y \rangle) = F'^\star(y) = t_2$. □

Theorem 5.5. *Let $\mathcal{R}$ be a TRS.*

1. *For a rational term s and regular set of positions Δ, the number of rational terms t such that $s \to^{\Delta}_{\mathcal{R}} t$ is finite and a regular representation of each term t can be constructed effectively.*
2. *For rational terms s and t and a regular set of positions Δ, it is decidable whether $s \to^{\Delta}_{\mathcal{R}} t$ or not.*
3. *For a given regular representation of rational term s, it is decidable whether s is in normal form with respect to $\mathcal{R}$. If not, a regular representation of a rational term t such that $s \to_{\mathcal{R}} t$ can be constructed effectively.*

Proof. (1) By Lemma 5.4, for a given s, Δ, and $l \to r$ there is at most one t such that $s \to_{l \to r}^{\Delta} t$. Thus, the number of terms t such that $s \to_{\mathcal{R}}^{\Delta} t$ is bounded by $|\mathcal{R}|$. Furthermore, we can effectively construct a term t with $s \to_{l \to r}^{\Delta} t$ by Lemma 4.17 if there is such a term. (2) Immediately by the previous. (3) A rational term s is $\mathcal{R}$-normal if there is no subterm u of s with $u = l\sigma$ for some $l \to r \in \mathcal{R}$ and σ. Since the (finite) set of subterms of any rational term is effectively constructed, this is decidable. Take a canonical representation E_x of s. Then for any subterm s' of s there is variable $y \in \mathcal{D}om(E)$ such that $s' = E^\star(y)$. We can check whether $E^\star(y) = l\sigma$ for some rule $l \to r \in \mathcal{R}$ and substitution σ. If there is such a rule and such a substitution, then there exists a variable substitution ρ such that $E^\star(y) = E^\star(l\rho)$ by Proposition 3.7. Define $E' = (E \setminus \{y = t\}) \cup \{y = l\rho\}$. By Lemma 3.2, $E'^\star_x$ is a representation of s. Moreover, we can effectively construct $F' = (E' \setminus \{y = l\rho\}) \cup \{y = r\rho\}$ and, hence, $s \to_{\mathcal{R}} t$ for $t = F'^\star(x)$.   □

Let s and t be finite terms with $\mathcal{V}(s) \cap \mathcal{V}(t) = \emptyset$. Then, s *overlaps* t (at position p) if there exists a non-variable subterm $u = t|_p$ of t such that u and s are unifiable in $\mathcal{T}_{\mathit{fin}}(\Sigma, \mathcal{V})$. Let $l_1 \to r_1$ and $l_2 \to r_2$ be rewrite rules. Suppose l_1 overlaps l_2 at position p (without loss of generality $\mathcal{V}(l_1) \cap \mathcal{V}(l_2) = \emptyset$). Let σ be the most general unifier of l_1 and $l_2|_p$. The pair $\langle l_2[r_1]_p\sigma, r_2\sigma \rangle$ is called a *critical pair*, where $l_2[r_1]_p\sigma$ is the result of rewriting $l_2\sigma$ by means of $l_1 \to r_1$ at position p. In case of self-overlap (i.e., when $l_1 \to r_1$ and $l_2 \to r_2$ are identical modulo renaming of variables), we do not consider $p = \epsilon$. The critical pair $\langle l_2[r_1]_p\sigma, r_2\sigma \rangle$ is *trivial* if $l_2[r_1]_p\sigma = r_2\sigma$. A TRS $\mathcal{R}$ is *orthogonal* if it is left-linear and there is no critical pairs and *weakly orthogonal* if all critical pairs are trivial instead.

Lemma 5.6 (Diamond Property). *Let $\mathcal{R}$ be an orthogonal TRS. For a rational term s and regular set of positions Δ and Γ such that $s \to_{\mathcal{R}}^{\Delta} t_1$ and $s \to_{\mathcal{R}}^{\Gamma} t_2$, a regular representation of a rational term u and regular sets of positions Δ' and Γ' such that $t_1 \to_{\mathcal{R}}^{\Delta'} u$ and $t_2 \to_{\mathcal{R}}^{\Gamma'} u$ can be constructed effectively.*

Proof. By assumption, there are representations E_x and F_y of s and representations E'_x and F'_x of t_1 and t_2, resp., with $V \sqsubseteq x$ and $W \sqsubseteq y$ such that $\Delta = \mathcal{SP}_{E_x}(V)$ and $\Gamma = \mathcal{SP}_{F_y}(W)$ and such that $x_i = l_1\rho_i$ for a rule $l_1 \to r_1$ and all $x_i \in V$ and $y_j = l_2\delta_j$ for a rule $l_2 \to r_2$ and all $y_j \in W$ with ρ_i and δ_j variable substitutions. Apply Proposition 3.8 to obtain canonical representations $\hat{E}_x$ and $\hat{F}_y$. Since $E^\star(x) = F^\star(y) = s$, we have that $(\hat{E} \times \hat{F})_{\langle x, y \rangle}$ is well-defined and a regular representation of $s \times s$. By Proposition 5.2(4), we have $\Delta = \mathcal{SP}_{\hat{E}_x}(V) = \mathcal{SP}_{(\hat{E} \times \hat{F})_{\langle x, y \rangle}}(V')$ with $V' = \{\langle x_i, y_j \rangle \in V \times \mathcal{D}om(\hat{F}) \mid \langle x_i, y_j \rangle \sqsubseteq \langle x, y \rangle\}$. Similarly, $\Gamma = \mathcal{SP}_{\hat{E}_y}(W) = \mathcal{SP}_{(\hat{E} \times \hat{F})_{\langle x, y \rangle}}(W')$, where $W' = \{\langle x_i, y_j \rangle \in \mathcal{D}om(\hat{E}) \times W \mid \langle x_i, y_j \rangle \sqsubseteq \langle x, y \rangle\}$.

Let $\hat{G} = \pi_1(\hat{E} \times \hat{F}) = \pi_2(\hat{E} \times \hat{F})$. Then, $\Delta = \mathcal{SP}_{\hat{G}_{\langle x, y \rangle}}(V')$ and $\Gamma = \mathcal{SP}_{\hat{G}_{\langle x, y \rangle}}(W')$. Moreover, by Proposition 5.2(3), we have for any $\langle x_i, y_j \rangle \sqsubseteq \langle x, y \rangle$ that $\hat{G}^\star(\langle x_i, y_j \rangle) = E^\star(x_i) = F^\star(y_j)$; in particular, $\hat{G}_{\langle x, y \rangle}$ is a canonical regular representation of s. Furthermore, for every $\langle x_i, y_j \rangle \in V'$, we have $\hat{E}^\star(x_i) = \hat{E}^\star(l_1\rho_i)$ and, for every $\langle x_i, y_j \rangle \in W'$, $\hat{F}^\star(y_j) = \hat{F}^\star(l_2\delta_j)$. Hence, by Lemma 3.2, $\hat{G}^\star = G_1^\star$ where $G_1 = \{\langle x_i, y_j \rangle = w \in \hat{G} \mid \langle x_i, y_j \rangle \notin V'\} \cup \{\langle x_i, y_j \rangle = l_1\rho_{i,j} \mid \langle x_i, y_j \rangle \in V'\}$ and $\rho_{i,j}$ is such that $\hat{G}^\star(\langle x_i, y_j \rangle) = \hat{G}^\star(l_1\rho_{i,j})$, where existence of

$\rho_{i,j}$ follows as $\hat{G}$ is canonical and as every rewrite rule is left-linear. Similarly, $\hat{G}^\star = G_2^\star$ where $G_2 = \{\langle x_i, y_j \rangle = w \in \hat{G} \mid \langle x_i, y_j \rangle \notin W'\} \cup \{\langle x_i, y_j \rangle = l_2 \delta_{i,j} \mid \langle x_i, y_j \rangle \in W'\}$ and $\delta_{i,j}$ is such that $\hat{G}^\star(\langle x_i, y_j \rangle) = \hat{G}^\star(l_2 \delta_{i,j})$.

Let $G_1' = \{\langle x_i, y_j \rangle = w \in \hat{G} \mid \langle x_i, y_j \rangle \notin V'\} \cup \{\langle x_i, y_j \rangle = r_1 \rho_{i,j} \mid \langle x_i, y_j \rangle \in V'\}$ and $G_2' = \{\langle x_i, y_j \rangle = w \in \hat{G} \mid \langle x_i, y_j \rangle \notin W'\} \cup \{\langle x_i, y_j \rangle = r_2 \delta_{i,j} \mid \langle x_i, y_j \rangle \in W'\}$. Then, $s = G_1^\star(\langle x, y \rangle) \to^{\Delta} G_1'^\star(\langle x, y \rangle)$ and, hence, $t_1 = G_1'^\star(\langle x, y \rangle)$ by Lemma 5.4. Similarly, we have $t_2 = G_2'^\star(\langle x, y \rangle)$. We now distinguish two cases, where we observe that l_1 and l_2 do not overlap, as $\mathcal{R}$ is orthogonal.

1. Let $l_1 \to r_1 = l_2 \to r_2 = l \to r$. For any $\langle x_i, y_j \rangle \in V' \cap W'$, we have $\hat{G}^\star(\langle x_i, y_j \rangle) = \hat{G}^\star(l \rho_{i,j}) = \hat{G}^\star(l \delta_{i,j})$, and, hence, $\rho_{i,j} = \delta_{i,j}$. Let

$$\begin{aligned}
G'' = \{&\langle x_i, y_j \rangle = w \in \hat{G} \mid \langle x_i, y_j \rangle \notin V' \cup W'\} \\
\cup \ &\{\langle x_i, y_j \rangle = r \rho_{i,j} \mid \langle x_i, y_j \rangle \in V' \cap W'\} \\
&\text{(which is equivalent to } \{\langle x_i, y_j \rangle = r \delta_{i,j} \mid \langle x_i, y_j \rangle \in V' \cap W'\}) \\
\cup \ &\{\langle x_i, y_j \rangle = r \rho_{i,j} \mid \langle x_i, y_j \rangle \in V' \setminus W'\} \\
\cup \ &\{\langle x_i, y_j \rangle = r \delta_{i,j} \mid \langle x_i, y_j \rangle \in W' \setminus V'\}.
\end{aligned}$$

Then, $t_1 = G_1'^\star(\langle x, y \rangle) \to^{\Delta'}_{l \to r} G''^\star(\langle x, y \rangle)$, where $\Delta' = \mathcal{SP}_{G_1'}(W' \setminus V')$, and $t_2 = G_2'^\star(\langle x, y \rangle) \to^{\Gamma'}_{l \to r} G''^\star(\langle x, y \rangle)$, where $\Gamma' = \mathcal{SP}_{G_2'}(V' \setminus W')$. Furthermore, Δ', Γ', and $G''_{\langle x, y \rangle}$ can be constructed effectively.

2. Let $l_1 \to r_1 \neq l_2 \to r_2$. Since $\hat{G}^\star(\langle x_i, y_j \rangle) = \hat{G}^\star(l_1 \rho_{i,j})$ for any $\langle x_i, y_j \rangle \in V'$ and $\hat{G}^\star(\langle x_i, y_j \rangle) = \hat{G}^\star(l_2 \delta_{i,j})$ for any $\langle x_i, y_j \rangle \in W'$ and since l_1 and l_2 are not variables, we obtain $V' \cap W' = \emptyset$. Let

$$\begin{aligned}
G'' = \{&\langle x_i, y_j \rangle = w \in \hat{G} \mid \langle x_i, y_j \rangle \notin V' \cup W'\} \\
\cup \ &\{\langle x_i, y_j \rangle = r_1 \rho_{i,j} \mid \langle x_i, y_j \rangle \in V'\} \\
\cup \ &\{\langle x_i, y_j \rangle = r_2 \delta_{i,j} \mid \langle x_i, y_j \rangle \in W'\}.
\end{aligned}$$

Then, $t_1 = G_1'^\star(\langle x, y \rangle) \to^{\Delta'}_{l_2 \to r_2} G''^\star(\langle x, y \rangle)$, where $\Delta' = \mathcal{SP}_{G_1'}(W')$, and $t_2 = G_2'^\star(\langle x, y \rangle) \to^{\Gamma'}_{l_1 \to r_1} G''^\star(\langle x, y \rangle)$, where $\Gamma' = \mathcal{SP}_{G_2'}(V')$. Furthermore, Δ', Γ', and $G''_{\langle x, y \rangle}$ can be constructed effectively. $\qquad \square$

Example 5.7. Let $E = \{x = \mathsf{f}(y), y = \mathsf{f}(x)\}$, $F = \{x' = \mathsf{f}(y'), y' = \mathsf{f}(z'), z' = \mathsf{f}(x')\}$, $s = E^\star(x) = F^\star(x')$, $\Delta = \{1 \cdot 1^{2n} \mid n \geq 0\}$, $\Gamma = \{11 \cdot 1^{3n} \mid n \geq 0\}$, and $\mathcal{R} = \{\mathsf{f}(x) \to \mathsf{g}(x)\}$. We have $s \to^{\Delta} t_1 = \{x = \mathsf{f}(y), y = \mathsf{g}(x)\}^\star(x)$ with $\Delta = \mathcal{SP}_{E_x}(\{y\})$ and $s \to^{\Gamma} t_2 = \{x' = \mathsf{f}(y'), y' = \mathsf{f}(z'), z' = \mathsf{g}(x')\}^\star(x')$ with $\Gamma = \mathcal{SP}_{F_x}(\{z'\})$. The term u with $t_1 \to u \leftarrow t_2$ is now constructed as follows: First we take product of E and F. We get

$$\begin{aligned}
(E \times F) {\upharpoonright} \langle x, x' \rangle = \\
\{&\langle x, x' \rangle = \mathsf{f}(\langle y, y' \rangle), \langle y, y' \rangle = \mathsf{f}(\langle x, z' \rangle), \langle x, z' \rangle = \mathsf{f}(\langle y, x' \rangle), \\
&\langle y, x' \rangle = \mathsf{f}(\langle x, y' \rangle), \langle x, y' \rangle = \mathsf{f}(\langle y, z' \rangle), \langle y, z' \rangle = \mathsf{f}(\langle x, x' \rangle)\}.
\end{aligned}$$

We now have $\Delta = (E \times F)_{\langle x, x' \rangle}(V)$ with $V = \{\langle y, x' \rangle, \langle y, y' \rangle, \langle y, z' \rangle\}$ and $\Gamma = (E \times F)_{\langle x, x' \rangle}(W)$ with $W = \{\langle x, z' \rangle, \langle y, z' \rangle\}$. Hence, we define

$$\begin{aligned}
u = \{&\langle x, x' \rangle = \mathsf{f}(\langle y, y' \rangle), \langle y, y' \rangle = \mathsf{g}(\langle x, z' \rangle), \langle x, z' \rangle = \mathsf{g}(\langle y, x' \rangle), \\
&\langle y, x' \rangle = \mathsf{g}(\langle x, y' \rangle), \langle x, y' \rangle = \mathsf{f}(\langle y, z' \rangle), \langle y, z' \rangle = \mathsf{g}(\langle x, x' \rangle)\}^\star(\langle x, x' \rangle)
\end{aligned}$$

and we obtain $t_1 \to^{\Delta'} u$ with $\Delta' = (E \times F)_{\langle x, x' \rangle}(W \setminus V)$ and $t_2 \to^{\Gamma'} u$ with $\Gamma' = (E \times F)_{\langle x, x' \rangle}(V \setminus W)$.

Lemma 5.4 allows us to define rewrite sequences $s_1 \to^*_{\mathcal{R}} s_n$ which are given by $\langle s_i, \Delta_i, l_i \to r_i \rangle_{1 \leq i \leq n-1}$ with $s_i \to^{\Delta_i}_{l_i \to r_i} s_{i+1}$ for each $1 \leq i \leq n-1$.

Theorem 5.8 (Confluence). *Let $\mathcal{R}$ is an orthogonal TRS. Given rewrite sequences $s \to^*_{\mathcal{R}} t_1$ and $s \to^*_{\mathcal{R}} t_2$, a rational term u and rewrite sequences $t_1 \to^*_{\mathcal{R}} u$ and $t_2 \to^*_{\mathcal{R}} u$ can be constructed effectively.*

Proof. Immediate by Lemma 5.6. □

The following example shows that weakly orthogonal TRS may be non-confluent in rational term rewriting. Hence, the above theorem cannot be extended to weakly orthogonal rewriting.

Example 5.9. Let $\mathcal{R} = \{\mathsf{p}(\mathsf{s}(x)) \to x, \mathsf{s}(\mathsf{p}(x)) \to x\}$. There are two critical pairs in $\mathcal{R}$, $\langle \mathsf{p}(x), \mathsf{p}(x) \rangle$ and $\langle \mathsf{s}(x), \mathsf{s}(x) \rangle$, which are both trivial. Hence, $\mathcal{R}$ is weakly orthogonal. Suppose $t = \{x = \mathsf{p}(\mathsf{s}(x))\}^\star(x)$. We have $t \to_{\mathcal{R}} \bot$ and $t \to_{\mathcal{R}} \mathsf{p}(\bot)$. Since $\bot$ and $\mathsf{p}(\bot)$ are distinct normal forms, they are not joinable. Hence, $\mathcal{R}$ is not confluent.

Non-confluence of weakly orthogonal systems may seem remarkable from the perspective of finitary rewriting, especially since confluence for weakly orthogonal systems does hold in a term graph rewriting formalism that allows for copying of subterms [1]. However, it is less remarkable from the infinitary perspective: A similar counterexample can be constructed within the approach to infinitary rewriting as taken by Kennaway et al.; that counterexample can be found in [8].

6 Conclusion

We have introduced rational term rewriting based on regular systems. Unlike [5,6], which is specialized to orthogonal rewrite systems, our definition is amenable to arbitrary rewrite systems. We have shown that our notion of rewrite steps is characterized by rational patterns. Moreover, we have shown decidable properties concerning our rewrite steps and constructive confluence for orthogonal rewrite systems, i.e., that for two rewrite sequences $s \to^* t_1$ and $s \to^* t_2$ we can effectively construct a regular system representing a rational term u and rewrite sequences $t_1 \to^* u$ and $t_2 \to^* u$. We have also given an example showing that rational term rewriting is in general not confluent for weakly orthogonal rewrite systems.

It is important to know for which classes of TRSs $\mathcal{R}$, $s \leftrightarrow^*_{\mathcal{R}} t$ can be decided. Even for an orthogonal TRS $\mathcal{R}$, our results do not immediately imply decidability of $s \leftrightarrow^*_{\mathcal{R}} t$. As future work, we plan to address this problem.

Acknowledgments. We would like to thank the reviewers for their valuable comments.

References

1. Ariola, Z.M., Klop, J.W.: Equational Term Graph Rewriting. Fundamenta Informaticae 26, 207–240 (1996)
2. Baader, F., Nipkow, T.: Term Rewriting and All That. Cambridge University Press (1998)
3. Barendsen, E.: Term Graph Rewriting. In: Terese (ed.) [17], ch. 13, pp. 712–743
4. Corradini, A.: Term Rewriting in CT_Σ. In: Gaudel, M.-C., Jouannaud, J.-P. (eds.) CAAP 1993, FASE 1993, and TAPSOFT 1993. LNCS, vol. 668, pp. 468–484. Springer, Heidelberg (1993)
5. Corradini, A., Drewes, F.: Term Graph Rewriting and Parallel Term Rewriting. In: TERMGRAPH 2011. EPTCS, vol. 48, pp. 3–18 (2011)
6. Corradini, A., Gadducci, F.: Rational Term Rewriting. In: Nivat, M. (ed.) FOSSACS 1998. LNCS, vol. 1378, pp. 156–171. Springer, Heidelberg (1998)
7. Courcelle, B.: Fundamental Properties of Infinite Trees. Theoretical Computer Science 25(2), 95–169 (1983)
8. Endrullis, J., Grabmayer, C., Hendriks, D., Klop, J.W., van Oostrom, V.: Unique Normal Forms in Infinitary Weakly Orthogonal Rewriting. In: RTA 2010. LIPIcs, vol. 6, pp. 85–102. Schloss Dagstuhl (2010)
9. Goguen, J.A., Thatcher, J.W., Wagner, E.G., Wright, J.B.: Initial Algebra Semantics and Continuous Algebras. Journal of the ACM 24(1), 68–95 (1977)
10. Inverardi, P., Zilli, M.V.: Rational Rewriting. In: Privara, I., Ružička, P., Rovan, B. (eds.) MFCS 1994. LNCS, vol. 841, pp. 433–442. Springer, Heidelberg (1994)
11. Kennaway, J.R., Klop, J.W., Sleep, M.R., de Vries, F.-J.: On the Adequacy of Graph Rewriting for Simulating Term Rewriting. ACM Transactions on Programming Languages and Systems 16(3), 493–523 (1994)
12. Kennaway, R., de Vries, F.-J.: Infinitary Rewriting. In: Terese (eds.) [17], ch. 12, pp. 668–711
13. Kennaway, R., Klop, J.W., Sleep, R., de Vries, F.-J.: Transfinite Reductions in Orthogonal Term Rewriting Systems. Information and Computation 119(1), 18–38 (1995)
14. Ketema, J., Simonsen, J.G.: Computing with Infinite Terms and Infinite Reductions (unpublished manuscript)
15. Plump, D.: Collapsed Tree Rewriting: Completeness, Confluence, and Modularity. In: Rusinowitch, M., Remy, J.-L. (eds.) CTRS 1992. LNCS, vol. 656, pp. 97–111. Springer, Heidelberg (1993)
16. Plump, D.: Term Graph Rewriting. In: Handbook of Graph Grammars and Computing by Graph Transformation. Applications, Languages and Tools, vol. 2, pp. 3–61. World Scientific (1999)
17. Terese (ed.): Term Rewriting Systems. Cambridge Tracts in Theoretical Computer Science, vol. 55. Cambridge University Press (2003)
18. Weihrauch, K.: Computable Analysis: An Introduction. Springer (2000)

A General Attribution Concept for Models in $\mathcal{M}$-Adhesive Transformation Systems

Ulrike Golas

Konrad-Zuse-Zentrum für Informationstechnik Berlin, Germany
golas@zib.de

Abstract. Attributes are an important concept for modeling data in practical applications. Up to now there is no adequate way to define attributes for different kinds of models used in $\mathcal{M}$-adhesive transformation systems, which are a special kind of graph transformation system based on $\mathcal{M}$-adhesive categories. Especially a proper representation and definition of attributes and their values as well as a suitable handling of the data does not fit well with other graph transformation formalisms.

In this paper, we propose a new method to define attributes in a natural, but still formally precise and widely applicable way. We define a new kind of adhesive category, called $\mathcal{W}$-adhesive, that can be used for transformations of attributes, while the underlying models are still $\mathcal{M}$-adhesive ones. As a result, attributed models can be used as they are intended to be, but with a formal background and proven well-behavior.

1 Introduction

Graph transformation is a well-known formalism for the rule-based derivation of graphs [1,2]. For different application areas, different graph types are necessary to express all the properties of the underlying models. An underlying category theoretical framework provides the possibility to once and for all prove different properties of graph transformation, like the local Church-Rosser property or local confluence, on the categorical level and then transfer it by instantiation to the different graphs. $\mathcal{M}$-adhesive categories [3] and their variants like adhesive [4] and weak adhesive HLR [5] categories form such a framework for graph transformations and their analysis. They are based on a distinguished morphism class $\mathcal{M}$, which is used for the rule morphisms and has to satisfy certain properties, and a special van Kampen (VK) property describing the compatibility of gluings and restrictions, i.e. pushouts and pullbacks in categorical terms. Based on this van Kampen property, many results can be proven that hold for all kinds of graphs.

Attributes play an important role for modeling, for example in object-oriented system models or network graphs. But up to now, no general method for the definition of attributes in arbitrary models has been found which represents attributes as understood in object-oriented systems. In theoretical contributions, often an algebraic approach combined with graphs is used which has disadvantages like allowing multiple attribute values for one attribute or integrating the complete data values into the graph structure leading to infinite graphs. While

H. Ehrig et al.(Eds.): ICGT 2012, LNCS 7562, pp. 187–202, 2012.

in other contexts, like conflict resolution of diverging transformations in model versioning [6], multiple attribute values may be useful, they are unexpected and lead to strange behavior in the general modeling of object-oriented systems.

The idea of this contribution is a separation of concerns, i.e. the separation of the graph structure and their attribution and data. The goal is to have a formalism that allows users to use attributes as in object-oriented models, but with a formal background. It should be applicable to all kinds of graphs and graph-like structures and compatible with well-known graph transformation approaches. We will not achieve a simple attribution concept – which does not seem to exist in a formal and wide-ranging way – but a theoretically well-founded one, which can be applied intuitively without the need to take care of unexpected side effects as necessary for current approaches. The formalization leads to the notion of W-adhesive categories, where the main idea is to restrict the VK property to those squares that actually appear in transformations.

This paper is organized as follows. In Section 2, we motivate our work by analyzing how attributes should work, how they are defined in current approaches and describe the general idea of our new attribution concept. This is defined formally in Section 3. In Section 4, we define transformations of attributed objects and show that the attribution concept does not lead to $\mathcal{M}$-adhesive categories. Instead, W-adhesive categories are defined in Section 5, where we show exemplarily the Local Church-Rosser Theorem to illustrate that this category is suitable for graph transformation. A conclusion and future work are given in Section 6.

We assume the reader to be familiar with graph transformation in the double pushout approach and the foundations of $\mathcal{M}$-adhesive transformation systems as, for example, introduced in [5]. The proofs of the theorems and facts, as far as they are not directly included in this paper, can be found in [7].

2 Motivation and Related Work

When working with attributes, most people expect attributes to behave like in object-oriented models. This means in particular that

- a model element has exactly one value for each attribute and
- model elements of different types may have the same attribute.

Unfortunately, this is not true for typed attributed graphs [5]. Typed attributed graphs are based on so-called E-graphs, which are graphs consisting of different node and edge types. The actual attribution of a node is done by an attribute edge of a certain type, representing this attribute's name, pointing from this node to the data value. Thus, also multiple attribute edges of the same type from a node form a valid graph, which means that the node has different values for the same attribute. Moreover, the attribute edges form a set which means that for different types we are not allowed to use the same attribute name. This situation is shown in Fig. 1. In the left, we see a class diagram-like graph modeling a cat with name = "Greebo" and a dog with name = "Gaspode", which is what we expect to see. In the middle, the corresponding type graph is shown. Since Cats

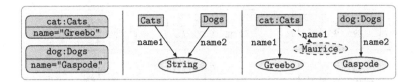

Fig. 1. Attribution in typed attributed graphs

and `Dogs` are different types, the attribute edge names have to be different, i.e. something like `name1` and `name2`. In the right, the typed attributed graph for our model is shown, where the typing of each element is denoted after the colon. While we would not expect a second name attribute for an object in the left representation, it could occur on the right, indicated by the dotted elements. To avoid this we had to define constraints forbidding the occurrence of this kind of double edges. Especially, we have to include application conditions for each rule to ensure that attribute edges can only be created if they do not already exist for the considered node. Another difficulty with this approach is that the data values are completely stored in the graph and are not only implicitly there, which means that the graph becomes infinite in general.

Symbolic graphs [8] are graphs labeled with variables and combined with formulas over these variables and a data algebra. Due to this concept, they allow for a separation of the graph and data part, such that the graphs themselves are not infinite only because of their data. In general, a symbolic graph represents a set of attributed graphs satisfying the formulas and allows for a logical reasoning about data. Nevertheless, the underlying graph structure is an E-graph, like for typed attributed graphs, and has the same behavior allowing multiple values for the same attribute, but needing different attribution names for different types.

Transformations of partially labelled graphs [9], i.e. graphs where nodes and edges may be labelled by (disjoint) label alphabets, can be treated somehow similar to attributes. But the overall setting in this approach is different from the standard version of double-pushout transformations. While we want to have injective rule morphisms and arbitrary matches, in [9] only the left rule morphism has to be injective, but in addition the match.

Other approaches also leave the context of $\mathcal{M}$-adhesive transformation systems like coding the graphs as algebras [10] or combining graphs with type theory for attribution [11], which needs a different transformation approach, since the attribution transformation is done via pullbacks instead of pushouts.

Basic Idea

For rule-based transformations, attribute values need to be preserved or refined. Note that actually changing attributes themselves would be a transformation in the meta model and is not our concern here. In addition to the creation or deletion of complete nodes including all attributes, our rules should be able to:

1. keep an attribute value, i.e. leave it unchanged (top of Fig. 2), or
2. change an attribute value, i.e. assign a new value to it (bottom of Fig. 2).

Fig. 2. Typical attribute behavior in rules

To allow this, we need "undefined" attributes, which are denoted by □. They should only appear in the interface of the rule and the corresponding intermediate graph models. For the morphisms used in rules, only certain mappings are allowed, such that all undefined attributes are concretized by a data value, and all other attribute values are identically mapped.

3   Attribution of Models

In this section, we define the category **AttC** of attributed objects with an underlying category **C**. To reach this goal, we have to formalize the ideas of the previous section. The key idea is to separate the (graphical) model and the data values used for attribution.

Attribution Values

Consider a finite set TYPES of available data types and for each $type \in$ TYPES we have a set A_{type} containing the data values. They may come, for example, from the sorts of a signature and their respective carrier sets in an algebra, an abstract data type, or a programming language. Moreover, not only concrete values (like all integers) but also abstract values (like all terms over variables of a certain signature [12]) can be used for attribution.

Definition 1 (Attribution values). *Given the data types* TYPES, *attribution values* $A = \dot{\bigcup}_{type \in \text{TYPES}} A_{type}$ *are defined by a set A_{type} of data values for each* $type \in$ TYPES.

Definition 2 (Attribution value morphism). *Consider attribution values* A_1 *and* A_2, *an attribution value morphism* $a : A_1 \to A_2$ *is defined by a family of functions* $(a_{type} : A_{1,type} \to A_{2,type})_{type \in \text{TYPES}}$ *such that* $a(x) = a_{type}(x)$ *for all* $x \in A_{1,type}$.

Example 1. For our running example, we only need one data type of strings, which means that TYPES = $\{string\}$. For models, we use as attribution values A_{string} the set of all strings. Moreover, we need attribution values for rules, where we use the set $T_{string}(\mathcal{V})$ of terms over variables $\mathcal{V} = \{*_1, *_2, *_3\}$, where certain standard string operations are defined. For an attribution value morphism $a : T(\mathcal{V}) \to A$, we use a variable assignment $\alpha : \mathcal{V} \to A$ such that a is the evaluation of terms according to α.

Attribution of Models

Consider different types of elements, which all should be attributed the same way, for example the types of a type graph. For each of these attribution types, we define a set of attribute names with a certain type. For readability, we do not allow the same attribute name with different type for one attribution type.

Definition 3 (Typed attribute). *Let $\mathcal{K}$ be a finite set whose elements are called attribution types and* Voc *a set (or vocabulary) of available attribute names. The mapping Atts* : $\mathcal{K} \to \mathcal{P}(\text{Voc} \times \text{TYPES})$ *defines a set of typed attributes for each $k \in \mathcal{K}$, if (s,t), $(s,t') \in Atts(k)$ for some $k \in \mathcal{K}$ implies that $t = t'$. The set $att_{name}(k)$ is the projection of $Atts(k)$ to its first component.*

For the attribution of the elements of a certain model, these have to be grouped into the attribution types $\mathcal{K}$. This is done by a so-called $\mathcal{K}$-functor. Note that the sets are not necessarily disjoint, which means that an element may have multiple attribution types. This can be used, for example, to express inheritance.

Definition 4 ($\mathcal{K}$-functor). *For a category $\mathbf{C}$, a pushout-preserving functor F : $\mathbf{C} \to \mathbf{Sets}^{\mathcal{K}}$ is said to be a $\mathcal{K}$-functor, where $\mathbf{Sets}^{\mathcal{K}}$ is the $|\mathcal{K}|$-fold product of $\mathbf{Sets}$. With $F_k : \mathbf{C} \to \mathbf{Sets}$ we denote the (also pushout-preserving) functor as projection of F to its k-th component.*

Using all the above definitions, we can now attribute the models. The $\mathcal{K}$-functor combined with *Atts* defines which attributes are mapped to which model element. We need an additional mapping of these attribute names to actual data values consistent with their defined type. This mapping may be partial, because we allow that certain attributes are not set. Note that for partial functions f and f', with $f(a) = f'(a)$ we denote that both $f(a)$ and $f'(a)$ are defined and equal, or both are undefined. Moreover, we write $f \leq f'$, if $f(d) = f'(d)$ for all $d \in Dom(f)$.

Definition 5 (Attributed object). *Given a category $\mathbf{C}$ and a $\mathcal{K}$-functor F : $\mathbf{C} \to \mathbf{Sets}^{\mathcal{K}}$ then $AO = (C, A, att)$ is an attributed object if $C \in \mathbf{C}$ with attribution values A and $att = (att_k : F_k(C) \times att_{name}(k) \rightarrowtail A)_{k \in \mathcal{K}}$ is a family of partial attribution functions such that for all $att_k(o, n) \in A_{type}$ we have that $(n, type) \in Atts(k)$.*

Example 2. We enrich our running example with humans. We define the attribution types $\mathcal{K} = \{pet, hum\}$ of pets and humans, where pets have a name, i.e. $Atts(pet) = \{(\mathbf{name}, String)\}$, while humans have a first and a last name, i.e. $Atts(hum) = \{(\mathbf{fname}, String), (\mathbf{lname}, String)\}$. Note that we are free to define also an attribute **name** for humans, which even may have a different type than *String* (if other types were available for our example). In this example, we only use node attribution, but of course edge attribution would also be possible – we only had to add an attribution type for the edges.

For the category $\mathbf{C}$ we choose the category of typed graphs using the type graph TG shown in the left of Fig. 3 defining cats, dogs, humans, and owns relations between humans and pets. An example typed graph G is depicted in the right of Fig. 3 defining two cats, one dog, and one human, which owns one of the cats. Note that actually we would need two different owns relations due to the uniqueness of edges in the type graph, but we omit this here. While we allow the same attributes for different elements within our attribution concept, we cannot solve the same problem for types in the underlying typed graphs. To model one owns relation for all pets we could implement an inheritance concept (see [5]) and extract a class pet.

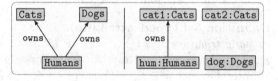

Fig. 3. Type graph TG and typed graph G

The $\mathcal{K}$-functor $F : \mathbf{Graphs_{TG}} \to \mathbf{Sets}^{\mathcal{K}}$ includes each node of type Cats or Dogs into the pet-set, and each human into the hum-set. This means that $F(G) = (\{n \mid type(n) \in \{\text{Cats}, \text{Dogs}\}\}, \{n \mid type(n) = \text{Humans}\})$, while morphisms are mapped to their corresponding components. For the example graph G in Fig. 3 this means that $F(G) = (\{\text{cat1}, \text{cat2}, \text{dog}\}, \{\text{hum}\})$.

An attributed object $AO' = (G, A, att')$ is shown in Fig. 4, where the underlying typed graph G is already given in Fig. 3. The attribution is defined by $att'_{pet}(\text{cat1}, \text{name}) = \text{"Greebo"}$, $att'_{pet}(\text{cat2}, \text{name}) = \text{"Maurice"}$, $att'_{pet}(\text{dog}, \text{name}) = \text{"Gaspode"}$, $att'_{hum}(\text{hum}, \text{fname}) = \text{"Nanny"}$, and $att'_{hum}(\text{hum}, \text{lname}) = \text{"Ogg"}$.

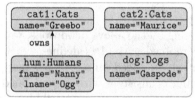

Fig. 4. An attributed graph

Morphisms

To express relations between attributed objects and apply graph transformation, we need to define attributed morphisms. We combine a valid morphism in the underlying category with an attribution value morphism to obtain attribution morphisms, where we allow to concretize undefined attribute values.

Definition 6 (Attributed morphism). *Consider a set $\mathcal{A}$ of available attribution value morphisms closed under composition.*

Given attributed objects $AO_1 = (C_1, A_1, att_1)$ and $AO_2 = (C_2, A_2, att_2)$,

$$\begin{array}{ccc} F_k(C_1) \times att_{name}(k) & \xrightarrow{\ att_{1,k}\ } & A_1 \\ {\scriptstyle F_k(g) \times id}\downarrow & \geq & \downarrow{\scriptstyle a} \\ F_k(C_2) \times att_{name}(k) & \xrightarrow{\ att_{2,k}\ } & A_2 \end{array}$$

an attributed morphism $f : AO_1 \to AO_2$ is a pair $f = (g, a)$ with $g : C_1 \to C_2 \in \mathbf{C}$ and $a : A_1 \to A_2 \in \mathcal{A}$ such that $a \circ att_{1,k} \leq att_{2,k} \circ (F_k(g) \times id)$.

We use the following convention: When given an attributed morphism f_i, we denote its components with g_i and a_i, i.e. $f_i = (g_i, a_i)$. Note that the set $\mathcal{A}$ restricts the available attribution value morphisms. This is useful if these

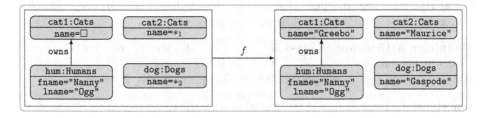

Fig. 5. An attributed morphism

morphisms shall preserve certain structure. In our example, we only want to use attribution value morphisms that stem from an evaluation based on a variable assignment.

Example 3. In Fig. 5, an attributed morphism $f = (g, a)$ from an attributed graph $AO = (G, T^{\mathcal{V}}(A), att)$ to $AO' = (G, A, att')$ is shown. The attribution of G in AO is defined by $att_{pet}(\text{cat2}, \text{name}) = *_1$, $att_{pet}(\text{dog}, \text{name}) = *_2$, $att_{hum}(\text{hum}, \text{fname}) = \text{"Nanny"}$, and $att_{hum}(\text{hum}, \text{lname}) = \text{"Ogg"}$, and $att_{pet}(\text{cat1}, \text{name})$ is undefined. AO' is already known from Fig. 4. g is the identical morphism on the graph G, while a is the evaluation with $\alpha(*_1) = \text{"Maurice"}$ and $\alpha(*_2) = \text{"Gaspode"}$.

To show that f is a valid attributed morphism we have to verify its properties for all $(o, n) \in Dom(a \circ att_k)$ with $k \in \{pet, hum\}$, which obviously holds.

Now attributed objects and attributed morphisms form a category.

Definition 7 (Category AttC). *Given a category* **C**, *a set* $\mathcal{A}$ *of attribution element morphisms, a finite set of attribution types* $\mathcal{K}$, *a* $\mathcal{K}$-*functor* F, *and a mapping Atts as above, then attributed objects and attributed morphisms, together with the component-wise composition and identities, form the category* $\textbf{AttC}^{F,\mathcal{A}}_{\textbf{Atts}}$. *If the setting is clear, we may also write* **AttC**.

Theorem 1. *The category* **AttC** *is well-defined, i.e. it is actually a category.*

Graphical Notation
Since the user does not want to know all the definitions and formal notations, attributed objects and morphisms are depicted in an UML-like notion, where the attributes are written down below the element they belong to as we already used in Fig. 4. Undefined attributes are denoted by $\square$.

4   Transformations of Attributed Models

To apply transformations in attributed objects, we need to define rules and in particular rule morphisms. For our rules, undefined attributes shall only occur in the gluing part, but not in the left or right hand side. Such undefined attributes have to be concretized. Moreover, the data should not be changed by a rule, i.e.

the attribution value morphisms should be isomorphic. The morphism class $\mathcal{R}$ represents these available morphisms.

Definition 8 (Morphism class $\mathcal{R}$). *Given an $\mathcal{M}$-adhesive category $(\mathbf{C}, \mathcal{M}_{\mathbf{C}})$, we define the morphism class $\mathcal{R}$ in $\mathbf{AttC}$ by $\mathcal{R} = \{f : AO_1 \to AO_2 \mid f = (g, a) \in \mathbf{AttC}, g \in \mathcal{M}_{\mathbf{C}}, a \text{ is isomorphism}, AO_2 = (C_2, A_2, att_2), att_2 \text{ is total}\}$.*

Unfortunately, the category $\mathbf{AttC}$ together with the morphism class $\mathcal{R}$ does not become an $\mathcal{M}$-adhesive category, because pushouts over $\mathcal{R}$-morphisms are not constructed preserving the morphism class $\mathcal{R}$.

Fact 1. *The category $(\mathbf{AttC}, \mathcal{R})$ is not an $\mathcal{M}$-adhesive category.*

Proof. The diagram in Fig. 6 shows that $\mathcal{M}$-morphisms are not stable under pushouts in $\mathbf{AttC}$. We look at the graph with one node $\mathbf{n}$ of type $\mathtt{Cats}$, where f_1 and f_2 are identities with total codomain attribution and thus in $\mathcal{R}$, and only the actual attribution of $\mathbf{n}$ is concretized. To obtain a pushout, we need a commuting square, i.e. its graph has to contain at least one node, where the nodes from B and C map to, which also has to be of type $\mathtt{Cats}$. For the attribution values, $\mathtt{"Greebo"}$ and $\mathtt{"Maurice"}$ have to be merged by a_3 and a_4 – otherwise, either f_3 or f_4 will not be a valid attributed morphism. Actually, the depicted attributed graph is the pushout of f_1 and f_2, but $f_3, f_4 \notin \mathcal{R}$. Therefore, $\mathbf{AttC}$ with the chosen $\mathcal{R}$ is no $\mathcal{M}$-adhesive category.

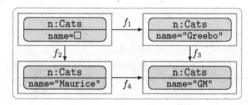

Fig. 6. Pushout in attributed graphs

Since our morphism class $\mathcal{R}$ is rather restrictive, one may argue that a more general morphism class may be suitable for $\mathbf{AttC}$ to become an $\mathcal{M}$-adhesive category. Thus we define the morphism class $\mathcal{M}$ as morphisms that are $\mathcal{M}_{\mathbf{C}}$-morphisms in the $\mathbf{C}$-component and isomorphisms for the attribution values.

Definition 9 (Morphism class $\mathcal{M}$). *Given an $\mathcal{M}$-adhesive category $(\mathbf{C}, \mathcal{M}_{\mathbf{C}})$, we define the morphism class $\mathcal{M}$ in $\mathbf{AttC}$ by $\mathcal{M} = \{f : AO_1 \to AO_2 \mid f = (g, a) \in \mathbf{AttC}, g \in \mathcal{M}_{\mathbf{C}}, a \text{ is isomorphism}\}$.*

But even $(\mathbf{AttC}, \mathcal{M})$ is not an $\mathcal{M}$-adhesive category. Fig. 6 represents a counterexample, because $\mathcal{R} \subseteq \mathcal{M}$, but $f_3, f_4 \notin \mathcal{M}$. Nevertheless, the morphism class $\mathcal{M}$ will be useful for our further analysis. The above example gives us an idea how pushouts in $\mathbf{AttC}$ along $\mathcal{M}$-morphisms are constructed. Basically, we use the pushout construction in $\mathbf{C}$, while we have to integrate different data values which are reached from the same elements by the given morphisms.

Fact 2 (Pushouts along $\mathcal{M}$). *The category $\mathbf{AttC}$ has pushouts along $\mathcal{M}$-morphisms.*

Construction. Consider the attributed morphisms $f_1 : AO_1 \to AO_3 \in \mathcal{M}$ and $f_2 : AO_1 \to AO_2$ with the single components as shown in the right diagrams, where due to the definition of $\mathcal{M}$ i and j are corresponding isomorphisms. We construct the pushout (2) in **C** and define a relation $\sim = \{(att_{2,k}(F_k(g_2)(o), n), a_2(j(att_{3,k}(F_k(g_1)(o), n)))) \in A_2 \times A_2 \mid k \in \mathcal{K}, n \in att_{name}(k), \exists o \in F_k(C_1), (o, n) \notin Dom(att_{1,k}), (F_k(g_2)(o), n) \in Dom(att_{2,k}), (F_k(g_1)(o), n) \in Dom(att_{3,k})\}$. Let $\equiv$ be the equivalence closure of $\sim$ and define $A_4 = A_2|_\equiv$. Now let $a_3(x) = [a_2(j(x))]$ and $a_4(x) = [x]$. Then $AO = (C_4, A_4, att_4)$ with $att_{4,k}(o, n) =$

$$\begin{cases} [a_2(j(att_{3,k}(o_3, n)))] & \exists o_3 \in F_k(C_3) : F_k(g_3)(o_3) = o, (o_3, n) \in Dom(att_{3,k}) \\ [att_{2,k}(o_2, n)] & \exists o_2 \in F_k(C_2) : F_k(g_4)(o_2) = o, (o_2, n) \in Dom(att_{2,k}) \\ \text{undefined} & \text{otherwise} \end{cases}$$

is the pushout object with morphisms $f_3 = (g_3, a_3)$ and $f_4 = (g_4, a_4)$.

Note that $\equiv$ is empty if each undefined attribute in AO_1 is also undefined in AO_2 or AO_3. In this case, A_4 is the pushout object of i and a_2 in **Sets**.

While the pushout construction is well-defined if one of the given morphisms is in $\mathcal{M}$, we do not want to glue attributes as done in Fig. 6 when applying a rule. Intuitively, gluing in the pushout construction has to be done for the following case: $\begin{smallmatrix} \square & \to & d \\ \downarrow & & \\ d' & & \end{smallmatrix}$. For transformations and their well-definedness, the key observation is the fact that such a situation could never occur. It is prevented because we have to construct a certain pushout complement first when applying a rule. Only the situations $\begin{smallmatrix} d & \leftarrow & \square & \to & d' \\ \downarrow & & \downarrow & & \downarrow \\ a(d) & \leftarrow & \square & \to & a(d') \end{smallmatrix}$ and $\begin{smallmatrix} d & \leftarrow & d & \to & d \\ \downarrow & & \downarrow & & \downarrow \\ a(d) & \leftarrow & a(d) & \to & a(d) \end{smallmatrix}$ can occur in the rule span and the corresponding application via a match. The construction of the pushout complement ensures that an undefined attribute in the intermediate object of the rule leads to an undefined attribute in the intermediate object of the rule application such that no data gluing has to occur. We describe such a well-behaviour as a special property and show that we can define a well-behaved pushout complement such that the double pushout of a direct transformation is actually well-behaved.

Definition 10 (Well-behaved pushouts). *The class $\mathcal{W}$ of morphism pairs with the same domain is defined by:* $\mathcal{W} = \{(f_1, f_2) \mid f_1 : AO_1 \to AO_3 \in \mathcal{R}, f_2 : AO_1 \to AO_2, \forall (o_1, n) \notin Dom(att_{1,k}) : (F_k(g_2)(o_1), n) \notin Dom(att_{2,k}), \forall o \in F_k(C_2) \backslash F_k(g_2)(C_1) : (o, n) \in Dom(att_{2,k})\}.$

A pushout (1) is called well-behaved, *or short a $\mathcal{W}$-pushout, if $(f_1, f_2) \in \mathcal{W}$.*

Note that the pushout over $(f_1, f_2) \in \mathcal{W}$ always exists, since **AttC** has pushouts along $\mathcal{M}$-morphisms (Fact 2). For $\mathcal{W}$-pushouts, in the construction of Fact 2 the relation $\sim$ is empty leading to $A_2 = A_4$, or $A_2 \cong A_4$ in general.

Now we can define rules and rule applications, which are called direct transformations, based on $\mathcal{W}$-pushouts. In contrast to rules for typed attributed graphs in [5], all attributes of the elements in the left or right hand side have to be defined, i.e. rules cannot be underspecified. While this requires some additional specification effort to at least assign a variable to each of these attributes, it clarifies the rule and is necessary for our approach. As a consequence, rules cannot be applied to attributed objects with only partial attribution.

Definition 11 (Rule and direct transformation). A rule is a span $p = (AO_L \xleftarrow{f_L} AO_K \xrightarrow{f_R} AO_R)$ of $\mathcal{R}$-morphisms in **AttC**.

Given a rule p, an attributed object AO_G, and an attributed morphism f_M :

$$\begin{array}{ccccc}
AO_L & \xleftarrow{f_L} & AO_K & \xrightarrow{f_R} & AO_R \\
f_M \downarrow & (1) & \downarrow f_K & (2) & \downarrow f_N \\
AO_G & \xleftarrow{f_G} & AO_D & \xrightarrow{f_H} & AO_H
\end{array}$$

$AO_L \to AO_G$, called match, a direct transformation $AG \xRightarrow{p, f_M} AH$ is given by two $\mathcal{W}$-pushouts (1) and (2) as in the diagram on the right.

Note that we could pragmatically adapt the rules of typed attributed graphs in [5] with conditions as expressed by $\mathcal{R}$-morphisms. These conditions can be checked statically to allow only intended attribute changes. Nevertheless, the underlying graph model still inhibits the disadvantages explained in Section 2.

Similar to the standard graph transformation, we define the gluing condition which characterizes valid situations, where the pushout complement exists.

Definition 12 (Gluing condition). Given f_1 and f_3 as in Def. 10, the gluing condition holds if the underlying pushout complement of $g_3 \circ g_1$ in **C** exists and for all $o_1, o_1' \in F_k(C_1)$ with $F_k(g_3 \circ g_1)(o_1) = F_k(g_3 \circ g_1)(o_1')$ we have that both $att_{1,k}(o_1, n).att_{1,k}(o_1', n)$ are either defined or undefined.

Since both $\begin{smallmatrix} d & \leftarrow & \square \\ \downarrow & & \downarrow \\ d & \leftarrow & d \end{smallmatrix}$ and $\begin{smallmatrix} d & \leftarrow & \square \\ \downarrow & & \downarrow \\ d & \leftarrow & \square \end{smallmatrix}$ are valid pushout complements, pushout complements in **AttC** are not unique, but only the second situation behaves well. Thus, we chose this one to construct pushout complements in **AttC**.

Fact 3 (Pushout complement along $\mathcal{M}$). The category **AttC** has pushout complements along $\mathcal{M}$-morphisms, if the gluing condition holds.

Construction. Given attributed morphisms $f_1 : AO_1 \to AO_3 \in \mathcal{M}$ and $f_3 : AO_3 \to AO_4$ as above, construct the pushout complement (2) in **C** (see Construction of Fact 2). Now define $A_2 = A_4$, $a_2 = a_3 \circ i$, $a_4 = id$, and $att_{2,k}(o, n) =$
$$\begin{cases} a_2(att_{1,k}(o_1, n)) & \exists o_1 \in F_k(C_1) : F_k(g_2)(o_1) = o, (o_1, n) \in Dom(att_{1,k}) \\ \text{undefined} & \exists o_1 \in F_k(C_1) : F_k(g_2)(o_1) = o, (o_1, n) \notin Dom(att_{1,k}) . \\ att_{4,k}(F_k(g_4)(o), n) & \text{otherwise} \end{cases}$$
Then $AO_2 = (C_2, A_2, att_2)$ together with the morphisms $f_2 = (g_2, a_2)$ and $f_4 = (g_4, a_4)$ is the pushout complement of f_1 and f_3.

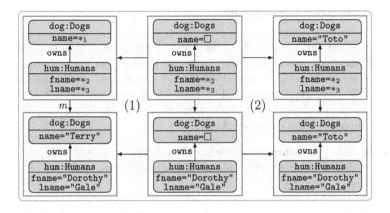

Fig. 7. Direct transformation of attributed graphs

A rule p is applicable to AO_G if the gluing condition holds and the attribution in AO_G is total. In this case, for the construction of a direct transformation we can first construct the pushout complement of Fact 3 and then the pushout of Fact 2, which are both unique and well-behaved.

Theorem 2 (Construction of direct transformation). *Given a situation as in Def. 11 with $f_L, f_R \in \mathcal{R}$, where the gluing condition holds and att_G is total, then the pushout complement (1) and the pushout (2) are uniquely defined $\mathcal{W}$-pushouts. Moreover, also $f_G, f_H \in \mathcal{R}$.*

Example 4. Now we can rename "Greebo" to "Maurice" using the second rule in Fig. 2 or an arbitrary dog that belongs to someone to "Toto" with the rule in the top row of Fig. 7. When applying such a rule, we construct first the pushout complement and then the pushout, as shown in Fig. 7. Note that the match m uses the variable assignment $\alpha(*_1) = $ "Terry", $\alpha(*_2) = $ "Dorothy", and $\alpha(*_3) = $ "Gale", and its codomain graph has total attribution. We can see that both pushouts are well-behaved because all attributes in the intermediate object that are undefined in AO_K are also undefined in AO_D.

As we will need this fact in the next section, we state that also pullbacks along $\mathcal{M}$-morphisms exist in **AttC**.

Fact 4 (Pullbacks over $\mathcal{M}$). *The category **AttC** has pullbacks along $\mathcal{M}$-morphisms. Moreover, $\mathcal{M}$ is closed under pullbacks.*

$$
\begin{array}{ccc}
AO_1 & \xrightarrow{f_1} & AO_3 \\
\downarrow{\scriptstyle f_2} & (1) & \downarrow{\scriptstyle f_3} \\
AO_2 & \xrightarrow{f_4} & AO_4
\end{array}
$$

Construction. Given $f_4 \in \mathcal{M}$ we construct the pullback in **C** with pullback object C_1 and morphisms g_1, g_2. Then $AO_1 = (C_1, A_3, att_1)$ with $f_1 = (g_1, id_{A_3})$, $f_2 = (g_2, a_4^{-1} \circ a_3)$, and $att_1(o, n) =$
$$
\begin{cases}
att_{3,k}(F_k(g_1)(o), n) & (F_k(g_1)(o), n) \in Dom(att_{3,k}), \\
& \qquad (F_k(g_2)(o), n) \in Dom(att_{2,k}) \\
\text{undefined} & \text{otherwise}
\end{cases}
$$
is the pullback of f_3 and f_4.

5 $\mathcal{W}$-Adhesive Categories

As we have analyzed in the last section, not all pushouts are important in the context of transformations, but only those we have called $\mathcal{W}$-pushouts. In this section, we generalize this idea to $\mathcal{W}$-adhesive categories, which are categories where the VK property is restricted to hold for $\mathcal{W}$-pushouts, which are defined by a class $\mathcal{W}$ of morphism spans. This restriction allows us to formulate a transformation theory for the category **AttC**.

Definition 13 ($\mathcal{W}$-adhesive category). *Given a category **C**, morphism classes $\mathcal{R} \subseteq \mathcal{M}$, and a class $\mathcal{W} \subseteq \mathcal{R} \times Mor_{\mathbf{C}}$ of morphism spans, $(\mathbf{C}, \mathcal{R}, \mathcal{M}, \mathcal{W})$ is a $\mathcal{W}$-adhesive category if:*

1. *$\mathcal{M}$ is a class of monomorphisms closed under isomorphisms, composition, and decomposition, with $id_A \in \mathcal{M}$ for all $A \in \mathbf{C}$,*
2. *$\mathbf{C}$ has pushouts along and pullbacks over $\mathcal{M}$-morphisms,*
3. *$\mathcal{M}$ is closed under pullbacks: Given a pullback (1) with $n \in \mathcal{M}$ then also $m \in \mathcal{M}$,*
4. *$\mathbf{C}$ has pushouts over $\mathcal{W}$-morphisms, called $\mathcal{W}$-pushouts,*
5. *$\mathcal{R}$ is closed under $\mathcal{W}$-pushouts: Given $\mathcal{W}$-pushout (1) with $m \in \mathcal{R}$ then also $n \in \mathcal{R}$,*
6. *$\mathcal{W}$ is closed under $\mathcal{R}$: $(m' : A' \to B', a) \in \mathcal{W}$, $f' : A' \to C' \in \mathcal{R}$ implies $(f', a) \in \mathcal{W}$,*
7. *$\mathcal{W}$-pushout composition and decomposition: Given pushout (1), (1) + (2) is a $\mathcal{W}$-pushout with $(f, h \circ m) \in \mathcal{W}$ if and only if (2) is a $\mathcal{W}$-pushout with $(g, h) \in \mathcal{W}$,*
8. *$\mathcal{W}$-pushouts fulfill the $\mathcal{W}$-van Kampen property: Given a commutative cube (3) with $\mathcal{W}$-pushout (1) in the bottom, $m, d \in \mathcal{R}$, $b, c \in \mathcal{M}$ and the back faces being pullbacks, it holds that the top is a pushout if and only if the front faces are pullbacks.*

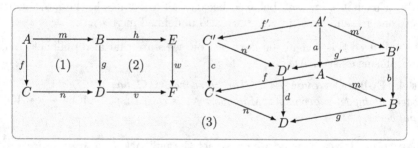

An example for a $\mathcal{W}$-adhesive category is the category of attributed objects and morphisms.

Theorem 3. AttC *with an underlying $\mathcal{M}$-adhesive category $(\mathbf{C}.\mathcal{M}_{\mathbf{C}})$, $\mathcal{R}$ as defined in Def. 8, $\mathcal{M}$ as defined in Def. 9, and $\mathcal{W}$ as defined in Def. 10 is a $\mathcal{W}$-adhesive category.*

Proof. Obviously, $\mathcal{M}$ is a class of monomorphisms because both components are monomorphisms with the required closure properties inherited from the components. **AttC** has pullbacks over $\mathcal{M}$-morphisms and $\mathcal{M}$ is closed under pullbacks as shown in Fact 4. Since **AttC** has pushouts along $\mathcal{M}$-morphisms, as shown in Fact 2, it has pushouts over $\mathcal{M}$-morphisms as well as $\mathcal{W}$-pushouts. The closure of $\mathcal{R}$ under $\mathcal{W}$-pushouts and the closure of $\mathcal{W}$ under $\mathcal{R}$ follow from Thm. 2. The $\mathcal{W}$-pushout composition and decomposition follows from the fact that the undefined attributes in A, B, and E are exactly the same. The proof of the $\mathcal{W}$-van Kampen property can be found in the appendix.

As specified in Def. 11 for the special case of **AttC**, in a $\mathcal{W}$-adhesive category rules are defined as spans of $\mathcal{R}$-morphisms, while transformations are defined by double $\mathcal{W}$-pushouts.

To prove important results for graph transformation, various so-called HLR properties have been used in [5]. Here, we show the corresponding variant for $\mathcal{W}$-categories for two of them, the $\mathcal{W}$-pushout-pullback decomposition property and that $\mathcal{W}$-pushouts are pullbacks.

Fact 5 ($\mathcal{W}$-pushout-pullback decomposition). *Given the above commutative diagram, where $(1) + (2)$ is a $\mathcal{W}$-pushout, (2) is a pullback, $v \in \mathcal{R}$, and $(f \in \mathcal{R}$ or $m, h \circ m \in \mathcal{R})$, then (1) is a pushout.*

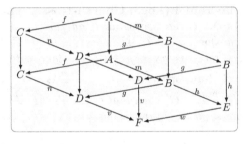

Proof. Consider the right cube, where all unnamed morphisms are identities. The bottom is the $\mathcal{W}$-pushout $(1) + (2)$ with $(f \in \mathcal{R}$ or $m, h \circ m \in \mathcal{R})$, $v \in \mathcal{R}$, and $h, id_C, id_D \in \mathcal{M}$ (Def. 13 Items 3 and 1). All back and front faces are pullbacks. Now the $\mathcal{W}$-van Kampen property (Item 8) implies that the top, i.e. the square (1), is a pushout.

Fact 6 ($\mathcal{W}$-pushouts are pullbacks). *Given the above $\mathcal{W}$-pushout (1) with $m \in \mathcal{R}$, then (1) is also a pullback.*

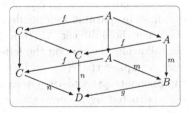

Proof. Consider the right cube, where all unnamed morphisms are identities. The bottom is the $\mathcal{W}$-pushout (1) with $m, n \in \mathcal{R}$ (Def. 13 Item 5), and $m, id_C \in \mathcal{M}$ (Item 1). All back faces are pullbacks and the top is a pushout. Now the $\mathcal{W}$-van Kampen property (Item 8) implies that the front faces, and in particular the square (1), are pullbacks.

In the following, we sketch for the example of (one direction of) the Local Church-Rosser Theorem that this result is also available in $\mathcal{W}$-adhesive categories. It is concerned with parallel and sequential independence of direct transformations. First, we define the notion of parallel and sequential independence. Then we state the Local Church-Rosser Theorem and prove it. The proof follows the one in [5] with certain adaptions for $\mathcal{W}$-adhesive categories.

Definition 14 (Parallel and sequential independence). *Two direct trans-formations* $G \xrightarrow{p_1,m_1} H_1$ *and* $G \xrightarrow{p_2,m_2} H_2$ *are parallel independent if there are morphisms* $i : L_1 \to D_2$ *and* $j : L_2 \to D_1$ *such that* $f_2 \circ i = m_1$ *and* $f_1 \circ j = m_2$.

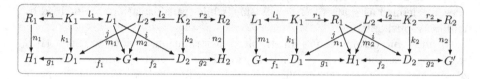

Two direct transformations $G \xrightarrow{p_1,m_1} H_1 \xrightarrow{p_2,m_2} G'$ *are sequentially indepen-dent if there are morphisms* $i : R_1 \to D_2$ *and* $j : L_2 \to D_1$ *such that* $f_2 \circ i = n_1$ *and* $g_1 \circ j = m_2$.

Theorem 4 (Local Church-Rosser Theo-rem). *Given two parallel independent direct transformations* $G \xrightarrow{p_1,m_1} H_1$ *and* $G \xrightarrow{p_2,m_2} H_2$ *there is an object* G' *together with di-rect transformations* $H_1 \xrightarrow{p_2,m_2'} G'$ *and* $H_2 \xrightarrow{p_1,m_1'} G'$ *such that* $G \xrightarrow{p_1,m_1} H_1 \xrightarrow{p_2,m_2'} G'$ *and* $G \xrightarrow{p_2,m_2} H_2 \xrightarrow{p_1,m_1'} G'$ *are sequentially independent.*

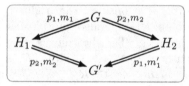

Proof. Consider the parallel independent direct transformations $G \xrightarrow{p_1,m_1} H_1$ and $G \xrightarrow{p_2,m_2} H_2$ depicted

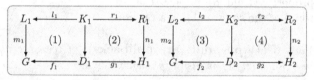

right. We combine the $\mathcal{W}$-pushouts (1) and (3) with the morphisms i_1 and i_2 obtained by parallel independence. Since $f_1, f_2 \in \mathcal{M}$ (Def. 13 Item 5) we can construct the pullback (5) (Item 2) and obtain morphisms j_1 and j_2 as shown in the following diagram on the left-hand side. Since $(1) = (6) + (5)$ with $f_2, l_1 \in \mathcal{R}$ Fact 5 implies that (6), and analogously (7), is a pushout. Now we construct the pushouts (8) and (9) (Item 2) along $r_1, r_2 \in \mathcal{M}$. Finally, the pushout (10) is constructed by decomposition of pushout (8) and the pushout over $r_1 \in \mathcal{M}$ and $h_1 \circ j_1$ (Item 2). From pushout (8) we obtain a morphism $s_1 : D_2' \to H_1$ such that $(2) = (8) + (11)$, and by Item 7 (11) is a $\mathcal{W}$-pushout. Analogously, we obtain the $\mathcal{W}$-pushout (12).

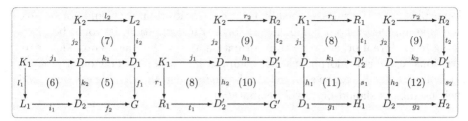

By Item 7, $(7) + (11)$ and $(6) + (12)$ are $\mathcal{W}$-pushouts, and using Item 6 we obtain the sequentially independent direct transformations $H_1 \xRightarrow{p_2} G'$ and $H_2 \xRightarrow{p_1} G'$.

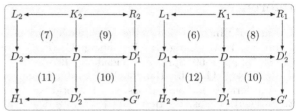

6   Conclusion and Future Work

In this paper, we proposed a new concept for attribution of objects in an arbitrary category based on a functor selecting attributable elements of objects and assigning attributes and values to them. We have then defined rules and transformations on attributed objects based on $\mathcal{M}$-adhesive categories, where transformations rely on the new concept of $\mathcal{W}$-pushouts. This concept leads to the notion of $\mathcal{W}$-categories, which are a suitable framework to show results for transformations of attributed objects, as we have demonstrated for the example of the Local Church–Rosser Theorem. Since the underlying objects come from an $\mathcal{M}$-adhesive category, we only have to prove the results for the attribution part and can rely on the underlying results for the pure, un-attributed transformations. In fact, we can use any suitable category for attribution. In this paper, we chose $\mathcal{M}$-adhesive categories, because a large number of results is available there, but there is actually no need for this restriction.

$\mathcal{W}$-adhesive categories have been introduced for attribution in this paper, but they may also fit for transformations in other non-$\mathcal{M}$-adhesive categories, where the definition of the proper pushout complement depends on both morphisms, like in RDF graphs (see [13]) or open Petri nets (see [14]). In this sense, they are more expressive then $\mathcal{M}$-$\mathcal{N}$-adhesive categories [15], which consider a special class $\mathcal{N}$ for vertical morphisms in transformations.

Future work includes to prove other important results and theorems for graph transformation, where additional requirements for $\mathcal{W}$-adhesive categories may have to be identified. For example, for the Parallelism Theorem some compatibility property of $\mathcal{W}$ with binary coproducts will be necessary. Moreover, rules should be extended with constraints and application conditions for data as done in [16] for triple rules. Another interesting extension would be to integrate an inheritance concept for the attribution types.

For language evolution, the deletion or addition of attributes, i.e. changing the meta-model, is an interesting field of work. Without the restrictions of the

$\mathcal{R}$-morphisms in the rules, especially the total attribution, we would not be able to define unique pushout complements for a transformation. It would be interesting to explore if other graph transformation approaches may require fewer restrictions to the rule morphisms. For computations and reasoning on attributes, symbolic graphs [8] can be adapted for the underlying $\mathcal{M}$-adhesive category, and their usefulness should be further elaborated.

References

1. Ehrig, H.: Introduction to the Algebraic Theory of Graph Grammars (A Survey). In: Ng, E.W., Ehrig, H., Rozenberg, G. (eds.) Graph Grammars 1978. LNCS, vol. 73, pp. 1–69. Springer, Heidelberg (1979)
2. Rozenberg, G. (ed.): Handbook of Graph Grammars and Computing by Graph Transformation. Foundations, vol. 1. World Scientific (1997)
3. Ehrig, H., Golas, U., Hermann, F.: Categorical Frameworks for Graph Transformation and HLR Systems based on the DPO Approach. BEATCS 102, 111–121 (2010)
4. Lack, S., Sobociński, P.: Adhesive Categories. In: Walukiewicz, I. (ed.) FOSSACS 2004. LNCS, vol. 2987, pp. 273–288. Springer, Heidelberg (2004)
5. Ehrig, H., Ehrig, K., Prange, U., Taentzer, G.: Fundamentals of Algebraic Graph Transformation. EATCS Monographs. Springer (2006)
6. Hermann, F., Ehrig, H., Ermel, C., Orejas, F.: Concurrent Model Synchronization with Conflict Resolution Based on Triple Graph Grammars. In: de Lara, J., Zisman, A. (eds.) FASE 2012. LNCS, vol. 7212, pp. 178–193. Springer, Heidelberg (2012)
7. Golas, U.: A General Attribution Concept for Models in M-adhesive Transformation Systems: Long Version. Technical Report 12-22, Zuse Institute Berlin (2012)
8. Orejas, F., Lambers, L.: Symbolic Attributed Graphs for Attributed Graph Transformation. ECEASST 30 (2010)
9. Habel, A., Plump, D.: Relabelling in Graph Transformation. In: Corradini, A., Ehrig, H., Kreowski, H.-J., Rozenberg, G. (eds.) ICGT 2002. LNCS, vol. 2505, pp. 135–147. Springer, Heidelberg (2002)
10. Löwe, M., Korff, M., Wagner, A.: An Algebraic Framework for the Transformation of Attributed Graphs. In: Term Graph Rewriting: Theory and Practice, pp. 185–199. Wiley (1993)
11. Rebout, M., Féraud, L., Soloviev, S.: A Unified Categorical Approach for Attributed Graph Rewriting. In: Hirsch, E.A., Razborov, A.A., Semenov, A., Slissenko, A. (eds.) CSR 2008. LNCS, vol. 5010, pp. 398–409. Springer, Heidelberg (2008)
12. Ehrig, H., Mahr, B.: Fundamentals of Algebraic Specification 1: Equations and Initial Semantics. EATCS Monographs. Springer (1985)
13. Braatz, B., Brandt, C.: Graph Transformations for the Resource Description Framework. ECEASST 10 (2008)
14. Baldan, P., Corradini, A., Ehrig, H., Heckel, R.: Compositional Modeling of Reactive Systems Using Open Nets. In: Larsen, K.G., Nielsen, M. (eds.) CONCUR 2001. LNCS, vol. 2154, pp. 502–518. Springer, Heidelberg (2001)
15. Habel, A., Plump, D.: M,N-Adhesive Transformation Systems. In: Ehrig, H., Engels, G., Kreowski, H.-J., Rozenberg, G. (eds.) ICGT 2012. LNCS, vol. 7562, pp. 218–233. Springer, Heidelberg (2012)
16. Anjorin, A., Varro, G., Schürr, A.: Complex Attribute Manipulation in TGGs with Constraint-Based Programming Techniques. ECEASST (2012)

DPO Transformation with Open Maps

Reiko Heckel

Department of Computer Science, University of Leicester, UK
reiko@mcs.le.ac.uk

Abstract. In graph transformation, a match just represents an occurrences of a rule's left-hand side in the host graph. This is expressed by a morphism preserving the graph structure. However, there are situations where occurrences are bound by additional constraints. These can either be implicit, such as the gluing conditions of the DPO, or explicit such as negative application conditions.

In this paper we study another type of implicit condition based on the reflection of structure. Morphisms reflecting some of the structures of their targets are abstractly characterised as open maps in the sense of Joyal, Nielsen, and Winskel. We show that under certain restrictions on the rules, DPOs preserve open maps. We establish an encoding of open maps into negative application conditions and study concurrency properties of the new approach.

Keywords: double pushout, open maps, negative application conditions, adhesive categories.

1 Introduction

Most graph transformation approaches control applicability of rules by means of subgraph embeddings or homomorphisms between rules' left-hand sides and the graphs they are meant to apply to. This has often been found insufficiently expressive in applications. Apart from adding explicit control structures or negative application conditions, more restrictive notions of matches can be used.

For example, in the context of document image analysis, Blostein [1] considers *induced subgraph matching*, where the left-hand side L has to be isomorphic to a complete subgraph of the host graph G. That means, if for any two nodes v, w in L an edge exists between their images in G, this edge must also be present in L. In other words, apart from preserving nodes and edges, the match is required to reflect edges between existing nodes. This approach provides more explicit control at the cost of having to specify a larger set of patterns [1].

The KAPPA approach uses graph rewriting for modelling biochemical reactions [3]. So-called site graphs, representing sets of molecules, consist of agents equipped with sites through which they can be linked with sites of other agents. Matches preserve agents, sites and links as expected, but they also reflect links outgoing from sites. That means, if in L an agent has a site who's image in G is linked to another site, such a link must be present in L also.

H. Ehrig et al.(Eds.): ICGT 2012, LNCS 7562, pp. 203–217, 2012.

While in these and similar examples negative applications conditions could be used, this may complicate the formalism, making it harder to understand and analyse. Also, as we will demonstrate, it complicates the concurrency properties of the approach. Instead, in this paper we develop a notion of graph transformation which allows to capture directly general reflection constraints. Such constraints on matches are specified relative to the type graph, similar to structural constraints on instance graphs. This makes it easier to design domain-specific languages based on graph transformation with specifically chosen classes of matches.

The approach is based on *open maps* [8], a categorical axiomatisation of morphisms in a category **C** reflecting structures specified by a subcategory **P**. Intuitively, this so-called *path category* represents a set of constraints capturing the specific reflection properties required. The name originates from the use of **P** to characterise reflection of certain paths in transition systems. Assuming **C** to be an adhesive category, we define a restricted form of the DPO approach where matches are open maps. We show that the constraints for open maps can be encoded as negative application conditions and study the concurrency properties of the new approach.

In particular, it turns out that using the standard definitions of independence, local Church-Rosser and switch equivalence for negative application conditions [7,10], independence of consecutive steps is not stable under switch equivalence. That means, causality is only well-defined over consecutive steps in the same sequence, but may be different for corresponding steps in an equivalent sequence. We show that, as in the classical DPO approach, independence of DPO transformations over open maps is preserved under switch equivalence.

The paper is organised as follows. After introducing basic notions below, Section 3 defines open maps and studies their relation with pullbacks. Section 4 presents DPO transformations of open maps and Section 5 considers its encoding by DPO with negative application conditions. The relation of independence and switch equivalence for both conditional DPO and DPO over open maps is analysed in Section 6 before Section 7 concludes the paper.

2   Basic Definitions

We use the double-pushout approach to (typed) graph transformation, occasionally with negative application conditions, but will state all definitions and results at the level of adhesive categories [9]. A category is *adhesive* if it has pullbacks as well as pushouts for all pairs of morphisms where one is a monomorphism, and where all such pushouts enjoy the van Kampen property. That means, when such a pushout is the bottom face of a commutative cube such as in the left of Fig. 1, whose rear faces are pullbacks, the top face is a pushout if and only if the front faces are pullbacks. In any adhesive category we have uniqueness of pushout complements, monomorphisms are preserved by pushouts and pushouts along a monomorphism are also pullbacks. Categories of typed graphs for a fixed type graph TG are adhesive [4].

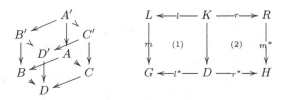

Fig. 1. van Kampen condition (left) and DPO diagram (right)

Rules are defined as spans of monomorphisms. Transformations follow the double-pushout approach [5]. A *rule* $p = (L \xleftarrow{l} K \xrightarrow{r} R)$ consists of two monomorphisms l and r in $\mathbf{C}$. Given a morphism $m : L \to G$ called the *match*, a *direct transformation* $G \xRightarrow{p,m} H$ from G to a H exists if a double-pushout (DPO) diagram can be constructed, where (1) and (2) in the right of Fig. 1 are pushouts in $\mathbf{C}$.

The applicability of rules can be restricted by specifying negative conditions stating the non-existence of certain structures in the context of the match. A *negative constraint* on a graph L is a morphism $n : L \to \hat{L}$ in $\mathbf{C}$. A morphism $m : L \to G$ satisfies n (written $m \models n$) iff there is no morphism $q : \hat{L} \to G$ such that $q \circ n = m$ in $\mathbf{C}$. A negative application condition (NAC) on L is a set of constraints N. A morphism $m : L \to G$ *satisfies* N (written $m \models N$) if and only if m satisfies every constraint in N, i.e., $\forall n \in N : m \models n$.

A graph transformation system (GTS) $\mathcal{G}$ consists of a set of rules. A derivation in $\mathcal{G}$ is a sequence of direct transformations $s = (G_0 \xRightarrow{p_1,m_1} G_1 \xRightarrow{p_2,m_2} \cdots \xRightarrow{p_n,m_n} G_n)$ such that all p_i are in $\mathcal{G}$. We write $s : G_0 \xRightarrow{*} G_n$ for a generic derivation and, given $s' : G_k \xRightarrow{*} G_m$ with $G_n = G_k$, we denote their composition by $s ; s' : G_0 \xRightarrow{*} G_m$.

3   Open Maps and Pullbacks

Open maps are a categorical characterisation for morphisms reflecting aspects of structure in a category of more general, structure preserving maps. The concept has been introduced in computer science when Joyal, Nielsen and Winskel [8] used it to characterise bisimulation functions in a category of labelled transition systems, where the standard morphisms preserve transitions but do not reflect them.

The idea is to use a subcategory $\mathbf{P}$ of an environment category $\mathbf{C}$ of "ordinary" morphisms to capture extensions of paths. An arrow $c : P \to Q$ in $\mathbf{P}$ can be seen as an implication saying that, for each occurrence of P in the source of an open morphism $m : X \to Y$, if a corresponding occurrence of Q can be found in Y, then this must give rise to a compatible occurrence of Q in X. If P represents a prefix of a path (e.g., in a transition system) and Q a possible extension, this amounts to demanding a reflection property of paths, specified by step-wise extension.

Let us illustrate this with the example discussed in the introduction, of matches as induced subgraphs. The constraint expressing that edges in G between nodes with a pre-image in L should be reflected is expressed by the constraint c below. The match m violates this constraint because the edge $1 \to 2$ in G is not reflected. This corresponds to the fact that there is no diagonal embedding of Q into L compatible with the embeddings of P into L and Q into G.

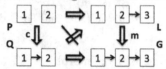

Next, after presenting the basic definitions, we are considering the relation of this notion with pushouts and pullbacks, preparing the ground for using open maps in DPO transformations.

Definition 1 (open maps [8]). *Let $\mathbf{C}$ be a category and $\mathbf{P} \subseteq \mathbf{C}$ be a subcategory, called* path category. *A morphism m is $\mathbf{P}$-open if all commutative squares based on $c : P \to Q \in \mathbf{P}$ as below have a fill-in, i.e., a morphism f such that the resulting triangles PQX and QXY commute. If $\mathbf{P}$ is understood from the context, we refer to m as* open.

For the rest of the paper, assume an adhesive category $\mathbf{C}$ and let $\mathbf{P}$ be a subcategory such that, for all objects X, Y in $\mathbf{P}$, all isomorphisms $i : X \to Y$ in $\mathbf{C}$ are also in $\mathbf{P}$.

Let us consider an example to illustrate the notions introduced so far.

Example 1 (paths and open maps). Figure 2 shows the type graph for a simple model of object structures, where Objects have Fields from which they refer to Handles representing the identities of other objects. The idea is that Fields and Handles act as ports for incoming and outgoing references.

The reflection properties are specified by the two arrows in Fig. 2: Constraint c_1 in the top states that, if a Handle is present its incoming references must be reflected. Constraint c_2 in the bottom states that, if a Field is present its outgoing references must be reflected. The path category is given by the closure of the constraints under composition, identities, and isomorphisms in $\mathbf{C}$.

Satisfaction of the condition of openness is illustrated by the morphisms shown in Fig 3. Here, m_1 satisfies c_1 but not c_2, because the latter has an embedding of the premise in the source and the conclusion into the target of m_1, but there is no embedding of the conclusion into the source of m_1 because the outgoing reference is not reflected. With the same justification, m_2 satisfies both constraints, so is an open map, while m_3 satisfies c_2 due to lack of occurrence of the premise, but does not satisfy c_1.

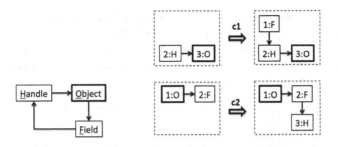

Fig. 2. Type graph for simple object structures (left) and constraints for reflection of references to Handles (top right) and from Fields (bottom right)

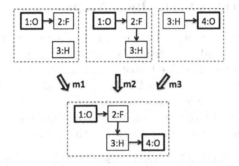

Fig. 3. Morphisms as candidate open maps

We continue analysing the relationship of open maps with pullbacks.

Lemma 1 (pullbacks preserve open maps). *Assume a pullback in* **C** *such as (1) below, where l is mono. Then, if m is open, so is d.*

$$
\begin{array}{ccccc}
L & \xleftarrow{\ l\ } & K & \longleftarrow & P \\
\downarrow{\scriptstyle m} & {\scriptstyle (1)} & \downarrow{\scriptstyle d} & {\scriptstyle g} \searrow & \downarrow{\scriptstyle c} \\
G & \longleftarrow & D & \longleftarrow & Q
\end{array}
$$

Proof. For every commuting diagram $PQKD$ as above on the right, we have to produce a fill-in. Since the outer diagram commutes and m is open, there exists a morphism $f : Q \to L$ commuting the resulting triangles. Now $Q \xrightarrow{f} L \xrightarrow{m} G = Q \to D \to G$, so by pullback property of (1) this induces g to commute with morphisms to D and L. Commutativity with the top-right triangle of the right-hand side pullback follows because l is mono.

Open maps are closed under composition and decomposition with monos.

Lemma 2 (composition and decomposition). *If morphisms $A \xrightarrow{f} B \xrightarrow{g} C$ are open, also $g \circ f$ is open. If $g \circ f$ is open and g is mono, also f is open.*

Proof. The first statement is obvious by composition of the diagram involved. For decomposition, assume $c \in \mathbf{P}$ as below such that the square commutes. Then, by openness of $g \circ f$ the fill-in exists such that the upper triangle commutes and $Q \to A \xrightarrow{f} B \xrightarrow{g} C = Q \to B \xrightarrow{g} C$. By monomorphism property of g this implies $Q \to A \xrightarrow{f} B = Q \to B$.

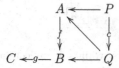

4   Transformations with Open Maps

Open maps are a class of morphisms enjoying the reflection properties specified by the path category. Used as matches they represent a global control condition, restricting the class of permissible transformations. While this is well-defined as it is, the inherent symmetry of the DPO approach suggests to investigate the conditions under which the resulting co-match of a transformation with an open match is open as well. In this section, after providing the basic definition, we investigate conditions on rules that guarantee this symmetry.

Definition 2 (transformation with open maps). *A DPO transformation* $G \overset{p,m}{\Longrightarrow}_{\mathbf{P}} H$ *with* $\mathbf{P}$*-open maps is a DPO step such that match m is* $\mathbf{P}$*-open.*

Example 2 (manipulation of objects). The rules in Fig 4 illustrate the power of open maps in controlling transformations. The first rule assigns a reference to a Field if no outgoing reference exists yet. The left-hand side is identical to the source of m_1 in Fig. 3, and as discussed there the constraints are not satisfied by the target of that match.

The remaining three rules implement a simple garbage collection mechanism. First, we disconnect outgoing references if there are no incoming ones. Then, empty Fields of unreferenced Objects are removed and finally, unreferenced Objects (without Fields, due to the dangling condition) are removed together with their Handles.

It is clear from phrases such as "empty Field" or "unreferenced Object" that we are conceptually in the territory of negative application conditions. This intuition is formalised in the following section.

Definition 3 (P-preserving, P-stable). *A rule* $p : L \xleftarrow{l} K \xrightarrow{r} R$ *is* $\mathbf{P}$*-preserving if for all transformations* $G \overset{p,m}{\Longrightarrow} H$*, the comatch m^* is open whenever m is. It is* $\mathbf{P}$*-stable if the same is true for the inverse rule* $p^{-1} : R \xleftarrow{r} K \xrightarrow{l} L$.

Using Lemma 1, it is not difficult to see that the left-hand side pushout of a DPO, which is also a pullback, translates open maps m into open maps d. The proposition below is therefore mostly concerned with the right-hand side.

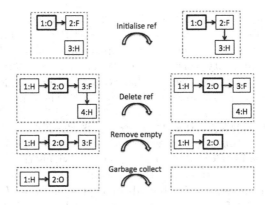

Fig. 4. Rules for manipulating object structures

Proposition 1 (P-preserving). *Rule* $p : L \xleftarrow{l} K \xrightarrow{r} R$ *is* **P**-*preserving if, in the diagram on the left below, for all* $c : P \to Q$ *in* **P** *and* $P \to R$ *with pushout (1) and epimorphism* q, *such that pushout complement (2) exists and* e *is* **P**-*open,*

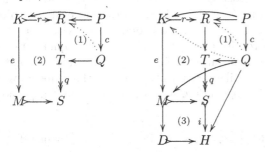

there exist (a) *morphism* $Q \to R$ *such that the upper right triangle of (1) commutes and* $Q \to R \to T \xrightarrow{q} S = Q \to T \xrightarrow{q} S$, *or* (b) *morphisms* $P \to K$ *and* $Q \to M$ *commuting all resulting diagrams.*

Proof. Assume that m is open. Due to the monic l, the left-hand side pushout is a pullback. Thus, by Lemma 1, d is open.

For the right-hand side, assume pushout (2+3) in the diagram on the right above. To show that $R \to H$ is open, let $c \in \mathbf{P}$ and assume $P \to R$ and $Q \to H$ commuting the square $PRQH$. Forming pushout (1), $R \to H$ decomposes over T such that triangle QTH commutes. Let $i \circ q$ be the epi-mono factorisation of the so-induced morphism. It follows from the pushout-pullback decomposition property [9] that we can decompose (2+3) into pushouts (2) and (3). In particular, r and i and all arrows parallel to them are monomorphisms. Since $K \to D$ is open, and $M \to D$ is mono, by Lemma 2 also e is open.

By assumption, there exist either $Q \to R$, or $P \to K$ and $Q \to M$, commuting the resulting diagrams. In the first case, this immediately delivers the fill-in for $PRQH$. In the second case, since e is open we have a fill-in $Q \to K$ which extends along r to a fill-in $Q \to R$.

While the notion of transformation does not formally require such restriction, we will usually assume rules to be **P**-stable or at least **P**-preserving. Conceptually, the process of verifying that a rule preserves open maps is similar to that of ensuring that it respects integrity constraints [6]: We create a counterexample without unnecessary context and Prop. 1 shows that every counterexample can be reduced to such a minimal one. Let us analyse the property for the rules in Fig. 4.

*Example 3 (**P**-stable rules).* Rule *Initialise ref* is **P**-stable because there is no embedding of c_1's premise P_1 into either left- or right-hand side, while the embedding of c_2's premise P_2 into the right-hand side yields the analysis depicted in Fig. 5 on the left. Constructing the pushout T of the embedding and c_2, there is pushout complement M, but also morphisms $P_2 \to K$ and $Q_2 \to M$ as required. Similarly, for quotient S_1 there is pushout complement M_1 with embedding $Q_2 \to M_1$. For the pushout complement M_2 of S_2 there is no such embedding, but $Q_2 \to R$ commuting with $R \to S_2$ and $T \to S_2$, thus invoking alternative (a). Rule *Delete reference* is analogous to the cases of T and the first quotient S_1 above.

Rule *Remove empty* illustrates a different case. There is an embedding of P_2 into its left-hand side that is not preserved. That means, the inverse rule may not be **P**-preserving. As shown in Fig. 5 on the right, constructing pushout object T, the morphism $L \to T$ does not have a pushout complement with respect to $K \to L$, and neither has $L \to S$ resulting from the only nontrivial quotient of T. That means, the premise of the implication is false, and thus *Remove empty*'s inverse is **P**-preserving, too. A similar argument holds for the embedding of P_1 into the left-hand side of *Garbage collect*, which does not admit a pushout complement either.

It is also worth considering examples of rules and path categories which do not enjoy this property.

*Example 4 (not **P**-preserving).* In order for a derived match $m^* : R \to H$ in a transformation $G \overset{p,m}{\Longrightarrow} H$ not to be **P**-open, we require a constraint $c : P \to Q \in$ **P** with morphisms $P \to R$ and $Q \to H$, but without a fill-in $Q \to R$. For m^* to be derived from a **P**-open match m, corresponding occurrences of premise P or conclusion Q must not exist into K and D, i.e., they are only created by the transformation. The two cases are explored in Fig. 6.

On the left, using constraint c_2 from Fig. 2, there is no embedding of the premise P_2 into K, which is only enabled by the creation of Object node $1 : O$. It is interesting to observe that a rule like this is meaningless if, as intended, fields depend on objects. That means, assuming an integrity constraints requiring that each Field node is pointed to by exactly one Object node, the rule would be violating this constraint.

On the right of Fig. 6 the same rule is shown with a different reflection constraint c_3, obtained from c_2 by dropping the Object node from its premise. In this case, the embedding of the conclusion is only created by the action of the rule. Also this example relies on graphs that do not satisfy the intended integrity constraint.

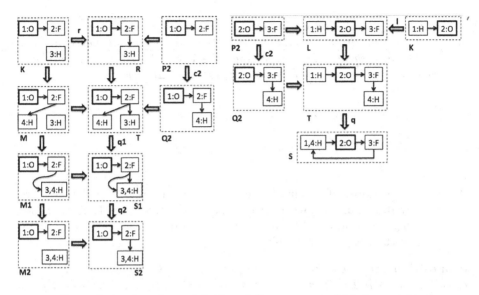

Fig. 5. Rules *Initialise ref* (left) and *Remove empty* (right) are stable

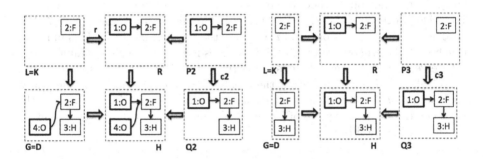

Fig. 6. New embedding created for premise (left) and conclusion (right) of constraint

The relationship between integrity and reflection constraints deserves further investigation. Both could be part of the definition of a domain-specific graph transformation language, integrity constraints over a type graph representing a metamodel defining the permissible graph structures and reflection constraints restricting the notion of matches.

5 Open Maps via Application Conditions

Open maps for a path category **P** can be encoded by negative application conditions, so that the conditions are satisfied if and only if the match is an open map. If the path category is finite (up to isomorphism), this is useful for implementing transformations with open maps based on existing tools.

Definition 4 (NACs for open maps). *For a rule* $p : L \xleftarrow{l} K \xrightarrow{r} R$ *we define negative application condition* $N(L, \mathbf{P})$ *as the set of all morphisms* $n : L \to L_Q$ *in the pushout square below, where* $c \in \mathbf{P}$ *and* $P \to L$ *is any morphism such that there is no* $f : Q \to L$ *commuting the upper triangle.*

$$
\begin{array}{ccc}
L & \xleftarrow{\ i\ } & P \\
{\scriptstyle n}\downarrow & \searrow{\scriptstyle f} & \downarrow{\scriptstyle c} \\
L_Q & \xleftarrow{\ i^*\ } & Q
\end{array}
$$

While most categories of graphical structures have all pushouts, adhesive categories only require pushouts where at least one of the given morphisms is mono. The following Proposition holds under the assumption that at least the pushout of c and i in the Def. 4 can be built.

Proposition 2 (NACs for open maps). *Assume* $\mathbf{C}$ *has pushouts with one given morphism in* $\mathbf{P}$. *Then, a match* $m : L \to G$ *satisfies* $N(L, \mathbf{P})$ *as defined above if and only if* m *is* $\mathbf{P}$-*open.*

Proof. If m is not open, there exists a commuting diagram like the one on the right below, without a fill-in f. By Def. 4 this implies a constraint $n \in N(L, \mathbf{P})$ obtained by the pushout of c and i, which exists because c is in $\mathbf{P}$. This induces o such that the triangle commutes, i.e., m does not satisfy the constraint.

Vice versa, assume that there is a constraint $n \in N(L, \mathbf{P})$ that is violated by m, i.e., there exists o commuting the triangle. By construction of n there are $c \in \mathbf{P}$ and $i : P \to L$ such that no f commutes the triangle. Composition with o extends the pushout square of Def. 4 to a commutative square over m and c, but still without a fill-in.

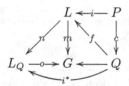

Example 5 (NACs ensuring openness). The result of the construction for the rules in Fig. 4 is illustrated in Fig. 7. The crossed-out parts in the rules' left-hand sides represent negative elements. The graph with positive and negative elements is the $\hat{L}$ of a constraint $L \to \hat{L}$ with L given by the positive elements only. Rule *Garbage collect* in Fig. 4 does not require any NAC because in its case, openness is subsumed by the dangling condition.

Reducing transformations with open maps to transformations with NACs, existing definitions and theorems for the second can be transferred to the first. This is demonstrated briefly in the following section for the notions of independence and the local Church Rosser theorem.

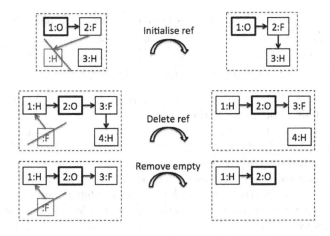

Fig. 7. NACs guaranteeing P-openness

6   Independence and Switch Equivalence

While open maps can be enforced into NACs, they provide a more uniform way of specifying reflection constraints, independently of individual rules. In addition to providing a more abstract style of specification, this uniformity has consequences for the concurrency properties of transformations, which we will explore in this section. We start with recalling some relevant definitions.

In the DPO approach, a derivation $s = G_0 \overset{p_1,m_1}{\Longrightarrow} G_1 \overset{p_2,m_2}{\Longrightarrow} G_2$ as in Fig. 8 is sequentially independent iff there exist morphisms $i : R_1 \to D_2$ and $j : L_2 \to D_1$ such that $r_1^* \circ j = m_2$ and $l_2^* \circ i = m_1^*$. Using the local Church-Rosser theorem ([4], theorem 3.20) it is possible to construct a derivation $s' = G_0 \overset{p_2,m_2'}{\Longrightarrow} G_1' \overset{p_1,m_1'}{\Longrightarrow} G_2$. We use $s \sim_{sw} s'$ to denote this relation. Switch-equivalence $\equiv_{sw} \subseteq \mathcal{G}^* \times \mathcal{G}^*$ over derivations of $\mathcal{G}$ is defined as the transitive and "context" closure of $\sim_{sw}$, i.e., the least equivalence relation containing $\sim_{sw}$ and such that if $s \equiv_{sw} s'$ then $s_1; s; s_2 \equiv_{sw} s_1; s'; s_2$.

The definitions of independence and switch equivalence carry over to transformations with NACs by requiring that the match for p_2 in G_0 given by $m_2' = l_1^* \circ j$ satisfies the NAC of p_2 and the induced match of p_1 into graph G_1' obtained by $G_0 \overset{p_2,m_2'}{\Longrightarrow} G_1'$ satisfies the NAC of p_1. The condition for independence of DPOs over open maps is analogous to that for steps with NACs.

$$L_1 \xleftarrow{l_1} K_1 \xrightarrow{r_1} R_1 \cdots\cdots\cdots L_2 \xleftarrow{l_2} K_2 \xrightarrow{r_2} R_2$$

$$G_0 \xleftarrow{l_1^*} D_1 \xrightarrow{r_1^*} G_1 \xleftarrow{l_2^*} D_2 \xrightarrow{r_2^*} G_2$$

Fig. 8. Sequential independence

Definition 5 (independence of P-open transformations). P-*open trans-formations* $G_0 \overset{p_1,m_1}{\Longrightarrow}_{\mathbf{P}} G_1 \overset{p_2,m_2}{\Longrightarrow}_{\mathbf{P}} G_2$ *are sequentially independent if the under-lying transformations* $G_0 \overset{p_1,m_1}{\Longrightarrow} G_1 \overset{p_2,m_2}{\Longrightarrow} G_2$ *are independent and morphisms* $m_1'^* = r_2^* \circ i : R_1 \to G_2$ *and* $m_2' = l_1^* \circ j : L_2 \to G_0$ *are* **P**-*open.*

Via the encoding of Prop. 2, this definition is equivalent to the one for steps with NACs. This is obvious for the requirement that m_2' needs to satisfy the application condition of p_2, but the requirement for p_1s NAC is expressed in terms of $m_1' : L_1 \to G_1'$ rather than $m_1'^*$. However, we know that DPOs preserve and reflect openness, so $m_1'^*$ is open iff m_1'.

The local Church-Rosser theorem for **P**-open transformations, like for those with NACs, follows directly from the definition of independence and the classical local Church-Rosser theorem. Therefore, the definition of switch-equivalence carries over as well. However, despite a considerable amount of work invested in developing the concurrency theory of transformations with NACs, the latter is not entirely satisfactory, as illustrated in the example.

Example 6 (context-dependency of independence with NACs). Figure 9 shows a sequence of transformations with NACs $t_1; t_2; t_3$ such that $t_1; t_2$ are inde-pendent and, with $t_2'; t_1'$ the steps after switching, $t_1'; t_3$ are independent, too. If $t_2'; t_3'; t_1'' \equiv_{sw} t_1; t_2; t_3$ is the result of switching $t_1'; t_3$, we might expect that independence of $t_2; t_3$ implies that of $t_2'; t_3'$, i.e., that switching preserves inde-pendence. However, this is not the case, because the match for p_3 into the first graph in the sequence does not satisfy p_3's NAC, as indicated by the dashed edges and node. This represents a single constraint ruling out the *joint* presence of the loop and the outgoing edge. Hence, independence can change depending on the derivation providing the context, even if derivations are equivalent.

Next we show that, under certain assumptions on the path category, this problem does not occur for transformations of open maps. The assumption is that, for arrows $c : P \to Q$ in **P**, Q does not extend P in two (or more) independent ways, where independence is used in the usual sense that structure can be added in any order. Consequently, if there are two different ways to decompose c, one has to be an extension of the other.

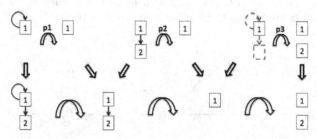

Fig. 9. Independence of steps $2; 3$ is not preserved by switching with 1

Definition 6 (incremental path category). *A path category* **P** *is incremental if for all arrows* $c : P \to Q$ *in* **P** *with decompositions* $P \overset{c_1}{\longrightarrow} O \overset{c_2}{\longrightarrow} Q = P \overset{c'_1}{\longrightarrow} O' \overset{c'_2}{\longrightarrow} Q$ *in* **C** *such that* c_2 *and* c'_2 *are monomorphism, there exists a morphism* $o : O \to O'$ *or* $o : O' \to O$ *commuting the resulting triangles.*

This is illustrated by the example below, where $c : P \to Q$ is not incremental, because Q extends P in two independent ways, by the loop on 1 in O_1 and the outgoing edge and node 2 in O_2. There is no compatible arrow relating these two additions. If instead we consider the addition of the loop and the outgoing edge as two separate constraints, they are incremental as shown on the right for $c_2 : P \to O_2$, which can only be decomposed in one way.

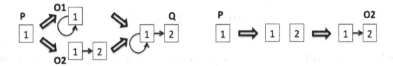

With this assumption, we show that independence is preserved by switching.

Proposition 3 (invariance of independence under switch equivalence).
Let **P** *be incremental and assume* **P***-open transformation sequences* $s = G_0 \overset{p_1,m_1}{\Longrightarrow}_{\mathbf{P}} G_1 \overset{p_2,m_2}{\Longrightarrow}_{\mathbf{P}} G_2 \overset{p_3,m_3}{\Longrightarrow}_{\mathbf{P}} G_3$ *and* $s' = G_0 \overset{p_2,m'_2}{\Longrightarrow}_{\mathbf{P}} G'_1 \overset{p_3,m'_3}{\Longrightarrow}_{\mathbf{P}} G'_2 \overset{p_1,m''_1}{\Longrightarrow}_{\mathbf{P}} G_3$ *using* **P***-stable rules* p_1, p_2, p_3 *such that* $s \equiv_{sw} s'$.

$$G_0 \overset{p_2,m'_2}{\Longrightarrow} G'_1 \overset{p_3,m'_3}{\Longrightarrow} G'_2$$
$$\| \qquad\qquad \| \qquad\qquad \|$$
$$p_1,m_1 \qquad p_1,m'_1 \qquad p_1,m''_1$$
$$\Downarrow \qquad\qquad \Downarrow \qquad\qquad \Downarrow$$
$$G_1 \overset{p_2,m_2}{\Longrightarrow} G_2 \overset{p_3,m_3}{\Longrightarrow} G_3$$

Then, $G_1 \overset{p_2,m_2}{\Longrightarrow}_{\mathbf{P}} G_2 \overset{p_3,m_3}{\Longrightarrow}_{\mathbf{P}} G_3$ *is sequentially independent if and only if* $G_0 \overset{p_2,m'_2}{\Longrightarrow}_{\mathbf{P}} G'_1 \overset{p_3,m'_3}{\Longrightarrow}_{\mathbf{P}} G'_2$ *is.*

Proof. According to Def. 5, $G_1 \overset{p_2,m_2}{\Longrightarrow}_{\mathbf{P}} G_2 \overset{p_3,m_3}{\Longrightarrow}_{\mathbf{P}} G_3$ implies that the match of p_3 extends to an open map to G_1. Using that m_3 and m'_3 are open, we show that the match into G_0 that exists by classical local Church-Rosser is open, too, which provides one half of the independence of $G_0 \overset{p_2,m'_2}{\Longrightarrow}_{\mathbf{P}} G'_1 \overset{p_3,m'_3}{\Longrightarrow}_{\mathbf{P}} G'_2$.

By inverting the two horizontal sequences in the diagram of Prop. 3, we obtain the proof for the other half, i.e., the comatch of p_2 into G'_2 is open. Inverting the vertical steps yields the reverse implication, that independence of the upper sequence implies independence of the lower.

The diagram in the left of Fig. 10 shows a deconstruction of the transformations $G_0 \overset{p_1,m_1}{\Longrightarrow} G_1 \overset{p_2,m_2}{\Longrightarrow} G_2$ and $G_0 \overset{p_2,m'_2}{\Longrightarrow} G'_1 \overset{p_1,m'_1}{\Longrightarrow} G_2$ according to the proof of the local Church-Rosser theorem ([4], theorem 3.20). Hence $D'_2 G_2 D^*_2 D_2$ is a

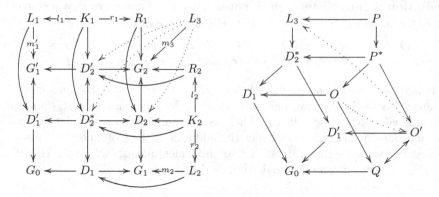

Fig. 10. Proof of Prop. 3

pullback while all other squares are pushouts, and all morphisms in the inner square $G_1'G_0G_2G_1$ are monos.

The match m_3 is also shown. Lets assume that $G_1 \overset{p_2,m_2}{\Longrightarrow}_{\mathbf{P}} G_2 \overset{p_3,m_3}{\Longrightarrow}_{\mathbf{P}} G_3$ are independent. Then, there exists $L_3 \to D_2$ commuting with m_3 such that $L_3 \to D_2 \to G_1$ is open. Also, $G_1' \overset{p_1,m_1'}{\Longrightarrow}_{\mathbf{P}} G_2 \overset{p_3,m_3}{\Longrightarrow}_{\mathbf{P}} G_3$ are independent because equivalence of s and s' requires to switch them, so there exists $L_3 \to D_2'$ commuting with m_3 such that $L_3 \to D_2' \to G_1'$ is open.

There exists a morphism $L_3 \to D_2^*$ commuting the resulting triangles induced by the pullback $D_2'G_2D_2^*D_2$. Morphisms $L_3 \to D_2^* \to D_1$ and $L_3 \to D_2^* \to D_1'$ are open by Lemma 2 because they are prefixes of open maps $L_3 \to D_2 \to G_1$ and $L_3 \to D_2' \to G_1'$, respectively, with $D_1 \to G_1$ and $D_1' \to G_1'$ monos.

To show that $m_3' = L_3 \to D_2^* \to D_1 \to G_0 = L_3 \to D_2^* \to D_1' \to G_0$ is open assume $c : P \to Q \in \mathbf{P}$ with morphisms $P \to L_3$ and $Q \to G_0$ commuting the square with m_3'. We can construct the diagram on the right by forming pullbacks D_1G_0OQ, $D_1'G_0O'Q$, $D_2^*D_1P^*O$, and $D_2^*D_1'P^*O'$. We obtain $P \to P^*$ because by pullback composition and decomposition the right-hand square $P^*OO'Q$ is a pullback, too. Further, by the van Kampen property of the left square, the right square is a pushout. Since $\mathbf{P}$ is incremental and $P \to O \to Q = P \to O' \to Q$ with $O \to Q$ and $O' \to Q$ monic, without loss of generality we have a morphism $O \to O'$ commuting the triangles. It can be shown that in this case $O' \leftrightarrow Q$ is an isomorphism, so $P \to O'$ is in $\mathbf{P}$. Since $L_3 \to D_2^* \to D_1'$ is open, there exists a fill-in $O' \to L_3$ which extends to a fill-in $Q \to L_3$ using the isomorphism.

7   Conclusion

We presented a general representation of reflection properties for matches in the DPO approach and studied its relationship with NACs. We found that, for DPO transformations with open maps, independence is invariant under switch equivalence if we restrict reflection constraints to be incremental. The counterexample for the same property of DPO with NACs suggested that an analogous restriction to "incremental NACs" might solve the problem, and this has indeed been confirmed now in [2].

Our original motivation is the representation of KAPPA [3] in the DPO approach, to study techniques for refinement and model reduction. The ability to model KAPPA-style rewriting using DPO with open maps is a first step in this direction.

References

1. Blostein, D.: Graph Transformation in Document Image Analysis: Approaches and Challenges. In: Brun, L., Vento, M. (eds.) GbRPR 2005. LNCS, vol. 3434, pp. 23–34. Springer, Heidelberg (2005)
2. Corradini, A., Heckel, R., Hermann, F., Gottmann, S., Nachtigall, N.: On the concurrent semantics of transformation systems with negative application conditions. In: Workshop on Algebraic Development Techniques, WADT 2012, Salamanca, Spain (2012) (Presentation and abstract)
3. Danos, V., Feret, J., Fontana, W., Harmer, R., Krivine, J.: Abstracting the differential semantics of rule-based models: Exact and automated model reduction. In: LICS, pp. 362–381. IEEE Computer Society (2010)
4. Ehrig, H., Ehrig, K., Prange, U., Taentzer, G.: Fundamentals of Algebraic Graph Transformation. EATCS Monographs in Theoretical Comp. Sci. Springer (2006)
5. Ehrig, H., Pfender, M., Schneider, H.: Graph grammars: an algebraic approach. In: 14th IEEE Symp. on Switching and Automata Theory, pp. 167–180. IEEE (1973)
6. Ehrig, H., Ehrig, K., Habel, A., Pennemann, K.H.: Theory of constraints and application conditions: From graphs to high-level structures. Fundam. Inf. 74(1), 135–166 (2006)
7. Habel, A., Heckel, R., Taentzer, G.: Graph grammars with negative application conditions. Fundamenta Informaticae 26(3,4), 287–313 (1996)
8. Joyal, A., Nielsen, M., Winskel, G.: Bisimulation from open maps. Inf. Comput. 127(2), 164–185 (1996)
9. Lack, S., Sobociński, P.: Adhesive Categories. In: Walukiewicz, I. (ed.) FOSSACS 2004. LNCS, vol. 2987, pp. 273–288. Springer, Heidelberg (2004)
10. Lambers, L., Ehrig, H., Prange, U., Orejas, F.: Parallelism and concurrency in adhesive high-level replacement systems with negative application conditions. ENTCS 203(6), 43–66 (2008)

$\mathcal{M}, \mathcal{N}$-Adhesive Transformation Systems

Annegret Habel[1] and Detlef Plump[2]

[1] Carl von Ossietzky Universität Oldenburg
annegret.habel@informatik.uni-oldenburg.de
[2] The University of York
detlef.plump@york.ac.uk

Abstract. The categorical framework of $\mathcal{M}$-adhesive transformation systems does not cover graph transformation with relabelling. Rules that relabel nodes are natural for computing with graphs, however, and are commonly used in graph transformation languages. In this paper, we generalise $\mathcal{M}$-adhesive transformation systems to $\mathcal{M}, \mathcal{N}$-adhesive transformation systems, where $\mathcal{N}$ is a class of morphisms containing the vertical morphisms in double-pushouts. We show that the category of partially labelled graphs is $\mathcal{M}, \mathcal{N}$-adhesive, where $\mathcal{M}$ and $\mathcal{N}$ are the classes of injective and injective, undefinedness-preserving graph morphisms, respectively. We obtain the Local Church-Rosser Theorem and the Parallelism Theorem for graph transformation with relabelling and application conditions as instances of results which we prove at the abstract level of $\mathcal{M}, \mathcal{N}$-adhesive systems.

1 Introduction

The double-pushout approach to graph transformation, which was invented in the early 1970's, is the best studied framework for graph transformation [20,5,10,4]. As applications of graph transformation come with a large variety of graphs and graph-like structures, the double-pushout approach has been generalised to the abstract settings of high-level replacement systems [9], adhesive categories [17] and $\mathcal{M}$-adhesive categories [8,6,7].

The categories of labelled graphs, typed graphs, and typed attributed graphs, for example, are known to be $\mathcal{M}$-adhesive categories if one chooses $\mathcal{M}$ to be the class of injective graph morphisms [8]. Each such category induces a class of $\mathcal{M}$-adhesive transformation systems for which several classical results of the double-pushout approach hold. Specifically, the Local Church-Rosser Theorem, the Parallelism Theorem, the Concurrency Theorem, the Amalgamation Theorem, the Embedding Theorem and the Local Confluence Theorem have been established for rules with nested application conditions [6,7].

However, $\mathcal{M}$-adhesive transformation systems do not cover graph transformation systems with rules that relabel nodes. Such rules are natural for computing with graphs and are used as a foundation for the graph transformation language GP [18,19]. The double-pushout approach can be extended with relabelling by

H. Ehrig et al.(Eds.): ICGT 2012, LNCS 7562, pp. 218–233, 2012.
© Springer-Verlag Berlin Heidelberg 2012

introducing rules with partially labelled interface graphs [14], providing a theoretical foundation for graph transformation languages that is much simpler than attributed graph transformation in the sense of [4]. In the latter approach, attributed graphs contain the algebra underlying the operations in the attributes as well as special edges which connect nodes and edges with their attributes. Hence they are (usually) complex infinite objects which are difficult to comprehend and which do not directly correspond to the graph data structures used to implement graph transformation languages.

In this paper, we study transformation systems over the category PLG of partially labelled graphs and the class $\mathcal{M}$ of injective graph morphims (which are used in rules). It turns out that PLG violates two of the properties required for $\mathcal{M}$-adhesive categories: pushouts along $\mathcal{M}$-morphisms do not always exist and, when they exist, need not be pullbacks. We therefore generalise $\mathcal{M}$-adhesive categories to $\mathcal{M}, \mathcal{N}$-adhesive categories, where $\mathcal{N}$ is a class of morphisms containing the vertical morphisms in double-pushouts. $\mathcal{M}$-adhesive categories are then the special case where $\mathcal{N}$ is the class of all morphisms.

For $\mathcal{M}, \mathcal{N}$-adhesive transformation systems with (nested) application conditions, we prove two classical results of the double-pushout approach: the Local Church-Rosser Theorem and the Parallelism Theorem. We then show that PLG is $\mathcal{M}, \mathcal{N}$-adhesive, where $\mathcal{N}$ is the class of injective morphisms that preserve unlabelled nodes and edges. As a result, we obtain both theorems for the setting of graph transformation with relabelling and application conditions.

The paper is structured as follows. In Section 2, we generalise $\mathcal{M}$-adhesive categories to $\mathcal{M}, \mathcal{N}$-adhesive categories, prove that they satisfy the so-called HLR properties, and identify two additional factorization properties. In Section 3, we present the Local Church-Rosser Theorem and the Parallelism Theorem for $\mathcal{M}, \mathcal{N}$-adhesive transformation systems with application conditions. In Section 4, we show that the category PLG is $\mathcal{M}, \mathcal{N}$-adhesive for suitable classes $\mathcal{M}$ and $\mathcal{N}$ of morphisms. As a consequence, we obtain the Local Church-Rosser Theorem and the Parallelism Theorem for graph transformation with relabelling. In Section 5, we conclude and give some topics for future work.

The proofs omitted in this paper are given in [15], as well as the Concurrency Theorem for $\mathcal{M}, \mathcal{N}$-adhesive transformation systems with application conditions.

2 $\mathcal{M}, \mathcal{N}$-Adhesive Categories

In [8] an overview is given on some categorical frameworks for double-pushout transformations. It is shown that adhesive categories [17], weak adhesive HLR categories [4], and partial map adhesive categories [16] are special cases of so-called $\mathcal{M}$-adhesive categories. A large number of results have been proved for $\mathcal{M}$-adhesive transformation systems, such as the Local Church-Rosser Theorem, the Parallelism Theorem, the Concurrency Theorem, the Amalgamation Theorem, the Embedding Theorem, and the Local Confluence Theorem [6,7].

In this section, we generalize $\mathcal{M}$-adhesive categories as defined in [8,6] to $\mathcal{M}, \mathcal{N}$-adhesive categories.

Definition 1 ($\mathcal{M},\mathcal{N}$-adhesive category). A category $\mathcal{C}$ is $\mathcal{M},\mathcal{N}$-*adhesive*, where $\mathcal{M}$ is a class of monomorphisms and $\mathcal{N}$ a class of morphisms, if the following properties are satisfied:

1. $\mathcal{M}$ and $\mathcal{N}$ contain all isomorphisms and are closed under composition and decomposition (see [6]). Moreover, $\mathcal{N}$ is closed under $\mathcal{M}$-decomposition, that is, $g \circ f \in \mathcal{N}$, $g \in \mathcal{M}$ implies $f \in \mathcal{N}$.
2. $\mathcal{C}$ has pushouts along $\mathcal{M},\mathcal{N}$-morphisms and pullbacks along $\mathcal{M}$-morphisms. Also, $\mathcal{M}$ and $\mathcal{N}$ are stable under $\mathcal{M},\mathcal{N}$-pushouts and $\mathcal{M}$-pullbacks (see below).
3. Pushouts along $\mathcal{M},\mathcal{N}$-morphisms are $\mathcal{M},\mathcal{N}$-van Kampen squares (see below).

Remark 1. A pushout *along* $\mathcal{M},\mathcal{N}$-morphisms, or $\mathcal{M},\mathcal{N}$-*pushout*, is a pushout where one of the given morphisms is in $\mathcal{M}$ and the other morphism is in $\mathcal{N}$. A pullback *along* an $\mathcal{M}$-morphism, or $\mathcal{M}$-*pullback*, is a pullback where at least one of the given morphisms is in $\mathcal{M}$. A class $\mathcal{X}$ of morphisms is *stable under* $\mathcal{M},\mathcal{N}$-*pushouts* if, given the $\mathcal{M},\mathcal{N}$-pushout (1) in the diagram below, $m \in \mathcal{X}$ implies $n \in \mathcal{X}$. Class $\mathcal{X}$ is *stable under* $\mathcal{M}$-*pullbacks* if, given the $\mathcal{M}$-pullback (1) in the diagram below, $n \in \mathcal{X}$ implies $m \in \mathcal{X}$.

A pushout along $\mathcal{M},\mathcal{N}$-morphisms is an $\mathcal{M},\mathcal{N}$-*van Kampen square* if for the commutative cube in the diagram below with the pushout as bottom square, $b, c, d, m \in \mathcal{M}$, $f \in \mathcal{N}$, and the back faces being pullbacks, we have that the top square is a pushout if and only if the front faces are pullbacks.

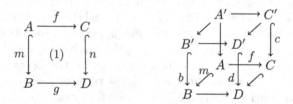

Fact 1. Let $\mathcal{C}$ be any category and let $\mathcal{N}$ be the class of all morphisms in $\mathcal{C}$. Then $\mathcal{C}$ is $\mathcal{M},\mathcal{N}$-adhesive if and only if $\mathcal{C}$ is $\mathcal{M}$-adhesive in the sense of [6].

Proof. This follows from the definition of an $\mathcal{M}$-adhesive category because if $\mathcal{N}$ contains all morphisms, then $\mathcal{M},\mathcal{N}$-pushouts and $\mathcal{M},\mathcal{N}$-van Kampen squares are precisely the $\mathcal{M}$-pushouts and $\mathcal{M}$-van Kampen squares of [6], respectively. □

In Section 4, we show that the category PLG of partially labelled graphs is $\mathcal{M},\mathcal{N}$-adhesive but not $\mathcal{M}$-adhesive. In this case, $\mathcal{M}$ is the class of injective graph morphisms and $\mathcal{N}$ is the class of injective, undefinedness preserving graph morphisms.

$\mathcal{M},\mathcal{N}$-adhesive categories satisfy generalised versions of the so-called HLR-properties [9] of $\mathcal{M}$-adhesive categories.

Theorem 1 (HLR-properties). Every $\mathcal{M}, \mathcal{N}$-adhesive category satisfies the following *HLR-properties*:

1. Pushouts along $\mathcal{M}, \mathcal{N}$-morphisms are pullbacks.
2. $\mathcal{M}, \mathcal{N}$-pushout-pullback decomposition: If (1)+(2) in the diagram below is a pushout, (2) is a pullback, $l \in \mathcal{M}$, and $k, w \in \mathcal{N}$, then (1) and (2) are pushouts as well as pullbacks.
3. Cube $\mathcal{M}, \mathcal{N}$-pushout-pullback decomposition: If in the commutative cube (3) of the diagram below, all morphisms in the top square and in the bottom square are in $\mathcal{M}$, all vertical morphisms are in $\mathcal{N}$, the top square is a pullback, and the front faces are pushouts, then the bottom square is a pullback if and only if the back faces are pushouts.
4. Uniqueness of pushout complements: Given morphisms $A \hookrightarrow B$ in $\mathcal{M}$ and $B \to D$ in $\mathcal{N}$, there is, up to isomorphism, at most one object C with morphisms $A \to C$ and $C \hookrightarrow D$ such that (4) in the diagram below is a pushout.

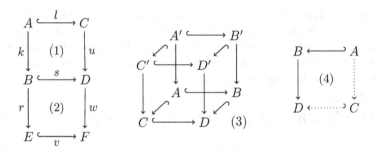

In order to prove the desired results for $\mathcal{M}, \mathcal{N}$-adhesive transformation systems, three more properties will be needed.

Definition 2 (HLR$^+$-properties). Let $\mathcal{C}$ be an $\mathcal{M}, \mathcal{N}$-adhesive category, $\mathcal{E}$ a class of morphisms, and $\mathcal{E}'$ a class of pairs of morphism with the same codomain. Then the following properties are the *HLR$^+$-properties* with respect to $\mathcal{M}$, $\mathcal{N}$, $\mathcal{E}$ and $\mathcal{E}'$.

1. $\mathcal{C}$ has binary coproducts.
2. $\mathcal{C}$ has an $\mathcal{E}$-$\mathcal{N}$ *factorization* if for each coproduct morphism $f : A_1 + A_2 \to C$ induced by morphisms $f_i : A_i \to C$ in $\mathcal{N}$ $(i = 1, 2)$, there is a decomposition, unique up to isomorphism, $f = n \circ e$ with $e \in \mathcal{E}$ and $n \in \mathcal{N}$.
3. $\mathcal{C}$ has an $\mathcal{E}'$-$\mathcal{M}$ *pair factorization* if, for each pair of morphisms $f_1 : A_1 \to C$ and $f_2 : A_2 \to C$, there exist a unique (up to isomorphism) object K and unique (up to isomorphism) morphisms $e_1 : A_1 \to K$, $e_2 : A_2 \to K$, and $m : K \hookrightarrow C$ with $(e_1, e_2) \in \mathcal{E}'$ and $m \in \mathcal{M}$ such that $m \circ e_1 = f_1$ and $m \circ e_2 = f_2$.

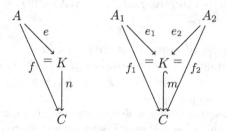

General Assumption. We assume that C is an $\mathcal{M}, \mathcal{N}$-adhesive category and that $\mathcal{E}$ and $\mathcal{E}'$ are classes of morphisms and morphisms pairs, respectively, such that C satisfies the HLR$^+$-properties.

The $\mathcal{E}$-$\mathcal{N}$ factorization is used in the proof of the Parallelism Theorem. The $\mathcal{E}'$-$\mathcal{M}$ pair factorization is used in the proof of a shift lemma for application conditions and in the construction of E-related transformations in [15].

Example 1. The category PLG considered in Section 4 satisfies the HLR$^+$-properties, where $\mathcal{M}$ is the class of injective morphisms, $\mathcal{N}$ is the class of injective, undefinedness preserving morphisms, $\mathcal{E}$ is the class of surjective, undefinedness preserving morphisms, and $\mathcal{E}'$ is the class of pairs of jointly surjective, undefinedness preserving morphisms.

3 $\mathcal{M}, \mathcal{N}$-Adhesive Transformation Systems

In this section, we introduce $\mathcal{M}, \mathcal{N}$-adhesive transformation systems and present the Local Church-Rosser Theorem and the Parallelism Theorem in this setting.
   We start by defining rules, direct transformations, and transformation systems.

Definition 3 (Rules, transformations, and systems). Given an $\mathcal{M}, \mathcal{N}$-adhesive category, a *rule* $\varrho = \langle p, \mathrm{ac}_L \rangle$ consists of a *plain rule* $p = \langle L \hookleftarrow K \hookrightarrow R \rangle$ with morphisms $l \colon K \hookrightarrow L$ and $r \colon K \hookrightarrow R$ in $\mathcal{M}$, and an application condition ac_L over L (see below). A *direct transformation* from an object G to an object H via the rule ϱ consists of two pushouts (1) and (2) as below where the vertical morphisms[1] are in $\mathcal{N}$ and $g \models \mathrm{ac}_L$. We write $G \Rightarrow_{\varrho,g} H$ if there exists such a direct transformation. For a set of rules $\mathcal{R}$, we write $G \Rightarrow_{\mathcal{R}} H$, if $G \Rightarrow_{\varrho} H$ with $\varrho \in \mathcal{R}$.

$$
\begin{array}{ccccc}
\mathrm{ac}_L \blacktriangleright L & \xleftarrow{\;l\;} & K & \xrightarrow{\;r\;} & R \\
\downarrow g & (1) & \downarrow d & (2) & \downarrow h \\
G & \longleftarrow & D & \longrightarrow & H
\end{array}
$$

An *$\mathcal{M}, \mathcal{N}$-adhesive transformation system* consists of an $\mathcal{M}, \mathcal{N}$-adhesive category and a set $\mathcal{R}$ of rules.

---

[1] By stability of $\mathcal{N}$ under $\mathcal{M}, \mathcal{N}$-pushouts, it is equivalent to require $d \in \mathcal{N}$.

Remark 2. Every $\mathcal{M}$-adhesive transformation system in the sense of [6] is an $\mathcal{M}, \mathcal{N}$-adhesive transformation system if we choose $\mathcal{N}$ as the class of all morphisms in $\mathcal{C}$. Our notion of transformation system is more flexible because it allows to restrict the class of morphisms that are used to match rules. For example, one can show that every $\mathcal{M}$-adhesive category is $\mathcal{M}, \mathcal{M}$-adhesive and hence gives rise to an $\mathcal{M}, \mathcal{M}$-adhesive transformation system. A concrete example for this is the category of totally labelled graphs together with the class of injective graph morphisms (see also [12] for this setting).

Application conditions are nested constructs which can be represented as trees of morphisms equipped with quantifiers and Boolean connectives.

Definition 4 (Application condition). *Application conditions are inductively defined as follows. For every object P, true is an application condition over P. For every morphism $a \colon P \to C$ and every application condition ac over C, $\exists(a, \mathrm{ac})$ is an application condition over P. For application conditions $\mathrm{ac}, \mathrm{ac}_i$ over P with $i \in I$ (for a given index set I), $\neg\mathrm{ac}$ and $\wedge_{i \in I}\mathrm{ac}_i$ are application conditions over P.*

Satisfiability of application conditions is also defined inductively. Every morphism satisfies true. A morphism $p \colon P \to G$ satisfies $\exists(a, \mathrm{ac})$ over P if there exists a morphism $q \colon C \hookrightarrow G$ in $\mathcal{M}$ such that $q \circ a = p$ and q satisfies ac.

$$\exists(\ P \xrightarrow{\ a\ } C, \blacktriangleleft \mathrm{ac}\ )$$
$$p \searrow \ \underset{=}{\ } \ \nearrow q$$
$$G$$

A morphism $p \colon P \to G$ satisfies $\neg\mathrm{ac}$ over P if p does not satisfy ac, and p satisfies $\wedge_{i \in I}\mathrm{ac}_i$ over P if p satisfies each ac_i ($i \in I$). We write $p \models \mathrm{ac}$ to express that p satisfies ac.

Next we state two important technical results. The first lemma allows to shift application conditions over arbitrary morphisms.

Lemma 1 (Shift of application conditions over morphisms [6]). There is a construction Shift such that, for each application condition ac over P and for each morphism $b \colon P \to P'$, Shift transforms ac via b into an application condition $\mathrm{Shift}(b, \mathrm{ac})$ over P' such that, for each morphism $n \colon P' \to H$, $n \circ b \models$ $\mathrm{ac} \iff n \models \mathrm{Shift}(b, \mathrm{ac})$.

$$\mathrm{ac} \blacktriangleright P \xrightarrow{\ b\ } P' \blacktriangleleft \mathrm{Shift}(b, \mathrm{ac})$$
$$n \circ b \searrow \underset{=}{\ } \nearrow n$$
$$H$$

The other technical result that we need is that application conditions can be shifted over rules.

Lemma 2 (Shift of application conditions over rules [13]). There is a construction L such that, for each rule ϱ and each application condition ac over R, L transforms ac via ϱ into an application condition $L(\varrho, ac)$ over L such that, for each direct transformation $G \Rightarrow_{\varrho, m, m^*} H$, we have $m \models L(\varrho, ac) \iff m^* \models ac$.

$$
\begin{array}{ccccccc}
L(\varrho, ac) \rhd & L & \hookleftarrow & K & \hookrightarrow & R & \lhd \text{ ac} \\
& \Big\downarrow m & (1) & \Big\downarrow & (2) & \Big\downarrow m^* & \\
& G & \hookleftarrow & D & \hookrightarrow & H &
\end{array}
$$

Remark 3. There is a construction R with $R(\varrho, ac) = L(\varrho^{-1}, ac)$ that transforms left application conditions ac via the rule ϱ into right application conditions.

Assumption. For $i = 1, 2$, let $\varrho_i = \langle p_i, ac_{L_i} \rangle$ be a rule with plain rule $p_i = \langle L_i \hookleftarrow K_i \hookrightarrow R_i \rangle$. Also, let $\varrho = \langle p, ac_L \rangle$ and $\varrho' = \langle p', ac_{L'} \rangle$ be rules with plain rules $p = \langle L \hookleftarrow K \hookrightarrow R \rangle$ and $p' = \langle L' \hookleftarrow K' \hookrightarrow R' \rangle$, respectively.

First, we formulate the notions of parallel and sequential independence and present the Local Church-Rosser Theorem.

Definition 5 (Parallel and sequential independence). Two direct transformations $H_1 \Leftarrow_{\varrho_1, g_1} G \Rightarrow_{\varrho_2, g_2} H_2$ are *parallelly independent* if in the diagram below there are morphisms $d_{ij} \colon L_i \to D_j$ such that $g_i = b_j \circ d_{ij}$, $g_i' = (c_j \circ d_{ij}) \in \mathcal{N}$, and $g_i' \models ac_{L_i}$ ($i, j \in \{1, 2\}$ and $i \neq j$).

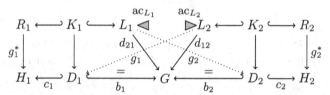

Two direct transformations $G \Rightarrow_{\varrho_1, g_1} H_1 \Rightarrow_{\varrho_2, g_2} M$ are *sequentially independent* if in the diagram below there are morphisms $d_{12} \colon R_1 \to D_2$ and $d_{21} \colon L_2 \to D_1$ such that $g_1^* = b_2 \circ d_{12}$, $g_2 = b_1 \circ d_{21}$, $g_2' = (c_1 \circ d_{21}) \in \mathcal{N}$, $g_1' = (c_2 \circ d_{12}) \in \mathcal{N}$, $g_2' \models ac_{L_2}$, and $g_1' \models R(\varrho_1, ac_{L_1})$.

$$
\begin{array}{ccccccccc}
ac_{L_1} \rhd L_1 & \hookleftarrow & K_1 & \hookrightarrow & R_1 & & ac_{L_2} \rhd L_2 & \hookleftarrow K_2 \hookrightarrow R_2 \\
\Big\downarrow g_1 & & \Big\downarrow & & & d_{21} \quad d_{12} & & \Big\downarrow g_2^* \\
G \hookleftarrow_{c_1} D_1 & \hookleftarrow_{b_1} & = & H_1 & \hookleftarrow_{b_2} & = & D_2 \hookrightarrow_{c_2} M
\end{array}
$$

The following Local Church-Rosser Theorem generalises the corresponding result in [6] from $\mathcal{M}$-adhesive transformation systems to $\mathcal{M}, \mathcal{N}$-adhesive transformation systems.

Theorem 2 (Local Church-Rosser Theorem). Given parallelly indepen-
dent direct transformations $H_1 \Leftarrow_{\varrho_1,g_1} G \Rightarrow_{\varrho_2,g_2} H_2$, there are an object M and
direct transformations $H_1 \Rightarrow_{\varrho_2,g_2'} M \Leftarrow_{\varrho_1,g_1'} H_2$ such that $G \Rightarrow_{\varrho_1,g_1} H_1 \Rightarrow_{\varrho_2,g_2'}$
M and $G \Rightarrow_{\varrho_2,g_2} H_2 \Rightarrow_{\varrho_1,g_1'} M$ are sequentially independent.

Given sequentially independent direct transformations $G \Rightarrow_{\varrho_1,g_1} H_1 \Rightarrow_{\varrho_2,g_2}$
M, there are an object H_2 and direct transformations $G \Rightarrow_{\varrho_2,g_2'} H_2 \Rightarrow_{\varrho_1,g_1'} M$
such that $H_1 \Leftarrow_{\varrho_1,g_1} G \Rightarrow_{\varrho_2,g_2'} H_2$ are parallelly independent:

Next we consider parallel rules, quotients rules, and parallel transformations.
The parallel rule $\varrho_1 + \varrho_2$ of the rules ϱ_1 and ϱ_2 is defined by using the binary
coproducts of the components of the rules (which exist by the General Assump-
tion).

Definition 6 (Parallel rule, quotient rule, parallel transformation). The
parallel rule of ϱ_1 and ϱ_2 is the rule $\varrho_1+\varrho_2 = \langle p, \mathrm{ac}_L \rangle$ where $p = \langle L_1+L_2 \hookleftarrow$
$K_1+K_2 \hookrightarrow R_1+R_2 \rangle$ is the parallel rule of p_1 and p_2 and $\mathrm{ac}_L = \wedge_{i=1}^2 \mathrm{Shift}(k_i, \mathrm{ac}_{L_i}) \wedge$
$\mathrm{L}(p, \mathrm{Shift}(k_i^*, \mathrm{R}(\varrho_i, \mathrm{ac}_{L_i})))$.

The rule ϱ' is a *quotient rule* of a parallel rule ϱ if there are two pushouts (1) and
(2) as in the figure above where $k\colon K \to K'$ is an epimorphism in the class of
coproduct morphisms induced by $\mathcal{N}$ and $\mathrm{ac}_{L'} = \mathrm{Shift}(l, \mathrm{ac}_L)$. The set of quotient
rules of ϱ is denoted by $\mathrm{Q}(\varrho)$.

A direct transformation via a quotient of a parallel rule is called *parallel* direct
transformation or *parallel transformation*, for short.

Fact 2 ([6]). $K_1+K_2 \hookrightarrow L_1+L_2$ and $K_1+K_2 \hookrightarrow R_1+R_2$ are in $\mathcal{M}$.

The connection between sequentially independent direct transformations and
parallel direct transformations is given in the Parallelism Theorem.

Theorem (Parallelism Theorem).
1. Synthesis. Given two sequentially independent direct transformations $G \Rightarrow_{\varrho_1,g_1}$
$H_1 \Rightarrow_{\varrho_2,g_2'} M$, there is a parallel transformation $G \Rightarrow_{\mathrm{Q}(\varrho_1+\varrho_2),g} M$.

2. Analysis. Given a parallel transformation $G \Rightarrow_{Q(\varrho_1+\varrho_2),m} M$, there are sequentially independent direct transformations $G \Rightarrow_{\varrho_1,g_1} H_1 \Rightarrow_{\varrho_2,g_2'} M$ and $G \Rightarrow_{\varrho_2,g_2} H_i \Rightarrow_{\varrho_1,g_1'} M$.

3. Bijective correspondence. The synthesis and analysis constructions are inverse to each other up to isomorphism:

$$
\begin{array}{ccc}
 & H_1 & \\
{}^{\varrho_1}\nearrow & & \searrow{}^{\varrho_2} \\
{}^{Q(\varrho_1+\varrho_2)} & & \\
G \Longrightarrow & & M \\
{}_{\varrho_2}\searrow & & \nearrow{}_{\varrho_1} \\
 & H_2 & \\
\end{array}
$$

We conclude this section by mentioning that the Concurrency Theorem for $\mathcal{M},\mathcal{N}$-adhesive transformation systems is established in [15].

4   Category PLG Is $\mathcal{M},\mathcal{N}$-Adhesive

In this section, we consider the category PLG of partially labelled graphs [14]. We first show that PLG is not $\mathcal{M}$-adhesive for the class $\mathcal{M}$ of injective graph morphisms. We then prove that PLG is $\mathcal{M},\mathcal{N}$-adhesive, though, and satisfies the HLR$^+$-properties if we choose $\mathcal{N}$ as a suitable class of morphisms. As a consequence, we obtain the Local Church-Rosser Theorem and the Parallelism Theorem as new results for the setting of graph transformation with relabelling and application conditions.

We start by recalling the basic notions of partially labelled graphs and their morphisms.

Definition 7 (Graphs and morphisms). A *(partially labelled) graph* is a system $G = (V_G, E_G, s_G, t_G, l_{G,V}, l_{G,E})$ consisting of finite sets V_G and E_G of *nodes* and *edges*, source and target functions $s_G, t_G \colon E_G \to V_G$, and partial labelling functions $l_{G,V} \colon V_G \to C_V$ and $l_{G,E} \colon E_G \to C_E$,[2] where C_V and C_E are fixed sets of node and edge labels. A graph G is *totally labelled* if $l_{G,V}$ and $l_{G,E}$ are total functions.

A *morphism* $g \colon G \to H$ between graphs G and H consists of two functions $g_V \colon V_G \to V_H$ and $g_E \colon E_G \to E_H$ that preserve sources, targets and labels, that is, $s_H \circ g_E = g_V \circ s_G$, $t_H \circ g_E = g_V \circ t_G$, and $l_H(g(x)) = l_G(x)$ for all x in $\mathrm{Dom}(l_G)$.[3] Such a morphism preserves *undefinedness* if it maps unlabelled items in G to unlabelled items in H. Morphism g is *injective* (*surjective*) if g_V and g_E are injective (surjective), and an *isomorphism* if it is injective, surjective and

---

[2] Given sets A and B, a partial function $f \colon A \to B$ is a function from some subset A' of A to B. The set A' is the *domain* of f and is denoted by $\mathrm{Dom}(f)$. We say that $f(x)$ is *undefined*, and write $f(x) = \bot$, if x is in $A - \mathrm{Dom}(f)$.

[3] We often do not distinguish between nodes and edges in statements that hold analogously for both sets.

preserves undefinedness. In the latter case G and H are *isomorphic*, which is denoted by $G \cong H$. Furthermore, g is an *inclusion* if $g(x) = x$ for all x in G (note that inclusions need not preserve undefinedness). The *composition* $h \circ g$ of g with a morphism $h \colon H \to M$ consists of the composed functions $h_V \circ g_V$ and $h_E \circ g_E$. We write PLG for the category having partially labelled graphs as objects and graph morphisms as arrows.

In pictures of graphs, nodes are drawn as circles with their labels (if existent) inside, and edges are drawn as arrows with their labels (if existent) placed next to them. Graph morphisms are graphically represented by attaching the same number to nodes and their images.

Example 2. Consider the partially labelled graphs G and H below. Nodes 4 and 5 in G, nodes 4 and 6 in H, and all edges are unlabelled. The graph morphism $g \colon G \hookrightarrow H$ is injective but not undefinedness preserving, because it maps the unlabelled node 5 in G to a labelled node in H.

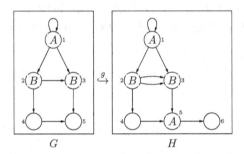

While the category of labelled graphs with arbitrary morphisms has pushouts [3], the category of partially labelled graphs with injective morphisms does not have all pushouts [14]. As a consequence, the category PLG with the class $\mathcal{M}$ of injective morphisms is not $\mathcal{M}$-adhesive.

Fact 3 (PLG is not $\mathcal{M}$-adhesive). Let $\mathcal{M}$ be the class of injective graph morphisms. Then PLG does not have pushouts along arbitrary $\mathcal{M}$-morphisms. Moreover, pushouts along $\mathcal{M}$-morphisms need not be pullbacks.

Example 3. The morphisms a and b in square (1) below are injective but their pushout does not exist: it is impossible to make both morphisms f and g label preserving. Square (2) is a pushout along $\mathcal{M}$, but not a pullback.

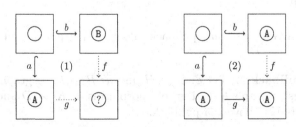

Assumption. For the rest of this section, we consider the category PLG and let $\mathcal{M}$ be the class of injective graph morphisms and $\mathcal{N}$ the class of injective, undefinedness preserving graph morphisms.

Theorem 3. The category PLG is $\mathcal{M}, \mathcal{N}$-adhesive.

To prove Theorem 3, we establish the properties required by Definition 1 in the following five lemmata.

Lemma 3 (Closure properties). $\mathcal{M}$ and $\mathcal{N}$ contain all isomorphisms and are closed under composition and decomposition. Moreover, $\mathcal{N}$ is closed under $\mathcal{M}$-decomposition.

Proof. Straightforward.   □

Lemma 4 (Pushouts along $\mathcal{M}, \mathcal{N}$-morphisms). Given graph morphisms $r \colon K \hookrightarrow R$ in $\mathcal{M}$ and $d \colon K \hookrightarrow D$ in $\mathcal{N}$, there exist a graph H and graph morphisms $c \colon D \hookrightarrow H$ and $h \colon R \to H$ such that square (2) below is a pushout.

$$
\begin{array}{ccc}
K & \overset{r}{\hookrightarrow} & R \\
d \downarrow & (2) & \downarrow h \\
D & \underset{c}{\hookrightarrow} & H
\end{array}
$$

Construction. The sets of nodes and edges are defined by $H = (D - d(K)) + R$. The source function s_H is defined by $s_H(e) = $ if $e \in E_R$ then $s_R(e)$ else $s_D(e)$; the target function t_D is defined analogously. The labelling functions l_H are defined by

$$
l_H(x) = \begin{cases}
l_R(x) & \text{if } x \in R \text{ and } l_R(x) \neq \bot, \\
l_D(d(x')) & \text{if } x \in R,\, l_R(x) = \bot,\, r(x') = x \text{ and } l_D(d(x')) \neq \bot, \\
\bot & \text{if } x \in R,\, l_R(x) = \bot,\, r(x') = x \text{ and } l_D(d(x')) = \bot, \\
l_D(x) & \text{if } x \in (D - d(K)).
\end{cases}
$$

Morphism $h \colon R \to H$ is the inclusion of R in H and $c \colon D \hookrightarrow H$ is defined by $c(x) = $ if $x \in D - d(K)$ then x else $r(k)$ for the unique $k \in K$ with $d(k) = x$.

Proof. See [14].   □

The category PLG has not only pullbacks along $\mathcal{M}$-morphisms but possesses all pullbacks.

Lemma 5 (Pullbacks). Let $c \colon D \to H$ and $h \colon R \to H$ be graph morphisms. Then there exist a graph K and graph morphisms $d \colon K \to D$ and $r \colon K \to R$ such that square (2) above is a pullback.

Construction. The sets of nodes and edges are defined by

$$K = \{\langle x, y \rangle \in D \times R \mid c(x) = h(y)\}.$$

The source function s_K is defined by $s_K(\langle x, y \rangle) = \langle s_D(x), s_R(y) \rangle$, the target function t_K is defined analogously. The labelling functions l_K are defined by

$$l_K(\langle x, y \rangle) = \text{ if } (l_D(x) = l_R(y) \neq \perp) \text{ then } l_R(x) \text{ else } \perp.$$

The morphisms $d \colon K \to D$ and $r \colon K \to R$ are the projections from $D \times R$ to D and R, that is, they are given by $d(\langle x, y \rangle) = x$ and $r(\langle x, y \rangle) = y$.

Proof. See [14]. $\qquad\qquad\square$

Lemma 6 ($\mathcal{M}$ and $\mathcal{N}$ are stable). The classes $\mathcal{M}$ and $\mathcal{N}$ are stable under $\mathcal{M}, \mathcal{N}$-pushouts and $\mathcal{M}$-pullbacks.

Proof. This follows from the construction of $\mathcal{M}, \mathcal{N}$-pushouts and pullbacks in Lemma 4 and Lemma 5, and the fact that pushouts and pullbacks are unique up to isomorphism. $\qquad\qquad\square$

Lemma 7 ($\mathcal{M}, \mathcal{N}$-van Kampen squares). Pushouts along $\mathcal{M}, \mathcal{N}$-morphisms are $\mathcal{M}, \mathcal{N}$-van Kampen squares.

Proof. We exploit the fact that the category ULG of unlabelled graphs is $\mathcal{M}$-adhesive. (This follows from Fact 4.1.6 for labelled graphs in [4], by restricting the label alphabet to a single label.)

Consider the pushout (1) below where $m \in \mathcal{M}$ and $f \in \mathcal{N}$. We have to show that, given a commutative cube (2) with (1) as bottom face, $b, c, d \in \mathcal{M}$, and pullbacks as back faces, the following holds:

the top face is a pushout $\Leftrightarrow$ the front faces are pullbacks.

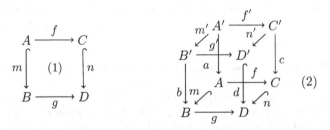

Part 1 ("$\Rightarrow$"). Assume that the top face of cube (2) is a pushout. Since pullback objects are unique up to isomorphism, it is sufficient to prove that B' and C' are isomorphic to the corresponding pullback objects. Let B'' be the pullback object of g and d with morphisms $b'' \colon B'' \to B$ and $g'' \colon B'' \to D'$. By the universal property of pullbacks, there is a unique morphism $u \colon B' \to B''$ such that $b'' \circ u = b$ and $g'' \circ u = g$. By forgetting all labels, cube (2) becomes a cube in ULG. Since ULG is $\mathcal{M}$-adhesive, every pushout in ULG is a van Kampen square.

Consequently, the morphism u is injective and surjective. It remains to show that u is $\perp$-preserving. Let $x \in B' - \mathrm{Dom}(1_{B'})$. Suppose that $u(x) \in \mathrm{Dom}(1_{B''})$. Then $b(x) \in \mathrm{Dom}(1_B)$ and $g'(x) \in \mathrm{Dom}(1_{D'})$.

Since the top is a pushout in PLG and $x \in B' - \mathrm{Dom}(1_{B'})$, there exists $y \in \mathrm{Dom}(1_{C'})$ with $g'(x) = n'(y)$. Since the bottom is a pushout in PLG and $m \in \mathcal{M}$, by Theorem 1, it is also a pullback, $b(x) \in \mathrm{Dom}(1_B)$, $c(y) \in \mathrm{Dom}(1_C)$, and the left front face commutes, $g(b(x)) = d(g'(x)) = d(n'(y))$ and there exists $z \in \mathrm{Dom}(1_A)$ such that $m(z) = b(x)$ and $f(z) = c(y)$. Since the back right face is a pullback, $y \in \mathrm{Dom}(1_{C'})$ and $z \in \mathrm{Dom}(1_A)$ with $c(y) = m(z)$, there is some $x' \in \mathrm{Dom}(1_{A'})$ with $m'(x') = x$. Then $x \in \mathrm{Dom}(1_{B'})$, a contradiction. Thus u is $\perp$-preserving and B' and B'' are isomorphic. Similarly, it is shown that C' and the pullback object C'' of d and n are isomorphic. Thus, the back faces of cube (2) are pullbacks.

Part 2 ("$\Leftarrow$"). Assume that the front faces of cube (2) are pullbacks in PLG. Since pushout objects are unique up to isomorphism, it is sufficient to prove that D' is isomorphic to the corresponding pushout object. Let D'' be the pushout object of m' and f' in PLG with morphisms $g'': B' \to D''$ and $n'': C' \to D''$. By the universal property of pushouts, there is a unique morphism $u: D'' \to D'$ such that $g' = u \circ g''$ and $n' = u \circ n''$. Consider now the underlying pushout in ULG. Since ULG is $\mathcal{M}$-adhesive, every pushout in ULG is a van Kampen square. Consequently, the morphism u is injective and surjective. It remains to show that u is $\perp$-preserving. Let $x \in D'' - \mathrm{Dom}(1_{D''})$. Suppose that $u(x) \in \mathrm{Dom}(1_{D'})$. Then $d(u(x)) \in \mathrm{Dom}(1_D)$. Since the bottom is a pushout, there are two cases. In the first case, there exists an item $y \in \mathrm{Dom}(1_B)$ such that $g(y) = d(u(x))$. Since the left front face is a pullback, $y \in \mathrm{Dom}(1_B)$ and $u(x) \in \mathrm{Dom}(1_{D'})$ with $g(y) = d(u(x))$, there is some $z \in \mathrm{Dom}(1_{B'})$ with $b(z) = y$ and $g'(z) = u(x)$. By commutativity of the left front face, $d(g'(z)) = g(b(z)) = g(y) = d(u(x))$. By $d \in \mathcal{M}$, $g'(z) = u(x) \in \mathrm{Dom}(1_{D''})$, a contradiction. In the second case, there exists an item $y \in \mathrm{Dom}(1_C)$ such that $n(y) = d(u(x))$. Since the right front face is a pullback, we obtain a contradiction. Thus, the morphism u is $\perp$-preserving and the top face is a pushout. Since the back faces are pullbacks and $\mathcal{M}$ and $\mathcal{N}$ are stable under $\mathcal{M}$-pullbacks, $m \in \mathcal{M}$ and $f \in \mathcal{N}$ imply $m' \in \mathcal{M}$ and $f' \in \mathcal{N}$, i.e. the top face is an $\mathcal{M}, \mathcal{N}$-pushout. $\qquad\square$

Proof of Theorem 3. See Lemma 3 to Lemma 7. $\qquad\square$

Lemma 8 (HLR$^+$-properties). PLG has binary coproducts, an $\mathcal{E}$-$\mathcal{N}$ factorization, and an $\mathcal{E}'$-$\mathcal{M}$ pair factorization, where $\mathcal{E}$ is the class of surjective, undefinedness preserving morphisms and $\mathcal{E}'$ is the class of pairs of jointly surjective, undefinedness preserving morphisms.

Proof. Routine. $\qquad\square$

By Theorem 3 and Lemma 8, we obtain the following corollary.

Corollary 1. The Local Church-Rosser Theorem and the Parallelism Theorem hold for $\mathcal{M}, \mathcal{N}$-adhesive tranformation systems over PLG.

Remark 4. $\mathcal{M}, \mathcal{N}$-adhesive transformation systems over PLG provide a foundation for the semantics of the graph programming language GP [18,19]. The graphs on which GP programs operate are totally labelled, and instances of GP's conditional rule schemata are rules with application conditions whose left- and right-hand graphs L and R are also totally labelled. The interface graph K consists of unlabelled nodes and hence enables relabelling of nodes. Moreover, the requirement that the vertical morphisms in double-pushouts must preserve unlabelled nodes guarantees that pushout complements are unique (see [14]).

In comparison with the approach of [14], $\mathcal{M}, \mathcal{N}$-adhesive tranformation systems over PLG are more restrictive in that unlabelled nodes in rules must not match labelled nodes in host graphs. However, to allow certain nodes in rules to match nodes with arbitrary labels, one can use rule schemata with label variables instead of unlabelled nodes. As in GP, rule schemata are instantiated to rules with totally labelled left- and right-hand graphs, while unlabelled nodes are solely used for relabelling. Indeed, label variables in left-hand graphs are more versatile than unlabelled nodes because they can be typed in order to match only subsets of labels.

5   Conclusion

Double-pushout graph transformation with relabelling is not covered by $\mathcal{M}$-adhesive transformation systems. Relabelling is natural for computing with graphs, though, and provides a foundation for graph transformation languages such as GP. We have generalised $\mathcal{M}$-adhesive transformation systems to $\mathcal{M}, \mathcal{N}$-adhesive transformation systems which do cover graph transformation with relabelling. We have proved the Local Church-Rosser Theorem and the Parallelism Theorem for $\mathcal{M}, \mathcal{N}$-adhesive transformation systems with application conditions, and hence these results hold for graph transformation with relabelling. The Concurrency Theorem is proved in the long version of this paper [15].

We hope to establish the Amalgamation Theorem, the Embedding Theorem and the Local Confluence Theorem in our new framework, too. These results have recently been proved for $\mathcal{M}$-adhesive transformation systems with application conditions [6,7].

In future work, we expect to be able to show that the category of term graphs is $\mathcal{M}, \mathcal{N}$-adhesive. This category is known to be not $\mathcal{M}$-adhesive, too, but has been shown to be quasi-adhesive [2]. Indeed the categories of term graphs and partially labelled graphs are similar in that PLG can also be shown to be quasi-adhesive. In PLG, the regular monomorphisms are precisely the undefinedness preserving injective morphisms.

An extension of $\mathcal{M}, \mathcal{N}$-adhesive transformation systems with rules that have a non-monomorphic right-hand morphism, allowing to merge items, may be possible. In the context of graph transformation with relabelling, the approach of [14] already includes such rules. Independently, in [1] a class of categories is identified for which the local Church-Rosser property holds for certain classes of rules with non-monomorphic right-hand morphisms.

Finally, the W-adhesive transformation systems introduced in [11] provide a general framework for attributed objects. They allow undefined attributes in the interface of a rule to change attributes, which is similar to relabelling. But the precise relationship to $\mathcal{M}, \mathcal{N}$-adhesive transformation systems remains to be worked out.

Acknowledgements. We are grateful to Berthold Hoffmann for drawing our attention to the problem that graph transformation with relabelling is not covered by $\mathcal{M}$-adhesive transformation systems, and for comments on a draft version of this paper. Thanks are also due to the anonymous referees for their helpful comments.

References

1. Baldan, P., Gadducci, F., Sobociński, P.: Adhesivity Is Not Enough: Local Church-Rosser Revisited. In: Murlak, F., Sankowski, P. (eds.) MFCS 2011. LNCS, vol. 6907, pp. 48–59. Springer, Heidelberg (2011)
2. Corradini, A., Gadducci, F.: On term graphs as an adhesive category. In: Fernández, M. (ed.) Proc. International Workshop on Term Graph Rewriting (TERMGRAPH 2004). Electronic Notes in Theoretical Computer Science, vol. 127(5), pp. 43–56 (2005)
3. Ehrig, H.: Introduction to the Algebraic Theory of Graph Grammars. In: Claus, V., Ehrig, H., Rozenberg, G. (eds.) Graph Grammars 1978. LNCS, vol. 73, pp. 1–69. Springer, Heidelberg (1979)
4. Ehrig, H., Ehrig, K., Prange, U., Taentzer, G.: Fundamentals of Algebraic Graph Transformation. Monographs in Theoretical Computer Science. Springer (2006)
5. Ehrig, H., Engels, G., Kreowski, H.-J., Rozenberg, G. (eds.): Handbook of Graph Grammars and Computing by Graph Transformation. Applications, Languages, and Tools, vol. 2. World Scientific (1999)
6. Ehrig, H., Golas, U., Habel, A., Lambers, L., Orejas, F.: $\mathcal{M}$-adhesive transformation systems with nested application conditions. Part 1: Parallelism, concurrency and amalgamation. Mathematical Structures in Computer Science (to appear, 2012)
7. Ehrig, H., Golas, U., Habel, A., Lambers, L., Orejas, F.: $\mathcal{M}$-adhesive transformation systems with nested application conditions. Part 2: Embedding, critical pairs and local confluence. Fundamenta Informaticae 118, 35–63 (2012)
8. Ehrig, H., Golas, U., Hermann, F.: Categorical frameworks for graph transformation and HLR systems based on the DPO approach. Bulletin of the EATCS 102, 111–121 (2010)
9. Ehrig, H., Habel, A., Kreowski, H.-J., Parisi-Presicce, F.: Parallelism and concurrency in high-level replacement systems. Mathematical Structures in Computer Science 1, 361–404 (1991)
10. Ehrig, H., Kreowski, H.-J., Montanari, U., Rozenberg, G. (eds.): Handbook of Graph Grammars and Computing by Graph Transformation. Concurrency, Parallelism, and Distribution, vol. 3. World Scientific (1999)
11. Golas, U.: A General Attribution Concept for Models in M-Adhesive Transformation Systems. In: Ehrig, H., Engels, G., Kreowski, H.-J., Rozenberg, G. (eds.) ICGT 2012. LNCS, vol. 7562, pp. 187–202. Springer, Heidelberg (2012)

12. Habel, A., Müller, J., Plump, D.: Double-pushout graph transformation revisited. Mathematical Structures in Computer Science 11(5), 637–688 (2001)
13. Habel, A., Pennemann, K.-H.: Correctness of high-level transformation systems relative to nested conditions. Mathematical Structures in Computer Science 19(2), 245–296 (2009)
14. Habel, A., Plump, D.: Relabelling in Graph Transformation. In: Corradini, A., Ehrig, H., Kreowski, H.-J., Rozenberg, G. (eds.) ICGT 2002. LNCS, vol. 2505, pp. 135–147. Springer, Heidelberg (2002)
15. Habel, A., Plump, D.: $\mathcal{M}, \mathcal{N}$-adhesive transformation systems (long version) (2012), http://formale-sprachen.informatik.uni-oldenburg.de/pub/index.html
16. Heindel, T.: Hereditary Pushouts Reconsidered. In: Ehrig, H., Rensink, A., Rozenberg, G., Schürr, A. (eds.) ICGT 2010. LNCS, vol. 6372, pp. 250–265. Springer, Heidelberg (2010)
17. Lack, S., Sobociński, P.: Adhesive and quasiadhesive categories. Informatique Théorique et Applications 39(3), 511–545 (2005)
18. Plump, D.: The Graph Programming Language GP. In: Bozapalidis, S., Rahonis, G. (eds.) CAI 2009. LNCS, vol. 5725, pp. 99–122. Springer, Heidelberg (2009)
19. Plump, D.: The design of GP 2. In: Escobar, S. (ed.) Proc. International Workshop on Reduction Strategies in Rewriting and Programming (WRS 2011). Electronic Proceedings in Theoretical Computer Science, vol. 82, pp. 1–16 (2012)
20. Rozenberg, G. (ed.): Handbook of Graph Grammars and Computing by Graph Transformation. Foundations, vol. 1. World Scientific (1997)

Generalised Compositionality in Graph Transformation

Amir Hossein Ghamarian and Arend Rensink

Department of Computer Science, Universiteit Twente
{a.h.ghamarian, rensink}@cs.utwente.nl

Abstract. We present a notion of composition applying both to graphs and to rules, based on graph and rule interfaces along which they are glued. The current paper generalises a previous result in two different ways. Firstly, rules do not have to form pullbacks with their interfaces; this enables graph passing between components, meaning that components may "learn" and "forget" subgraphs through communication with other components. Secondly, composition is no longer binary; instead, it can be repeated for an arbitrary number of components.

1 Introduction

We believe that, for graph transformation to become a practicable specification technique, its native strengths should be complemented with a notion of compositionality which allows the user to specify and analyse a system modularly. Failing that, graphs always have to be specified monolithically, which for large graphs quickly becomes prohibitive and causes the advantage of visualisation to be lost. Moreover, if graph transformations are used to specify the dynamic behaviour of systems, having a large monolithic graph as a state introduces the dreaded problem of state space explosion.

The issue of compositionality (or dually, modularity) has indeed been identified and addressed in a number of different approaches over the years — see Sect. 4 for an overview. Several of these, such as borrowed contexts [6], transformation units [16] and synchronised hyperedge replacement [7] have been inspired to some degree or another by notions of composition from process algebra. In this paper we continue an investigation started in [19] based on the following initial requirements:

- Composition should make it possible to construct large graphs (describing the global system in context) from smaller graphs (describing individual components).
- Composition should act as an operator over graph production systems: given a number of production systems describing individual components, the result should be a production system describing the global system. We want to introduce as little additional structure as possible.
- The behaviour of the composed system, in terms of rule applications, should likewise arise out of the composition of local system behaviour. This means that local rule applications need to be synchronised and exchange information.

In [19], we proposed to use graphs and rules with interfaces for the local systems; composition glues graphs and rules together over their interfaces using a categorical construction called *pushout*. A limitation of that setup is that components cannot exchange node or edge identities; i.e., it is impossible for one component to "publish"

H. Ehrig et al.(Eds.): ICGT 2012, LNCS 7562, pp. 234–248, 2012.

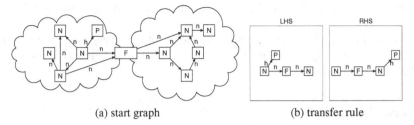

(a) start graph   (b) transfer rule

Fig. 1. Running example: Firewall between local networks

part of its internal structure and share it with others. (Instead, shared structure can only arise through simultaneous creation.)

In the current paper we lift this restriction by using a different notion of rule: rather than relying on the usual spans of morphisms we resort to *cospans*. Though ordinary cospan rules and span rules have been shown to be equally expressive, cospan rules turn out to be advantageous in the presence of interfaces and composition. Another important difference with the usual concept of rule interface (called *kernel* in [10]) is that the relation between rule and interface is less strict (we do not insist on pullbacks).

Motivating example. To illustrate our setup, we use a running example based on two local networks which are connected via a firewall. Such a network is depicted in Fig. 1a. Each cloud represents a local network and each local network has its own network nodes, represented by $\boxed{N}$s. Network nodes are connected via next edges (denoted by $\boxed{N}\!\xrightarrow{n}\!\boxed{N}$) to their neighbours. Nodes can also have packets which are denoted by ($\boxed{N}\!\xrightarrow{h}\!\boxed{P}$). The firewall node ($\boxed{F}$) is the only interface between the local networks, through which they can communicate by sending and receiving packets. The firewall node passes safe packets through and deletes the infected packets.

Each local network may have dedicated rules to reflect its packet generation and transmission protocols. In order to avoid the state space explosion that ensues in the global network, it is desirable to specify and analyse each local network separately and obtain the global analysis by composition.

In this example, all the local structure (nodes and edges) of each network can be hidden from the other, except the firewall node which serves as the interface between the two networks. Similarly, the dynamic behaviour of the networks, given by graph rules, can also be considered local except the rule dealing with the transmission of packets from one network to another.

In the following we do not consider the local network behaviour; instead we focus on the transmission of a packet through the firewall. This behaviour is captured as a graph rule in Fig. 1b. To obtain the local effect on each network, we have to split this into two rules. An ad hoc decomposition can consist of a rule which deletes the packet node in one local network and the counterpart of this rule that creates the packet node in the other network. However, this has one major drawback: packets in the network usually have content, which is lost by the ad hoc rule decomposition. The approach of this paper enables us to pass the node itself between the decomposed systems.

Roadmap. In the next section, we give the basic definitions, especially the composition of graphs. Sect. 3 contains the main results. In Sect. 4 we review related approaches and summarise the contribution. Due to space limitations proofs are omitted and can be found in [8].

The basic ideas of this paper were presented for a concrete category of graphs in [9]. With respect to that paper, we have ironed out a number of technical issues and lifted the theory to the algebraic level.

2 Basic Definitions

In the grand tradition of algebraic graph transformation, we develop our theory in the setting of adhesive categories. It has been shown in [17] that adhesive categories form a nice, general framework in which properties of graph transformation systems can be proved abstractly; they generalise in some part the High-Level Replacement systems studied in, e.g., [4].

Definition 1 (adhesive category). *A category* $\mathbf{C}$ *is* adhesive *if it satisfies the following properties:*

1. $\mathbf{C}$ *has pushouts along monomorphisms (monos);*
2. $\mathbf{C}$ *has pullbacks;*
3. *Pushouts along monos are Van Kampen squares.*

For those that are not familiar with this theory, the following intuitions may be helpful:

– A mono $f\colon A \hookrightarrow B$ identifies a subobject of B that is isomorphic to A;
– The pushout of $B \xleftarrow{f} A \xrightarrow{g} C$ may be thought of as the union of B and C, where the shared subset is given by A and its "embedding" in B and C;
– The pullback of $B \xrightarrow{h} D \xleftarrow{k} C$ may be thought of as the intersection of B and C, where their "embedding" in D determines which elements they have in common.

As an example concrete category, one may think of edge-labelled directed graphs $\langle N, E, L, s, t, l \rangle$ with $s, t\colon E \to N$ the source and target function from the edges E to nodes N, and $l\colon E \to L$ a labelling function to a set of labels L. This is the context in which our running example is formulated.

In contrast with the usual setup, we take a transformation rule p not to be a span but a cospan of morphisms. In a cospan rule, creation occurs before deletion instead of the other way around. Cospan rules have been studied in [5], where the following is shown:

– A cospan rule is equivalent to the span rule arising from the pullback of the cospan;
– A span rule is equivalent to the cospan rule arising from the pushout of the span.

As a corollary, it follows that for every cospan rule there exists an equivalent cospan rule in which the morphisms are jointly epimorphic. Curiously, as we will see, cospan rules that are *not* jointly epi do play an essential role in our notion of rule composition.

Definition 2 (rule). *Let* $\mathbf{C}$ *be an adhesive category.*

– *A rule* p *consists of a cospan of monos* $L \xrightarrow{l} U \xleftarrow{r} R$ *where* L *is the left hand side, R the right hand side and U the* upper object.

– *The application of p on G is defined by the following diagram, where* $m: L \hookrightarrow G$
is a mono (in this paper we consider monomorphic matches [12]).

$$
\begin{array}{ccccc}
L & \overset{l}{\hookrightarrow} & U & \overset{r}{\hookleftarrow} & R \\
\scriptstyle m \downarrow & \textit{PO} & \scriptstyle k \uparrow & \textit{PO} & \uparrow \scriptstyle m' \\
G & \underset{g}{\hookrightarrow} & K & \underset{h}{\hookleftarrow} & H
\end{array}
$$

We write $G \overset{p,m}{\Longrightarrow} H$ to denote the existence of a rule application of p to G under match
m, with result H.

2.1 Marked Objects

We now define the general notion of a marked object, as a monomorphism from an inner
object to an outer object. The inner object serves as an interface used to glue marked
objects together: gluing two marked objects with the same interface comes down to
taking the pushout of the corresponding span. This extends to arrows naturally.

Definition 3 (marked object and arrow). *Let* **C** *be a category.*
- *A* marked object X *is a monomorphism* $e_X: \underline{X} \hookrightarrow \overline{X}$. $\underline{X}$ *is called the inner object*
  and $\overline{X}$ *the outer object. Two marked objects* X, Y *are* compatible *if* $\underline{X} = \underline{Y}$. *If this*
  is the case, we will use $X+Y$ *to refer to the marked object defined by* $c_Y \circ e_X: \underline{X} \hookrightarrow$
  Z *in the diagram*

$$
\begin{array}{ccc}
\overline{X} & \overset{c_Y}{\hookrightarrow} & Z \\
\scriptstyle e_X \uparrow & \textit{PO} & \uparrow \\
\underline{X}{=}\underline{Y} & \hookrightarrow & \overline{Y}
\end{array}
$$

  We will refer to the composed object as global *and each of the original objects as*
  local.
- *Given two marked objects* X, Y, *a* marked arrow $f: X \to Y$ *is a pair of morphisms*
  $\underline{f}: \underline{X} \to \underline{Y}$ *and* $\overline{f}: \overline{X} \to \overline{Y}$ *such that the resulting (left hand) diagram commutes:*

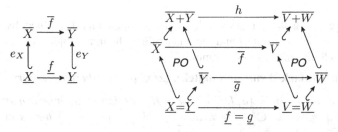

  Two marked arrows $f: X \to V, g: Y \to W$ *are* compatible *if* $\underline{f} = \underline{g}$. *If this is the*
  case, we will use $f + g$ *to refer to the marked arrow consisting of* $\overline{f}$ *and the medi-*
  ating morphism h, *connected by* e_{X+Y} *and* e_{V+W} *(right hand diagram above).*
- *A marked arrow* f *is called* strict *if* $\underline{f}$ *and* $\overline{f}$ *are monos and the left hand diagram*
  above is a pullback in **C**.

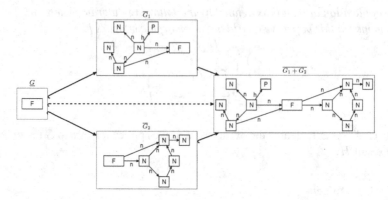

Fig. 2. Running example: composition of marked objects

Example. In our running example, the graphs associated to the local networks can be specified as compatible marked objects with the firewall node as their inner object. These marked graphs and their composition is illustrated in Fig. 2.

Given a category $\mathbf{C}$, we use $\mathbf{C}^M$ to denote the cateory of marked $\mathbf{C}$-objects and -arrows. The following properties are important.

Proposition 4 (properties of marked arrows). *Let $\mathbf{C}$ be an adhesive category.*

1. *Monos in $\mathbf{C}^M$ correspond to pairs of monos in $\mathbf{C}$, stacked on top of one another. (Hence, for instance, strict arrows are monos in $\mathbf{C}^M$.)*
2. *Pushouts over strict arrows in $\mathbf{C}^M$ correspond to pairs of pushouts in $\mathbf{C}$, stacked on top of one another.*

2.2 Marked Rules

In the remainder of this paper, we will mainly deal with transformation in $\mathbf{C}^M$. As expected (given the above), marked rules will be cospans of marked monos. We do *not* require that the monos are strict (i.e., pullbacks). This makes the definition quite a bit more general than similar notions in [19,10], a fact which is at the core of this paper's contribution.

The intention is that a marked rule should act upon a marked object by applying the outer rule to the outer object and the inner rule to the inner object. To make this work, we have to limit matches to strict monos.

Definition 5 (marked rule and match). *Let $\mathbf{C}$ be an adhesive category.*

- *A marked rule $p = (a, L \hookrightarrow U \hookleftarrow R)$ consists of a name a and a cospan of marked monos in $\mathbf{C}^M$. We write $\underline{p} = (\underline{L} \hookrightarrow \underline{U} \hookleftarrow \underline{R})$ for its inner rule and $\overline{p} = (\overline{L} \hookrightarrow \overline{U} \hookleftarrow \overline{R})$ for its outer rule.*
- *A marked match is a strict arrow in $\mathbf{C}^M$.*

Example. An example marked rule, transfer-1, is shown in Fig. 3 where all L_1, U_1 and R_1 are marked objects with $\overline{L}_1, \overline{U}_1, \overline{R}_1$ and $\underline{L}_1, \underline{U}_1, \underline{R}_1$ as their outer and inner graphs respectively. There are several things to be noted:

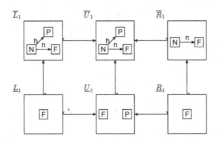

Fig. 3. Example: a marked rule transfer-1

- The inner rule $\underline{L}_1 \hookrightarrow \underline{U}_1 \hookleftarrow \underline{R}_1$ is not jointly epimorphic: the P-node in $\underline{U}_1$ is in the image of neither morphism. In fact, in this inner rule, the node is created and then immediately deleted, which does not appear to be very useful. However, it is precisely this feature that allows the node to be communicated to any other local rule with which transfer-1 is composed.
- The $\mathbf{C}^M$-mono $L_1 \hookrightarrow U_1$ is not strict; i.e., the corresponding square is not a pull-back. Thus, deletion and creation in the inner rule do not strictly follow the outer rule. This reflects the fact that previously private parts of the graph may be (temporarily) "published" to the interface.

Under the assumption that $\mathbf{C}$ is adhesive, the category of marked objects is known to be *quasi-adhesive* [15], which is a weaker notion that still retains all the nice properties of graph transformation, provided the rules are made up of strict arrows only. However, since our marked rules are not made up of strict monos, we cannot benefit from this result. In the remainder of the paper we ignore the link to quasi-adhesive categories.

The following states the first important result of this paper: the application of a marked rule to a marked object is fully characterised by the applications of the inner and outer rules. Informally speaking, the embedding morphisms of the intermediate and target (marked) objects take care of themselves.

Theorem 6 (marked transformation). *Let $\mathbf{C}$ be an adhesive category. If p is a marked rule, G a marked object and $m: L \rightarrow G$ a marked match, then $\overline{G} \xRightarrow{\overline{p},\overline{m}} H_1$ and $\underline{G} \xRightarrow{\underline{p},\underline{m}} H_2$ (in $\mathbf{C}$) if and only if $G \xRightarrow{p,m} H$ (in $\mathbf{C}^M$) with $\overline{H} = H_1$ and $\underline{H} = H_2$.*

This is related to the fact (Prop. 4.2) that pushout squares in $\mathbf{C}^M$ are precisely stacks of pushout squares in $\mathbf{C}$ for the inner and outer objects. The relevant properties are stated in the following two propositions.

Proposition 7 (pushout of marked objects). *In the following diagram, if $U \hookleftarrow L \hookrightarrow G$ is a span of marked monos, the top and bottom faces are pushouts and $m: L \hookrightarrow G$ is a marked match, then (a) there is a unique monomorphism $e_K: \underline{K} \hookrightarrow \overline{K}$ that makes the diagram commute (and which therefore makes K a marked object); and (b) the back right rectangle, $U \hookrightarrow K$, is a pullback.*

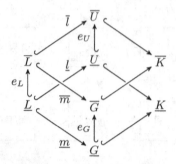

Proposition 8 (pushout complement of marked objects). *In the following diagram, if $R \hookrightarrow U$ is a marked mono and $U \hookrightarrow K$ a marked match, and the top and bottom faces are pushouts, then there is a unique monomorphism $e_H : \underline{H} \hookrightarrow \overline{H}$ that makes the diagram commute (and which therefore makes H a marked object).*

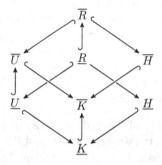

Proof (Th. 6).

If. The marked transformation $G \xrightarrow{p,m} H$ consists of a double pushout in $\mathbf{C}^M$, which according to Prop. 4.2 are stacked double pushouts for the inner and outer objects.

Only if. Given transformations $\underline{G} \xrightarrow{\underline{p},\underline{m}} H_1$ and $\overline{G} \xrightarrow{\overline{p},\overline{m}} H_2$ with derived cospans $\underline{G} \hookrightarrow K_1 \hookleftarrow H_1$ and $\overline{G} \hookrightarrow K_2 \hookleftarrow H_2$, respectively, we know by Prop. 7 that K_1 and K_2 form a marked object K and the intermediate morphism $U \to K$ is strict; and hence by Prop. 8 that H_1 and H_2 also form a marked object H. Moreover, the resulting stacked double pushouts form a double pushout in $\mathbf{C}^M$ according to Prop. 4.2.

Example. The application of the transfer-1 (see Fig. 3) on marked graph G_1 shown in Fig. 2 is illustrated in Fig. 4. The front faces represent the marked rule and the back face shows the marked graphs. Note that the outer rule is applied to the outer graph and the inner rule is applied to the inner graph. The network nodes that are not involved in the rule applications are omitted for simplification.

Joint epimorphism revisited. As an aside, the proposition below states that a marked cospan can be jointly epi in $\mathbf{C}^M$ *even if the inner cospan is not jointly epi in* $\mathbf{C}$: the property only depends on the outer cospan. Moreover, joint epimorphism of rules is preserved under composition. Thus, we suspect that the results of [5] might after all still hold in this setting.

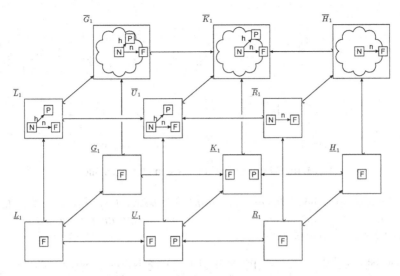

Fig. 4. Running example: application of a marked rule

Proposition 9 (joint epimorphism in $\mathbf{C}^M$). *Let $\mathbf{C}$ be an arbitrary category.*

1. *A marked cospan $X \to Z \leftarrow Y$ is jointly epimorhpic in $\mathbf{C}^M$ if and only if $\overline{X} \to \overline{Z} \leftarrow \overline{Y}$ is jointly epimorphic in $\mathbf{C}$.*
2. *If p_1, p_2 are two compatible jointly epimorphic rules in $\mathbf{C}^M$, then $p_1 + p_2$ is also jointly epimorphic.*

3 Rule Composition

We now come to the actual topic of the paper, namely the composition of rules and rule applications. First we define how rules are composed. This is entirely in line with the composition of objects and arrows in Def. 3, except that we need an additional compatibility condition.

Definition 10 (rule composition). *Consider a category $\mathbf{C}^M$ of marked objects.*

- *A marked rule $p = (a, L \hookrightarrow U \hookleftarrow R)$ consists of a name a and a cospan of marked monos. We write $\underline{p} = (\underline{L} \hookrightarrow \underline{U} \hookleftarrow \underline{R})$ for the inner rule and $\overline{p} = (\overline{L} \hookrightarrow \overline{U} \hookleftarrow \overline{R})$ for the outer rule.*
- *Two marked rules p, q are compatible if $a_p = a_q$, $\underline{p} = \underline{q}$, and $\underline{L}_p$ resp. $\underline{R}_p$ are the limits of the following diagrams:*

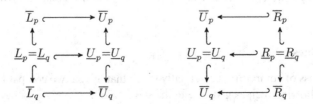

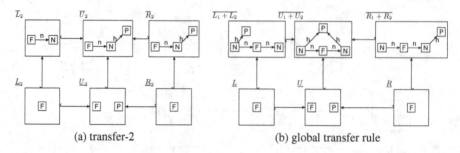

Fig. 5. Running example: composition of marked rules

If p and q are compatible, their composition is defined by $p + q = (a, L_p + L_q \rightarrow U_p + U_q \leftarrow R_p + R_q)$, where the arrows are the composition of the left and right morphisms of p and q.

It should be noted that the limit property of the inner left hand side is automatically fulfilled if one of the local left hand side morphisms is strict; and similarly for the right hand side. Thus, in a sense, this is the price we pay for relaxing our rules to allow non-strict morphisms.

Under this notion of compatibility, rule composition is well-defined, i.e., always yields a rule. In particular, we have to establish that the composed rule morphisms are monic.

Proposition 11 (marked rule). *The composition of two compatible marked rules is a marked rule.*

Example. The marked rule, transfer-2, depicted in Fig. 5a is compatible with transfer-1 (see Fig. 3): they share the same inner rule. For transfer-2 it is $R_2 \hookrightarrow U_2$ that is not a pullback: after the P-node has been received, it is removed again from the interface and becomes local to this component. The composition of transfer-1 and transfer-2 is illustrated in Fig. 5b. Note that the outer rule of Fig. 5b is the same as our original transfer rule given in Fig. 1b.

Finally, we extend composition to rule applications. When compatible rules are applied to compatible graphs under compatible matches, we also call the entire transformations compatible.

Definition 12 (transformation compatibility). *Let* $\mathbf{C}$ *be an adhesive category. We call two marked transformations* $G_1 \xrightarrow{p_1, m_1} H_1$ *and* $G_2 \xrightarrow{p_2, m_2} H_2$ *(in* $\mathbf{C}^M$*) compatible if (1)* p_1 *and* p_2 *are compatible rules, (2)* G_1 *and* G_2 *are compatible objects and (3)* $\underline{m_1} = \underline{m_2}$*.*

3.1 Soundness

The soundness of composition essentially states that, given two compatible transformations, the following recipes give rise to the same result:

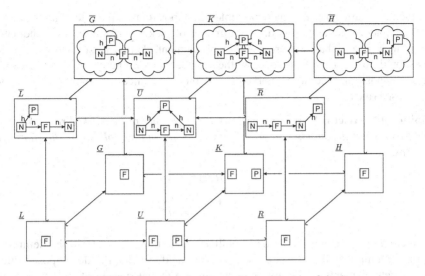

Fig. 6. Running example: composition of the transfer rule

- Compose the local rules into a single global rule and the local objects to a global object, then apply the rule (under the composed match);
- Apply the local rules to the local objects, then compose the target objects.

This can be succinctly summarised as "local behaviour generates global behaviour".

Theorem 13 (soundness). *Let* $\mathbf{C}$ *be an adhesive category. If* $G_i \xrightarrow{p_i, m_i} H_i$ *(in* $\mathbf{C}^M$*) for* $i = 1, 2$ *are compatible marked transformations, then* $G_1 + G_2 \xrightarrow{p_1 + p_2, m_1 + m_2} H_1 + H_2$ *(in* $\mathbf{C}^M$*).*

The proof is essentially due to the fact that all compositions are by pushout, and pushouts commute. Moreover, due to Th. 6 we can separately concentrate on the inner and outer part of the global transformation; and since the inner part is identical for the local and global rules, there is nothing to be shown.

Example. In Fig. 4 we have seen that K_1 and H_1 obtained by applying the transfer-1 rule (Fig. 3) to G_1 (Fig. 2). Similarly, by applying transfer-2 (Fig. 5a) to G_2 (Fig. 2) we can obtain K_2 and H_2. It is not difficult to verify that K and H in Fig. 6 obtained by application of the composed rule (Fig. 5b) to G (Fig. 2) are in fact equivalent to $K_1 + K_2$ and $H_1 + H_2$ respectively. Moreover, the application of transfer-1 and transfer-2 allows node P to be transferred from G_1 to H_2, while its content is preserved.

3.2   Completeness

Completeness is the dual of soundness, and can be summarised as "all global behaviour arises from local behaviour". The proof entails showing that every global graph transformation can be decomposed into local transformations, for an arbitrary decomposition of the host graph into local graphs.

In fact, there are generally many possible decompositions of the rules, and the proof of the completeness theorem is mainly a matter of picking an appropriate candidate. For instance, if a global rule creates a graph element, then at least one of the local rules should do so. Therefore, to show the completeness property, we define an specific decomposition of a global marked rule, which guarantees the applicability of the decomposed rules.

Definition 14 (strict rule decomposition). *Let p, p_1, and p_2 be marked rules. We call p_1 and p_2 a strict decomposition of p if $p = p_1 + p_2$ and the following diagrams are pullbacks.*

$$
\begin{array}{ccccc}
\overline{R}_1 & \longhookrightarrow & \overline{R} & \longleftarrow & \overline{R}_2 \\
\downarrow & PB & \downarrow & PB & \downarrow \\
\overline{U}_1 & \longhookrightarrow & \overline{U} & \longleftarrow & \overline{U}_2
\end{array}
$$

The intention of insisting on the extra pullback condition, $U \hookleftarrow R$, in the definition of strict decomposition is to state the conditions under which the decomposed rules do not delete elements which are preserved by the global rule. In other words, if an element is preserved by the application of a global rule then it is also preserved by the application of both strictly decomposed ones. Note that this decomposition does not capture all possible rule decompositions. For instance, the decomposition of the transfer rule in our example is not strict: node P is deleted in transfer-1 while it is preserved by the global rule. For completeness, however, it turns out to be sufficient to use strict decompositions only.

This is convenient because strict decomposition guarantees the applicability of the decomposed rules whenever the original global marked rule is applicable. To show this property, first we prove the following lemma.

Lemma 15. *Let $\mathbf{C}$ be an adhesive category, p a marked rule where $U \hookleftarrow R$ is strict, G a marked object and m a marked matching (in $\mathbf{C}^M$). If there exists a transformation $\overline{G} \xrightarrow{\overline{p},\overline{m}} \overline{H}$ (in $\mathbf{C}$), then there is a marked transformation $G \xrightarrow{p,m} H$ (in $\mathbf{C}^M$).*

Now we show the conditions where a marked transformation can be decomposed to two compatible marked transformations.

Lemma 16. *Let $\mathbf{C}$ be an adhesive category, $G \xrightarrow{p,m} H$ (in $\mathbf{C}^M$) a marked transformation, G_1 and G_2 a decomposition of G, and p_1 and p_2 a strict rule decomposition of p, such that $\overline{L}_1$ and $\overline{L}_2$ are the pullbacks of the following diagrams.*

$$
\begin{array}{ccccccc}
\overline{L}_1 & \longhookrightarrow & \overline{L} & & \overline{L} & \longleftarrow & \overline{L}_2 \\
\downarrow & PB & \downarrow & & \downarrow & PB & \downarrow \\
\overline{G}_1 & \longhookrightarrow & \overline{G} & & \overline{G} & \longleftarrow & \overline{G}_2
\end{array}
$$

Then there are two compatible marked transformations $G_i \xrightarrow{p_i, m_i} H_i$ for $i = 1, 2$ such that $H = H_1 + H_2$.

In fact Lemma 16 states that a marked transformation $G \xrightarrow{p,m} H$ can be decomposed to two compatible transformations for any strict decomposition of p as long as they have valid matches. Now we have come to another main contribution of the paper.

The completeness theorem states that given a global transformation, and a decomposition of its start object, we can always decompose the transformation to two compatible marked transformations according to the given start object decomposition. To prove the completeness theorem, we only need to show the existence of two such marked transformations.

Theorem 17 (completeness). *Let $\mathbf{C}$ be an adhesive category, $G = G_1 + G_2$ be a decomposition of a marked object G. A transformation $\overline{G} \xrightarrow{p,m} H$ (in $\mathbf{C}$) can be decomposed to two compatible marked transformations $G_1 \xrightarrow{p_1,m_1} H_1$ and $G_2 \xrightarrow{p_2,m_2} H_2$ (in $\mathbf{C}^M$) such that $\overline{p_1 + p_2} = p$ and $H = \overline{H_1 + H_2}$.*

4 Conclusion

We have defined a notion of composition for graphs and graph transformation rules, in the setting of adhesive categories, which allows passing subgraphs between components. This was done by equipping every graph and graph transformation rule with an interface, which declare the parts that are exposed to the environment. Graphs and rules can be composed when they have compatible interfaces. The contributions involved are:

- Rule composition both preserves transformations from the local to the global level (soundness, see Th. 13) and reflects them from the global to the local level (completeness, see Th. 17). There are no futher restrictions on the rules: the results are completely general.
- Our rule interfaces, in contrast to what we have seen elsewhere, do not have to form pullbacks with the main rule. This means that deletion and creation in the interface does not strictly follow that in the main rule. This is essential to the framework, since it enables rules to publish part of their inner structure to the outside workd (via the interface).
- With respect to [19] we have changed from span rules to cospan rules, which makes the framework quite a bit more expressive; in particular, the notion of graph passing answers one of the items identified as future work in that paper.
- With respect to [9], where the concept of graph passing was presented in a concrete category of graphs, we have lifted the framework to the prevailing algebraic setting.

4.1 Related Work

The concepts of graph and rule composition, with the appropriate notions of soundness and completeness, were introduced in [19] and later generalised in [13]. With respect to those papers, the variation studied here offers a more powerful notion of composition, in which nodes and edges can be deleted in one component and simultaneously created in the other.

In addition, there are a number of other approaches to introduce aspects of compositionality into graph transformation.

Synchronised Hyperedge Replacement. This is a paradigm in which graph transformation rules (more specifically, hyperedge replacement rules) can be synchronised based one the adjacency of their occurrences within a graph; see [14,7]. The synchronised rules are not themselves understood as graph transformation rules, and consequently the work does not address the type of compositionality issues that we have studied here. Still, it is interesting to see whether SHR synchronisation can be understood as a special type of composition in our sense.

History-Dependent Automata. This is a behavioural model in which states are enriched with a set of *names* (see [18] for an overview). Transitions expose names to the environment, and can also record the deletion, creation and permutation of names. HD-automata can be composed while synchronising their transitions: this provides a model for name passing. Transition systems induced by graph transformation rules can be understood as a variant of HD-automata where the states are enriched with graphs rather than just sets, and the information on the transitions is extended accordingly.

Rule amalgamation and distributed graph transformation. Studied in [3] and later, more extensively, in [20], the principle of rule amalgamation provides a general mechanism for rule (de)composition. This is a sub-problem of the one we have addressed here, as we study composition of the graphs as well as the rules. Our notion of rule composition is actually a generalisation of rule amalgamation, as local rules do not have to synchronise on deletions and creations.

Borrowed contexts. Like our paper, the work on borrowed contexts [6,1] uses a setting where only part of a graph is available, and studies the application of rules to such subgraphs in a way that is compatible with the original, reductive semantics. In contrast to our approach, however, they do not decompose rules: instead, when a rule is applied to a graph in which some of the required structure ("context") for the match is missing, this is imported ("borrowed") as part of the transformation. As a result, in this paradigm the subgraphs grow while being transformed, incorporating ever more context information. This is quite different from the basic intuitions behind our approach.

Summarising, where only rules are (de)composed in rule amalgamation, and only graphs in borrowed contexts, in our approach both rules and graphs are subject to (de)composition.

Compositional model transformation. [2] studies a notion of compositionality in model transformation. Though on the face of it this sounds similar, in fact they study a different question altogether, namely whether a transformation affects the *semantics* of a model (given as a separate mapping to a semantic domain) in a predictable (compositional) manner. This is in sharp contrast with our work, which rather addresses the compositionality of the graph transformation framework itself.

Graph Transformation Units. The graph transformation units exemplified in [16], also provide a notion of composition. However, this work takes the form of an explicit structuring mechanism of local graph transformation systems, called Units. The question of equivalence of a monolithic graph transformation system and a composition of local units is not addressed in this approach.

4.2 Future Work

Though with this paper we have addressed a major outstanding question of [19], there is still a lot of work to be done before the compositional framework can be used in practice.

For instance, negative application conditions (NACs) as introduced in [11] have shown to be very useful in practice. It will be interesting to extend our notion of compositionality to rules with NACs, in particular with respect to the soundness and completeness properties.

Another important problem is finding an automatic mechanism for splitting the start graph, and decomposing the rule system such that both the number of states and the number of required rules for the local systems stays minimal. This was also discussed in some detail in [9], where we proposed to use *partial graphs* for this purpose; hoewever, so far these lack a good definition on the algebraic level.

Finally, composition as introduced here is only part of the story. Again inspired by process algebra, in particular the *hiding* operator, it makes sense to think of ways in which to restrict the interface of a marked rule or graph, thus making part of the previously published interface structure private. Also, if we want to compose rules whose interfaces to not quite match, one may think of a *partial composition* operator, using ideas of borrowed contexts [6,1] (see above).

Acknowledgement. This paper could not have been written without the invaluable help of Barbara König and members of her group, in particular Matthias Hülsbusch, who provided essential hints in the proof of Th. 6.

References

1. Baldan, P., Ehrig, H., König, B.: Composition and Decomposition of DPO Transformations with Borrowed Context. In: Corradini, A., Ehrig, H., Montanari, U., Ribeiro, L., Rozenberg, G. (eds.) ICGT 2006. LNCS, vol. 4178, pp. 153–167. Springer, Heidelberg (2006)
2. Bisztray, D., Heckel, R., Ehrig, H.: Compositionality of model transformations. In: Aldini, A., ter Beek, M., Gadducci, F. (eds.) 3rd International Workshop on Views On Designing Complex Architectures (VODCA). ENTCS, vol. 236, pp. 5–19 (2009)
3. Boehm, P., Fonio, H.R., Habel, A.: Amalgamation of graph transformations: A synchronization mechanism. J. Comput. Syst. Sci. 34(2/3), 377–408 (1987)
4. Ehrig, H., Habel, A., Kreowski, H.J., Parisi-Presicce, F.: From Graph Grammars to High Level Replacement Systems. In: Ehrig, H., Kreowski, H.-J., Rozenberg, G. (eds.) Graph Grammars 1990. LNCS, vol. 532, pp. 269–291. Springer, Heidelberg (1991)
5. Ehrig, H., Hermann, F., Prange, U.: Cospan DPO approach: An alternative for DPO graph transformations. Bulletin of the EATCS 98, 139–149 (2009)
6. Ehrig, H., König, B.: Deriving bisimulation congruences in the DPO approach to graph rewriting with borrowed contexts. Mathematical Structures in Computer Science 16(6), 1133–1163 (2006)
7. Ferrari, G.L., Hirsch, D., Lanese, I., Montanari, U., Tuosto, E.: Synchronised Hyperedge Replacement as a Model for Service Oriented Computing. In: de Boer, F.S., Bonsangue, M.M., Graf, S., de Roever, W.-P. (eds.) FMCO 2005. LNCS, vol. 4111, pp. 22–43. Springer, Heidelberg (2006)
8. Ghamarian, A.H., Rensink, A.: Generalised compositionality in graph transformation. Tech. Rep. TR-CTIT-12-17, Centre for Telematics and Information Technology, University of Twente (2012)
9. Ghamarian, A., Rensink, A.: Graph passing in graph transformation. In: Fish, A., Lambers, L. (eds.) Graph Transformation and Visual Modelling Techniques. ECEASST (to be published, 2012)

10. Golas, U., Ehrig, H., Habel, A.: Multi-Amalgamation in Adhesive Categories. In: Ehrig, H., Rensink, A., Rozenberg, G., Schürr, A. (eds.) ICGT 2010. LNCS, vol. 6372, pp. 346–361. Springer, Heidelberg (2010)
11. Habel, A., Heckel, R., Taentzer, G.: Graph grammars with negative application conditions. Fundam. Inform. 26(3/4), 287–313 (1996)
12. Habel, A., Müller, J., Plump, D.: Double-pushout graph transformation revisited. Mathematical Structures in Computer Science 11(5), 637–688 (2001)
13. Heindel, T.: Structural decomposition of reactions of graph-like objects. In: Aceto, L., Sobocinski, P. (eds.) Structural Operational Semantics (SOS). EPTCS, vol. 32, pp. 26–41 (2010)
14. Hirsch, D., Montanari, U.: Synchronized Hyperedge Replacement with Name Mobility. In: Larsen, K.G., Nielsen, M. (eds.) CONCUR 2001. LNCS, vol. 2154, pp. 121–136. Springer, Heidelberg (2001)
15. Johnstone, P.T., Lack, S., Sobociński, P.: Quasitoposes, Quasiadhesive Categories and Artin Glueing. In: Mossakowski, T., Montanari, U., Haveraaen, M. (eds.) CALCO 2007. LNCS, vol. 4624, pp. 312–326. Springer, Heidelberg (2007)
16. Kreowski, H.-J., Kuske, S., Rozenberg, G.: Graph Transformation Units – An Overview. In: Degano, P., De Nicola, R., Meseguer, J. (eds.) Concurrency, Graphs and Models. LNCS, vol. 5065, pp. 57–75. Springer, Heidelberg (2008)
17. Lack, S., Sobociński, P.: Adhesive Categories. In: Walukiewicz, I. (ed.) FOSSACS 2004. LNCS, vol. 2987, pp. 273–288. Springer, Heidelberg (2004)
18. Montanari, U., Pistore, M.: History-Dependent Automata: An Introduction. In: Bernardo, M., Bogliolo, A. (eds.) SFM-Moby 2005. LNCS, vol. 3465, pp. 1–28. Springer, Heidelberg (2005)
19. Rensink, A.: A first study of compositionality in graph transformation. Tech. Rep. TR-CTIT-10-08, Centre for Telematics and Information Technology, University of Twente (2010)
20. Taentzer, G.: Parallel high-level replacement systems. TCS 186(1-2), 43–81 (1997)

Towards Automatic Verification
of Behavior Preservation
for Model Transformation via Invariant Checking*

Holger Giese and Leen Lambers

Hasso Plattner Institute at the University of Potsdam, Germany
[holger.giese,leen.lambers]@hpi.uni-potsdam.de

Abstract. The correctness of model transformations is a crucial element for model-driven engineering of high quality software. In particular, behavior preservation is the most important correctness property avoiding the introduction of semantic errors during the model-driven engineering process. Behavior preservation verification techniques either show that specific properties are preserved, or more generally and complex, they show some kind of bisimulation between source and target model of the transformation. Both kinds of behavior preservation verification goals have been presented with automatic tool support for the instance level, i.e. for a given source and target model specified by the model transformation. However, up until now there is no automatic verification approach available at the transformation level, i.e. for all source and target models specified by the model transformation. In this paper, we present a first approach towards automatic behavior preservation verification for model transformations specified by triple graph grammars and semantic definitions given by graph transformation rules. In particular, we show that the behavior preservation problem can be reduced to invariant checking for graph transformation. We discuss today's limitations of invariant checking for graph transformation and motivate further lines of future work in this direction.

1   Introduction

The correctness of model transformations is a crucial element for model-driven engineering of high quality software. Many quality related activities are obtained using the source models of the transformations rather than the results of a single transformation or chains of transformations. Therefore, only if the model transformation works correctly and introduces no additional faults, the full benefits of working with the higher-level source models can be realized.

In this context in particular *behavior preservation* is the most important correctness property avoiding the introduction of semantic errors during the model-driven engineering process. Behavior preservation verification techniques either show that specific properties are preserved, or more generally and complex, they show some kind of bisimulation [1] between source and target model of the transformation.

---

* This work was funded by the Deutsche Forschungsgemeinschaft in the course of the project - Correct Model Transformations - see http://www.hpi.uni-potsdam.de/giese/ projekte/kormoran.html?L=1 .

H. Ehrig et al.(Eds.): ICGT 2012, LNCS 7562, pp. 249–263, 2012.

For both kinds of behavior preservation, verification goals have been presented with automatic tool support for the instance level [2,3,4], i.e. for a given source and target model specified by the model transformation. Nevertheless, up until now there is no automatic verification approach available at the transformation level, i.e. for *all* source and target models specified by the model transformation. However, as usually the transformation development and the application development that employs the developed transformation are separate activities that are addressed by different people or even different organizations, detecting that the transformation is not correct during application development time is thus usually too late.

Consequently, ensuring behavior preservation for the transformation in general already during the development of the transformation is highly desirable, but to our best knowledge so far no work exists presenting a generic proof scheme for this problem, allowing to tackle it in an automated manner. We presented a first approach [5] attacking this problem in a semi-automated manner in form of a verification technique based on interactive theorem proving. Hülsbusch et. al [6] presented and compared different proof strategies for manual proofs on the transformation level without solving the problem of automation. Some first approaches tackling the problem for the special case of model refactorings are present [7,8], not covering complete model transformations.

In this paper, we present a first approach towards automatic verification of behavior preservation for model transformations specified by triple graph grammars [9] (TGG) and semantic definitions given by graph transformation systems (GTS). In particular, we show that the behavior preservation problem can be reduced to invariant checking for GTS, which in restricted cases can be automatically verified using our existing verification technique [10]. We reduced the problem of consistency preservation of refactorings [11] accordingly. Due to a mapping of TGGs on specially typed graph transformations [12,13] both the transformation and the semantics are captured in a homogeneous manner, which greatly facilitates mapping the problem on invariants for GTS. We demonstrate with a simple example which degree of automation can be achieved today and further discuss today's limitations of invariant checking for graph transformation and motivate further lines of future work in this direction.

The paper is structured as follows: In Section 2 we review the foundations required to tackle the behavioral preservation problem in the form of graph constraints, GTSs and TGGs. In Section 3 the problem of behavior preservation in general and for the particular setting is introduced. How the problem in this setting can be reduced to invariant checking GTSs is outlined in Section 4. Afterwards, the automation for a restricted class of the problem, current limitations, and directions for future work are reviewed in Section 5. The paper closes with a final conclusion.

2   Foundations

We reintroduce graph conditions, graph transformation and invariants, and TGGs rather informally and refer to [13,14,15] for formal definitions.

Graph Conditions. Nested *graph conditions* [14,15] generalize the corresponding notions in [16], where a negative (positive) application condition, NAC (PAC) for short, over a graph P, denoted $\neg\exists a$ ($\exists a$) is defined in terms of a graph morphism. We use $\rightarrow$ to denote a graph morphism in general and $\hookrightarrow$ to denote an injective graph morphism in particular. Informally, a morphism $p : P \rightarrow G$ satisfies $\neg\exists a$ ($\exists a$) if there does not exist a morphism $q : C \rightarrow G$ extending p (if there exists q extending p). Then, a *(nested) graph condition AC* is either the special condition true or a pair of the form $\neg\exists(a, ac_C)$ or $\exists(a, ac_C)$, where the first case corresponds to a NAC and the second to a PAC, and in both cases ac_C is an additional AC on C. Intuitively, a morphism $p : P \rightarrow G$ satisfies $\exists(a, ac_C)$ if p satisfies a and the corresponding extension q satisfies ac_C. ACs (and also NACs and PACs) may be combined with the usual logical connectors. A morphism $p : P \rightarrow G$ satisfies $\neg c$ if p does not satisfy c and satisfies $\wedge_{i \in I} c_i$ if it satisfies each c_i ($i \in I$).

$$\exists(\ P \xrightarrow{\ \ a\ \ } C, \blacktriangleleft^{ac_C}\ )$$
$$p \searrow \underset{=}{\ } \nearrow q \ \not\models$$
$$G$$

Graph conditions over the empty graph I are also called *graph constraints*. A graph G satisfies a graph constraint ac_I, written $G \models ac_I$, if the initial morphism $i_G : I \rightarrow G$ satisfies ac_I. This means that if a constraint simply should state that a match for a graph C must exist, we have the graph constraint $\exists(i_C, \text{true})$.

Notation: Note that $\exists a$ abbreviates $\exists(a, \text{true})$, $\forall(a, ac_C)$ abbreviates $\neg\exists(a, \neg ac_C)$ and $\exists(i_C, ac_C)$ with the initial morphism $i_C : I \rightarrow C$ abbreviates $\exists(C, ac_C)$.

Graph Transformation and Invariants. In this paper, we assume the double-pushout approach (DPO) to graph transformation with injective matching [17]. A *plain rule* $p = \langle L \hookleftarrow I \hookrightarrow R \rangle$ consists of a left-hand side (LHS) L and a right-hand side (RHS). Additionally, we allow rules to be equipped with LHS application conditions [14,15], allowing to apply a given rule to a graph G only if the corresponding match morphism satisfies the AC of the rule. Thus, a *rule* $\rho = \langle p, ac_L \rangle$ consists of a *plain* rule $p = \langle L \hookleftarrow I \hookrightarrow R \rangle$ and an application condition ac_L over L.

$$ac_L \ \blacktriangleright\!\!\blacktriangleright \ L \hookleftarrow I \hookrightarrow R$$
$$m \downarrow \ (1) \ \downarrow \ (2) \ \downarrow m^*$$
$$G \hookleftarrow D \hookrightarrow H$$

A *direct transformation* via rule $\rho = \langle p, ac_L \rangle$ consists of two pushouts (1) and (2), called DPO, with injective match m and comatch m^* such that $m \models ac_L$. If there exists a direct transformation from G to G' via rule ρ and match m, we write $G \Rightarrow_{m,\rho} G'$. If we are only interested in the rule ρ, we write $G \Rightarrow_\rho G'$. If a rule ρ in a set of rules $\mathcal{R}$ exists such that there exists a direct transformation via rule ρ from G to G', we write $G \Rightarrow_{\mathcal{R}} G'$. A *graph transformation*, denoted as $G_0 \overset{*}{\Rightarrow}_{\mathcal{R}} G_n$, is a sequence $G_0 \Rightarrow_{\mathcal{R}} G_1 \Rightarrow_{\mathcal{R}} \cdots \Rightarrow_{\mathcal{R}} G_n$ of $n \geq 1$ direct transformations.

Graph conditions, rules and transformations as described before can be equipped with *typing* over a given type graph TG as usual [17] by adding typing morphisms from each graph to TG and by requiring type-compatibility with respect to TG for each graph morphism. We denote with $\mathscr{L}(TG)$ the set of all graphs G typed over TG.

Definition 1 (graph transformation system). *A* graph transformation system *(GTS)* gts = $(\mathscr{R}, TG)$ *consists of a set of rules* $\mathscr{R}$ *typed over a type graph* TG. *A graph transformation system may be equipped with an initial graph* G_0 *or a set of initial graphs* I *being graphs typed over* TG. *For a GTS* gts = $(\mathscr{R}, TG)$ *and a set of initial graphs* I *the set of reachable graphs* $REACH(\text{gts}, I)$ *is defined as* $\{G \mid G_0 \overset{*}{\Rightarrow}_{\mathscr{R}} G, G_0 \in I\}$.

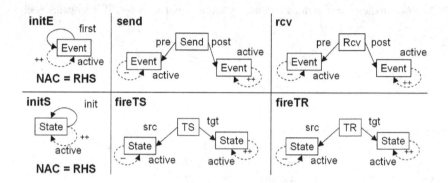

Fig. 1. gts_s and gts_t - Operational semantics

Example 1 (graph transformation system). In the upper part of Fig. 1, three GT rules are depicted that are typed over S_{RT}, the left part of the graph depicted on the right of Fig. 2, building a GTS $\text{gts}_s = (\{initE, send, rcv\}, S_{RT})$. We use a notation marking elements that are created or deleted by the rule with "++" or "- -", respectively. The rule *initE* holds a NAC identical to its RHS, expressing that it should not be applied to the same match twice.

The applicability of a rule can be expressed as a graph constraint. We exploited this feature already in [11,18] for the consistency preservation verification of rule-based refactorings and for consistency verification of integrated behavior models, respectively. The rule applicability constraint for a rule $\rho = \langle p, \text{ac}_L \rangle$ with $p = \langle L \overset{l}{\hookleftarrow} I \overset{r}{\hookrightarrow} R \rangle$, expresses that an injective match m exists such that the application condition ac_L and the so-called deletable condition $Deletable(p)^1$, guaranteeing the existence of a PO-complement for $m \circ l$, are fulfilled. Then it is obvious that the rule $\rho = \langle p, \text{ac}_L \rangle$ is applicable with injective matching to a graph G if and only if G fulfills the rule applicability constraint.

Definition 2 (rule applicability constraint). *Given a rule* $\rho = \langle p, \text{ac}_L \rangle$ *with* plain *rule* $p = \langle L \hookleftarrow I \hookrightarrow R \rangle$, *then* $\text{App}(\rho) = \exists(i_L, \text{ac}_L \wedge Deletable(p))$ *is the rule applicability constraint of* ρ.

---

[1] In Lemma 5.9 of [19], it is described how to construct $Deletable(l)$ (we write $Deletable(p)$ instead of $Deletable(l)$). Basically, it prohibits the existence of additional adjacent edges, making use of additional NACs, for nodes that are to be deleted.

Example 2 (rule applicability as constraint). The applicability of the rule *send*, depicted in Fig. 1, can be expressed as graph constraint $\exists(i_L, \text{true})$, or abbreviated $\exists L$, with L the LHS of *send*. This is because ac_L is true and *Deletable*(p) is true, since the rule does not delete any nodes. The same holds for the applicability of rule *fireTS*. In Fig. 5, the graph constraint App(*send*) $\Rightarrow$ App(*fireTS*), equivalent to $\neg$App(*send*) $\vee$ App(*fireTS*), is depicted.

The satisfaction of graph constraints can be invariant with respect to a GTS. In our verification approach, we reduce the problem of behavior preservation to invariant checking. In Section 5, we explain how and with which restrictions automatic invariant checking can be performed statically.

Definition 3 (inductive invariant [20]). *A graph constraint* ac_I *is an* inductive invariant *of the GTS* gts $= (\mathscr{R}, \text{TG})$, *if for all graphs G in* $\mathscr{L}(TG)$, *and for all rules* $\rho \in \mathscr{R}$, *it holds that* $G \models ac_I \wedge G \Rightarrow_\rho G'$ *implies* $G' \models ac_I$.

Triple Graph Grammars Triple graph grammars (TGGs) define model transformations in a relational (declarative) way. We use [12,13] a TGG formalization more suitable for the current practice for TGGs than the one introduced originally in [9]. Thereby, the main idea is to use a distinguished, fixed graph TRIPLE which all triple graphs, including the type triple graph $S_{TT}C_{TT}T_{TT}$, are typed over.

$$\text{TRIPLE} \overset{l_s}{\underset{s}{\curvearrowright}} \xleftarrow{\;\;e_{cs}\;\;} c \xrightarrow{\;\;e_{ct}\;\;} t \overset{l_t}{\curvearrowleft}$$

We say that TRIPLE$_S$, TRIPLE$_C$, and TRIPLE$_T$, as shown below,

$$\text{TRIPLE}_S \overset{l_s}{\underset{s}{\curvearrowright}} \qquad \text{TRIPLE}_C \;\; s \xleftarrow{\;\;e_{cs}\;\;} c \xrightarrow{\;\;e_{ct}\;\;} t \qquad \text{TRIPLE}_T \;\; t \overset{l_t}{\curvearrowleft}$$

are the *source, correspondence*, and *target component* of TRIPLE, respectively. Analogously to the aforementioned case, the projection of a graph G typed over TRIPLE to TRIPLE$_S$, TRIPLE$_C$, or TRIPLE$_T$ selects the corresponding component of this graph.

We denote a triple graph as a combination of three indexed capitals, as for example $G = S_G C_G T_G$, where S_G denotes the *source* and T_G denotes the *target component* of G, while C_G denotes the *correspondence component*, being the smallest subgraph of G such that all c-nodes as well as all e_{cs}- and e_{ct}-edges are included in C_G. Note that C_G has to be a proper graph, i.e. all target nodes of e_{cs} and e_{ct}-edges have to be included. The category of triple graphs and triple graph morphisms is called **TripleGraphs**.

Analogously to typed graphs, *typed triple graphs* are triple graphs typed over a distinguished triple graph $S_{TT}C_{TT}T_{TT}$, called type triple graph. The category of typed triple graphs and morphisms is called **TripleGraphs$_{TT}$**. In the remainder of this paper, we assume every triple graph $S_G C_G T_G$ and triple graph morphism f to be typed over $S_{TT}C_{TT}T_{TT}$, even if not explicitly mentioned. In particular, this means that S_G is typed over S_{TT}, C_G is typed over C_{TT}, and T is typed over T_{TT}. We say that S_G (T_G or C_G) is a *source graph* (*target graph* or *correspondence graph*, respectively) belonging to the language $\mathscr{L}(S_{TT})$ ($\mathscr{L}(T_{TT})$ or $\mathscr{L}(C_{TT})$, respectively).

Notation: Note that each source graph (target graph) corresponds uniquely to a triple graph with empty correspondence and target (source and correspondence) components, respectively. Therefore, if it is clear from the context that we are dealing with triple graphs, we denote triple graphs $S_G \varnothing \varnothing$ ($\varnothing \varnothing T_G$) with empty correspondence and target components (source components) also as S_G (T_G), respectively.

A *triple graph rule* $p : S_L C_L T_L \xrightarrow{r} S_R C_R T_R$ consists of a triple graph morphism r, which is an inclusion. A *direct triple graph transformation* $S_G C_G T_G \Rightarrow_{p,m} S_H C_H T_H$ from $S_G C_G T_G$ to $S_H C_H T_H$ via p and m consists of the pushout *(PO)* in **TripleGraphs$_{TT}$**.

$$
\begin{array}{ccc}
S_L C_L T_L & \xrightarrow{\ \ r\ \ } & S_R C_R T_R \\
\downarrow{\scriptstyle m} & (PO) & \downarrow{\scriptstyle n} \\
S_G C_G T_G & \xrightarrow{\ \ h\ \ } & S_H C_H T_H
\end{array}
$$

A *triple graph transformation*, denoted as $S_{G_0} C_{G_0} T_{G_0} \overset{*}{\Rightarrow} S_{G_n} C_{G_n} T_{G_n}$, is a sequence $S_{G_0} C_{G_0} T_{G_0} \Rightarrow S_{G_1} C_{G_1} T_{G_1} \Rightarrow \cdots \Rightarrow S_{G_n} C_{G_n} T_{G_n}$ of direct triple graph transformations. As in the context of classical triple graphs, we consider triple graph grammars (TGGs) with non-deleting rules. Moreover, we allow grammars to be equipped with a so-called *TGG constraint* $\mathscr{C}_{tgg}$ typed over $S_{TT} C_{TT} T_{TT}$, restricting the language of triple graphs generated by the TGG to a subset of triple graphs satisfying $\mathscr{C}_{tgg}$.

Definition 4 (Triple graph grammar, $\mathscr{L}(\text{tgg}, \mathscr{C}_{tgg})$). *A triple graph grammar (TGG)* tgg $= ((\mathscr{R}, S_{TT} C_{TT} T_{TT}), S_A C_A T_A)$ *consists of a set of triple graph rules $\mathscr{R}$ typed over $S_{TT} C_{TT} T_{TT}$ and a triple start graph $S_A C_A T_A$, called axiom, also typed over $S_{TT} C_{TT} T_{TT}$. Given a TGG constraint $\mathscr{C}_{tgg}$ for* tgg, *being a graph constraint typed over $S_{TT} C_{TT} T_{TT}$ such that $S_A C_A T_A \models \mathscr{C}_{tgg}$, then the* triple graph language $\mathscr{L}(\text{tgg}, \mathscr{C}_{tgg})$ *consists of $S_A C_A T_A$ and all triple graphs $S_G C_G T_G \models \mathscr{C}_{tgg}$ such that $S_A C_A T_A \overset{*}{\Rightarrow} S_G C_G T_G$ via rules in $\mathscr{R}$.*

Example 3 (tgg). In Fig. 2, an example TGG tgg is depicted with an axiom $S_A C_A T_A$ and two rules typed over the type graph $S_{TT} C_{TT} T_{TT}$. The type graph $S_{TT} C_{TT} T_{TT}$ is the subgraph of the type graph $S_{RT} C_{TT} T_{RT}$ shown in the upper right part of Fig. 2, obtained by deleting the active loops in the source and target component. The rules describe a model transformation between a sequence chart with one lifeline being able to send and receive messages and an automaton with two different types of transitions, one for the sending and one for the receiving of messages. The events before and after a send/receive message on the lifeline correspond to states before and after send/receive transitions in the automaton. On the lifeline, there is one first event which corresponds to an initial state of the automaton. We also have a TGG constraint $\mathscr{C}_{tgg}$, for which a fragment is shown in the lower right part of the figure. It expresses that each event should be connected with at most one previous message (subsequent message) of type Send/Rcv. An analogous condition holds for states and transitions in the automaton.

3   Behavior Preservation for Model Transformation

Assuming that a model transformation is defined by some relation over the source and target language, we can formulate the general problem of behavior preservation on the transformation level as follows:

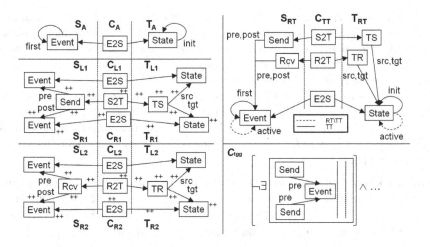

Fig. 2. tgg with axiom and two rules, type graph $S_{RT}C_{TT}T_{RT}$, fragment of $\mathscr{C}_{\text{tgg}}$

Problem Statement 1 (Behavior preservation). *Given a model transformation* MT $\subseteq$ $\mathscr{L}(S_{TT}) \times \mathscr{L}(T_{TT})$ *and operational semantic definitions* sem$_S$ *and* sem$_T$ *for source and target language* $\mathscr{L}(S_{TT})$ *and* $\mathscr{L}(T_{TT})$, *respectively, we say that* MT *is* behavior preserving *if for each pair of source and target graphs* $(S, T) \in$ MT, *it holds that* sem$_S$ *of S is somehow equivalent to* sem$_T$ *of T.*

For our specific setting we explain now how we define the elements and the equivalence mentioned in Problem Statement 1. At first, analogous to [5,6], the model transformation MT $\subseteq \mathscr{L}(S_{TT}) \times \mathscr{L}(T_{TT})$ is derived in our case from a given TGG tgg typed over $S_{TT}C_{TT}T_{TT}$. Additionally, we allow tgg to be equipped with a TGG constraint $\mathscr{C}_{\text{tgg}}$ such that MT can be derived from the language $\mathscr{L}(\text{tgg}, \mathscr{C}_{\text{tgg}})$.

Definition 5 (MT(tgg, $\mathscr{C}_{\text{tgg}}$)). *Given a TGG* tgg *with TGG constraint* $\mathscr{C}_{\text{tgg}}$, MT(tgg, $\mathscr{C}_{\text{tgg}}$) $\subseteq \mathscr{L}(S_{TT}) \times \mathscr{L}(T_{TT})$ *consists of pairs of source and target graphs* (S, T) *such that there exists some triple graph SCT* $\in \mathscr{L}(\text{tgg}, \mathscr{C}_{\text{tgg}})$ *having S and T as source and target component, respectively.*

*Example 4 (*MT(tgg, $\mathscr{C}_{\text{tgg}}$)*).* In Fig. 3, a source graph S and target graph T is depicted in concrete syntax belonging to MT(tgg, $\mathscr{C}_{\text{tgg}}$) with tgg and $\mathscr{C}_{\text{tgg}}$ as described in Example 3 and depicted in Fig. 2. A triple graph *SCT* fulfilling $\mathscr{C}_{\text{tgg}}$ can be generated by tgg, being the subgraph of S_2CT_2 depicted in Fig. 3 obtained by deleting the two *active* loops.

Secondly, analogous to the work of Hülschbusch et al. [6], we define the operational semantics sem$_S$ and sem$_T$ of source models and target models as graph transformation systems gts$_s$ and gts$_t$, respectively. In order to be able to encode runtime information into the source and target language $\mathscr{L}(S_{TT})$ and $\mathscr{L}(T_{TT})$ the according type graphs can be enriched with so-called *run-time types* allowing to define the operational semantics of both languages as the possible changes in instances of this enhanced type graph (similar to the dynamic metamodeling approach [3]). We denote these enhanced type graphs for the source (target) language as S_{RT} (T_{RT}), respectively. Accordingly, the type graph $S_{TT}C_{TT}T_{TT}$ enriched with run-time types for source and target languages is denoted as $S_{RT}C_{TT}T_{RT}$. In this context, we say that a type (or corresponding instance element) is *static* if it belongs to $S_{TT}C_{TT}T_{TT}$. We assume that operational semantic rules have the property that they do not change elements with static type, since they merely model the change of run-time information.

Note that if some graph $S_G C_G T_G$, morphism m, rule ρ, or condition ac is typed over a subgraph $S_{SG}C_{SG}T_{SG}$ of $S_{RT}C_{TT}T_{RT}$, then it is straightforward to extend the codomain of the corresponding typing morphisms to $S_{RT}C_{TT}T_{RT}$ such that $S_G C_G T_G$, m, ac, or ρ are actually typed over $S_{RT}C_{TT}T_{RT}$. We therefore do not explicitly mention this anymore in the rest of this paper.

Definition 6 (gts$_s$ **and** gts$_t$). *Given a source and target enhanced type graph S_{RT} and T_{RT}, respectively, we have a source GTS* gts$_s = (\mathcal{R}_s = \{\rho_s^i \mid i \in I\}, S_{RT})$ *for the source language $\mathcal{L}(S_{TT})$ and a target GTS* gts$_t = (\mathcal{R}_t = \{\rho_t^j \mid j \in J\}, T_{RT})$ *for the target language $\mathcal{L}(T_{TT})$, consisting of rules that preserve all elements with static type.*

Example 5 (gts$_s$ *and* gts$_t$). The GTS gts$_s = (\{initE, send, rcv\}, S_{RT})$ depicted in the upper part of Fig. 1 typed over S_{RT} (see Fig. 2) provides a semantics for the source models. The semantics for the target models (depicted in the lower part of Fig. 1) is provided by the GTS gts$_t = (\{initS, fireTS, fireTR\}, T_{RT})$ typed over T_{RT} (see Fig. 2).

Finally, we also have to define what it means for two operational semantics to be somehow equivalent. For behavioral models that describe a reactive behavior, the external visible interactions rather than the usually encapsulated states are relevant. This can be captured by considering the labeled transitions systems induced by the source and target GTS, where the labeling describes the externally visible interactions, defined by the corresponding rule names leading to operational semantics rule applications.

Definition 7 (induced LTS(gts, G_0)). *A labeled transition systems (LTS) lts = $\langle i, \rightarrow, Q, L \rangle$ consists of the initial state i, the labeled transition relation $\rightarrow \subseteq Q \times L \times Q$ over the label alphabet L and the set of states Q. Given a relabeling mapping $l : L \rightarrow L'$ with L' a new label alphabet, then $l(lts)$ is the labeled transition system where each label α in lts has been replaced by $l(\alpha)$. The labeled transition system LTS(gts, G_0) induced by gts = $(\mathcal{R}, TG)$ and the initial graph G_0 equals $\langle G_0, \rightarrow_{gts}, Q_{gts}, \mathcal{R} \rangle$ with $\rightarrow_{gts} = \{(G, \rho, G') \mid G, G' \in Q_{gts}, \rho \in \mathcal{R} \wedge G \Rightarrow_\rho G'\}$, and $Q_{gts} = REACH(gts, \{G_0\})$.*

Given two relabeling mappings $l_s : \mathcal{R}_s \rightarrow A$ and $l_t : \mathcal{R}_t \rightarrow A$ for LTS(gts$_s$, S) and LTS(gts$_t$, T), respectively, we can then obtain two transition systems $l_s(\text{LTS}(\text{gts}_s, S))$ and $l_t(\text{LTS}(\text{gts}_t, T))$ over a common alphabet A.[2]

Example 6 ($l_s(\text{LTS}(\text{gts}_s, S))$ *and* $l_t(\text{LTS}(\text{gts}_t, T))$). For the semantic definitions gts$_s$ and gts$_t$ from Example 5 and the induced LTS LTS(gts$_s$, S) and LTS(gts$_t$, T) we define bijective relabeling mappings $l_s : \mathcal{R}_s \rightarrow A$ and $l_t : \mathcal{R}_t \rightarrow A$ with $A = \{init, s, r\}$ as $l_s = \langle initE \rightarrow init, send \rightarrow s, rcv \rightarrow r \rangle$ and $l_t = \langle initS \rightarrow init, fireTS \rightarrow s, fireTR \rightarrow r \rangle$. The application of the relabelings to the LTSs of the example source and target model is depicted in Fig. 3, where the original labels α are depicted on the left side of a transition, while the mapped labels $l_s(\alpha)$ are depicted on the right side of a transition.

---

[2] It is important that the relabeling is not trivial, e.g. mapping every label to the same element in A, as otherwise the fact that the LTSs are bisimilar is not very significant.

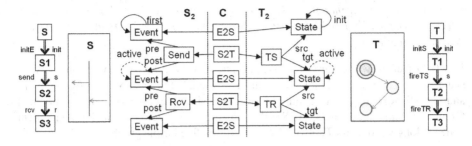

Fig. 3. S and T in concrete syntax, $SCT \subseteq S_2C_2T_2$, LTS(gts_s, S) and LTS(gts_t, T) with relabeling

Now we can compare source and target behavior looking only at the labeling of transitions by requiring that a source S and target graph T have bisimilar LTSs $l_s(\text{LTS}(\text{gts}_s, S))$ and $l_t(\text{LTS}(\text{gts}_t, T))$. Bisimilarity [1] of two LTSs over the same alphabet is defined as follows:

Definition 8 (bisimulation relation, bisimilarity [1]). *A bisimulation relation between two labeled transition systems* $\text{lts}_1 = \langle i_1, \rightarrow, Q_1, A \rangle, \text{lts}_2 = \langle i_2, \rightarrow, Q_2, A \rangle$ *over the same alphabet A is a relation* $B \subseteq Q1 \times Q2$ *such that whenever* $(q1, q2) \in B$

1. *If* $q_1 \xrightarrow{\alpha} q_1'$, *then* $q_2 \xrightarrow{\alpha} q_2'$ *and* $(q_1', q_2') \in B$.
2. *If* $q_2 \xrightarrow{\alpha} q_2'$, *then* $q_1 \xrightarrow{\alpha} q_1'$ *and* $(q_1', q_2') \in B$.

We say that lts_1 *and* lts_2 *are* bisimilar *if there exists a bisimulation relation between them such that* $(i_1, i_2) \in B$.

We still need to specify for our specific setting how the bisimulation relation $B \subseteq$ REACH$(\text{gts}_s, S) \times$ REACH(gts_t, T) for the LTSs $l_s(\text{LTS}(\text{gts}_s, S))$ and $l_t(\text{LTS}(\text{gts}_t, T))$ is given. We derive it from a given graph constraint $\mathscr{C}_{\text{Bis}}$ typed over $S_{RT}C_{TT}T_{RT}$, called the *bisimulation constraint* and introduced more in detail in Section 4. Since we know that for each (S, T) in MT$(\text{tgg}, \mathscr{C}_{\text{tgg}})$, there exists some SCT in $\mathscr{L}(\text{tgg}, \mathscr{C}_{\text{tgg}})$, we say that (S, T) belongs to the bisimulation relation if SCT fulfills the bisimulation constraint $\mathscr{C}_{\text{Bis}}$. Concluding, the bisimulation relation B in our specific setting, denoted Bis$(\mathscr{C}_{\text{Bis}}, SCT)$, is derived from the bisimulation constraint $\mathscr{C}_{\text{Bis}}$ and SCT.

Definition 9 (induced bisimulation relation Bis$(\mathscr{C}_{\text{Bis}}, SCT)$**).** *Given a pair of source and target graphs* (S, T) *in* MT(tgg) *such that the triple graph SCT typed over $S_{TT}C_{TT}T_{TT}$ belongs to* $\mathscr{L}(\text{tgg}, \mathscr{C}_{\text{tgg}})$ *with two LTSs* $l_s(\text{LTS}(\text{gts}_s, S))$ *and* $l_t(\text{LTS}(\text{gts}_t, T))$, *and given a graph constraint $\mathscr{C}_{\text{Bis}}$ typed over $S_{RT}C_{TT}T_{RT}$, called the* bisimulation constraint, *then the* induced bisimulation relation Bis$(\mathscr{C}_{\text{Bis}}, SCT) \subseteq$ REACH$(\text{gts}_s, S) \times$ REACH(gts_t, T) *consists of all* (S', T') *such that $S'CT'$ fulfills $\mathscr{C}_{\text{Bis}}$.*

The above definition for the induced bisimulation relation is well-defined because of the following lemma:

Lemma 1 (semantic rules preserve static types and correspondences). *Given two LTSs* $l_s(\text{LTS}(\text{gts}_s, S))$ *and* $l_t(\text{LTS}(\text{gts}_t, T))$ *for a pair of source and target graphs* (S, T)

in $\mathrm{MT}(\mathrm{tgg}, \mathscr{C}_{\mathrm{tgg}})$ *such that the triple graph SCT typed over* $S_{TT}C_{TT}T_{TT}$ *belongs to* $\mathscr{L}(\mathrm{tgg}, \mathscr{C}_{\mathrm{tgg}})$, *then* $S'CT'$ *with* $S' \in REACH(\mathrm{gts}_s, S)$ *and* $T' \in REACH(\mathrm{gts}_t, T)$ *is a well-defined triple graph.*

Proof. In Def. 6, we required for gts_s and gts_t that the rules preserve elements with static types. Therefore, it holds that S and T is included in each $S' \in REACH(\mathrm{gts}_s, S)$ and $T' \in REACH(\mathrm{gts}_t, T)$, respectively. Moreover, since the correspondence component C consists of all edges connecting S and T via correspondence nodes including incident nodes belonging to S and T, also $S'CT'$ is a well-defined triple graph.   □

Summarizing, we can refine the more general Problem Statement 1 to the following statement for our specific setting:

Problem Statement 2 (Behavior preservation in our setting). *Given a model transformation* $\mathrm{MT}(\mathrm{tgg}, \mathscr{C}_{\mathrm{tgg}}) : \mathscr{L}(S_{TT}) \times \mathscr{L}(T_{TT})$ *for a* $\mathrm{tgg} = ((\mathscr{R}, S_{TT}C_{TT}T_{TT}), S_A C_A T_A)$ *with TGG constraint* $\mathscr{C}_{\mathrm{tgg}}$, *operational semantic definitions* $\mathrm{gts}_s = (\mathscr{R}_s, S_{RT})$ *and* $\mathrm{gts}_t = (\mathscr{R}_t, T_{RT})$ *for source and target language* $\mathscr{L}(S_{TT})$ *and* $\mathscr{L}(T_{TT})$, *resp., relabeling mappings* $l_s : \mathscr{R}_s \to A$ *and* $l_t : \mathscr{R}_t \to A$ *and a bisimulation constraint* $\mathscr{C}_{\mathrm{Bis}}$, *then* $\mathrm{MT}(\mathrm{tgg}, \mathscr{C}_{\mathrm{tgg}})$ *is behavior preserving if for each triple graph SCT typed over* $S_{TT}C_{TT}T_{TT}$ *belonging to* $\mathscr{L}(\mathrm{tgg}, \mathscr{C}_{\mathrm{tgg}})$ *such that* $(S, T) \in \mathrm{MT}(\mathrm{tgg}, \mathscr{C}_{\mathrm{tgg}})$, *it holds that* $l_s(\mathrm{LTS}(\mathrm{gts}_s, S))$ *and* $l_t(\mathrm{LTS}(\mathrm{gts}_t, T))$ *are bisimilar via the bisimulation relation* $\mathrm{Bis}(\mathscr{C}_{\mathrm{Bis}}, SCT)$.

Example 7 (problem statement). The complete example for the problem statememt 2 consists of the TGG of Example 3 shown in Fig. 2, the operational semantics definitions $\mathrm{gts}_s = (\mathscr{R}_s, S_{RT})$ and $\mathrm{gts}_t = (\mathscr{R}_t, T_{RT})$ of Example 5 depicted in Fig. 1 for the source and target model of the TGG tgg, and the relabeling mappings $l_s : \mathscr{R}_s \to A$ and $l_t : \mathscr{R}_t \to A$ from Example 6. In addition, we still require the bisimulation constraint $\mathscr{C}_{\mathrm{Bis}}$ that will be constructed in the following section.

4   Behavior Preservation Verification

Given a model transformation $\mathrm{MT}(\mathrm{tgg}, \mathscr{C}_{\mathrm{tgg}})$ with $\mathrm{tgg} = ((\mathscr{R}, S_{TT}C_{TT}T_{TT}), S_A C_A T_A)$, operational semantic definitions gts_s and gts_t, relabeling mappings l_s and l_t and a bisimulation constraint $\mathscr{C}_{\mathrm{Bis}}$, then we prove in Theorem 3 that the *problem of behavior preservation* of $\mathrm{MT}(\mathrm{tgg}, \mathscr{C}_{\mathrm{tgg}})$ in the sense of Problem Statement 2 can be *reduced to invariant checking* of $\mathscr{C}_{\mathrm{Bis}}$ for specific GTSs inherent to the problem. The bisimulation constraint $\mathscr{C}_{\mathrm{Bis}}$ is defined as the conjunction of a so-called runtime constraint $\mathscr{C}_{\mathrm{RT}}$, a pairwise constraint $\mathscr{C}_{\mathrm{Pair}}$, and the TGG constraint $\mathscr{C}_{\mathrm{tgg}}$. First, the *runtime constraint* $\mathscr{C}_{\mathrm{RT}}$ is a constraint typed over $S_{RT}C_{TT}T_{RT}$ that has to be provided, expressing how the runtime structure of source and target language are related to each other via the correspondences between them defined by the tgg.

Example 8 (runtime constraint $\mathscr{C}_{\mathrm{RT}}$). The runtime constraint for our Example 7, as depicted in Fig. 4, expresses that if an active loop on some event in the sequence chart domain occurs, then this event should be connected to a corresponding state with active loop in the automaton domain and the other way round. This runtime constraint is typed over the type graph $S_{RT}C_{TT}T_{RT}$ shown in Fig. 2.

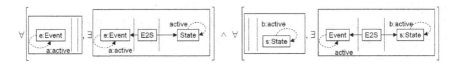

Fig. 4. $\mathscr{C}_{\mathrm{RT}}$ for tgg, gts$_s$, gts$_t$, l_s, and l_t of Example 7

Secondly, the *pairwise constraint* $\mathscr{C}_{\mathrm{Pair}}$, expresses that the applicability of a rule ρ_s of the source semantics gts$_s$ implies the applicability of a rule ρ_t of the target semantics gts$_t$ and the other way round, whenever ρ_s and ρ_t are mapped by the relabeling mappings l_s and l_t to the same label in the common alphabet A. Trivially speaking, rules "with the same label" should be applied pairwise, since this is exactly what we need for proving bisimulation.

Definition 10 (pairwise constraint $\mathscr{C}_{\mathrm{Pair}}$, set of parallel rules $\mathscr{P}(l_s, l_t)$). *Given* gts$_s$ = $(\mathscr{R}_s, S_{RT})$ *and* gts$_t$ = $(\mathscr{R}_t, T_{RT})$ *as well as the relabeling mappings* $l_s : \mathscr{R}_s \to A$ *and* $l_t : \mathscr{R}_t \to A$, *the pairwise constraint*

$$\mathscr{C}_{\mathrm{Pair}} = \wedge_{(\rho_s, \rho_t) \in \mathrm{Pair}(l_s, l_t)}((\mathrm{App}(\rho_s) \Rightarrow \mathrm{App}(\rho_t)) \wedge (\mathrm{App}(\rho_t) \Rightarrow \mathrm{App}(\rho_s)))$$

typed over $S_{RT}C_{TT}T_{RT}$ *with the set of pairs* $\mathrm{Pair}(l_s, l_t) = \{(\rho_s, \rho_t) | l_s(\rho_s) = l_t(\rho_t) \wedge \rho_s \in \mathscr{R}_s, \rho_t \in \mathscr{R}_t\}$. *We define* $\mathscr{P}(l_s, l_t) = \{\rho_s + \rho_t | (\rho_s, \rho_t) \in \mathrm{Pair}(l_s, l_t)\}$ *as the set of parallel rules [21,15] induced by* $\mathrm{Pair}(l_s, l_t)$.

Example 9 ($\mathscr{C}_{\mathrm{Pair}}$). For the pair $(send, fireTS) \in Pair(l_s, l_t)$ of semantics rules of Example 7 the constraint $\mathrm{App}(send) \Rightarrow \mathrm{App}(fireTS)$ depicted in Fig. 5 results.

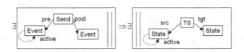

Fig. 5. $\mathrm{App}(send) \Rightarrow \mathrm{App}(fireTS)$ in $\mathscr{C}_{\mathrm{Pair}}$

The following theorem expresses that behavior preservation is given if runtime states in source and target models are always corresponding, and source and target operational rules with equivalent labels are always applicable together.

Theorem 3 (behavior preservation verification). *Given a model transformation* $\mathrm{MT}(\mathrm{tgg}, \mathscr{C}_{\mathrm{tgg}}) : \mathscr{L}(S_{TT}) \times \mathscr{L}(T_{TT})$ *for a* $\mathrm{tgg} = ((\mathscr{R}, S_{TT}C_{TT}T_{TT}), S_A C_A T_A)$ *with TGG constraint* $\mathscr{C}_{\mathrm{tgg}}$, *operational semantic definitions* gts$_s$ = $(\mathscr{R}_s, S_{RT})$ *and* gts$_t$ = $(\mathscr{R}_t, T_{RT})$ *for source and target language* $\mathscr{L}(S_{TT})$ *and* $\mathscr{L}(T_{TT})$, *respectively, relabeling mappings* $l_s : \mathscr{R}_s \to A$ *and* $l_t : \mathscr{R}_t \to A$, *a bisimulation constraint* $\mathscr{C}_{\mathrm{Bis}} = \mathscr{C}_{\mathrm{RT}} \wedge \mathscr{C}_{\mathrm{Pair}} \wedge \mathscr{C}_{\mathrm{tgg}}$ *typed over* $S_{RT}C_{TT}T_{RT}$ *with* $\mathscr{C}_{\mathrm{RT}}$ *a given runtime constraint and* $\mathscr{C}_{\mathrm{Pair}}$ *the pairwise constraint derived according to Def. 10, then* $\mathrm{MT}(\mathrm{tgg}, \mathscr{C}_{\mathrm{tgg}})$ *is behavior preserving in the sense of Problem Statement 2 if the following conditions are fulfilled:*

1. $S_A C_A T_A \models \mathscr{C}_{RT} \wedge \mathscr{C}_{Pair}$.
2. $\mathscr{C}_{RT} \wedge \mathscr{C}_{Pair}$ is an inductive invariant (see Def. 3) of $(\mathscr{R}, S_{RT} C_{TT} T_{RT})$.
3. $\mathscr{C}_{RT} \wedge \mathscr{C}_{Pair}$ is an inductive invariant (see Def. 3) of $(\mathscr{P}(l_s, l_t), S_{RT} C_{TT} T_{RT})$ with $\mathscr{P}(l_s, l_t)$ as given in Def. 10.

Proof. We have to show that for any $(S, T) \in M(\text{tgg}, \mathscr{C}_{\text{tgg}})$ it holds that $l_s(\text{LTS}(\text{gts}_s, S))$ and $l_t(\text{LTS}(\text{gts}_t, T))$ are bisimilar via the induced bisimulation relation $\text{Bis}(\mathscr{C}_{RT} \wedge \mathscr{C}_{Pair} \wedge \mathscr{C}_{\text{tgg}}, SCT)$. We therefore prove (1) that $\text{Bis}(\mathscr{C}_{RT} \wedge \mathscr{C}_{Pair} \wedge \mathscr{C}_{\text{tgg}}, SCT)$ is indeed a bisimulation relation according to conditions 1 and 2 of Def. 8 and (2) that the pair of initial states (S, T) of $l_s(\text{LTS}(\text{gts}_s, S))$ and $l_t(\text{LTS}(\text{gts}_t, T))$ is always in $\text{Bis}(\mathscr{C}_{RT} \wedge \mathscr{C}_{Pair} \wedge \mathscr{C}_{\text{tgg}}, SCT)$.

(1) $\text{Bis}(\mathscr{C}_{RT} \wedge \mathscr{C}_{Pair} \wedge \mathscr{C}_{\text{tgg}}, SCT)$ *is a bisimulation relation:* We first have to show for condition 1 of Def. 8 that for all $(S_1, T_1) \in \text{Bis}(\mathscr{C}_{RT} \wedge \mathscr{C}_{Pair} \wedge \mathscr{C}_{\text{tgg}}, SCT)$ (equivalent to $S_1 C_1 T_1 \models \mathscr{C}_{RT} \wedge \mathscr{C}_{Pair} \wedge \mathscr{C}_{\text{tgg}}$ according to Def. 9), if $S_1 \underset{\rightarrow}{\alpha} S_2$, then $T_1 \underset{\rightarrow}{\alpha} T_2$ and $(S_2, T_2) \in \text{Bis}(\mathscr{C}_{RT} \wedge \mathscr{C}_{Pair} \wedge \mathscr{C}_{\text{tgg}}, SCT)$ for $l_s(\text{LTS}(\text{gts}_s, S))$ and $l_t(\text{LTS}(\text{gts}_t, T))$, respectively. This holds if $S_1 \Rightarrow_{\rho_s} S_2$ implies $T_1 \Rightarrow_{\rho_t} T_2$ with $l_s(\rho_s) = l_t(\rho_t)$ and $(S_2, T_2) \in \text{Bis}(\mathscr{C}_{RT} \wedge \mathscr{C}_{Pair} \wedge \mathscr{C}_{\text{tgg}}, SCT)$ for ρ_s in gts_s and ρ_t in gts_t. We first prove that $T_1 \Rightarrow_{\rho_t} T_2$ if $S_1 \Rightarrow_{\rho_s} S_2$ with $l_s(\rho_s) = l_t(\rho_t)$. If we have $S_1 \Rightarrow_{\rho_s} S_2$, then we also have $S_1 C T_1 \Rightarrow_{\rho_s} S_2 C T_1$. Because $S_1 C T_1 \models \mathscr{C}_{Pair}$, applicability of ρ_s to $S_1 C T_1$ implies applicability of ρ_t to $S_1 C T_1$ such that $S_1 C T_1 \Rightarrow_{\rho_t} S_1 C T_2$ with $l_s(\rho_s) = l_t(\rho_t)$. This means, in particular, that $T_1 \Rightarrow_{\rho_t} T_2$. We still need to prove that $(S_2, T_2) \in \text{Bis}(\mathscr{C}_{RT} \wedge \mathscr{C}_{Pair} \wedge \mathscr{C}_{\text{tgg}}, SCT)$. This is because, as ρ_s and ρ_t consist of disjoint types, they can only be applied in a parallel independent way to $S_1 C T_1$. Due to the Parallelism Theorem [21], then it follows that $S_1 C T_1 \Rightarrow_{\rho_s + \rho_t} S_2 C T_2$ with $\rho_s + \rho_t \in \mathscr{P}(l_s, l_t)$. Because of condition 3 of the Theorem and the fact that $S_1 C T_1 \models \mathscr{C}_{RT} \wedge \mathscr{C}_{Pair}$, then it follows that $S_2 C T_2 \models \mathscr{C}_{RT} \wedge \mathscr{C}_{Pair}$. As gts_s and gts_t preserve static types, $\mathscr{C}_{\text{tgg}}$ typed over $S_{TT} C_{TT} T_{TT}$ is by construction an inductive invariant for $\mathscr{P}(l_s, l_t)$ implying $S_2 C T_2 \models \mathscr{C}_{\text{tgg}}$. Thus, we have $S_2 C T_2 \models \mathscr{C}_{RT} \wedge \mathscr{C}_{Pair} \wedge \mathscr{C}_{\text{tgg}}$ and according to Def. 9, this means that $(S_2, T_2) \in \text{Bis}(\mathscr{C}_{RT} \wedge \mathscr{C}_{Pair} \wedge \mathscr{C}_{\text{tgg}}, SCT)$. Condition 2 of Def. 8 follows analogously to condition 1 as the roles of S and T are symmetric.

(2) $(S, T) \in \text{Bis}(\mathscr{C}_{RT} \wedge \mathscr{C}_{Pair} \wedge \mathscr{C}_{\text{tgg}}, SCT)$: Each triple graph SCT in $\mathscr{L}(\text{tgg}, \mathscr{C}_{\text{tgg}})$ fulfills $\mathscr{C}_{\text{tgg}}$ by construction. We further prove by induction over the number of TGG rule applications that each triple graph SCT in $\mathscr{L}(\text{tgg}, \mathscr{C}_{\text{tgg}})$ fulfills also $\mathscr{C}_{RT} \wedge \mathscr{C}_{Pair}$ such that according to Def. 9 $(S, T) \in \text{Bis}(\mathscr{C}_{RT} \wedge \mathscr{C}_{Pair} \wedge \mathscr{C}_{\text{tgg}}, SCT)$. The *base clause* for the axiom $S_A C_A T_A \models \mathscr{C}_{RT} \wedge \mathscr{C}_{Pair}$ follows directly from condition 1 of the Theorem. Condition 2 of the Theorem then provides the *induction step* that for any TGG rule application $S_n C_n T_n \Rightarrow_{\mathscr{R}} S_{n+1} C_{n+1} T_{n+1}$ such that $S_{n+1} C_{n+1} T_{n+1}$, it holds that $S_{n+1} C_{n+1} T_{n+1} \models \mathscr{C}_{RT} \wedge \mathscr{C}_{Pair}$ assuming the *induction hypothesis* that $S_n C_n T_n \models \mathscr{C}_{RT} \wedge \mathscr{C}_{Pair}$. □

In [12] we showed for TGGs (showing this for TGGs with constraints is ongoing work) that they are conform with our TGG implementation. Thus we can guarantee that forward and backward transformation implementations are indeed behavior preserving.

Besides bisimulation also preorders may be employed to establish behavioral consistency. For *simulation* [1], we simply have to weaken the *pairwise constraint* to obtain a similar result as Thm. 3 into: $\mathscr{C}'_{Pair} = \wedge_{(\rho_s, \rho_t) \in Pair}(\text{App}(\rho_t) \Rightarrow \text{App}(\rho_s))$. In order to support e.g. weak simulation or source rule applications being equivalent to several target rule applications, the labeling function can be adjusted accordingly.

5 Automation of Behavior Preservation Verification

We can perform inductive invariant checking automatically [10] for the constraints $\mathscr{C}_{RT}$ and $\mathscr{C}_{Pair}$, if they can be written as a conjunction of constraints of the basic form $\neg \exists P$ or more complex form $\forall (P, \exists n)$ with $n : P \to N$, or equivalently $\neg \exists (P, \neg \exists n)$, and if rules are restricted to the form $\rho = (\langle L \hookleftarrow I \hookrightarrow R \rangle, \wedge_{i \in I} \neg \exists n_i)$ with $n_i : L \to N_i$ a negative application condition. While this restriction is usually fulfilled by the runtime constraint $\mathscr{C}_{RT}$, the rules $\mathscr{R}_s$ and $\mathscr{R}_t$ for the source and target semantics, resp., and the TGG rules $\mathscr{R}$, the pairwise constraint $\mathscr{C}_{Pair}$ will usually not fulfill it. Only if the rules $\mathscr{R}_s$ and $\mathscr{R}_t$ have no NACs and a trivially true Deletable condition, the resulting $\mathscr{C}_{Pair}$ can be formulated as a conjunction of constraints of the form $\forall (P, \exists n)$. This is because $(\text{App}(\rho_s) \Rightarrow \text{App}(\rho_t)) \wedge (\text{App}(\rho_t) \Rightarrow \text{App}(\rho_s))$ can be written as $(\neg (\text{App}(\rho_s) \wedge \neg \text{App}(\rho_t))) \wedge (\neg (\text{App}(\rho_t) \wedge \neg \text{App}(\rho_s)))$ and $\text{App}(\rho_s) \wedge \neg \text{App}(\rho_t)$ is equivalent in the restricted case to $\exists (S_L, \neg \exists (S_L \hookrightarrow S_L + T_L))$ with S_L and T_L the LHS of ρ_s and ρ_t, resp.

The presented Example 7 fulfills the required restrictions with the exception of the *initE* and *initS* rules which have a NAC each. However, we can slightly relax the conditions of Def. 6 such that not only runtime, but also static edges can be deleted. This does not break Lemma 1, since static nodes and correspondences are preserved anyway. We can then emulate the NACs by instead testing that the first resp. init loop are present and delete them afterwards. In addition, we then have to check that $\mathscr{C}_{tgg}$ is an inductive invariant for $(\mathscr{P}(l_s, l_t), S_{RT} C_{TT} T_{RT})$ resp. exclude that erasing the init edge can invalidate the constraint $\mathscr{C}_{tgg}$. Condition 1 of Theorem 3 can be checked by any GTS tool that is able to check constraints of the form $\mathscr{C}_{RT} \wedge \mathscr{C}_{Pair}$ for a given graph. For the slightly adjusted but semantically equivalent semantic rules, we can use our invariant checker to show that behavior preservation with an interactively strengthened[3] bisimulation constraint in the sense of Problem Statement 2 holds.

Increasing the expressiveness of the constraints for which invariant checking can be performed, would allow improved automatic verification of behavior preservation, motivating the further investigation of invariant checking. In [22], it is described how the invariant checking (or constraint preservation) problem can be reduced to the implication problem. As proven in [14], in the case of graphs, nested conditions are expressively equivalent to first order graph formulas. Consequently, the implication problem for ACs is undecidable, in general. However, in [19,23], techniques are presented to tackle this problem in practice. Also [24] is concerned with invariant checking by studying Myhill-Nerode quasi orders. Finally, it is possible to translate the invariant checking problem to the input for a suitable constraint solver, see for example [25,26].

The check enforced by $\mathscr{C}_{Pair}$ is sufficient but not necessary. In particular, when alternative rule applications are possible, e.g., when the models relate to a non-deterministic induced LTS, the check may deliver false negatives, since source and target steps with equivalent labels may lead to a non-bisimilar combined state, where at the same time

---

[3] This becomes necessary because the invariant checker is otherwise lacking knowledge to rule out false negatives, i.e. situations that would actually never occur in the problem context, when verifying condition 2 and 3 in Theorem 3. The strengthening of the bisimulation constraint consists of additional invariants satisfied by the axiom and guaranteed by the TGG and parallel rules such that they can be assumed as extra knowledge during verification. It is part of ongoing work to support as much as possible the automation of this step.

equivalently labeled steps may lead to a bisimilar combined state. A more fine grain labeling of the LTS based on the rule and the match that allows to better distinguish the different rule applications can help here. More advanced GTS concepts such as attributes would be required, but this is beyond the scope of this paper.

6   Conclusion

We presented a first verification scheme promising that behavior preservation verification for model transformations at the transformation level can be automated. For model transformations specified by TGGs and semantic definitions for input and output models given by GTS rules, we can reduce the behavior preservation problem to an invariant checking problem for GTSs derived from the TGG and semantics rules, and constraints encoding the bisimilarity and the applicability of equivalent steps in the source and target models. Furthermore, we described which degree of automation can be achieved today using the existing verification technique [10].

Acknowledgement. We thank Basil Becker and Johannes Dyck for their continuous development work on the invariant checker [10].

References

1. Milner, R.: Communication and Concurrency. Prentice Hall International (UK) Ltd., Hertfordshire (1995)
2. Varró, D., Pataricza, A.: Automated Formal Verification of Model Transformations. In: Jürjens, J., Rumpe, B., France, R., Fernandez, E.B. (eds.) CSDUML 2003: Critical Systems Development in UML; Proceedings of the UML 2003 Workshop. Number TUM-I0323 in Technical Report, Technische Universitat Munchen, pp. 63–78 (September 2003)
3. Engels, G., Kleppe, A., Rensink, A., Semenyak, M., Soltenborn, C., Wehrheim, H.: From UML Activities to TAAL - Towards Behaviour-Preserving Model Transformations. In: Schieferdecker, I., Hartman, A. (eds.) ECMDA-FA 2008. LNCS, vol. 5095, pp. 94–109. Springer, Heidelberg (2008)
4. Narayanan, A., Karsai, G.: Verifying Model Transformations by Structural Correspondence. Electronic Communications of the EASST: Graph Transformation and Visual Modeling Techniques 2008 10 (2008)
5. Giese, H., Glesner, S., Leitner, J., Schäfer, W., Wagner, R.: Towards Verified Model Transformations. In: Hearnden, D., Süß, J., Baudry, B., Rapin, N. (eds.) Proc. of the 3rd International Workshop on Model Development, Validation and Verification (MoDeV²a). Genova, Italy, Le Commissariat à l'Energie Atomique - CEA, pp. 78–93 (October 2006)
6. Hülsbusch, M., König, B., Rensink, A., Semenyak, M., Soltenborn, C., Wehrheim, H.: Showing Full Semantics Preservation in Model Transformation - A Comparison of Techniques. In: Méry, D., Merz, S. (eds.) IFM 2010. LNCS, vol. 6396, pp. 183–198. Springer, Heidelberg (2010)
7. Rangel, G., Lambers, L., König, B., Ehrig, H., Baldan, P.: Behavior Preservation in Model Refactoring using DPO Transformations with Borrowed Contexts. In: Ehrig, H., Heckel, R., Rozenberg, G., Taentzer, G. (eds.) ICGT 2008. LNCS, vol. 5214, pp. 242–256. Springer, Heidelberg (2008)

8. Bisztray, D., Heckel, R., Ehrig, H.: Compositional Verification of Architectural Refactorings. In: de Lemos, R., Fabre, J.-C., Gacek, C., Gadducci, F., ter Beek, M. (eds.) Architecting Dependable Systems VI. LNCS, vol. 5835, pp. 308–333. Springer, Heidelberg (2009)

9. Schürr, A.: Specification of Graph Translators with Triple Graph Grammars. In: Mayr, E.W., Schmidt, G., Tinhofer, G. (eds.) WG 1994. LNCS, vol. 903, pp. 151–163. Springer, Heidelberg (1995)

10. Becker, B., Beyer, D., Giese, H., Klein, F., Schilling, D.: Symbolic Invariant Verification for Systems with Dynamic Structural Adaptation. In: Proc. of the 28th International Conference on Software Engineering (ICSE), Shanghai, China. ACM Press (2006)

11. Becker, B., Lambers, L., Dyck, J., Birth, S., Giese, H.: Iterative Development of Consistency-Preserving Rule-Based Refactorings. In: Cabot, J., Visser, E. (eds.) ICMT 2011. LNCS, vol. 6707, pp. 123–137. Springer, Heidelberg (2011)

12. Giese, H., Hildebrandt, S., Lambers, L.: Bridging the Gap Between Formal Semantics and Implementation of Triple Graph Grammars - Ensuring Conformance of Relational Model Transformation Specifications and Implementations. Software and Systems Modeling (2012)

13. Golas, U., Lambers, L., Ehrig, H., Giese, H.: Toward Bridging the Gap between Formal Foundations and Current Practice for Triple Graph Grammars. In: Ehrig, H., Engels, G., Kreowski, H.-J., Rozenberg, G. (eds.) ICGT 2012. LNCS, vol. 7562, pp. 141–155. Springer, Heidelberg (2012)

14. Habel, A., Pennemann, K.H.: Correctness of high-level transformation systems relative to nested conditions. Mathematical Structures in Computer Science 19, 1–52 (2009)

15. Ehrig, H., Golas, U., Habel, A., Lambers, L., Orejas, F.: M-adhesive transformation systems with nested application conditions, part 1: Parallelism, concurrency and amalgamation. Mathematical Structures in Computer Science (to appear, 2012)

16. Habel, A., Heckel, R., Taentzer, G.: Graph Grammars with Negative Application Conditions. Fundamenta Informaticae 26(3/4), 287–313 (1996)

17. Ehrig, H., Ehrig, K., Prange, U., Taentzer, G.: Fundamentals of Algebraic Graph Transformation. Springer (2006)

18. Ermel, C., Gall, J., Lambers, L., Taentzer, G.: Modeling with Plausibility Checking: Inspecting Favorable and Critical Signs for Consistency between Control Flow and Functional Behavior. In: Giannakopoulou, D., Orejas, F. (eds.) FASE 2011. LNCS, vol. 6603, pp. 156–170. Springer, Heidelberg (2011)

19. Pennemann, K.H.: Development of Correct Graph Transformation Systems. PhD thesis, Department of Computing Science, University of Oldenburg, Oldenburg (2009)

20. Charpentier, M.: Composing Invariants. In: Araki, K., Gnesi, S., Mandrioli, D. (eds.) FME 2003. LNCS, vol. 2805, pp. 401–421. Springer, Heidelberg (2003)

21. Ehrig, H., Habel, A., Lambers, L.: Parallelism and Concurrency Theorems for Rules with Nested Application Conditions. In: Festschrift dedicated to Hans-Jorg Kreowski at the Occasion of his 60th Birthday. EC-EASST, vol. 26 (2010)

22. Lambers, L.: Certifying Rule-Based Models using Graph Transformation. PhD thesis, Technische Universität Berlin (2010)

23. Orejas, F., Ehrig, H., Prange, U.: A Logic of Graph Constraints. In: Fiadeiro, J.L., Inverardi, P. (eds.) FASE 2008. LNCS, vol. 4961, pp. 179–198. Springer, Heidelberg (2008)

24. Blume, C., Bruggink, H.S., König, B.: Recognizable Graph Languages for Checking Invariants. In: Proc. of GT-VMT 2010 (Workshop on Graph Transformation and Visual Modeling Techniques). Electronic Communications of the EASST, vol. 29 (2010)

25. Cabot, J., Clarisó, R., Guerra, E., Lara, J.: Verification and validation of declarative model-to-model transformations through invariants. J. Syst. Softw. 83(2), 283–302 (2010)

26. Frias, M.F., Galeotti, J.P., Pombo, C.L., Aguirre, N.: DynAlloy: Upgrading Alloy with actions. In: Proc. of Int. Conf. of Software Engineering, pp. 442–451. ACM (2005)

Efficient Symbolic Implementation of Graph Automata with Applications to Invariant Checking*

Christoph Blume, H. J. Sander Bruggink, Dominik Engelke, and Barbara König

Universität Duisburg-Essen, Germany
{christoph.blume,sander.bruggink,barbara_koenig}@uni-due.de

Abstract. We introduce graph automata as a more automata-theoretic view on (bounded) automaton functors and we present how automaton-based techniques can be used for invariant checking in graph transformation systems. Since earlier related work on graph automata suffered from the explosion of the size of the automata and the need of approximations due to the non-determinism of the automata, we here employ symbolic BDD-based techniques and recent antichain algorithms for language inclusion to overcome these issues. We have implemented techniques for generating, manipulating and analyzing graph automata and perform an experimental evaluation.

1 Introduction

Regular languages and (word) automata are the cornerstone of several verification techniques (for example [9]). Similarly, tree automata [11] have been used in regular model-checking [8]. Challenges in the analysis of dynamic graph-like structures, such as pointer structures on the heap, object graphs or evolving networks, naturally lead to the question whether graph languages and graph automata can serve the same purpose. There is indeed an established theory of recognizable graph languages by Courcelle [12], although substantial work needs to be done before this theory can be put to good use in complex verification scenarios.

In order to close this gap, we here give a very concrete variant of graph automata accepting a subclass of the recognizable graph languages à la Courcelle. Furthermore we reformulate our own earlier work on invariant checking [7] in this setting. However, our main motivation is to fight state explosion, which is a major problem when working with graph automata. Graph automata cannot input all graphs but only graphs up to a certain width (in our case: path width), which is a restriction on the interface size of the alphabet of "building blocks" of graphs. However the size of the automaton typically grows exponentially when this bound is raised. This is a major problem that forced earlier work such as [16], based on the algorithms in [14], [20] and [13] to restrict to very small interface

---

* This work is supported by the DFG-project GaReV.

H. Ehrig et al.(Eds.): ICGT 2012, LNCS 7562, pp. 264–278, 2012.

sizes. Recent work abstains from a representation of the automaton, but pursues a game-based approach, obtaining much better runtime results [18].

However, all these approaches have a different focus than ours, in that they concentrate on solving the membership problem: given a description of the language (often specified by a formula in monadic second-order graph logic) and a graph, check whether the graph is in the language. Courcelle's theorem shows that for a fixed formula this can be done in linear time for graphs of bounded treewidth (or path width). However the large constants involved lead to severe efficiency problems when the automata are represented directly.

Here we are less interested in solving the membership problem: with the applications that we have in mind we are interested in designing an automaton tool suite that treats automata as representatives of languages that can be suitably manipulated and analyzed. However, we have to face the same problem as the other approaches: the sheer size of the automata involved. Hence we are using symbolic BDD-based techniques to represent the set of states and the transition function, which enable us to generate non-deterministic automata for large interface sizes. To avoid determinization, our earlier work [7] used an approximation, but this approach will not be used in this paper. In order to perform useful analyses, needed for instance for invariant checking as mentioned above, we have to solve the language inclusion problem, which is PSPACE-complete. Our new approach uses recent methods based on antichains as introduced in [21,1]. We have implemented our techniques and we perform an extensive experimental evaluation, which shows a clear improvement over earlier work.

The structure of the paper is as follows. In Sect. 2 we will introduce preliminary definitions such as cospans, hypergraphs and binary decision diagrams. In Sect. 3 we will take a look at graph automata and the connection between them and automaton functors of bounded size. Then in Sect. 4 we will show how techniques for solving the language inclusion problem can be used to perform invariant checking and in Sect. 5 we will present implementation details about the *Raven* tool suite which implements language inclusion algorithms for invariant checking. Furthermore we will present results about our case studies. Finally, we will conclude in Sect. 6.

2   Preliminaries

By $\mathbb{N}_k$ we denote the set $\{1, \ldots, k\}$. The set of finite sequences over a set A is denoted by A^*. If $f: A \to B$ is a function from A to B, we will implicitly extend it to subsets and sequences; for $A' \subseteq A$ and $\boldsymbol{a} = a_1 \ldots a_n \in A^*$: $f(A') = \{f(a) \mid a \in A'\}$ and $f(\boldsymbol{a}) = f(a_1) \ldots f(a_n)$. By $|\boldsymbol{a}|$ we denote the length of $\boldsymbol{a} \in A^*$. By $\wp(A)$ we denote the powerset of A.

Categories and Cospans. We presuppose a basic knowledge of category theory. For an arrow f from A to B we write $f: A \to B$ and define $dom(f) = A$ and $cod(f) = B$. For arrows $f: A \to B$ and $g: B \to C$, the composition of f and g is denoted $(f \,;\, g): A \to C$. The category **Rel** has sets as objects and relations as arrows.

Let $\mathbf{C}$ be a category in which all pushouts exist. A cospan in $\mathbf{C}$ is a pair $c = (c_L, c_R)$ of $\mathbf{C}$-arrows $J -c_L\!\rightarrow G \leftarrow c_R- K$. Two cospans c, d are isomorphic if their middle objects are isomorphic (such that the isomorphism commutes with the component morphisms of the cospan). In this case we write $c \simeq d$. Isomorphism classes of cospans are the arrows of so-called cospan categories. That is, for a category $\mathbf{C}$ with pushouts, the category $Cospan(\mathbf{C})$ has the same objects as $\mathbf{C}$. The isomorphism class of a cospan $c\colon J -c_L\!\rightarrow G \leftarrow c_R- K$ in $\mathbf{C}$ is an arrow from J to K in $Cospan(\mathbf{C})$ and will be denoted by $c\colon J \dashrightarrow K$. Composition of two cospans $(c_L, c_R), (d_L, d_R)$ is computed by taking the pushout of the arrows c_R and d_L. A cospan is called *output linear* if the right leg of the cospan is a monomorphism.

Graphs and Output Linear Cospans. Let Λ be a set of labels and let $ar\colon \Lambda \to \mathbb{N}$ be the function that maps each alphabet symbol to its arity.

A *hypergraph* over a set of labels Λ (in the following also simply called *graph*) is a structure $G = (V, E, att, lab)$, where V is a finite set of nodes, E is a finite set of edges, $att\colon E \to V^*$ maps each edge to a finite sequence of nodes attached to it, such that $|att(e)| = ar(lab(e))$, and $lab\colon E \to \Lambda$ assigns a label to each edge. A *discrete graph* is a graph without edges; the discrete graph with node set $\mathbb{N}_k$ is denoted by D_k. We denote the *empty graph* by $\emptyset$ instead of D_0.

A graph morphism is a structure preserving map between two graphs. The category of graphs and graph morphisms is denoted by **Graph**. Recall, that the *monomorphisms* (monos) and *epimorphisms* (epis) of the category **Graph** are the injective and surjective graph morphisms, respectively.

A cospan $J -c_L\!\rightarrow G \leftarrow c_R- K$ (over a set of labels Λ) in **Graph** can be viewed as a graph (G over Λ) with two interfaces (J and K), called the *inner interface* and *outer interface* respectively. Informally said, only elements of G which are in the image of one of the interfaces can be "touched". By $[G]$ we denote the trivial cospan $\emptyset \to G \leftarrow \emptyset$, the graph G with two empty interfaces.

The *category of output linear cospans* $\mathbf{OLCG}_n$ has discrete graphs (of size at most n) as objects and output linear cospans of graphs with discrete interfaces (of size at most n) as arrows. Note that the middle objects of the cospans of the category $\mathbf{OLCG}_n$ can still be arbitrary graphs. The idea for using this category is that we want to be able to fuse nodes via cospan composition, but we want to avoid that nodes of the middle graph are shared in the outer interface.

Binary Decision Diagrams. A *binary decision diagram* (BDD) is a rooted, directed, acyclic graph which serves as a representation of a boolean function. Every BDD has two distinguished terminal nodes, called *one* and *zero*, representing the logical constants *true* and *false*. The inner nodes are labeled by the variables of the boolean formula represented by the BDD, such that on each path from the root to the terminal nodes, every variable of the boolean formula occurs at most once. Each inner node has exactly two distinguished outgoing edges, called *high* and *low*, which represent the case that the variable of the inner node has been set to *true* or *false* respectively. A boolean formula $f(x_1, \ldots, x_n)$ can be evaluated by following the path from the root node to a terminal node.

We will use a special class of BDDs, called *reduced and ordered BDDs* (ROBDDs), in which the order of the variables occuring in the BDD is fixed and redundancy is avoided, i.e. if both child nodes of a parent node are identical, the parent node is dropped from the BDD and isomorphic parts of the BDD are merged. The great advantage of ROBDDs is that each boolean formula can be uniquely represented by an ROBDD (if the order of the variables is fixed). For a detailed introduction of these BDDs see [2].

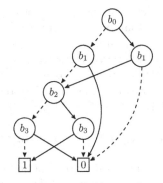

Fig. 1. BDD for the set $\{0000, 0011, 1100, 1111\}$

As an example, we consider the following set of 4-bit vectors: $\{0000, 0011, 1100, 1111\}$. We assume that the bits of the bit vectors are numbered from b_0 to b_3 with b_0 the least significant bit. The ROBDD representing this set of bit vectors is shown in Fig. 1. Variables are depicted as rounded nodes, terminals as rectangular nodes. The *high* and *low* edges are depicted as solid and dashed lines respectively.

3   Bounded Graph Automata

Recognizable graph languages are a generalization of regular (word) languages to graph languages which were first investigated by Courcelle [3,12]. In this section we define bounded graph automata, which accept a subclass of the recognizable graph languages due to the bound. Similar to word languages we define graph languages based on an alphabet. Each letter of the alphabet represents an output linear cospan such that the concatenation of these letters (or cospans respectively) yields a graph (seen as a cospan with empty interfaces).

Let $n \in \mathbb{N}$ and a *doubly-ranked alphabet* $\Sigma = (\Sigma_{i,j})_{i,j \leq n}$ be given. The set of *(doubly-ranked) sequences* $S_\Sigma = (S_{i,j})_{i,j \leq n}$ over a doubly-ranked alphabet Σ is defined inductively:

- for every $i \leq n$ the *empty sequence* ε_i is in $S_{i,i}$
- for every $i, j \leq n$ every letter $\sigma \in \Sigma_{i,j}$ is in $S_{i,j}$
- for every $i, j, k \leq n$ and for every $\boldsymbol{\sigma} \in S_{i,j}$, $\boldsymbol{\sigma}' \in S_{j,k}$ the *concatenation* $\boldsymbol{\sigma} ; \boldsymbol{\sigma}'$ of $\boldsymbol{\sigma}$ and $\boldsymbol{\sigma}'$ is in $S_{i,k}$

The *width* of a sequence is the maximum rank of its letters. We will also write S instead of S_Σ if the underlying alphabet is clear from the context.

Let Λ be a set of labels. By $\Gamma(\Lambda)$ we denote the doubly-ranked alphabet containing the following letters:

Letter:	$connect^i_A$	$fuse^i$	$perm^i$	res^i	$trans^i$	$vertex^i$
Type:	(i, i)	$(i, i-1)$	(i, i)	$(i, i-1)$	(i, i)	$(i, i+1)$
Constraint:	$A \in \Lambda$, $ar(A) \leq i$	$i \geq 2$	$i \geq 3$	$i \geq 1$	$i \geq 2$	–

The meaning of these letters is given by the evaluation function defined below. Note that *res* is a restriction of the interface, *perm* permutes the interface and *trans* transposes the first two interface nodes. Due to the fact that for two elements permutation and transposition are identical operations the constraint of the letter $perm^n$ is $n \geq 3$.

Now we define an evaluation function which maps each letter of the alphabet $\Gamma(\Lambda)$ to an output linear cospan.

Definition 1 (Evaluation function). *Let Λ be a set of labels.*

(i) *The* evaluation function $\eta \colon \Gamma(\Lambda) \to \mathbf{OLCG}_i$ *maps each letter to an output linear cospan as shown below:*

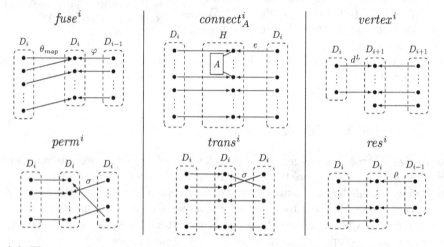

(ii) *The* extended evaluation function $\hat{\eta} \colon S_{\Gamma(\Lambda)} \to \mathbf{OLCG}_i$ *is defined as*

$$\hat{\eta}(\sigma) = \begin{cases} D_j \to D_j \leftarrow D_j, & \text{if } \sigma = \varepsilon_j \in S_{j,j} \\ \eta(\sigma), & \text{if } \sigma = \sigma \in \Gamma(\Lambda) \\ \hat{\eta}(\sigma_1) \,;\, \hat{\eta}(\sigma_2), & \text{if } \sigma = \sigma_1 \,;\, \sigma_2 \end{cases}$$

We call the cospans which correspond to the six letters above *atomic cospans*.

Let c be an output-linear cospan. The *width* of c is the minimal width of all σ such that $\hat{\eta}(\sigma) = c$.

The following lemma shows that every graph (seen as an output linear cospan with two empty interfaces) can be constructed by the alphabet $\Gamma(\Lambda)$. Hence, we will restrict ourselves to this alphabet in the following:

Lemma 1 ([7]). *Let c be an output linear cospan over Λ. Then it holds that:*

1. c *can be constructed by a sequence $c_1, \ldots, c_m$ of atomic cospans, i.e. c can be obtained as the decomposition $c = c_1 \,;\, \ldots \,;\, c_m$.*
2. *There exists a sequence $\sigma \in S_{\Gamma(\Lambda)}$ such that $\hat{\eta}(\sigma) = c$.*

In the following we are considering graphs with an arbitrary inner interface and an empty outer interface. We need the arbitrary inner interface in order to state Theorem 2 below. We could, without major problems, also parametrize over the outer interface, but this is not necessary for the theory.

Definition 2 (Bounded graph automaton). *Let $n \in \mathbb{N}$ and $k \leq n$ be given. An n-bounded graph automaton $\mathcal{A} = (Q, \Sigma, \delta, I, F)$ from k, where $\Sigma = \Gamma(\Lambda)$, consists of*

- *$Q = (Q_i)_{i \leq n}$ the family of finite state sets,*
- *$\Sigma = (\Sigma_{i,j})_{i,j \leq n}$ the doubly-ranked input alphabet,*
- *$\delta = (\delta_{i,j})_{i,j \leq n}$ is a family of transition functions, where $\delta_{i,j} \colon Q_i \times \Sigma_{i,j} \to \wp(Q_j)$*
- *$I \subseteq Q_k$ the set of initial states and*
- *$F \subseteq Q_0$ the set of final states*

such that the following condition holds for all $q \in Q$ and $\sigma_1, \sigma_2 \in S_{i,j}$:

$$\text{if } \hat{\eta}(\sigma_1) \simeq \hat{\eta}(\sigma_2) \text{ then } \hat{\delta}_{i,j}(\{q\}, \sigma_1) = \hat{\delta}_{i,j}(\{q\}, \sigma_2), \qquad (\star)$$

where $\hat{\delta}_{i,j} \colon \wp(Q_i) \times S_{i,j} \to \wp(Q_j)$ is defined as follows:

$$\hat{\delta}_{i,j}(R, \sigma) := \begin{cases} R & \text{if } \sigma = \epsilon_i \in \Sigma_{i,i} \text{ and } i = j \\ \delta(R, \sigma) & \text{if } \sigma = \sigma \in \Sigma_{i,j} \\ \hat{\delta}_{k,j}(\hat{\delta}_{i,k}(R, \sigma_1), \sigma_2) & \text{if } \sigma = (\sigma_1 \, ; \, \sigma_2), \sigma_1 \in S_{i,k}, \sigma_2 \in S_{k,j} \end{cases}.$$

A sequence $\sigma \in S_{k,0}$ over Σ is accepted by $\mathcal{A}$ if and only if $\hat{\delta}_{k,0}(I, \sigma) \cap F \neq \emptyset$.

The idea behind a graph automaton is to get a decomposition of an input graph and to process it "piece by piece". The condition $(\star)$ guarantees that the graph automaton accepts an input graph independently of the decomposition of the graph. Showing that this condition holds for some prospective graph automaton is not trivial in general. A solution would be to automatically translate formulas of monadic second-order logic to correct graph automata.

Definition 3 (Accepted language). *Let an n-bounded graph automaton $\mathcal{A}$ from k be given. The language accepted by $\mathcal{A}$, denoted by $\mathcal{L}(\mathcal{A})$, contains exactly the sequences accepted by $\mathcal{A}$. The cospan language accepted by $\mathcal{A}$ is*

$$\mathcal{G}(\mathcal{A}) = \{c \mid \hat{\eta}(\sigma) = c \text{ for some } \sigma \in \mathcal{L}(\mathcal{A})\}.$$

The cospan language of a bounded graph automaton contains cospans. When we want to accept graphs, we can interpret the cospan $[G]$ as the graph G.

Since a graph automaton is bounded, it is a kind of non-deterministic finite automaton (NFA). Therefore, we can apply standard algorithms from formal language theory, such as the subset construction and constructing the cross product of two automata. It can be shown that these constructions preserve the condition $(\star)$ of graph automata. Thus, the languages accepted by n-bounded graph automata are closed under boolean operations, and many important decision problems (such as the membership, emptiness and language inclusion problems) are decidable. Note that the language inclusion algorithm for NFA is PSPACE-complete, and thus no efficient algorithms for the problem exist yet.

Example 1. First we consider the language L_U of all graphs which contain a fixed subgraph U. The bounded graph automaton $\mathcal{A}_U$ accepting this language works as follows: Every state in each of the state sets Q_i contains two pieces of information. The first piece of information says which parts of the subgraph have already been recognized. The second piece of information is a function which maps every outer node to a node which has already been recognized or to some "bottom element" to indicate that the interface node is not mapped to a node of the wanted subgraph U. The transition function "updates" this information according to the letter which is currently processed. Since the input graph might contain several parts which are isomorphic to the wanted subgraph U, the bounded graph automaton is highly non-deterministic. More details about the construction of this graph automaton can be found in [5].

Example 2. Now we consider the language $C_{(k)}$ of all k-colorable graphs (for some $k \in \mathbb{N}$). A k-coloring of a graph G is a function $f: V_G \to \mathbb{N}_k$ such that for all edges $e \in E_G$ and for all nodes $v_1, v_2 \in att_G(e)$ it holds that $f(v_1) \neq f(v_2)$ if $v_1 \neq v_2$. The question whether a graph is k-colorable is essential in many applications, for example in scheduling. The idea of the graph automaton $\mathcal{A}_{(k)}$ accepting all k-colorable graphs (as defined in [10]) is as follows: Every state is a valid k-coloring of D_i, that is $Q_i = \{f: V_{D_i} \to \mathbb{N}_k \mid f$ is a valid k-coloring of $D_i\}$. The transition function $\delta_{i,j}$ maps a coloring $f \in Q_i$ and a letter $\sigma \in \Sigma$ to a coloring $f' \in Q_j$ if and only if the coloring of the inner nodes of $\eta(\sigma)$ according to f and the coloring of the outer nodes of $\eta(\sigma)$ according to f' leads to a valid coloring of $\eta(\sigma)$. More details on graph automata for coloring can be found in Sect. 5.1.

In the rest of the section we compare bounded graph automata to automaton functors, which were introduced in [10], in particular to automaton functors for the category **OLCG**$_i$ (bounded automaton functors). We show that they accept the same class of language. The main difference between the two is that bounded automaton functors are defined on *all* cospans of bounded size (of which there are infinitely many), while graph automata are only defined for the letters of the input alphabet, which correspond to only the atomic cospans (of which there are finitely many).

Definition 4 (Bounded Automaton Functor). *Let* $n \in \mathbb{N}$. *An* n-bounded automaton functor from k *is a structure* $\mathcal{A} = (\mathcal{A}_0, I, F)$, *where*

- $\mathcal{A}_0: \mathbf{OLCG}_n \to \mathbf{Rel}$ *is a functor which maps every discrete graph* D_i *to a finite set* $\mathcal{A}_0(D_i)$ *(the state set of* D_i*) and every output linear cospan* $c: D_i \looparrowright D_j$ *to a relation* $\mathcal{A}_0(c) \subseteq \mathcal{A}_0(D_i) \times \mathcal{A}_0(D_j)$ *(the transition relation of c),*
- $I \subseteq \mathcal{A}_0(D_k)$ *is the set of initial* states *and*
- $F \subseteq \mathcal{A}_0(\emptyset)$ *is the set of final states.*

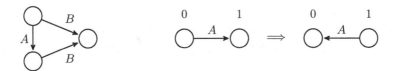

Fig. 2. Wanted subgraph D **Fig. 3.** Transformation rule ρ_A

For a discrete graph G or a output linear cospan c we will, in the following, usually write $\mathcal{A}(G)$ and $\mathcal{A}(c)$ instead of $\mathcal{A}_0(G)$ and $\mathcal{A}_0(c)$, respectively. A cospan $c\colon D_k \dashrightarrow \emptyset$ is accepted by $\mathcal{A}$, if $(q, q') \in \mathcal{A}(c)$ for some $q \in I$ and $q' \in F$.

Definition 5 (Accepted language). *Let $\mathcal{A}$ be an n-bounded automaton functor. The language accepted by $\mathcal{A}$, denoted by $\mathcal{G}(\mathcal{A})$, contains exactly the cospans accepted by $\mathcal{A}$.*

Theorem 1. *Let L be a language of cospans from D_k to $\emptyset$. Then L is the cospan language of an n-bounded graph automaton from k if and only if it is the language of an n-bounded automaton functor from k.*

4   Invariant Checking and Language Inclusion

One of the applications of our approach is to automatically check invariants of graph transformation systems (GTSs). The following definition of graph transformation is equivalent to the well-known double-pushout approach [19], where we have injective rule spans and not necessarily injective matches.

Definition 6 (Graph transformation).

(i) *Let $\ell\colon \emptyset \dashrightarrow D_i$ and $r\colon \emptyset \dashrightarrow D_i$ be two output linear cospans (called left-hand and right-hand side). The pair $\rho = (\ell, r)$ is called a (graph) transformation rule. A graph transformation system is a finite set of transformation rules.*

(ii) *Let $\rho = (\ell, r)$ be a transformation rule. The rule ρ is applicable to a graph G if and only if $[G] = \ell\,;\,c$ for some output linear cospan $D_i \dashrightarrow \emptyset$. In this case we write $G \Rightarrow_{\rho,c} H$, where H is the graph obtained from $[H] = r\,;\,c$.*

A language L is an invariant according to a graph transformation rule ρ if it holds for all graphs G and H with $G \Rightarrow_\rho H$ that $[G] \in L$ implies $[H] \in L$.

Example 3. As an example we take the graph D (which is depicted in Fig. 2) as wanted subgraph. The language L_D of all graphs containing D as a subgraph is an invariant for the rule ρ_A (shown in Fig. 3) which "switches" an A-labeled edge. Obviously, every graph which contains D as subgraph before the application of ρ_A does contain D also after the rule application.

Example 4. The next example we consider is the language $C_{(2)}$ of all 2-colorable graphs (see Ex. 2 for details about $C_{(2)}$). This language is an invariant for the transformation rule α_n depicted in Fig. 4 which adds two new nodes between two adjacent nodes on a path. That the language $C_{(2)}$ is an invariant for this rule is clear since every path with an even number of nodes is 2-colorable.

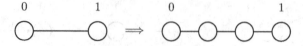

Fig. 4. Transformation rule α_n

For an output linear cospan $c\colon D_k \dashrightarrow D_m$ and a n-bounded graph automaton $\mathcal{A} = (Q, \Sigma, \delta, I, F)$ from k we obtain a new n-bounded graph automaton $\mathcal{A}[c] = (Q, \Sigma, \delta, I', F)$ from m with $I' = \hat{\delta}_{k,m}(I, \sigma)$, where σ is some word from S_Σ such that $\hat{\eta}(\sigma) = c$. (If the width of c is larger than n, such a σ does not exist, and we take $I' = \emptyset$, such that $\mathcal{L}(\mathcal{A}[c]) = \emptyset$.) The new automaton has as new initial states all states reachable from the original initial states by processing c. Note that I' is independent of the specific decomposition of c into a sequence σ.

The following theorem easily follows from the observation that $\sigma_\ell \,;\, \sigma_c \in \mathcal{L}(\mathcal{A})$ if and only if $\sigma_c \in \mathcal{L}(\mathcal{A}[\ell]))$, where σ_ℓ and σ_c are sequences such that $\hat{\eta}(\sigma_\ell) = \ell$ and $\hat{\eta}(\sigma_c) = c$.

Theorem 2 (Invariant checking). *Let $\mathcal{A}$ be an n-bounded graph automaton (from 0) accepting the cospan language L, and let $\rho = (\ell, r)$ be a transformation rule. The cospan language L is an invariant of ρ if and only if $\mathcal{L}(\mathcal{A}[\ell]) \subseteq \mathcal{L}(\mathcal{A}[r])$.*

5   Implementation and Results

We implemented a language inclusion algorithm and invariant checking in the Java-based tool *Raven*. In this section we examine some implementation details of the tool and present results of case studies.

5.1   Representation of Automata with BDDs

Graph automata are represented by means of BDDs. First, states of the automaton are represented by a bit string, and secondly the transition relations for the various atomic cospans (or letters respectively) are stored as a BDD which encodes a relation on these bit strings.

As an example, we look at the encoding for the automaton which accepts all k-colorable graphs (see Ex. 2). The state encoding has to take care of the following information: the interface size (of the outer interface of the graph seen so far) and the color of each node currently occurring in the outer interface.

A good ordering of the bits holding the information is essential to construct compact BDDs. We have experimented with different orderings, and found the following to be the best. Let n be the maximum interface size and k the number of colors. Furthermore, let $m = \lceil \log_2 n \rceil$ be the number of bits required to store the interface size, and $\ell = \lceil \log_2(k+1) \rceil$ the number of bits to store one color (we need an extra value to represent uncolored or unused nodes). A state is encoded by the bit sequence $b\,c_1 \ldots c_n = b_1 \ldots b_m (c_{1,1} \ldots c_{1,\ell}) \ldots (c_{n,1} \ldots c_{n,\ell})$, where $b = b_1 \ldots b_m$ encodes the current interface size as a binary number and $c_i = (c_{i,1} \ldots c_{i,\ell})$ (for $1 \leq i \leq n$) represents the color of the i-th interface node.

For each of the letters of $\Gamma(\Lambda)$ we define a propositional formula describing the transition relations – for all permitted interfaces – of the graph automaton. These formulas can then be easily transformed into BDDs which describe the transition functions. (As usual with BDD representations of relations, the bits of the domain and codomain states are interleaved.)

We present the formula $f_{connect_A^i}$ as an example. To distinguish between the bits for the current state and the bits for the successor state we indicate the successor state encoding by $b'c_1' \ldots c_n'$. The formula consists of four parts (where $p = i - ar(A) + 1$ is the index of the first node attached to the new edge):

$$f_1 := (ar(A) \leq i) \wedge (b = i) \wedge (b = b') \qquad f_3 := \bigwedge_{j=1}^{n} (c_j = c_j')$$

$$f_2 := \bigwedge_{j=i+1}^{n} (c_j = 0) \qquad\qquad f_4 := \bigwedge_{j=p}^{i} \bigwedge_{j'=p}^{i} (j \neq j') \to (c_j \neq c_{j'})$$

The subformula f_1 expresses that the arity of the added edge is less than or equal to the current interface and that the interface size of both the current state and the successor state is i. The subformula f_2 expresses that the nodes of the encoding which do not belong to the current interface, that is the last $n-i+1$ nodes in the encoding, have not been colored. Next, f_3 expresses that all nodes have the same color in the source and the target state. Finally, f_4 expresses that the nodes which are connected by the new edge have different colors. Now, we take $f_{connect_A} := f_1 \wedge f_2 \wedge f_3 \wedge f_4$, that is, a transition $q - connect_A^i \to q'$ is allowed if and only if the above four conditions hold.

Example 5. We consider the 3-colorability automaton with a maximum interface size of 5. The size of the state encoding is $3 + (2 \cdot 5) = 13$ bits. Consider the state q depicted in Fig. 5 (on the left): we have five nodes in the current interface, colored with color 1, 2, 3, 2 and 3, respectively. The bit string which encodes this state is given in Fig. 5 on the right.

Fig. 5. State q and its representation as bit string

Suppose that the graph automaton is currently in state q, and that the next letter is $connect_A^5$, where A is a label with arity 2. Since the last two nodes of the interface are colored differently, none of the nodes connected by the A-edge have the same color. Hence the transition $q - connect_A^5 \to q$ is in the transition relation. Suppose on the other hand that the graph automaton is in state q and the next operation is $connect_B^5$, where the arity of B is 3. Since the third and fifth node of q's interface have the same color, no state can be reached from q.

Apart from a graph automaton which accepts k-colorable graphs and one which accepts graphs with a specific subgraph (see Ex. 1), we also implemented graph automata for vertex cover and dominating set. A *vertex cover* of graph G is a set C of nodes of G such that each edge is incident to at least one node of C. A *dominating set* of a graph G is a set D of nodes of G such that each node of G is either in D or adjacent to a node in D. The states of automata checking if the input graph has a vertex cover of size k or if the input graph has a dominating set of size k respectively need to encode the following pieces of information:

- *Vertex cover:* the interface size of the outer interface of the graph seen so far, which nodes of the current interface are part of the vertex cover and the size of the vertex cover (where nodes in the vertex cover are counted when they are removed from the interface).
- *Dominating set:* the interface size of the outer interface of the graph seen so far, which nodes of the current interface are part of the dominating set, which nodes of the current interface are dominated by some node of the dominating set and the size of the dominating set (where nodes in the dominating set are counted when they are removed from the interface).

Note that we use BDDs in a different way than other tools. In our case, the alphabet is small and the state set is huge, and we use BDDs to encode a transition relation for each symbol. In other tools, such as MONA [17], the state set is relatively small and the alphabet is huge. Thus MONA uses BDDs not to encode the transition relation for each symbol, but to encode the possible transitions of each single state, that is for each state there is a BDD encoding all transitions for each alphabet symbol starting at that specific state.

5.2   Checking Language Inclusion

In [7] we presented a technique for checking invariants based on the Myhill-Nerode quasi-order. The main disadvantage of this approach is that the algorithm for computing the Myhill-Nerode quasi-order applies only to deterministic (graph) automata, whereas in general our graph automata are highly non-deterministic. Determinization is not an option because it would lead to an exponential blow-up of already huge automata. Therefore we had to settle for an approximation.

To overcome this problem, here we use the antichain-based algorithm from [21] to check for language inclusion, which can be used to check invariants via Theorem 2. In the worst case this approach can still need exponential time, but in practise one can often achieve very good runtimes.

An antichain is a set of elements which are uncomparable with respect to some ordering. What the elements look like and what ordering is used depends on the application; here we present an antichain-based algorithm to decide language inclusion. In this subsection, we forget typing information of the states and consider bounded automata as regular finite automata.

Let $\mathcal{A} = (Q_\mathcal{A}, \Sigma, \delta_\mathcal{A}, I_\mathcal{A}, F_\mathcal{A})$ and $\mathcal{B} = (Q_\mathcal{B}, \Sigma, \delta_\mathcal{B}, I_\mathcal{B}, F_\mathcal{B})$ be n-bounded graph automata. Let $\overline{F_\mathcal{B}} = Q_\mathcal{B} \setminus F_\mathcal{B}$, that is, the set of $\mathcal{B}$'s non-accepting states. We want to decide whether $\mathcal{L}(\mathcal{A}) \subseteq \mathcal{L}(\mathcal{B})$. In particular, we are trying to falsify that claim by finding a state $q \in I_\mathcal{A}$ and a set of states $S \subseteq I_\mathcal{B}$ such that $\hat{\delta}_\mathcal{A}(\{q\}, \sigma) \cap F_\mathcal{A} \neq \emptyset$ and $\hat{\delta}_\mathcal{B}(S, \sigma) \subseteq \overline{F_\mathcal{B}}$, for some word σ.

Let $\mathcal{U} = Q_\mathcal{A} \times \wp(Q_\mathcal{B})$. For $(q_1, S_1), (q_2, S_2) \in \mathcal{U}$, we define $(q_1, S_1) \leq (q_2, S_2)$ if $q_1 = q_2$ and $S_1 \subseteq S_2$. Now, an *antichain* (for language inclusion) is a set $K \subseteq \mathcal{U}$ such that for all $p_1, p_2 \in K$ with $p_1 \neq p_2$, it holds that neither $p_1 \leq p_2$ nor $p_2 \leq p_1$. A pair $p \in K$ is called maximal, if there is no $p' \in K$ such that $p \leq p'$; by $\lceil K \rceil$ we denote the set of maximal elements of K. Minimal elements and the set $\lfloor K \rfloor$ of minimal elements are defined symmetrically.

The algorithm searches through the automaton backwards. We define:

$$\mathsf{Pre}_{\mathcal{A},\mathcal{B}}(K) = \left\{ (q, S) \mid \exists \sigma \in \Sigma : \exists (q', S') \in K : q' \in \delta_\mathcal{A}(q, \sigma) \wedge \hat{\delta}_\mathcal{B}(S, \sigma) \subseteq S' \right\}.$$

The function does the following: For each $(q', S') \in K$, we take the pairs (q, S) such that, for some symbol σ, q is an σ-predecessor of q' and S is the set of states, from which a state in S' is surely reached when reading σ.

Formally, the basic version of the algorithm, which returns *true* if and only if $\mathcal{L}(\mathcal{A}) \not\subseteq \mathcal{L}(\mathcal{B})$, works as follows:

> **input:** $\mathcal{A} = (Q_\mathcal{A}, \Sigma, \delta_\mathcal{A}, I_\mathcal{A}, F_\mathcal{A})$ and $\mathcal{B} = (Q_\mathcal{B}, \Sigma, \delta_\mathcal{B}, I_\mathcal{B}, F_\mathcal{B})$
> $K \leftarrow F_\mathcal{A} \times \{\overline{F_\mathcal{B}}\}$
> **repeat**
> $K' \leftarrow K$
> $K \leftarrow \lceil K \cup \mathsf{Pre}_{\mathcal{A},\mathcal{B}}(K) \rceil$
> **until** $K = K'$
> **return** there exist $q \in I_\mathcal{A}$ and $S \supseteq I_\mathcal{B}$ such that $(q, S) \in K$

The line $K \leftarrow \lceil K \cup \mathsf{Pre}(K) \rceil$ adds new elements to the current antichain and removes all but the maximal ones. At all times it holds that for all $(q, S) \in K$ there is a word σ such that $\hat{\delta}_\mathcal{A}(\{q\}, \sigma) \cap F_\mathcal{A} \neq \emptyset$ and $\hat{\delta}_\mathcal{B}(S), \sigma) \subseteq \overline{F_\mathcal{B}}$.

The basic algorithm can be optimized in various ways. First, only new elements need to be processed in each step instead of all the elements in K. Second, since the function is monotone, the algorithm can return *true* as soon as the final condition is satisfied (meaning that $\mathcal{L}(\mathcal{A}) \not\subseteq \mathcal{L}(\mathcal{B})$). For a correctness proof of the algorithm, we refer to [21].

Note that in the implementation that we used in the tool, both the automata and the pairs in the antichains are represented symbolically as BDDs. We also tested a forward search variant of the algorithm, but do not include it here due to poor runtimes.

5.3   Results

In this section we present results for several case studies. All tests were performed on a 64-bit Linux machine with a Xeon Dualcore 5150 processor and 8 GB of available main memory.

In the following, we briefly describe the several examples for which we have computed results for different interface sizes using our tool suite. For each of

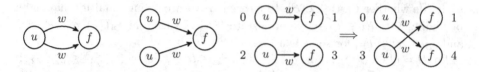

Fig. 6. Forbidden subgraphs "Double Access" and "Two Users"

Fig. 7. Operation "Switch Write Access" as transformation rule

these examples we used the backwards language inclusion algorithm to compute our results.

3-Colorability and 4-Colorability. We checked $C_{(3)} \subseteq C_{(4)}$ and $C_{(4)} \not\subseteq C_{(3)}$ (in the case of non-inclusion a counter example is generated).

Triangle subgraph and *2-Colorability with path extension*. These are the invariants from Ex. 3 and Ex. 4, respectively.

Multi-user file system. We validate the file system example from [7]. In this example, a system state is modelled as a graph: users and files are nodes, access permissions (either "read" or "write") are labelled, directed edges. The system behaviour (add new user, change access permissions, ...) is modelled as transformation rules. The problem is to check whether the file system can reach at least one of two forbidden states. These forbidden states are modelled as the subgraphs depicted in Fig. 6: "Double Access" models the situation where a user obtains double write access to a system resource and "Two Users" models the situation where two users both have write access to the same system resource.

To validate this system we perform a "backwards invariant check": we swap the left- and right-hand sides of the rules and check whether the language of all graphs which contain certain "forbidden subgraphs" is an invariant of this reversed system. The idea is that a forbidden state is reachable (in the original system) only if the system already started in a forbidden state.

Because in [5] a simulation relation was used to approximate the Myhill-Nerode quasi-order, validating the operation "Switch Write Access" (see Fig. 7), which switches the write access of two users, was unsuccessful, although the language is an invariant w.r.t. this operation. Now we succesfully verified it.

Dominating Set and Vertex Cover. We computed results for the inclusion of the language $NonIso \cap D_{(k)}$ of all graphs without isolated nodes which have a dominating set of size at most k in the language $V_{(k)}$ of all graphs which have a vertex cover of size at most k and the non-inclusion of the opposite direction.

In Table 1 the runtime results for the case studies are presented. We can handle some non-trivial examples up to relatively large interface size (note that in practical applications the width, and thus the interface size of graphs, is in general relatively small). For example, the "triangle subgraph automaton" has 37 440 states in case of maximum interface 3 and 19 173 952 states in case of interface size 6. From the first two and last two case studies, it is also apparent that

Table 1. Case study runtimes (in seconds); TO: timed out, OM: out of memory, n.a.: not applicable

Case study	Maximum Interface Size							
	3	4	5	6	7	8	9	10
$C_{(3)} \subseteq C_{(4)}$	< 1	3	14	410	28 713	TO	–	–
$C_{(4)} \not\subseteq C_{(3)}$	n.a.	9	270	63 065	TO	–	–	–
Triangle subgraph	4	15	123	1 978	OM	–	–	–
$C_{(2)}$ and path extension	2	2	3	5	13	53	385	4 193
Multi-user file system	n.a.	19	217	OM	–	–	–	–
$NonIso \cap D_{(2)} \subseteq V_{(2)}$	n.a.	432	26 337	TO	–	–	–	–
$V_{(2)} \not\subseteq NonIso \cap D_{(2)}$	n.a.	2	12	14	154	4 701	TO	–

the runtimes are better when the first automaton is small (the automaton for $C_{(3)}$ and $V_{(2)}$, respectively). This is unsurprising, because the states of the first automaton are explicitly represented (more formally, as a BDD representing a singleton set), whereas the (sets of) states of the second automaton are collectively represented by a BDD.

6   Conclusion

We gave a concrete variant of graph automata accepting recognizable languages. The languages such graph automata can accept contain cospans which have a bounded width, which means that we can only accept graphs with a bounded path width [6]. We applied the approach to automatically checking whether the language of one automaton is included in the language of the other and whether a language is an invariant of a graph transformation system. Case studies show that we can handle non-trivial examples in a relatively short time. However, it seems that the size of the generated automata and the running times grow exponentially with the interface size of the automaton.

Note that our approach differs from the approach in MONA [17], another tool based on recognizable languages. MONA is suitable when the alphabet is large (since BDDs are used to encode the alphabet), whereas in our case the state space is huge.

Another related work [4] considers graph patterns consisting of negative and positive components and shows that they are invariants via an exhaustive search.

For further research, we would like to try more algorithms; in particular we want to implement the simulation-based algorithm of [1] in our tool to see if better results can be obtained. Also, an algorithm that can translate formulas of monadic second-order logic into automata would be helpful. Finally, it is ongoing research to see whether graph automata can help in proving termination of graph transformation systems, much like in the case of string rewrite systems [15].

References

1. Abdulla, P.A., Chen, Y.-F., Holík, L., Mayr, R., Vojnar, T.: When Simulation Meets Antichains (On Checking Language Inclusion of Nondeterministic Finite (Tree) Automata). In: Esparza, J., Majumdar, R. (eds.) TACAS 2010. LNCS, vol. 6015, pp. 158–174. Springer, Heidelberg (2010)
2. Andersen, H.R.: An introduction to binary decision diagrams. Course Notes (1997), http://www.configit.com/fileadmin/Configit/Documents/bdd-eap.pdf
3. Bauderon, M., Courcelle, B.: Graph expressions and graph rewritings. Mathematical Systems Theory 20(2-3), 83–127 (1987)
4. Becker, B., Beyer, D., Giese, H., Klein, F., Schilling, D.: Symbolic invariant verification for systems with dynamic structural adaptation. In: Proc. of ICSE 2006 (International Conference on Software Engineering), pp. 72–81. ACM (2006)
5. Blume, C.: Graphsprachen für die Spezifikation von Invarianten bei verteilten und dynamischen Systemen. Master's thesis, Universität Duisburg-Essen (2008)
6. Blume, C., Bruggink, S., Friedrich, M., König, B.: Treewidth, pathwidth and cospan decompositions. In: Proc. of GT-VMT 2011 (2011)
7. Blume, C., Bruggink, S., König, B.: Recognizable graph languages for checking invariants. In: Proc. of GT-VMT 2010. ECEASST, vol. 29 (2010)
8. Bouajjani, A., Habermehl, P., Rogalewicz, A., Vojnar, T.: Abstract Regular Tree Model Checking of Complex Dynamic Data Structures. In: Yi, K. (ed.) SAS 2006. LNCS, vol. 4134, pp. 52–70. Springer, Heidelberg (2006)
9. Bouajjani, A., Jonsson, B., Nilsson, M., Touili, T.: Regular Model Checking. In: Emerson, E.A., Sistla, A.P. (eds.) CAV 2000. LNCS, vol. 1855, pp. 403–418. Springer, Heidelberg (2000)
10. Bruggink, H.J.S., König, B.: On the Recognizability of Arrow and Graph Languages. In: Ehrig, H., Heckel, R., Rozenberg, G., Taentzer, G. (eds.) ICGT 2008. LNCS, vol. 5214, pp. 336–350. Springer, Heidelberg (2008)
11. Comon, H., Dauchet, M., Gilleron, R., Löding, C., Jacquemard, F., Lugiez, D., Tison, S., Tommasi, M.: Tree automata techniques and applications (October 12, 2007), http://www.grappa.univ-lille3.fr/tata
12. Courcelle, B.: The monadic second-order logic of graphs I. Recognizable sets of finite graphs. Inf. Comput. 85(1), 12–75 (1990)
13. Courcelle, B., Durand, I.: Verifying monadic second order graph properties with tree automata. In: European Lisp Symposium (May 2010)
14. Flum, J., Frick, M., Grohe, M.: Query evaluation via tree-decompositions. Journal of the ACM 49(6), 716–752 (2002)
15. Geser, A., Hofbauer, D., Waldmann, J.: Match-bounded string rewriting. Applicable Algebra in Engineering, Communication and Computing 15(3-4), 149–171 (2004)
16. Gottlob, G., Pichler, R., Wei, F.: Bounded treewidth as a key to tractability of knowledge representation and reasoning. Journal of Artificial Intelligence 174(1), 105–132 (2010)
17. Klarlund, N., Møller, A., Schwartzbach, M.I.: MONA implementation secrets. International Journal of Foundations of Computer Science 13(4), 571–586 (2002)
18. Kneis, J., Langer, A., Rossmanith, P.: Courcelle's theorem – a game-theoretic approach. Discrete Optimization (2011)
19. Sassone, V., Sobociński, P.: Reactive systems over cospans. In: Proc. of LICS 2005, pp. 311–320. IEEE (2005)
20. Soguet, D.: Génération automatique d'algorithmes linéaires. Ph.D. thesis, Université Paris-Sud (2008)
21. De Wulf, M., Doyen, L., Henzinger, T.A., Raskin, J.-F.: Antichains: A New Algorithm for Checking Universality of Finite Automata. In: Ball, T., Jones, R.B. (eds.) CAV 2006. LNCS, vol. 4144, pp. 17–30. Springer, Heidelberg (2006)

Testing against Visual Contracts: Model-Based Coverage

Tamim Ahmed Khan[1], Olga Runge[2], and Reiko Heckel[1]

[1] Department of Computer Sciences, Leicester University, UK
{tak12,reiko}@mcs.le.ac.uk
[2] Department of Software Engineering and Theoretical Computer Science, TU-Berlin, Germany
o.runge@mailbox.tu-berlin.de

Abstract. Testing service-oriented or component-based systems poses new challenges due to the non-availability of code and the distributed nature of the applications being tested. Structural coverage criteria, traditionally used to assess test suites, require access to code. As an alternative we consider model-based criteria based on interface specifications using visual contracts.

Formally represented as graph transformation rules, visual contracts are analysed for potential dependencies and conflicts and dependency graphs are derived for defining the criteria. In order to assess the coverage of a given set of tests, AGG is used for simulating the model while tests are executed. In the course of the simulation, which also serves as a test oracle, conflicts and dependencies are observed and recorded. This allows us to see if the statically detected potential dependencies and conflicts are exercised at runtime. For evaluation purposes, we compare coverage with respect to model-based criteria and traditional structural ones.

1 Introduction

To assess the quality of a test suite we traditionally rely on coverage criteria measuring the proportion of features of a certain type (statement, branch, data or call dependency, etc.) exercised by the tests [2]. Services or components hide their implementations, so client-side testing is limited to information available from interface descriptions. Visual contracts were proposed for interface specification of services in [13].

Their use for model-based testing [18,12] is supported by a formal interpretation as graph transformation rules, which are executable and hence suitable for the generation of test cases [12] and test oracles [17]. The diagram below shows the overall setup for the latter, where a test driver implements calls to the system under test (SUT) and the oracle provided by AGG. Results returned from model and SUT are compared to determine success.

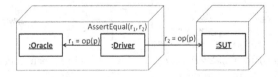

Independently of how test cases are generated or executed, we have to assess their effectiveness. Our approach to coverage combines static and dynamic analysis of models.

H. Ehrig et al.(Eds.): ICGT 2012, LNCS 7562, pp. 279–293, 2012.
© Springer-Verlag Berlin Heidelberg 2012

First, static analysis provides a dependency graph where coverage criteria can be defined [14]. Second, while executing the tests, the model is simulated and coverage is recorded and measured against the criteria. The original contribution of this paper is in the formalisation of the criteria over dependency graphs and the dynamic detection of conflicts and dependencies. That requires keeping track of occurrences and overlaps of pre- and post-conditions, their enabling and disabling, in successive model states, and interpreting these in terms of the static dependency graph.

Next, Sect. 2 introduces visual contracts and the case study. Sect. 3 describes our coverage criteria based on dependency graphs obtained from (static) critical pair and dependency analysis. Dynamic analysis of dependencies and conflicts is presented in Sect. 4, while Sect. 5 evaluates, in particular, the relation to code-based coverage. Related work and conclusion are presented in Sect. 6 and 7, respectively.

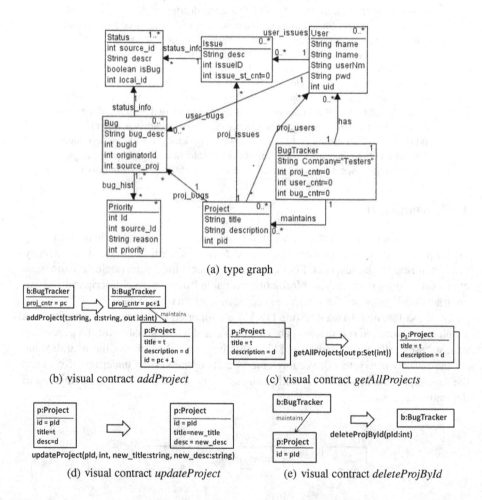

(a) type graph

(b) visual contract *addProject*

(c) visual contract *getAllProjects*

(d) visual contract *updateProject*

(e) visual contract *deleteProjById*

Fig. 1. Interface specification consisting of data model (type graph (a)) and visual contracts (rules (b), (c), (d) and (e))

2   Visual Contracts

A visual contract [13] is a pair of object diagrams specifying the pre- and post-
conditions of an operation, formally a graph transformation rule with an operation signature as in Figure 1. The signature distinguishes input/output and output parameters.
The latter, designated by "out", are not intended to be instantiated as part of invocations,
in analogy to OO return types.

We use a case study of a Bug Tracker service derived from a desktop application[1] in C#. The application provides operations for adding, updating and deleting
projects, users, faults, and issues, distinguishing roles such as administrators, users, and
testers.

Example 1 (signatures of service operations). The service has more than 30 operations
overall. Signatures for a subset are shown below.

```
public Int32 addProject(String title, String description)
public String updateProject(Int32 pId, String new_t, String new_d)
public String deleteProjById(Int32 pId)
public List<ProjectInfo> getAllProjects()
```

In Figure 1(b), signature *addProject(t: String, d: String, out Id: int)* is associated with
the corresponding contract. Parameters *t* and *d* are used to provide title and description
of a new project. Output parameter *Id* represents the project id returned by the system.
The signature *getAllProjects(out p: Set(int))* associated with the visual contract in Figure 1(c) has a multi object as output. The visual contracts are typed over the type graph
in Fig. 1(a) representing the interface data model. Fig. 2 provides a sample instance
graph representing the initial state.

Formally, such a model is represented by a typed attributed graph transformation
system with rule signatures, consisting of an attributed type graph, rule names with
parameter declarations, and for each such signature a set of rules, each representing a
different outcome of the operation [14].

Definition 1 (TAGTS with rule signatures). *A typed attributed graph transformation
system with rule signatures is a tuple* $G = (ATG, P, X, \pi, \sigma)$ *where*

- *ATG is an attributed type graph,*
- *P is a countable set of rule names,*
- *X is an S-indexed family* $(X_s)_{s \in S}$ *of sets of variables,*
- π *assigns each rule name p a finite set of rules* $\pi(p)$ *of the form* $L \xleftarrow{l} K \xrightarrow{r} R$
  over ATG and with local attribute variables in X,
- $\sigma : P \to (\{\epsilon, out\} \times X)^*$ *assigns to each rule name p a list of formal input and output
  parameters* $\sigma(p) = \bar{x} = (q_1 x_1 : s_1, \ldots, q_n x_n : s_n)$ *were* $q_i \in \{\epsilon, out\}$ *and* $x_i \in X_{s_i}$ *for*
  $1 \leq i \leq n$. *We write p's rule signature* $p(\bar{x})$ *and refer to the set of all rule signatures
  as signature of* G.

---

[1] Available at http://btsys.sourceforge.net/

The signatures' main purpose is to provide us with labels of transformations that represent invocations. Semantically, visual contracts are seen as potentially incomplete specifications [15]. In general, both pre- and post-conditions can be under-specified. For their use as oracles we allow under-specified pre-conditions, but insist on complete specification of the operations' effects. Otherwise the states of the model and the SUT would get out of sync [17]. That means, the operational semantics can be described by double-pushout transformations of attributed typed graphs [9]. We allow multiple contracts per operation to represent alternative actions inside the same operation, chosen by different input values and the system's internal state. However, these visual contracts are currently created manually.

3   Coverage Criteria Based on Visual Contracts

A dependency graph over visual contracts, which is the basis for defining coverage, captures the potential dependencies and conflicts of the underlying graph transformation rules. We use an asymmetric version of dependencies and conflicts, captured abstractly by directed relations $\prec$ and $\nearrow$ over rules.

Definition 2 (asymmetric dependencies and conflicts). *Given two rules p_1, p_2 we say that p_2 may disable p_1, written $p_1 \nearrow p_2$, if there are steps $G \overset{p_1,m_1}{\Longrightarrow} H_1$ and $G \overset{p_2,m_2}{\Longrightarrow} H_2$ without $k : L_1 \to D_2$ such that $m_1 = l_2^* \circ k$.*

$$R_1 \xleftarrow{r_1} K_1 \xrightarrow{l_1} L_1 \qquad\qquad L_2 \xleftarrow{l_2} K_2 \xrightarrow{r_2} R_2$$

$$m_1^* \downarrow \quad k_1 \downarrow \quad\quad m_1 \quad m_2 \quad k \quad\quad \downarrow k_2 \quad\quad \downarrow m_2^*$$

$$H_1 \xleftarrow{r_1^*} D_1 \xrightarrow{l_1^*} G \xleftarrow{l_2^*} D_2 \xrightarrow{r_2^*} H_2$$

We say that p_1 may enable p_2, written $p_1 \prec p_2$, if there are steps $G_0 \overset{p_1,m_1}{\Longrightarrow} G_1 \overset{p_2,m_2}{\Longrightarrow} G_2$ without $j : L_2 \to D_1$ such that $m_2 = r_1^* \circ j$.

$$L_1 \xleftarrow{l_1} K_1 \xrightarrow{r_1} R_1 \qquad\qquad L_2 \xleftarrow{l_2} K_2 \xrightarrow{r_2} R_2$$

$$m_1 \downarrow \quad k_1 \downarrow \quad j \quad m_1^* \quad m_2 \quad\quad \downarrow k_2 \quad\quad \downarrow m_2^*$$

$$G \xleftarrow{l_1^*} D_1 \xrightarrow{r_1^*} G_1 \xleftarrow{l_2^*} D_2 \xrightarrow{r_2^*} G_2$$

Example 2 (dependencies and conflicts). The rules in Fig. 1 and the start graph in Fig. 2 allow the following sequences of rule applications.

addProject("proj", "desc", 19); updateProject(19, "CMS", "GeneralPurpose").

updateProject(18, "MIS", "version 1"); delProjById(18).

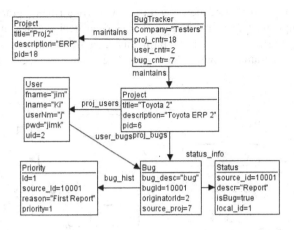

Fig. 2. Start graph

The first sequence exhibits an asymmetric dependency *addProject("proj", "desc", 19)* < *updateProject(19, "CMS", "General Purpose")* between the first and the second step. Similarly, there exists an asymmetric conflict *updateProject(18, "MIS", "version 1") ↗ delProjById(18)* in the second sequence. Hence either *updateProject(18, "MIS", "version 1")* must precede *delProjById(18)*, or only the latter can occur in a sequence.

Definition 3 (dependency graph). *A dependency graph* $DG = \langle G, P, op, lab \rangle$ *is a structure where*

- $G = \langle V, E, src, tar \rangle$ *is a graph.*
- *P is a set of operation names*
- $op : V \to P$ *labels vertices by operation names*
- $lab : E \to \{c, r, d\} \times \{<, \nearrow\} \times \{c, r, d\}$ *is a labeling function distinguishing source and target types* <u>c</u>reate, <u>r</u>ead, <u>d</u>elete *and dependency types* $<, \nearrow$.

The labeling of edges by < or ↗ refers to the "may enable" and "may disable" relations on rules, but the graphs also allow a distinction between different kinds of dependencies or conflicts. These are summarized in Table 1, which also shows that not all combinations are possible.

Table 1. Label combinations indicating conflicts and dependencies

Labels	Conflict	Dependency
cr	×	√
cd	×	√
rd	√	×
dd	√	×

The construction of a dependency graph from a typed attributed graph transformation system is implemented as an extension of AGG's critical pair and minimal dependency analysis [1]. The dependency graph for the case study introduced in Sect. 2 is presented in two diagrams, showing the $\prec$ and $\nearrow$ relations in Fig. 3(a) and Fig. 3(b), respectively.

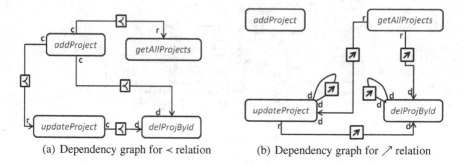

(a) Dependency graph for $\prec$ relation   (b) Dependency graph for $\nearrow$ relation

Fig. 3. Partial view of the dependency graph for $\prec$ (a) and $\nearrow$ (b) relations

As basic coverage criteria we consider the combinations of labels shown in Table 1. The labels cr in the first row mean that the first step creates an item, e.g., a graph element or an attribute, which is read by the second step. Labels cd represent the fact that the first step creates an item deleted by the second step. Labels rd and dd in the third and forth rows denote an item read or deleted by the first step and deleted by the second step. The first two rows mark dependencies, while the last two represent conflicts. For a given dependency graph, a criterion therefore defines a set of edges of that graph that are to be covered by dependencies or conflicts encountered when executing transformation sequences. This is described in more detail in the following section.

Example 3 (edge labels). Consider Figure 3(a) and rules *addProject* and *getAllProjects*. The edge between them is labeled cr, because *addProject* creates a project node read by *getAllProjects*. We use c to represent both the creation of a node as well as the update of an attribute and d to represent the deletion of a node or the update of an attribute. This is justified by the fact that attribute updates are realized by the deletion and creation of edges between nodes and values.

Consider Figure 3(b) and rules *deleteProjByID* and *updateProject*. The loop labeled dd on the *deleteProjByID* rule marks the fact that a project cannot be deleted twice. A similar loop dd on the *updateProject* rule is a symmetric conflict where attributes are updated.

A test case consists of a start graph and a sequence of invocations. An invocation is a rule name instantiated with respect to input parameters. Output parameters are not initialized until the execution reaches the point where their values are computed.

4   Dynamic Analysis of Dependencies and Conflicts

As outlined in the introduction, we use AGG [1] as an oracle simulating the model while the tests are executing. Monitoring the transformation sequences created, we can detect conflicts and dependencies at runtime and therefore measure the coverage of the dependency graph with respect to a given set of criteria. We consider dependencies and conflicts separately.

Definition 4 (coverage of dependencies). *A dependency edge* $p \prec q$ *is covered by a transformation sequence* $G_0 \overset{p_1,m_1}{\Longrightarrow} G_1 \overset{p_2,m_2}{\Longrightarrow} \cdots \overset{p_n,m_n}{\Longrightarrow} G_n$ *if there are* $i < j \leq n$ *such that* $p = p_i, q = p_j$ *and the residual comatch* m_i^{j-1} *of* p_i *into* G_{j-1} *overlaps with the match* m_j *of* p_j *in accordance with the source and target types of the relation. That means, there exist a node or edge* x *or an attribute* a *in* $m_i^{j-1}(L_i) \cap m_j(L_j) \subseteq G_{j-1}$ *such that*

cr: *x is created by* p_i *and read by* p_j
cd: *x is created by* p_i *and deleted by* p_j

The residual comatch m_i^{j-1} *of* p_i *into graph* G_{j-1} *is obtained by composing comatch* m_i^* *with the tracking morphism* $p_{j-1}^* \circ \ldots \circ p_{i+1}^*$ *between* G_i *and* G_{j-1} *as illustrated in the diagram of Fig. 4.*

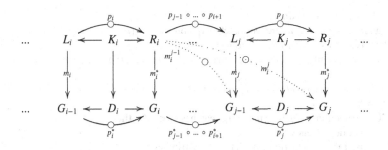

Fig. 4. Dependencies in a transformation sequence

This definition is implemented by Algorithm 1, whose input is a sequence of invocations s as generated by the test as well as the start graph of the grammar. For each step in s we apply the corresponding rule schema or basic rule. AGG stores (co-)matches as mappings into a pool of graph elements. If an element is deleted it is removed with its details, leaving a hash value assigned upon creation of the element. We use these hash values to represent matches and comatches and to calculate their difference and intersection after each step to find out what was created, preserved, and deleted. We detect dependencies of type *cr* and *cd* by maintaining a list of partial comatches into all subsequent steps of the sequence, computing the overlaps between matches and partial comatches. We process s one step at a time. We first investigate if the invocation is related to a rule or a rule scheme, as the process to find and apply the match are significantly different. Once we have collected the information about matches and co-matches, we check for each rule if any of the subsequent rules have a dependency with the considered rule. We do this using two nested loop such that we first process the nodes and edges and then the effect on attribute values.

The output of the algorithm is stored in the dynamic dependency matrix and compared to the static dependency matrix created by AGG when the model was first loaded. Note that the host graph can change without requiring to recalculate the stored static information, saving considerable execution time. The comparison of the two matrices provides the coverage data.

Algorithm 1. Algorithm for marking dependencies

Input: s where size(s)>= 2
  set host graph to start graph of *GraphGrammar*
  for (i=0; i<size(s); i++) **do**
    if Rule r_i *instanceof* RuleScheme **then**
      apply Rule Scheme
    else
      apply Rule r_i
    end if
    store hash values of involved graph elements in arrays matches and comatches
  end for
  for (i=0; i<size(s); i++) **do**
    createdElements[i] = difference(matches[i], comatches[i])
    deletedElements[i] = difference(comatches[i], matches[i])
    preservedElements[i] = intersection(matches[i], comatches[i])
  end for
  for (i=size(s)-1; i>0; i- -) **do**
    for (j=0; j<size(s); j++) **do**
      if (intersection(createdElements[j], matches[i] <> ϕ)) **then**
        mark dependency between Rule r[i] and r[j]
        repeat the intersection calculation for attributes lists
      end if
      if (intersection(preservedElements[j], matches[i] <> ϕ)) **then**
        mark dependency between Rule r[i] and r[j]
        repeat the intersection calculation for attributes lists
      end if
    end for
  end for

We make use of rule schemes to implement rules containing multi-objects, such as *getAllProjects* as shown in Fig. 1(c). The concept of amalgamated transformations [6,11] is implemented in AGG [22]. An amalgamated transformation returns the set of nodes corresponding to the multi-object on the right-hand side of the rule. Rule schemes implement all-quantified operations on recurring graph patterns. The kernel rule is a common subrule of a set of multi-rules. It is matched only once, while multi-rules are matched as often as suitable matches are found. In AGG an amalgamated rule is constructed from all matches found for multi-rules that share the match of the kernel rule.

Example 4 (Coverage of dependencies). Consider the dependency graph in Fig. 3(a) and the following test sequences.

> *addProject("p1", "d", 19); updateProject(19, "CMS", "desc"); deleteProjById(19).*
>
> *addProject("p1", "d", 19); updateProject(6, "CMS", "desc"); deleteProjById(18).*

The first sequence exercises a direct *cr* dependency between steps *addProject(...)* and *updateProject(...)* since the first produces a project and the second updates the project created in the previous step. Steps *addProject(...)* and *deleteProjById(...)* have an indirect *cd* dependency since the project node created by the first operation is deleted by the third. Observe that these dependencies rely on the matches of the steps as determined by the parameters of the operations. The second sequence does not exercise the dependency between *addProject* and *updateProject* because the update is done to a previously existing project different from what was created by *addProject("p1", "d", 19)*. Similarly, steps *addProject("proj", "desc", 19)* and *deleteProjById(18)* are unrelated since the latter deletes a different project than the one produced by the first.

The dynamic detection of conflicts is based on finding, for each graph in the sequence, all matches for all rules and comparing them via the tracking morphisms. If, for a given step $G \stackrel{p,m}{\Longrightarrow} H$, rule q has a match into graph G which is not present in H, this match is disabled by p. In this case, we have observed a conflict $q \nearrow p$.

Definition 5 (coverage of conflicts). *A conflict edge $q \nearrow p$ is covered by a transformation sequence $G_0 \stackrel{p_1,m_1}{\Longrightarrow} G_1 \stackrel{p_2,m_2}{\Longrightarrow} \cdots \stackrel{p_n,m_n}{\Longrightarrow} G_n$ if there exists a step $G_{i-1} \stackrel{p_i,m_i}{\Longrightarrow} G_i$ such that $p = p_i$ and for any match m of q into G_{i-1} there is no match m' of q in G_i such that $p_i^* \circ m = m'$.*

The source and target labels of the edge are determined according to one of the following cases, for a node or edge x or an attribute a in $m_i(L_i) \cap m(L) \subseteq G_{j-1}$.

rd: *x is read by q and deleted by p_i*
dd: *x is deleted by both q and p_i*

The implementation is presented in Algorithm 2 with the same input as before and executing the same sequence of steps on invocation. At each step we find and store all matches for all rules in the grammar, computing the difference between the sets of matches into graph i and graph $i + 1$ to find out those that were disabled by that step. Each disabled match represents an asymmetric conflict, which is recorded in the dynamic conflict matrix. Like for dependencies, this is compared to the result of the static analysis to calculate the coverage.

Example 5 (Coverage of conflicts). Consider the following two test sequences.

> *addProject("p1", "d1", 19); updateProject(19, "CMS", "desc"); deleteProjById(19).*
>
> *addProject("p1", "d1", 19); deleteProjById(19).*

Algorithm 2. Algorithm for marking conflicts

Input: s where size(s)>= 2
  set host graph to start graph of *GraphGrammar*
  for (i=0; i<size(s); i++) **do**
    if Rule r_i *instanceof* RuleScheme **then**
      apply Rule scheme
    else
      apply Rule r_i
    end if
    store the hash value of graph elements in an array
    for all (Rule r in *GraphGrammar*) **do**
      find all possible matches and store in an array
    end for
  end for
  for (i=0; i<size(s)-1; i++) **do**
    select all the matches found for i^{th} row
    select all the matches found for $(i + 1)^{th}$ row
    for j=0; j<size(row);j++ **do**
      analyze matches details to mark conflict between rule[i] and rule[j]
    end for
  end for

The *rd* conflict between *updateProject(...)* and *deleteProjById(...)* in the first sequence means that these operations are not executable in the reverse order. Step
deleteProjById(...) has a *dd* conflict with itself, meaning that it can only occur once
in a sequence. The second sequence results from choosing *deleteProjById(...)* before,
and instead of, *updateProject(...)* in the *rd* conflict.

5   Evaluation

In this section, we evaluate the relation of our model-based approach with traditional
code-based coverage criteria. Let us repeat that, since the code is not available for services or components, the tester would not have access to code-based coverage data.
Therefore, using model-based criteria instead, we are interested in finding out how good
a substitute they may be. We also evaluate the scalability of our approach in terms of
the size of the specifications, the length of a test case, and number of test cases that can
be executed in a given period of time. Finally, we discuss some threats to the validity of
the evaluation as well as limitations of our approach.

We evaluate coverage with respect to model-based coverage criteria in relation to
code-based coverage. Using our own AGG-based tool to measure coverage with respect
to the selected criteria on the model, we determine code-based coverage with respect to
the most common criteria using the *NCover* tool.[2] In particular, we consider symbol and
branch coverage. The first is essentially a more fine-grained version of statement coverage, including elements in expressions. The second requires that, for each condition

---

[2] See http://www.ncover.com/

Table 2. Evaluation results

S/N	Criteria	# of test cases	average length of test case	SUT Symbol Coverage	SUT Branch Coverage
1.	*cd*	10	3	49.19%	45.07%
2.	*cr*	8	5	52.10%	56.34%
3.	*cr + cd*	10	7	83.50%	87.32%
4.	*cr + cd +rd + dd*	14	9	91.91%	92.96%

of a branch (such as in an *if, while, do while*, etc.) both positive and negative outcome should be tested.

Results are reported in Table 2, where each row represents a selection of basic or combined model-based coverage criteria. We report the number of test cases necessary to achieve this coverage, the average length of these test cases, and the corresponding code-based coverage achieved with respect to the two criteria. Model-based coverage is based on the dependency graphs in Fig. 3(a) and 3(b). Given a test suite derived manually, the numbers of test cases reported are based on minimal subsets of test cases that are able to achieve the required coverage.

For example, considering coverage criterion *cd* we require eight test cases of average length 3 achieving 49.19% symbol and 45.07% branch coverage. For criterion *cr* the values are slightly higher, while combing the two coverage jumps above 80%. Obviously, some of the nodes are required by both criteria, such as *addProject* which is involved in both *cr* and *cd* edges.

The forth row represents the results for complete coverage of dependencies and conflicts in the dependency graph, but fails to provide complete coverage with respect to code-based criteria. Further analysis reveals that the remaining $8 - 9\%$ correspond to code that is not executable as part of the normal behaviour. This includes exception handling code triggered by technical errors outside the specification, e.g., a failure to connect to the data base, or glue code added by the IDE. An example is shown in Figure 5 where the code fragment in the rectangle is triggered by a technical failure. Since the approach is concerned with testing against functional specifications, technical exception handling is out of scope and the failure to cover it based on functional testing is not surprising.

We have evaluated the scalability of our approach by considering the size of the specifications in terms of the number of rules and the size of the start graph as well as the length of the test sequence. Our case study has 31 rules with the start graph having 12 projects, 9 users, and 1 bug, plus priority and issue objects. We compute the static information, i.e., critical-pairs and minimal dependencies, only once using AGG's API and store them locally for repeated use. The time taken for this calculation is 783.53 seconds, while loading the stored data for subsequent use takes 1.72 seconds. This is acceptable given that the effort is only incurred once, but we have to be aware that the runtime is quadratic in the number of rules and exponential in the size of the rules themselves. That means, large numbers of complex operations will continue to pose

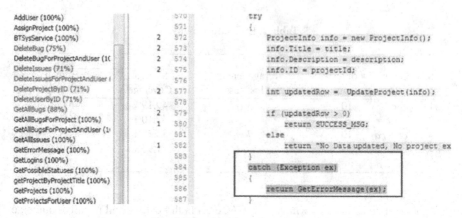

Fig. 5. Unreachable code example

challenges. On the other hand, service interfaces should be high-level, and split up if they become too large.

We also conducted experiments with different lengths of test sequences. We produced a routine to automatically provide inputs for test sequence repeating only two rules, i.e., for adding and deleting a project repeatedly. That means, the size of the graph remains stable. The time taken for executing test sequences of lengths 25, 50 and 100 was 4.579, 12.844 and 65.189 seconds, respectively, while the system crashed with an *out of memory* exception for a sequence of 106 steps. The problem here is the maintenance of partial matches for all earlier rules into later graphs of the sequence, which incurs a cost quadratic in the length of the sequence in terms of the memory used. It should be noted that in practice, most test sequences will be short (e.g., [5] states that the majority of faults are revealed by test sequences of size 2), but the size of the longest possible sequence is still a limiting factor.

Executing 14 test sequences of average length 9, as required but the 4th combination of criteria in Table 2, takes about 15 seconds. The effort is in fact linear in the number of test cases, so there is no significant barrier to executing large test suites.

The validity of the evaluation is limited by several factors. First, the implementation of the case study, if non-trivial, is relatively small, although the interface (and model) are of reasonable size. The sequences for evaluating scalability in terms of the length of test cases are clearly artificial, but based on our knowledge of the implementation we can say that the actions of the rules in the sequence are marginal for the effort, which is caused by maintaining and comparing matches into a large number of graphs. The start graph of 25 nodes used for the evaluation is probably realistic for tests with specifically created data, but tests using real data will require much larger graphs. These may represent not only a challenge to scalability, but also call for automation of the generation of start graphs from initial test data.

6   Related Work

We follow other approaches to coverage in creating an abstract representation of dependencies in the system based on specifications [2,4]. In our case, this takes the form of a graph representing data dependencies over visual contracts, rather than that of a finite state machine or control-flow diagram. Dependency graphs for software testing are considered in [21] capturing direct and indirect data dependencies [24]. Approaches to model-based testing using data dependencies are also known for object-oriented systems [8,7]. Our graphs are limited to direct data dependencies.

There are several approaches to testing web services based on dataflow graphs extracted from semantic information [3,23,16]. The approach discussed in [3] is aimed at testing service composition using BPEL specifications. BPEL is also considered in [16], where dependency analysis is carried out over variables acquired from WSDL interface description to extract paths through the graph of the BPEL specification. Criteria for data flow testing, originally established in [10], are applied by [19] to the functional testing of services using BPEL and WSDL. The authors of [20] have made use of call-based dependence graphs for coverage in object-oriented systems. They incorporate both control and data dependence. In our approach, the combination of data and control flow analysis could be interesting when considering service specifications complementing visual contracts with orchestrations. However, our handling of data dependencies is more advanced than what can be extracted from operation signatures in WSDL.

Visual contracts have been used in [18] not only for testing individual operations but also for operation sequences. The work proposes a mapping between visual contracts and JML assertions that can be considered as providing an oracle. Visual contracts are also used in [12] for formalizing pre- and post-conditions of use cases to be used as test models for the generation of logical test cases. This work provides the basis for establishing a relation between UML specifications and visual contracts and proposes a test suites generation mechanism for required and provided interfaces. The generation of test cases is not our concern in this paper, so work on generating test cases from visual contracts can be considered complementary.

7   Conclusion

We have proposed an approach to model-based coverage based on a two-step process combing static and dynamic analysis. Statically, we use AGG's critical pair and minimal dependency analysis to create a dependency graph over rules representing visual contracts. These graphs, which distinguish different types of dependencies and conflicts, are the basis for coverage criteria. The evaluation of a set of tests based on the criteria is performed dynamically while executing the model as an oracle.

The approach requires further evaluation in particular in terms of scalability to longer sequences. It is clear that improvements are possible by reducing the number of matches kept and compared, using information from the static analysis which provides a conservative approximation of the real dependencies and conflicts. We also plan to consider negative application conditions to strengthen the visual contracts specification. This will create additional types of dependency, such as create-forbid, where one rule creates part of the structure preventing the application of another, resulting in new coverage criteria. We also plan to investigate how we can construct test case directly from the dependency graph.

Another line of investigation is to evaluate the approach with developers in order to assess the benefit of a model-based approach, where oracle and test coverage are provided, against a more informal documentation of the service interface where no such help is given. Seeding faults in the service implementation, this would allow us to assess the added value of the model-based approach. The major cost factor is of course the creation and maintenance of the models. If and when these costs are outweighed by the benefits can only be evaluated in a more realistic industrial setting. However, scenarios where specifications are created once and used repeatedly, e.g., as part of standards, are likely to provide good tradeoffs.

References

1. AGG: AGG - Attributed Graph Grammar System Environment (2007),
   http://tfs.cs.tu-berlin.de/agg
2. Ammann, P., Offutt, J.: Introduction to Software Testing. Cambridge University Press, New York (2008)
3. Bartolini, C., Bertolino, A., Marchetti, E., Parissis, I.: Data Flow-Based Validation of Web Services Compositions: Perspectives and Examples. In: de Lemos, R., Di Giandomenico, F., Gacek, C., Muccini, H., Vieira, M. (eds.) Architecting Dependable Systems V. LNCS, vol. 5135, pp. 298–325. Springer, Heidelberg (2008)
4. Beizer, B.: Black-box testing: techniques for functional testing of software and systems. John Wiley & Sons, Inc., New York (1995)
5. Belli, F., Guandler, N., Linschulte, M.: Are longer test sequences always better? - a reliability theoretical analysis. In: Fourth International Conference on Secure Software Integration and Reliability Improvement Companion (SSIRI-C), pp. 78–85. IEEE (June 2010)
6. Biermann, E., Ehrig, H., Ermel, C., Golas, U., Taentzer, G.: Parallel Independence of Amalgamated Graph Transformations Applied to Model Transformation. In: Engels, G., Lewerentz, C., Schäfer, W., Schürr, A., Westfechtel, B. (eds.) Nagl Festschrift. LNCS, vol. 5765, pp. 121–140. Springer, Heidelberg (2010)
7. Briand, L., Labiche, Y., Lin, Q.: Improving the coverage criteria of UML state machines using data flow analysis. Software Testing, Validation, and Reliability 20(3) (2010)
8. Chen, Y., Liu, S., Nagoya, F.: An Approach to Integration Testing Based on Data Flow Specifications. In: Liu, Z., Araki, K. (eds.) ICTAC 2004. LNCS, vol. 3407, pp. 235–249. Springer, Heidelberg (2005)
9. Ehrig, H., Ehrig, K., Prange, U., Taentzer, G.: Fundamentals of Algebraic Graph Transformation (Monographs in Theoretical Computer Science. An EATCS Series). Springer (2006)
10. Frankl, P.G., Weyuker, E.J.: An applicable family of data flow testing criteria. IEEE Trans. Softw. Eng. 14(10), 1483–1498 (1988)
11. Golas, U., Biermann, E., Ehrig, H., Ermel, C.: A Visual Interpreter Semantics for Statecharts Based on Amalgamated Graph Transformation. In: Proceedings of Int. Workshop on Graph Computation Models (GCM 2010). Electronic Communications of the EASST, vol. 39 (2011)
12. Güldali, B., Mlynarski, M., Wübbeke, A., Engels, G.: Model-based system testing using visual contracts. In: Proceedings of Euromicro SEAA Conference 2009, Special Session on "Model Driven Engineering", pp. 121–124. IEEE Computer Society, Washington, DC (2009)
13. Hausmann, J.H., Heckel, R., Lohmann, M.: Model-based development of web services descriptions enabling a precise matching concept. Int. J. Web Service Res. 2(2), 67–84 (2005)
14. Heckel, R., Khan, T.A., Machado, R.: Towards test coverage criteria for visual contracts. In: Proceedings of Graph Transformation and Visual Modeling Techniques, GTVMT 2011. Electronic Communications of the EASST, vol. 41 (2011)

15. Heckel, R., Llabrés, M., Ehrig, H., Orejas, F.: Concurrency and loose semantics of open graph transformation systems. Mathematical Structures in Computer Science 12(4), 349–376 (2002)
16. Hou, J., Xu, B., Xu, L., Wang, D., Xu, J.: A testing method for web services composition based on data-flow. Wuhan University Journal of Natural Sciences 13, 455–460 (2008)
17. Khan, T.A., Runge, O., Heckel, R.: Visual contracts as test oracle in AGG 2.0. In: Proceedings of Graph Transformation and Visual Modeling Techniques, GTVMT 2012. Electronic Communications of the EASST, vol. 47 (2012)
18. Lohmann, M., Mariani, L., Heckel, R.: A model-driven approach to discovery, testing and monitoring of web services. In: Test and Analysis of Web Services, pp. 173–204. Springer (2007)
19. Mei, L., Chan, W., Tse, T., Kuo, F.C.: An empirical study of the use of Frankl-Weyuker data flow testing criteria to test BPEL web services. In: 33rd Annual IEEE International Computer Software and Applications Conference, COMPSAC 2009, vol. 1, pp. 81–88 (July 2009)
20. Najumudheen, E., Mall, R., Samanata, D.: A dependence representation for coverage testing of object-oriented programs. Journal of Object Technology 9(4), 1–23 (2010)
21. Podgurski, A., Lori, C.A.: A formal model of program dependences and its implications for software testing, debugging, and maintenance. IEEE Transactions on Software Engineering 16, 965–979 (1990)
22. Runge, O., Ermel, C., Taentzer, G.: AGG 2.0 – new features for specifying and analyzing algebraic graph transformations. In: Proceedings of the 4th International Symposium on Applications of Graph Transformation with Industrial Relevance, AGTIVE 2011. LNCS, vol. 7233, Springer (2012)
23. Sinha, A., Paradkar, A.: Model-based functional conformance testing of web services operating on persistent data. In: Proceedings of the 2006 Workshop on Testing, Analysis, and Verification of Web Services and Applications, TAV-WEB 2006, pp. 17–22. ACM, New York (2006)
24. Zhu, H., Hall, P.A.V., May, J.H.R.: Software unit test coverage and adequacy. ACM Comput. Surv. 29, 366–427 (1997)

A Truly Concurrent Semantics for the $\mathbb{K}$ Framework Based on Graph Transformations[*]

Traian Florin Șerbănuță[1,2] and Grigore Roșu[1,2]

[1] Alexandru Ioan Cuza University of Iasi
[2] University of Illinois at Urbana-Champaign
traian.serbanuta@info.uaic.ro,
grosu@illinois.edu

Abstract. This paper gives a truly concurrent semantics with sharing of resources for the $\mathbb{K}$ semantic framework, an executable (term-)rewriting-based formalism for defining programming languages and calculi. Akin to graph rewriting rules, the $\mathbb{K}$ (rewrite) rules explicitly state what can be concurrently shared with other rules. The desired true concurrency is obtained by translating the $\mathbb{K}$ rules into a novel instance of term-graph rewriting with explicit sharing, and then using classical concurrency results from the double-pushout (DPO) approach to graph rewriting. The resulting parallel term-rewriting relation is proved sound, complete, and serializable with respect to the jungle rewriting flavor of term-graph rewriting, and, therefore, also to term rewriting.

1 Introduction

There are several reasons for defining a truly concurrent semantics for a given model of computation. One reason is that specification languages based on truly concurrent models are more informative; e.g., testing sequences defining partial orderings may carry the same information as an exponentially larger number of interleaving traces. Another reason is that in truly concurrent models the existing fine parallelism of the application is fully specified. It is left to the implementer to take advantage of it by allocating concurrent events to different processors, or to partition events into coarser classes performed by a few concurrent processes. Finally, truly concurrent semantics carries extra information, being usually straightforward to recover interleaving semantics from it [17].

The $\mathbb{K}$ semantic framework [20] is a programming language definitional framework based on rewriting which attempts to bring together the strengths of existing frameworks (e.g., the chemical abstract machine (CHAM) [3], evaluation contexts [22], or continuations [8]) while avoiding their weaknesses. The $\mathbb{K}$ framework relies on computations, configurations, and $\mathbb{K}$ rules in giving semantics to programming language constructs. So far, $\mathbb{K}$ has been successfully used for defining several real-life programming languages like C [7] and Scheme [14].

Currently, computations and configurations are described as algebraic terms over a first-order signature, and the semantics of $\mathbb{K}$ definitions is given through

[*] This work is supported by Contract 161/15.06.2010, SMISCSNR 602-12516 (DAK).

H. Ehrig et al.(Eds.): ICGT 2012, LNCS 7562, pp. 294–310, 2012.

$$\left\langle \begin{array}{c} \langle \mathtt{set}(2,9)\ \cdots\rangle_{\mathsf{thrd}}\ \langle \mathtt{set}(3,0)\ \cdots\rangle_{\mathsf{thrd}} \\ \langle \cdots\ 2 \mapsto 5\ \cdots\ 3 \mapsto 1\ \cdots\rangle_{\mathsf{mem}} \end{array} \right\rangle \Rightarrow \left\langle \begin{array}{c} \langle 5\ \cdots\rangle_{\mathsf{thrd}}\ \langle 1\ \cdots\rangle_{\mathsf{thrd}} \\ \langle \cdots\ 2 \mapsto 9\ \cdots\ 3 \mapsto 0\ \cdots\rangle_{\mathsf{mem}} \end{array} \right\rangle \qquad (a)$$

$$\left\langle \begin{array}{c} \langle \mathtt{set}(3,0)\ \cdots\rangle_{\mathsf{thrd}}\ \langle \mathtt{set}(3,2)\ \cdots\rangle_{\mathsf{thrd}} \\ \langle \cdots\ 3 \mapsto 1\ \cdots\rangle_{\mathsf{mem}} \end{array} \right\rangle$$

$$\Rightarrow \left\langle \begin{array}{c} \langle 1\ \cdots\rangle_{\mathsf{thrd}}\ \langle \mathtt{set}(3,2)\ \cdots\rangle_{\mathsf{thrd}} \\ \langle \cdots\ 3 \mapsto 0\ \cdots\rangle_{\mathsf{mem}} \end{array} \right\rangle \Rightarrow \left\langle \begin{array}{c} \langle 1\ \cdots\rangle_{\mathsf{thrd}}\ \langle 0\ \cdots\rangle_{\mathsf{thrd}} \\ \langle \cdots\ 3 \mapsto 2\ \cdots\rangle_{\mathsf{mem}} \end{array} \right\rangle \quad (b)$$

$$\Rightarrow \left\langle \begin{array}{c} \langle \mathtt{set}(3,0)\ \cdots\rangle_{\mathsf{thrd}}\ \langle 1\ \cdots\rangle_{\mathsf{thrd}} \\ \langle \cdots\ 3 \mapsto 2\ \cdots\rangle_{\mathsf{mem}} \end{array} \right\rangle \Rightarrow \left\langle \begin{array}{c} \langle 2\ \cdots\rangle_{\mathsf{thrd}}\ \langle 1\ \cdots\rangle_{\mathsf{thrd}} \\ \langle \cdots\ 3 \mapsto 0\ \cdots\rangle_{\mathsf{mem}} \end{array} \right\rangle$$

Fig. 1. Synchronous access of memory in a multithreaded environment: (a) concurrent writes, and (b) interleaving dataraces

their translation in rewriting logic [15] theories. Structuring execution configurations as terms is quite convenient, as first order signatures are quite intuitive, and there is plenty of tool support for reasoning about first-order terms. Moreover, rewriting logic is generally appealing for defining truly concurrent systems, since rewrite rules can independently match and apply anywhere, unconstrained by the context [15]. However, although rewriting logic has proved successful in defining sequential programming languages as well as actor-like languages [16], it enforces that "the same object cannot be shared by two simultaneous rewrites" [16], i.e., rule instances are not allowed to overlap. Although there are good reasons for this choice, such as sufficing to capture concurrent synchronization like that of Petri Nets, this limitation enforces an interleaving semantics in situations where one may not want it, especially when describing systems which allow sharing of resources.

Consider a running configuration of a program where two threads are both ready to set the value of different memory locations, as in the left-hand side (lhs) of Fig. 1(a). Assume the (single-threaded) semantics of the memory update construct $\mathtt{set}$ is to update the value in the memory and return the old value, like $\langle \mathtt{set}(X, V')\ \cdots\rangle_{\mathsf{thrd}}\langle \cdots\ X \mapsto V\ \cdots\rangle_{\mathsf{mem}} \longrightarrow \langle V\ \cdots\rangle_{\mathsf{thrd}}\langle \cdots\ X \mapsto V'\ \cdots\rangle_{\mathsf{mem}}$. Then two instances of this rule (i.e., two threads attempting to concurrently update *distinct* memory locations) should be allowed to advance concurrently in one transition step as in Fig. 1(a). In fact, two instances of such memory access rules should be forced to interleave *only if trying to concurrently access the same location, one of the accesses being a* $\mathtt{set}$, as exemplified in Fig. 1(b).

However, using the rewriting logic semantics associated to the rule above, the concurrent rewriting transition from Fig. 1(a) would need to be interleaved, too. The reason is that the two instances of the rule overlap on the mem cell and on the algebraic constructs representing the cell composition operators. Nevertheless, it is worthwhile noting that the operators on which the rule instances overlap are only mentioned as context needed for the transition to apply, playing the same "gluing" role that interface graphs play in the DPO approach to graph transformations [4].

The special rewrite rules employed by the $\mathbb{K}$ framework, from here on named $\mathbb{K}$ rules, make this sharing of context explicit: rewrite rules are extended to allow specifying which parts of the matching pattern are effectively changed by a rule, allowing the rest to be shared with other rules. For example, the $\mathbb{K}$ rule corresponding to the set rewrite rule presented above is:

$$\langle \underbrace{\mathsf{set}(X, V')}_{V} \cdots \rangle_{\mathsf{thrd}} \quad \langle \cdots X \mapsto \underbrace{V}_{V'} \cdots \rangle_{\mathsf{mem}}$$

The intuition for the above rule is that, while the entire pattern of the top needs to be matched for the rule to apply, only the underlined set instruction and the value in the memory are actually changed by the rule, being replaced with the corresponding values below the line. This furthermore implies that the thrd and mem containers, along with the ellipses (specifying potential additional content), are only needed to specify the context in which the local transformations would apply, and thus can be shared with other concurrent instances of $\mathbb{K}$ rules.

The process of rewriting terms using $\mathbb{K}$ rules and following the intuition above will be called $\mathbb{K}$ *(term-)rewriting.*

Contributions. This paper gives semantics to $\mathbb{K}$ rewriting through the help of graph rewriting, adapting existing representations of terms and rules as graph and graph rewrite rules to maximize their potential for concurrent application. The main result, Theorem 4, shows that $\mathbb{K}$ rewriting is *sound and complete* w.r.t. standard term rewriting, and that the concurrent application of $\mathbb{K}$ rules is *serializable.* Soundness means that applying one $\mathbb{K}$ rule can be simulated by applying its corresponding direct representation as a rewrite rule. Completeness means the converse, i.e., that one application of a term rewriting rule can be simulated by applying the corresponding $\mathbb{K}$ rule directly. Finally, the serialization result ensures that applying multiple $\mathbb{K}$ rules in parallel can be simulated by applying them one by one, obtaining an interleaving semantics for $\mathbb{K}$ rewriting through standard rewriting, which is one of the desirable goals for all truly concurrent models of computation [17]. Interestingly, a novel and unexpected acyclicity condition (presented in Section 5) was required to ensure serializability.

The remainder of the article is organized as follows. Section 2 provides background and related work. Section 3 formally defines $\mathbb{K}$ rules and relates them to term-rewrite rules. Section 4 formalizes the encoding of terms as graphs used by $\mathbb{K}$ graph rewriting. Section 5 presents the encoding of $\mathbb{K}$ rules as graph-rewriting rules, defines $\mathbb{K}$ graph rewriting as a term-graph rewriting formalism, and shows it admits parallel derivations which are serializable. Finally, Section 6 reflects $\mathbb{K}$ graph rewriting upon term rewriting proving soundness, completeness, and serialization of parallel derivations w.r.t. term-rewriting, and Section 7 concludes. Proofs for all claimed results are available in the companion technical report [21].

2   Background and Related Work

We briefly recall here some basic notions from the theory of term rewriting [1], graph grammars and graph transformations—the double-pushout (DPO) approach—[4], as well as jungle (term-graph) rewriting [9], and relate them to the approach presented in this paper.

Rewriting logic [15] provided the first inference system for concurrent term rewriting allowing sideways and nested parallelism for rule applications.

A signature Σ is a pair (S, F) where S is a set of *sorts* and F is a set of operations $f : w \to s$, where f is an operation symbol, $w \in S^*$ is its arity, and $s \in S$ is its result sort. If w is the empty word ϵ then f is called a constant. T_Σ is the universe of (ground) terms over Σ and $T_\Sigma(\mathcal{X})$ is that of Σ-terms with variables from the S-sorted set $\mathcal{X}$. Given term $t \in T_\Sigma(\mathcal{X})$, let $vars(t)$ be the variables from $\mathcal{X}$ appearing in t. Given an ordered set of variables, $\mathcal{W} = \{\Box_1, \ldots, \Box_n\}$, named *context variables*, or *holes*, a $\mathcal{W}$-*context* over $\Sigma(\mathcal{X})$ (assume that $\mathcal{X} \cap \mathcal{W} = \emptyset$) is a term $C \in T_\Sigma(\mathcal{X} \cup \mathcal{W})$ in which each variable in $\mathcal{W}$ occurs once. The instantiation of a $\mathcal{W}$-context C with an n-tuple $\bar{t} = (t_1, \ldots, t_n)$, written $C[\bar{t}]$ or $C[t_1, \ldots, t_n]$, is the term $C[t_1/\Box_1, \ldots, t_n/\Box_n]$. One can regard $\bar{t}$ as a substitution $\bar{t} : \mathcal{W} \to T_\Sigma(\mathcal{X})$, defined by $\bar{t}(\Box_i) = t_i$, in which case $C[\bar{t}] = \bar{t}(C)$.

A (term) rewrite rule ρ over signature Σ with variables from $\mathcal{X}$ is a tuple $\rho : (\forall \mathcal{X})l \longrightarrow r$, where l and r are terms in $T_\Sigma(\mathcal{X})$. If variables are clear from the context, the quantification can be omitted. A set of such rewrite rules is called a term rewrite system (TRS).

Running Example. Our running example consists of a four-rule rewrite system, where h is a ternary operation, g is binary, f is unary, 0, 1, a, b are constants, and x, y are variables:

$$(1)\ h(x, y, 1) \longrightarrow h(g(x, x), y, 0) \qquad (3)\ a \longrightarrow b$$
$$(2)\ h(x, 0, y) \longrightarrow h(x, 1, y) \qquad\qquad (4)\ f(x) \longrightarrow x$$

together with the term $h(f(a), 0, 1)$ to be rewritten.

We say that a rule $\rho : (\forall \mathcal{X})l \longrightarrow r$ *matches* a Σ-term t at the position given by $\Box$-context C, if there exist a substitution $\sigma : T_\Sigma(\mathcal{X}) \to T_\Sigma$ such that $t = C[\sigma(l)]$. If that is the case, we can apply ρ on t to obtain Σ-term $t' = C[\sigma(r)]$; we say that t *rewrites to* t', written $t \stackrel{\rho, C}{\Longrightarrow} t'$. For example, $h(f(a), 0, 1) \stackrel{(4), h(\Box, 0, 1)}{\Longrightarrow} h(a, 0, 1) \stackrel{(3), h(\Box, 0, 1)}{\Longrightarrow} h(b, 0, 1) \stackrel{(2), \Box}{\Longrightarrow} h(b, 1, 1) \stackrel{(1), \Box}{\Longrightarrow} h(g(b, b), 1, 0)$.

Rewriting logic allows for a high degree of parallelism. For example, one can apply either rules (1), (3), and (4), or rules (2), (3), and (4) concurrently on the term $h(f(a), 0, 1)$, to obtain either $h(g(b, b), 0, 0)$ or $h(b, 1, 1)$, respectively. However, executing both rules (1) and (2) in parallel (which would amount to achieving concurrency with sharing of resources) is impossible within rewriting logic because rule instances are not allowed to overlap [16].

$\mathbb{K}$ rewriting can be viewed as orthogonal to rewriting logic, as one could envision using $\mathbb{K}$ rules to augment its concurrency.

Graph Rewriting. exhibited parallelism with sharing of resources since its formalization as an algebraic theory [6,5]. However, despite several theoretical approaches to defining concurrent programming languages [18,12,2], graph rewriting found more use in giving semantics to modeling languages.

$\mathbb{K}$ uses the idea of interfaces from graph-rewriting to "borrow" the potential for concurrency with sharing of resources from graph rewriting to term rewriting. The definitions and results presented here are used in Sections 4 and 5 to give semantics to $\mathbb{K}$ graph rewriting.

Assuming fixed sets $\mathcal{L}_V$ and $\mathcal{L}_E$ for node and for edge labels, respectively, a *graph G over labels* $(\mathcal{L}_V, \mathcal{L}_E)$ is a tuple $G = \langle V, E, \text{source}, \text{target}, \text{lv}, \text{le}\rangle$, where V is the set of *vertices* (or *nodes*), E is a set of *edges*, source, target : $E \to V$ are the *source* and the *target* functions, and lv : $V \to \mathcal{L}_V$ and le : $E \to \mathcal{L}_E$ are the node and the edge *labeling functions*, respectively.

A *graph morphism* $f : G \to G'$ is a pair $f = \langle f_V : V_G \to V_{G'}, f_E : E_G \to E_{G'}\rangle$ of functions preserving sources, targets, and labels. Let $\mathbf{Graph}(\mathcal{L}_V, \mathcal{L}_E)$ denote the category of graphs over labels $(\mathcal{L}_V, \mathcal{L}_E)$. Given graph G, let $\prec_G \subseteq V \times V$ be its *path relation*: $v_1 \prec_G v_2$ iff there is a path from v_1 to v_2 in G. G is *cyclic* iff there is some $v \in V_G$ s.t. $v \prec_G v$. Given $v \in V_G$, let $G\!\restriction_v$ be the subgraph of G (forwardly) reachable from v.

A *graph rewrite rule* $p : (L \overset{l}{\leftarrow} K \overset{r}{\to} R)$, is a pair of graph morphisms $l : K \to L$ and $r : K \to R$, where l is injective. The graphs L, K, and R are called the *left-hand-side* (lhs), the *interface*, and the *right-hand-side* (rhs) of p, respectively.

Given a graph G, a graph rule $p : (L \overset{l}{\leftarrow} K \overset{r}{\to} R)$, and a *match* $m : L \to G$, a *direct derivation from G to H using p (based on m)* exists iff the diagram to the right can be constructed, where both squares are pushouts in the category of graphs. In this case, C is called the *context* graph, and we write $G \overset{p,m}{\Longrightarrow} H$ or $G \overset{p}{\Rightarrow} H$. Whenever l or r is an inclusion, the corresponding l^* or r^* can be chosen to also be an inclusion. If it exists, H is unique up to graph isomorphism.

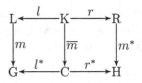

A direct derivation $G \overset{p,m}{\Longrightarrow} H$ exists iff the following *gluing conditions* hold [4]: *(Dangling condition)* no edge in $E_G \setminus m_E(E_L)$ is incident to any node in $m_V(V_L \setminus l_V(V_K))$; and *(Identification condition)* there are no $x, y \in V_L \cup E_L$ with $x \neq y$, $m(x) = m(y)$ and $x, y \notin l(V_K \cup E_K)$. The gluing conditions say that whenever a transformation deletes a node, it should also delete all its edges (dangling condition), and that a match is only allowed to identify elements coming from K (identification condition).

Given a family of graph-rewrite rules $p_i : (L_i \overset{l_i}{\leftarrow} K_i \overset{r_i}{\to} R_i)$, $i = \overline{1,n}$, not necessarily distinct, their *composed graph-rewrite rule* $p_1 + \cdots + p_n$ is a rule $p : (L \overset{l}{\leftarrow} K \overset{r}{\to} R)$ where L, K, and R are the direct sums of the corresponding components from $(p_i)_{i=\overline{1,n}}$ and, similarly, l and r are the canonical morphisms induced by $(l_i)_{i=\overline{1,n}}$ and $(r_i)_{i=\overline{1,n}}$, respectively. Given a graph G, matches $(m_i : L_i \to G)_{i=\overline{1,n}}$

induce a combined match $m : L \to G$ defined as the (unique) arrow amalgamating all individual matches. Matches $(m_i : L_i \to G)_{i=\overline{1,n}}$ have the *parallel independence* property iff for all $1 \leq i < j < n$, $m_i(L_i) \cap m_j(L_j) \subseteq m_i(K_i) \cap m_j(K_j)$. The following result, which we will use in our subsequent development, is proved in [10, Theorem 7.3], recasting previous results from [5,13]:

Theorem 1 (Existence and serializability of parallel derivations). *If $(m_i : L_i \to G)_{i=\overline{1,n}}$ have the parallel independence property and each m_i satisfies the gluing conditions for rule p_i, then the combined match m satisfies the gluing conditions for the composed rule $p_1 + \cdots + p_n$, and thus there exists a graph H such that $G \xrightarrow{p_1 + \cdots + p_n, m} H$. Moreover, this derivation is serializable, i.e., $G \xrightarrow{p_1 + \cdots + p_{n-1}, m'} H_{n-1} \xrightarrow{p_n} H$, where m' is the composition of $(m_i)_{i=\overline{1,n-1}}$.*

Term Graph Rewriting [19] was conceived as an embedding of term rewriting into graph rewriting, having the benefit of using terms and regular rewrite rules as a front end, while using graphs and graph transformations as implementation means. Nevertheless, although subterm sharing allowed by graph rewriting increases the efficiency of the rewriting process, the existing embeddings do not take advantage of the additional potential for concurrency available through graph rewriting. The semantics for $\mathbb{K}$ graph rewriting (Sections 4 and 5) is given by extending the term graph rewriting formalism known as Jungle evaluation [9,11,19] (described below) to allow sharing of resources.

Jungle evaluation uses (directed) hypergraphs to encode terms and rules. Hypergraphs generalize graphs by allowing edges to have multiple (or zero) sources and targets; more precisely, the source and target mappings now yield words over nodes instead of just a node (words, rather than sets, to maintain an order among nodes). A jungle represents a term as an acyclic hypergraph whose nodes are labeled by sort names, and whose edges are labeled by names of operations in the signature; the figure on the right depicts the jungle representation of term $h(f(a), 0, 1)$. Constants are edges without any target. Variables are represented as nodes that are not sources of any edge. Non-linear terms are represented by identifying the nodes corresponding to the same variable.

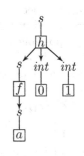

Let VAR_G denote the *variables of G*; we have that $VAR_G = \{v \in V_G \mid outdegree_G(v) = 0\}$. The *term represented by some node v* in a jungle G, $term_G(v)$, is obtained by descending along hyperedges and collecting the hyperedge labels:

$$term_G(v) = \begin{cases} v & \text{if } v \in VAR_G \\ le(e)(term_G^*(target(e))) & \text{otherwise,} \end{cases} \quad \text{where } \{e\} = source^{-1}(v).$$

A *root* of a jungle is a node v such that $indegree(v) = 0$. Let $ROOT_G$ denote the set of roots of G. Given a term t (with variables), a *variable-collapsed* tree representing t is a jungle G with a single root $root_G$ which is obtained from the tree representing t by identifying all nodes corresponding to the same variable, that is, $term_G(root_G) = t$, and for all $v \in V$, $indegree(v) > 1$ implies that $v \in VAR_G$.

A term rewrite rule *left* → *right* is encoded as a *jungle evaluation rule* $L \hookleftarrow K \xrightarrow{r} R$ in the following way: L is a variable-collapsed tree corresponding to *left*; K is obtained from L by removing the hyperedge corresponding to the top operation of *left*; if *right* is a variable (i.e., the rule is collapsing, then R is obtained from K by identifying $root_L$ with *right*; otherwise, R is the disjoint union of K and a variable collapsing tree R' corresponding to *right*, where $root_{R'}$ is identified with $root_L$ and each variable of R' is identified with its counterpart from VAR_L; $L \xleftarrow{l} K$ and $K \xrightarrow{r} R$ are inclusions with the exception that r maps $root_L$ to *right* if *right* is a variable. Rules (3) and (4) in Fig. 2 exemplify this encoding.

Among the many interesting results relating term rewriting with jungle evaluation, we will build our results on the one presented below [11,19]:

Theorem 2 (Soundness and completeness w.r.t. term rewriting). *Let p be an evaluation rule for a rewrite rule ρ, and let G be a jungle.*

Soundness. *If $G \overset{p}{\Rightarrow} H$, then for each $v \in V_G$ $term_G(v) \overset{\rho^*}{\Rightarrow} term_H(r^*(v))$*
Completeness. *If ρ is left-linear and $term_G(v) \overset{\rho}{\Rightarrow} t'$ for some $v \in V_G$, then there exists jungle H such that $G \overset{p}{\Rightarrow} H$.*

In the sequel, we will use the bipartite graph representation of jungles. In this representation, sort nodes and operation edges of the hypergraph both become graph nodes, belonging to the sorts or the operations partitions of the graph nodes, respectively. Graph edges are added to link them similarly to how they were linked in the hypergraph. To maintain the order given by the target word, corresponding edges are labeled with their position in that word. For example, the bipartite graph representation of the jungle representing $h(f(a), 0, 1)$ is graph G in Fig. 2.

3 $\mathbb{K}$ Rules

$\mathbb{K}$ rules describe how a term can be transformed into another term by altering some of its parts. They share the idea of match-and-replace of standard term rewriting, but each $\mathbb{K}$ rule also specifies which part of the pattern is read-only. For example, the $\mathbb{K}$ system below presents a possible encoding as $\mathbb{K}$ rules of the rewrite rules from our running example in Section 2:

(1) $h(\underset{g(x,x)}{\underline{x}}, y, 1)$ (2) $h(x, \underset{1}{\underline{0}}, y)$ (3) $a \to b$ (4) $f(x) \to x$

These read-only patterns (obtained by replacing the underlined subterms with holes) are akin to the interfaces in graph rewriting [4], being used to glue together read-write patterns (underlined subterms and their replacements), that is, subparts to be rewritten. Moreover, through their variables, the read-only patters also provide information which can be used and shared by the read-write patterns. Formally,

Definition 1. A $\mathbb{K}$ **rule** $\rho : (\forall \mathcal{X})\ k[\ L \Rightarrow \mathcal{R}\ ]$ over a signature $\Sigma = (S, F)$ is a tuple $(\mathcal{X}, k, L, \mathcal{R})$, where:

- $\mathcal{X}$ is an S-sorted set, called the **variables** of the rule ρ;
- k is a $\mathcal{W}$-context over $\Sigma(\mathcal{X})$, called the **rule pattern**, where $\mathcal{W}$ are the **holes** of k; k can be thought of as the "read-only" part of ρ;
- $L, \mathcal{R} : \mathcal{W} \to T_\Sigma(\mathcal{X})$ associate to each hole in $\mathcal{W}$ its **original** and **replacement term**; $L, \mathcal{R}$ can be thought of as the "read/write" part of ρ.

We may write $(\forall \mathcal{X})\ k[\ \underset{r_1}{\underline{l_1}}, \ldots, \underset{r_n}{\underline{l_n}}\ ]$ instead of $(\forall \mathcal{X})\ k[\ L \Rightarrow \mathcal{R}\ ]$ whenever $\mathcal{W} = \{\square_1, \cdots, \square_n\}$ and $L(\square_i) = l_i$ and $\mathcal{R}(\square_i) = r_i$; this way, the holes are implicit and need not be mentioned.

A set of $\mathbb{K}$ rules is called a $\mathbb{K}$ **system**.

The variables in $\mathcal{W}$ are only used to identify the positions in k where rewriting takes place; in practice we typically use the compact notation above, that is, underline the to-be-rewritten subterms in place and write their replacement underneath. When the set of variables $\mathcal{X}$ is clear, it can be omitted.

Given a $\mathbb{K}$ rule $\rho : (\forall \mathcal{X})\ k[\ L \Rightarrow \mathcal{R}\ ]$, its associated 0-*sharing* $\mathbb{K}$ *rule* is $\rho_0 : (\forall \mathcal{X})\ \square[\ \underset{\mathcal{R}(k)}{\underline{L(k)}}\ ]$, which is a $\mathbb{K}$ rule specifying the same transformation but without sharing anything. It is relatively easy to see that one can associate to any $\mathbb{K}$ rule ρ as above a regular rewrite rule $K2R(\rho) = (\forall \mathcal{X})L(k) \longrightarrow \mathcal{R}(k)$. This is to account for the fact that, when applied in a non-concurrent fashion, $\mathbb{K}$ rules must obey the standard rewriting semantics.

Conversely, given a rewrite rule $\rho : (\forall \mathcal{X})l \longrightarrow r$, let $R2K(\rho)$ denote the 0-sharing $\mathbb{K}$ rule for which $K2R(R2K(\rho)) = \rho$, that is $(\forall \mathcal{X})\ \square[\ \underset{r}{\underline{l}}\ ]$. For this reason, we take the liberty to denote a 0-sharing $\mathbb{K}$ rule $\rho : (\forall \mathcal{X})\ \square[\ \underset{r}{\underline{l}}\ ]$ by $l \longrightarrow r$.

In the remainder of this paper we will formalize $\mathbb{K}$ (term) rewriting, i.e., rewriting using $\mathbb{K}$ rules, through an embedding into graph rewriting theory, called $\mathbb{K}$ graph rewriting. The reasons for our choice are: (1) the intuition that the pattern k of a $\mathbb{K}$ rule is meant to be "shared" with competing concurrent rule instances is conceptually captured by the notion of interface graphs of graph rewrite rules in the double-pushout (DPO) algebraic approach to graph rewriting [4]; (2) (term) graph rewriting [9,19] was shown to be sound and complete for term rewriting, which we want to preserve for $\mathbb{K}$; and (3) the results in the DPO theory of graph rewriting show that if graph rule instances only overlap on the interface graphs, then they can be concurrently applied and the obtained rewrite step is serializable [10], which is also the desirable semantics for $\mathbb{K}$.

4 $\mathbb{K}$ Term-Graphs

$\mathbb{K}$ *graph rewriting* uses the same mechanisms and intuitions of jungle rewriting, but adapts the jungle term-graphs and graph-rewrite rules to increase the potential for concurrency, both with sharing and without sharing of context. Therefore, $\mathbb{K}$ *term-graphs* are close to the bipartite graph representation of jungles (they actually coincide for ground terms). The difference is that the $\mathbb{K}$ term-graph representation allows certain variables (the anonymous and the pattern-hole variables) to be omitted from the graph. By reducing the number of nodes that need to be shared (i.e., by not forcing these variable nodes to be shared in the interface graph), this "partiality" allows terms at those positions to be concurrently rewritten by other rules.

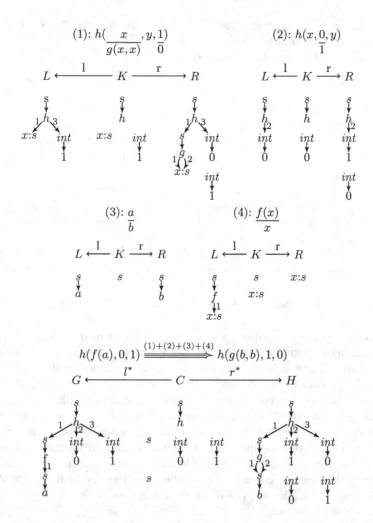

Fig. 2. Graph representations for the $\mathbb{K}$ rules (1)–(4) from the running example and their concurrent application

The top-half of Fig. 2 shows the $\mathbb{K}$ term-graphs involved in the graph representations of the $\mathbb{K}$ rules (1)–(4) of our running example. For example, the representation of variable x can be observed as the (singleton) graph R for rule (4), the constants a and b as graphs L and R from rule (3), and the term $f(x)$ as graph L in rule (4); all these $\mathbb{K}$ term-graphs are also graph jungles [9]. The bottom-half of Fig. 2 shows the $\mathbb{K}$ term-graphs involved in the graph transformation which uses all four rules combined to rewrite the graph representation of $h(f(a), 0, 1)$ (graph G) to one that can be used to retrieve $h(g(b, b), 1, 0)$ (graph H).

The novel aspect of our representation is that, unlike the graph jungles, the $\mathbb{K}$ term-graphs are *partial*: they do not require each operation node to have outward edges for all sorts in its arity. This partiality plays a key role in "abstracting away" the anonymous variables and the holes of the pattern. For example, the number of outward edges specified for the nodes labeled with h have all possible values between 3 (its normal arity) in graphs G and H, to 0, e.g., in graph K for rule (1). This flexibility is crucial for enhancing concurrency; only through it rules (1) and (2) can apply in parallel, as it allows the outward edge of h labeled with 1 to be rewritten by rule (1), while h is still shared with rule (2). This is achieved by relaxing the properties of the graph representation of jungles to allow partially specified operations.

Definition 2. *Given a signature $\Sigma = (S, F)$, a $\mathbb{K}$ Σ-term-graph is a graph G over labels $(S \cup F, \{\epsilon\} \cup Nat)$ satisfying:*

*0. G is bipartite, partitions given by nodes with labels in S—**sort nodes**—, and F—**operation nodes**—;*

1. every operation node labeled by $f : s_1 \cdots s_n \to s$ is
   (i) the target of exactly one edge, labeled with 0 and having its source labeled with s, and
   *(ii) the source of **at most** n edges having distinct labels in $\{1, \cdots, n\}$, such that $\mathrm{lv}(\mathrm{target}(e)) = s_{\mathrm{le}(e)}$ for each such edge e;*

2. every sort node has at most one outward edge; and

3. G is acyclic.

Let $\mathbf{KGraph}_\Sigma$ *denote the full subcategory of* $\mathbf{Graph}(\mathbf{S} \cup \mathbf{F}, \{\epsilon\} \cup Nat)$ *having $\mathbb{K}$ Σ-term-graphs as objects.*

As any graph jungle is a $\mathbb{K}$ term-graph, most of the definitions given for graph jungles in Section 2 can be easily extended for term-graphs. For simplicity $\mathbb{K}$ term-graphs will be referred to as just term-graphs.

Given a set of anonymous variables $A \subseteq \mathcal{X}$, an A-anonymizing variable-collapsed tree representation of a term $t \notin A$ with variables from $\mathcal{X}$ is obtained from a variable-collapsed tree representing t by removing the variable nodes corresponding to variables in A and their adjacent edges.

Let G be a term-graph over $\Sigma = (S, F)$. VAR_G is the set variables of G, that is, sort-nodes of G such that $outdegree(v) = 0$. Note that this definition only captures the non-anonymous variables. To capture all variables, we need to additionally identify partially specified operation nodes.

The set $OPEN_G$ of *open (or incomplete) operation nodes* of G, consists of the operation nodes whose outward edges are incompletely specified. Formally, $OPEN_G = \{v \in \mathrm{lv}^{-1}(S) \mid |s^{-1}(v)| < arity(\mathrm{lv}(v))\}$. The set of *term variables* of G, $TVARS_G$ consists of the variables of G and the positions of the unspecified outward edges for open operation nodes (which stand for anonymous variables). Formally, $TVARS_G = VAR_G \cup \{x_{v,i} \mid v \in OPEN_G, 1 \le i \le arity(\mathrm{lv}(v)) \land i \notin \mathrm{le}(source^{-1}(v))\}$.

The *term represented by some sort node v in a term-graph G*, $term_G(v)$, is obtained by descending along operation nodes and collecting their labels:

$$term_G(v_s) = \begin{cases} v_s, \text{ if } v_s \in VAR_G \\ \sigma(t_1, \ldots, t_n), \text{ if } \{v_e\} = target(source^{-1}(v_s)), \mathrm{le}(v_e) = \sigma : s_1 \ldots s_n \to s, \\ \qquad \text{and } t_i = subterm_G(v_e, i), 1 \le i \le n \end{cases}$$

where $subterm_G$ is defined on pairs of operation nodes with integers by

$$subterm_G(v_e, i) = \begin{cases} x_{v_e,i}, \text{ if } x_{v_e,i} \in TVARS_G \\ term_G(target(e)), \text{ if } source(e) = v_e \text{ and } \mathrm{le}(e) = i \end{cases}$$

5   𝕂 Graph Rewriting

As we want 𝕂 graph rewriting to be a conservative extension of graph jungle evaluation, every 0-sharing 𝕂 rule $(\forall \mathcal{X})\, \Box[\ \underline{left}\ ]$ is encoded as the graph jungle $\overline{right}$ evaluation rule corresponding to the rewrite rule $left \to right$—see, for example the encodings of rules (3) and (4) in Fig. 2. However, if the rule pattern k is non-empty, then the rule is encoded so that the variable-collapsed tree representing k would not be modified by the rule. To be more precise, instead of obtaining K by removing the outgoing edge from the root of L, we will instead only remove the edges connecting the hole variables to their parent operations. Moreover, to further increase concurrency, the variables which appear in the read only pattern k but not in the left substitution are anonymized. However, departing from the definition of jungle rules, we relax the requirement that the order between the nodes of K and variables of R should be the same as in L, to allow rules such as reading or writing the value of a variable from a store.

Consider the representation of the 𝕂 rule (1) in Fig. 2, namely $h(\ \underline{x}\ , y, 1)$. $\overline{g(x,x)}\quad 0$ The left-hand-side is represented as a $\{y\}$-anonymized variable collapsed tree representing $h(x, y, 1)$; variable y is anonymized as only appearing in the pattern k. The interface K is obtained from L by severing (through the removal of edges labeled by 1 and 3) the part of L representing the read-only pattern $h(\Box_1, y, \Box_2)$ (which is the $\{y, \Box_1, \Box_2\}$-anonymized variable collapsed tree representing $h(\Box_1, y, \Box_2)$) from the parts of L representing the left substitution (namely, x and 1). Thus, the l morphism from K to L is clearly an inclusion. R is obtained by taking the disjoint union between K and the variable-collapsed trees corresponding to terms $g(x, x)$ and 0 given by the right substitution, identifying the variables, and "gluing" them to the part representing the read-only pattern through edges from operation node h labeled 1 and 3, respectively. Like l, the r morphism can also be chosen to be an inclusion.

The graph rules in Fig. 2 are obtained using the definition below. To avoid clutter, we do not depict node or edge names (except for variables). Also, the actual morphisms are not drawn (they are either inclusions or obvious collapsing morphisms).

Definition 3. *Let $\rho :(\forall \mathcal{X}) k[L \Rightarrow R]$ be a $\mathbb{K}$ rule.*

If ρ is 0-sharing, then the $\mathbb{K}$ graph rewrite rule representing ρ coincide with the jungle evaluation rule corresponding to the rewrite rule associated to ρ.

*Otherwise, a $\mathbb{K}$ **graph rewrite rule** representing ρ is a graph rewrite rule $(L_\rho \xleftarrow{l_\rho} K_\rho \xrightarrow{r_\rho} R_\rho)$ such that:*

L_ρ *is an A-anonymized variable collapsed tree representation of $L(k)$, where $A = vars(k) \setminus vars(L)$ are the anonymous variables of ρ;*

K_ρ *Let K_0 be the subgraph of L_ρ which is a A-anonymized variable collapsed tree representing k; then $K_\rho = (V_{K_\rho}, E_{K_\rho})$ is given by $V_{K_\rho} = V_{L_\rho}$ and $E_{K_\rho} = E_{L_\rho} \setminus \{e \in E_{L_\rho} \mid \text{source}(e) \in V_{K_0} \text{ and } \text{target}(e) \notin V_{K_0}\}$. l_ρ is the inclusion morphism.*

R_ρ *Let R_0 be an A-anonymized variable collapsed tree representation of $R(k)$ containing K_0 as a subgraph. Then R_ρ is obtained as the pushout between the inclusions of $K_0 \cup VAR_{R_0}$ into K_ρ and R_0, respectively.*

The nodes from K_0 will be called pattern nodes.

Note that the edges removed from L_ρ to obtain K_ρ are those whose target corresponds to the hole variables of k.

Similarly to the graph jungle rules, the (basic) $\mathbb{K}$ graph rules defined above ensure that the gluing conditions are satisfied for any matching morphism. For the remainder of this section, let us fix G to be a term-graph, $\rho_i : (L_i \xleftarrow{l_i} K_i \xrightarrow{r_i} R_i)$, $i = \overline{1,n}$ to be $\mathbb{K}$ graph-rewrite rules, and $m_i : L_i \to G$ to be parallel independent matches. Let $\rho : (L \xleftarrow{l} K \xrightarrow{r} R)$ be the composed rule of $(\rho_i)_{i=\overline{1,n}}$, and let $m : L \to G$ be the composition of the individual matches. It follows that m satisfies the gluing conditions for ρ, and thus (ρ, m) can be applied as a graph transformation. Let us now provide a concrete construction for the derivation of (ρ, m) in **Graph** which is used in proving the subsequent results.

The pushout complement object of m and l can be defined in **Graph** as $C = G \setminus m(L \setminus K)$ where the difference is taken component-wise. That C is a graph is ensured by the gluing conditions. The standard construction of the pushout object H is to factor the disjoint union of C and R through the equivalence induced by the pushout morphism $\overline{m} : K \to C$ and r. We do this directly, by taking preference for elements in C, and thus choosing representatives from $\overline{m}(K)$ and by choosing as representatives variables for the equivalence classes induced by the parts of r belonging to collapsing rules.

Suppose G is a $\mathbb{K}$ graph representation of term t, i.e., that $ROOT_G = \{root_G\}$, $G = G\lceil_{root_G}$, and $term_G(root_G) = t$. When applying a (composed, or not) $\mathbb{K}$ graph rewrite rule to graph G, $root_G$ must be preserved in the context C, because K contains all nodes of L. Therefore, let us define the top of the obtained graph H as being $root_H = r^*(root_G)$. Note that $root_H$ might not be equal to $root_G$, because $root_G$ could be identified with a variable node by a collapsing rule; moreover, $root_H$ might not be the only element of $ROOT_H$, because of the potential "junk" left by the application of the rule. Nevertheless, the term $term_H(root_H)$ would be the one to which $term_G(root_G)$ was rewritten.

To show that **KGraph$_\Sigma$** admits similar constructions for (composed) $\mathbb{K}$ graph-rewrite rules as **Graph**, that is, that the graphs described above are in fact term-graphs, we need to strengthen the constraints on the matching morphisms.

Indeed, without further constraints, applying $\mathbb{K}$ graph rules on term-graphs can produce cyclic graphs. Consider $\mathbb{K}$ rules $f(g(\underline{a}), x)$ and $f(y, h(\underline{b}))$ together with the term to rewrite $f(g(a), h(b))$. Upon formalizing terms as term-graphs and $\mathbb{K}$ rules as $\mathbb{K}$ graph rewrite rules, the result of applying the composed $\mathbb{K}$ graph rewrite rule on the graph representing $f(g(a), h(b))$ is the graph H in Fig. 3, *which has a cycle* and thus it is not a term-graph.

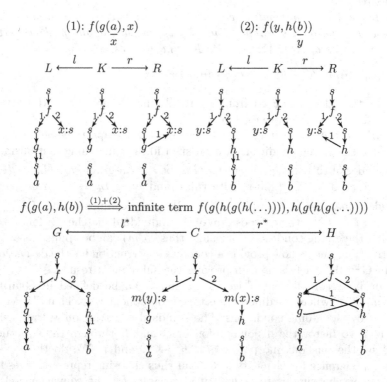

Fig. 3. Parallel $\mathbb{K}$ graph rewriting can introduce cycles

The reason for the cycle being introduced is that the matches overlap, allowing variable nodes to precede operation nodes in the path order of G, while r reorders the mapping of the variables to create a cycle. In jungle rewriting this issue is prevented by imposing a statically checkable condition on the rules, namely that the path relation between the nodes preserved from L should not be changed by R. Formally, we say that a rule $\rho : (L \overset{l}{\leftarrow} K \overset{r}{\rightarrow} R)$ is *cycle-free* if whenever $v \prec_R x$ with $v \in V_K$ and $x \in VAR_L \cap V_K$, it must be that $v \prec_L x$. This condition is sufficient to prevent the introduction of cycles; however, we find it rather strong in our programming language context—in particular, this condition would disallow rules like the one for reading the value of a variable from the store. We propose below a more relaxed sufficient condition (on the matching morphism m) for avoiding the introduction of cycles.

Given a (composed) term-graph rewrite rule $\rho : (L \overset{l}{\leftarrow} K \overset{r}{\rightarrow} R)$, r induces on K a (partial) replacement order $\prec_r = r^{-1}(\prec_R)$, i.e., $v_1 \prec_r v_2$ in K iff $r(v_1) \prec_R r(v_2)$ (there is a path from $r(v_1)$ to $r(v_2)$ in R). Moreover, given match m of p into G, m induces on K a (partial) matching order $\prec_m = l^{-1}(m^{-1}(\prec_G))$, i.e., $v_1 \prec_r v_2$ in K iff $m(v_1) \prec_G m(v_1)$ (l is an inclusion). Although both these (partial) orders are strict, their combination is not guaranteed to remain strict. We say that the match m is *cycle-free* w.r.t. p if the transitive closure of $\prec_m \cup \prec_r$ is also a strict (partial) order.

Proposition 1. *If any matching morphism for a $\mathbb{K}$ graph rewriting rule ρ is cycle-free, then ρ is a jungle graph rewriting rule. If ρ is a $\mathbb{K}$ graph rule, G is a term-graph, $G \overset{(\rho,m)}{\Longrightarrow} H$, and m is cycle-free w.r.t. ρ, then H is acyclic.*

The following result guarantees that if the original graph is a tree, then cycle-freeness of the matching morphism characterizes acyclicity of the resulting graph.

Proposition 2. *Let G be a tree term-graph. If ρ is a simple $\mathbb{K}$ graph rule and m is a match for ρ into G, then m is cycle-free. If ρ is a composed $\mathbb{K}$ graph rule and $G \overset{(\rho,m)}{\Longrightarrow} H$, then H is acyclic iff m is cycle-free w.r.t. ρ.*

Next result uses Theorem 1 to prove a similar result for the restricted category of $\mathbb{K}$ term graphs. Namely, it shows that, under cycle-freeness conditions, **KGraph$_\Sigma$** is closed under (parallel) derivations using $\mathbb{K}$ graph rewrite rules.

Theorem 3. *Let G, $(\rho_i)_{i=\overline{1,n}}$, $(m_i)_{i=\overline{1,n}}$, ρ, m, C, and H be defined as above. If m is cycle-free w.r.t. p, then:*

(Parallel) Derivation: $G \overset{\rho,m}{\underset{\mathbf{KGraph}_\Sigma}{\Longrightarrow}} H$;

Serialization: *There exist $(G_i)_{i=\overline{0,n}}$ such that $G_0 = G$, $G_n = H$, and $G_{i-1} \overset{\rho_i}{\underset{\mathbf{KGraph}_\Sigma}{\Longrightarrow}} G_i$ for each $1 \leq i \leq n$.*

6 $\mathbb{K}$ Term Rewriting

This section formally defines $\mathbb{K}$ rewriting on terms and shows that $\mathbb{K}$ rewriting is a conservative extension of the standard term rewriting relation.

Theorem 3 allows us to define $\mathbb{K}$ rewriting as follows:

Definition 4. *Let t be a Σ-term and $\rho_1, \cdots, \rho_n$ be $\mathbb{K}$ rules (not necessarily distinct). Then $t \overset{\rho_1+\cdots+\rho_n}{\Longrightarrow} t'$ iff there is a term-graph H s.t. $G \xrightarrow[\mathbf{KGraph_\Sigma}]{K2G(\rho_1)+\cdots+K2G(\rho_n)} H$ and $term_H(root_H) = t'$, where G is the tree term-graph representing t. We say that $t \Rightarrow t'$ iff there is a (composed) $\mathbb{K}$ rule ρ s.t. $t \overset{\rho}{\Rightarrow} t'$.*

We can give a straightforward definition for what it means for a $\mathbb{K}$ rule to match a term: *one* $\mathbb{K}$ rule $\rho : (\forall \mathcal{X})\; k[\; L \Rightarrow R\;]$ matches a term t at the position given by the $\square$-context C, yielding substitution σ, iff its corresponding rewrite rule $K2R(\rho) : (\forall X)L(k) \to R(k)$ matches t at the position given by C, yielding σ, that is, iff $t = C[\sigma(L(k))]$. This conforms to the intuition that, when applied sequentially, $\mathbb{K}$ rules behave exactly as their corresponding rewrite rules.

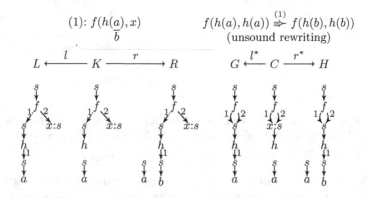

Fig. 4. Subterm sharing might lead to unsound $\mathbb{K}$ graph rewriting

However, it turns out that, although preserving the term-graph structure (under cycle-freeness assumptions), $\mathbb{K}$ rewriting on graphs might not be sound w.r.t. term rewriting in the presence of subterm sharing. Consider the example in Fig. 4. We want to apply rule $f(h(\underset{b}{a}), x)$, corresponding to the regular rewrite rule $f(h(a), x) \to f(h(b), x)$, to the term $f(h(a), h(a))$. If we would represent $f(h(a), h(a))$ as a tree, then the $\mathbb{K}$ graph rewriting step would be sound, leading to a graph depicting $f(h(b), h(a))$; however, if we decide to collapse the tree representing $h(a)$ then we obtain $f(h(b), h(b))$, as depicted in Fig. 4 which cannot be obtained through regular rewriting. The reason for this unsound rewriting is that part of the read-only pattern of the rule is shared. To overcome this, we will restrict the read-only pattern of the rule to only match against a tree in the graph to be rewritten. We say that a match $m : L \to G$ of a $\mathbb{K}$ graph rewrite rule $\rho : (L \overset{l}{\leftarrow} K \overset{r}{\to} R)$ is *safe* if $m(K\lceil_{root_L})$ is a tree in G, that is, if $indegree_G(m_V(v)) = 1$ for any $v \in V_{K\lceil_{root_L}} \setminus \{root_L\}$. Note that, if G is a tree then all matching morphisms on G are safe.

Proposition 3. *Let ρ be a proper $\mathbb{K}$ rewrite rule, let ρ_0 be its associated 0-sharing $\mathbb{K}$ rewrite rule, and let m be a cycle-free safe matching morphism for $K2G(\rho)$ in G. Let H be such that $G \xrightarrow[\textbf{KGraph}_\Sigma]{K2G(\rho),m} H$, and let H' be such that $G \xrightarrow[\textbf{KGraph}_\Sigma]{K2G(\rho_0),m} H'$. Then for any $v \in ROOT_G$, $term_H(v) = term_H(v)$.*

Since, as previously mentioned, the $\mathbb{K}$ graph representation of a term t without anonymous variables is precisely the bipartite graph representation of the jungle representing the same term, and since the $\mathbb{K}$ term-graph representation of a 0-sharing $\mathbb{K}$ rewrite rule is the graph representation of the jungle rule representing the rewrite rule associated to it, we can use the soundness and completeness of jungle rewriting (Theorem 2) to prove the sequential soundness and completeness of $\mathbb{K}$ graph rewriting w.r.t. standard term rewriting, and, by combining that with Theorem 3, to prove the serializability result for $\mathbb{K}$ rewriting.

Theorem 4. *Let ρ, ρ_1, $\ldots$, ρ_n be $\mathbb{K}$ rules. Then:*

Completeness. *If $t \xrightarrow{K2R(\rho)} t'$ then $t \overset{\rho}{\Rightarrow} t'$.*

Soundness. *If $t \overset{\rho}{\Rightarrow} t'$ then $t \xrightarrow{K2R(\rho)^*} t'$.*

Serializability. *If $t \xrightarrow{\rho_1 + \cdots + \rho_n} t'$, then there exists a sequence of terms $t_0, \cdots, t_n$, such that $t_0 = t$, $t_n = t'$, and $t_{i-1} \overset{\rho_i^*}{\Rightarrow} t_i$.*

Therefore, $\mathbb{K}$ rewriting is sound and complete for term rewriting, while providing a higher degree of concurrency in one step than existing approaches.

7   Conclusion

This paper presents a truly concurrent semantics with sharing of resources for the $\mathbb{K}$ framework, a term-rewriting-based semantic framework specialized for defining programming languages and calculi. The distinguishing aspect of the $\mathbb{K}$ rewrite rules is that they explicitly state what portions of the term can be concurrently shared with other rules. This sharing information allows one to increase the potential for concurrent rewriting, but it may also lead to inconsistencies if not used properly. We showed that, under reasonable conditions, $\mathbb{K}$ rewriting is actually sound, complete, and serializable w.r.t. term rewriting. Moreover, although being motivated by the $\mathbb{K}$ framework, $\mathbb{K}$ rewriting is not confined to it; it rather is an extension of rewriting which allows additional concurrency for any rewriting-based formalism.

Future work. Although we have found a sufficient condition for sound and serializable concurrent executions, this condition is rather semantical, and might be non-trivial to check. However, all of the rule combinations in our current definitions of programming languages seem to generate cycle-free executions. An interesting research problem would be to find generic enough syntactic conditions which would guarantee that cycle-freeness is satisfied for all possible combinations of matches.

References

1. Baader, F., Nipkow, T.: Term rewriting and all that. Cambridge University Press, New York (1998)
2. Baldan, P., Gadducci, F., Montanari, U.: Modelling calculi with name mobility using graphs with equivalences. In: TERMGRAPH. ENTCS, vol. 176(1), pp. 85–97 (2007)
3. Berry, G., Boudol, G.: The chemical abstract machine. J. of Theoretical Computer Science 96(1), 217–248 (1992)
4. Corradini, A., Montanari, U., Rossi, F., Ehrig, H., Heckel, R., Löwe, M.: Algebraic approaches to graph transformation: Basic concepts and double pushout approach. In: Handbook of Graph Grammars, vol. 1, pp. 163–246. World Sci. (1997)
5. Ehrig, H., Kreowski, H.-J.: Parallelism of Manipulations in Multidimensional Information Structures. In: Mazurkiewicz, A. (ed.) MFCS 1976. LNCS, vol. 45, pp. 284–293. Springer, Heidelberg (1976)
6. Ehrig, H., Pfender, M., Schneider, H.J.: Graph-grammars: An algebraic approach. In: SWAT (FOCS), pp. 167–180 (1973)
7. Ellison, C., Rosu, G.: An executable formal semantics of C with applications. In: POPL, pp. 533–544 (2012)
8. Felleisen, M., Friedman, D.P.: Control operators, the SECD-machine, and the lambda-calculus. In: 3rd Working Conference on the Formal Description of Programming Concepts, Ebberup, Denmark, pp. 193–219 (August 1986)
9. Habel, A., Kreowski, H.-J., Plump, D.: Jungle Evaluation. In: Sannella, D., Tarlecki, A. (eds.) Abstract Data Types 1987. LNCS, vol. 332, pp. 92–112. Springer, Heidelberg (1988)
10. Habel, A., Müller, J., Plump, D.: Double-pushout graph transformation revisited. Mathematical Structures in Computer Science 11(5), 637–688 (2001)
11. Hoffmann, B., Plump, D.: Implementing term rewriting by jungle evaluation. RAIRO—Theoretical Informatics and Applications 25, 445–472 (1991)
12. Kastenberg, H., Kleppe, A., Rensink, A.: Defining Object-Oriented Execution Semantics Using Graph Transformations. In: Gorrieri, R., Wehrheim, H. (eds.) FMOODS 2006. LNCS, vol. 4037, pp. 186–201. Springer, Heidelberg (2006)
13. Kreowski, H.-J.: Transformations of Derivation Sequences in Graph Grammars. In: Karpinski, M. (ed.) FCT 1977. LNCS, vol. 56, pp. 275–286. Springer, Heidelberg (1977)
14. Meredith, P., Hills, M., Rosu, G.: An executable rewriting logic semantics of K-Scheme. In: SCHEME, Tech. Rep. DIUL-RT-0701, pp. 91–103. U. Laval (2007)
15. Meseguer, J.: Conditional rewriting logic as a unified model of concurrency. J. of Theoretical Computer Science 96(1), 73–155 (1992)
16. Meseguer, J.: Rewriting Logic as a Semantic Framework for Concurrency. In: Sassone, V., Montanari, U. (eds.) CONCUR 1996. LNCS, vol. 1119, pp. 331–372. Springer, Heidelberg (1996)
17. Montanari, U.: True Concurrency: Theory and Practice. In: Bird, R.S., Wing, J.M., Morgan, C.C. (eds.) MPC 1992. LNCS, vol. 669, pp. 14–17. Springer, Heidelberg (1993)
18. Montanari, U., Pistore, M., Rossi, F.: Modeling concurrent, mobile and coordinated systems via graph transformations. In: Handbook of Graph Grammars, vol. 3, pp. 189–268. World Sci. (1999)
19. Plump, D.: Term graph rewriting. In: Handbook of Graph Grammars, vol. 2, pp. 3–61. World Sci. (1999)
20. Rosu, G., Șerbănuță, T.F.: An overview of the K semantic framework. J. of Logic and Algebraic Programming 79(6), 397–434 (2010)
21. Șerbănuță, T.F., Rosu, G.: KRAM—extended report. Technical Report, UIUC (September 2010), http://hdl.handle.net/2142/17337
22. Wright, A.K., Felleisen, M.: A syntactic approach to type soundness. Information and Computation 115(1), 38–94 (1994)

Probabilistic Graph Transformation Systems

Christian Krause* and Holger Giese

Hasso Plattner Institute at the University of Potsdam
Prof.-Dr.-Helmert-Str. 2-3, 14482 Potsdam, Germany
{christian.krause,holger.giese}@hpi.uni-potsdam.de

Abstract. In the recent years, extensions of graph transformation systems with quantitative properties, such as real-time and stochastic behavior received considerable attention. In this paper, we describe the new quantitative modeling approach of *probabilistic graph transformation systems* (PGTSs) which incorporate probabilistic behavior into graph transformation systems. Among other applications, PGTSs permit to model randomized protocols in distributed and mobile systems, and systems with on-demand probabilistic failures, such as message losses in unreliable communication media. We define the semantics of PGTSs in terms of Markov decision processes and employ probabilistic model checking for the quantitative analysis of finite-state PGTS models. We present tool support using HENSHIN and PRISM for the modeling and analysis and discuss a probabilistic broadcast case study for wireless sensor networks.

1 Introduction

Graph transformation systems (GTSs) provide a natural and expressive formalism for modeling dynamic distributed and mobile systems. In the recent past, extensions of graph transformation systems with quantitative properties such as real-time [1,2] and stochastic behavior [3] have been developed to increase their expressiveness further. However, many protocols used in distributed systems also employ randomization in the form of discrete probabilistic behavior to ensure liveness properties or to optimize quality of service properties without introducing a centralized authority. Probabilistic behavior is also a key ingredient for describing on-demand random failures, such as message losses in unreliable communication media. However, such discrete probabilistic decisions are not supported by any of the existing quantitative graph transformation based modeling approaches. Furthermore, since the employed models are always abstractions of the real systems, they inevitably contain nondeterminism for which no probabilistic assumption can be made. Consequently, a modeling approach is required that also permits to combine probabilistic and nondeterministic behavior.

As a case study for probabilistic and nondeterministic behavior in distributed systems, we consider a probabilistic broadcast protocol for wireless sensor networks (WSNs) described and formally analyzed in [4]. WSNs are decentralized

---

* Supported through a research grant by the Hasso Plattner Institute (HPI).

H. Ehrig et al.(Eds.): ICGT 2012, LNCS 7562, pp. 311–325, 2012.

and spatially distributed networks that do not rely on an existing infrastructure, such as routers or access points. To acquire or distribute information in such networks, often a simple form of a flooding protocol is employed, where *flooding* means that a node that receives a message forwards it to all its neighbors by a broadcast. However, the nodes in a WSN typically have to work with very limited resources such that unnecessary communications should be kept at a minimum in order to save energy. So-called gossiping protocols use randomization in order to reduce this overhead. In a gossiping protocol, every node decides with a certain probability whether to forward a received message or not, which reduces the communication costs. While the local decision whether to forward a received message or not requires probabilistic behavior, the asynchronous nature of the message delivery in such a network requires nondeterministic behavior.

In this paper, we introduce *probabilistic graph transformation systems* (PGTSs) which permit to describe both probabilistic and nondeterministic phenomena, and develop methods for their quantitative analysis. Transformation rules in PGTSs can have multiple right-hand sides, each of them annotated with a probability. The choice for an applicable rule and a particular match is nondeterministic, whereas the effect of a rule is probabilistic. We define the semantics of PGTSs in terms of Markov decision processes (MDPs) and employ probabilistic model checking for the quantitative analysis of finite-state PGTS models. We present tool support for the modeling and analysis of PGTSs using the HEN-SHIN [5] graph transformation tool and the probabilistic model checker PRISM [6]. We discuss some of the advantages of PGTSs over component-based modeling approaches using the WSN case study presented in [4]. Briefly, PGTSs provide a better modeling scalability as the complexity of the model does not grow with the complexity of the topology. Also, models in the graph transformation-based approach can be more easily adjusted to reflect topology or protocol changes.

Organization Section 2 and 3 recall the formal foundations, specifically typed graph transformation systems, Markov decision processes and the probabilistic logic PCTL. We use the case study to illustrate their particular capabilities. In Section 4 we introduce probabilistic graph transformation systems as a modeling language and define their semantics. In Section 5, we present a probabilistic model of our case study and compare it to existing models. In Section 6 we present our tool support and discuss the obtained analysis results for the case study. Section 7 contains related work and Section 8 conclusions and future work.

2   Typed Graph Transformation

We follow the double pushout (DPO) approach for typed graph transformation [7,8], which builds on category theory. Note that our probabilistic extensions could be also applied to the single pushout (SPO) approach.

Definition 1 (Typed graphs and graph morphisms).

-   *A graph $G = \langle V, E, s, t \rangle$ consists of a set of nodes V, a set of edges E and source and target functions $s, t : E \to V$.*

- A graph morphism $f : G_1 \to G_2$ is a pair of functions $f = \langle f_V, f_E \rangle$ with $f_V : V_1 \to V_2$, $f_E : E_1 \to E_2$, such that $f_V \circ s_1 = s_2 \circ f_E$, $f_V \circ t_1 = t_2 \circ f_E$.
- Let T be a graph, called a type graph. A typed graph $\langle G, \tau \rangle$ consists of a graph G and a graph morphism $\tau : G \to T$.
- For two typed graphs $\langle G_i, \tau_i \rangle$ with $i \in \{1, 2\}$ over the same type graph, a typed graph morphism is a graph morphism $f : G_1 \to G_2$ with $\tau_2 = f \circ \tau_1$.

Definition 2 (Rule). *A rule $p = \langle L \xleftarrow{\ell} K \xrightarrow{r} R \rangle$ is a span of injective typed graph morphisms. The graph L is called the left-hand side (LHS), and R the right-hand side (RHS) of p.*

Definition 3 (Transformation). *Given a rule $p = \langle L \xleftarrow{\ell} K \xrightarrow{r} R \rangle$, a typed graph M, and a typed graph morphism $m : L \to G$, called a match. A transformation $M \overset{p,m}{\Longrightarrow} N$ is defined by the double pushout diagram in Fig. 1.*

Operationally, the graph M is transformed by (1) removing the occurrence of $L \backslash \ell(K)$ in M, yielding the graph C, and (2) adding a copy of $R \backslash r(K)$ to C. A rule is applicable w.r.t. a given match, if the so-called gluing condition is satisfied. Informally, all dangling edges must be explicitly removed by the rule and all non-injectively matched nodes and edges must be consistently removed or preserved by the rule.

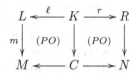

Fig. 1. DPO diagram

Negative Application Conditions. To increase the expressiveness of rules, several extensions of the basic format in Definition 2 are available. For instance, negative application conditions (NACs) provide a means to restrict the applicability of rules. Formally, a NAC is a pair $\langle N, c \rangle$ with N a typed graph and $c : L \to N$ a typed graph morphism from the rule's LHS into N. The applicability of the rule is restricted to those matches which cannot be extended to any of its NACs.

Nested Rules and Amalgamation. Nested rules provide a concept to extend a match of a basic rule to an unbounded number of substructures and to perform modifications to all these structures in an atomic step. Formally, a nested rule is modeled by a possibly nested embedding of one rule into another. The application of a nested rule can be carried out by constructing an *amalgamated* rule and applying it as a normal transformation as in Definition 3. For a comprehensive discussion of the formal foundations of parallel rule applications we refer to [9]. Regarding tooling, we use nested rules as supported by the approaches in [10,5]. For the examples in this paper, we require only nested rules of depth 1.

As our case study, we model the variant of the gossiping protocol with nondeterministic execution order and message collisions as presented in [4] using graph transformation. The type graph for this example is depicted in Fig. 2. Wireless sensors are modeled as nodes and network topologies using edges between such nodes. We usually assume bidirectional connections which we formally model

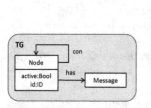

Fig. 2. Type graph

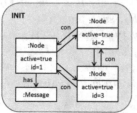

Fig. 3. Initial graph for a simple topology

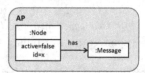

Fig. 4. Atomic proposition *received*(x)

using two edges in opposite directions. Additionally, every node can hold references to an unbounded number of messages. To identify nodes and to model their status, we use an attribute id of the finite type $ID = \{1, \ldots, n\}$ and a Boolean attribute active. Note that we did not formally introduce attributes. However, attributes over finite data domains can be easily encoded in graphs.

The behavior of the gossiping protocol is modeled using three rules. Fig. 5 depicts the rule $send_1$ which models the situation where a node decides to broadcast its message to all its neighbor nodes. We depict only the LHS and the RHS and indicate the partial mapping between them using indices. A node becomes inactive if it correctly received a message. The broadcasting is possible only as long as the sender is active and has exactly one message (ensured using a NAC). If a node has more than one message, a collision occurred. Informally, if multiple neighbors send messages to the same node, it can happen that the communication is disturbed and that the node receives only noise. We use a nested rule to model the synchronous broadcast to all neighbors, i.e., every connected node receives a copy of the original message. Fig. 6 depicts the rule $send_2$ which models that the node decides *not* to broadcast the message to its neighbors. The node can become inactive only if it correctly received the message (ensured using a NAC). Fig. 7 depicts the rule *reset* which allows a node to reset itself by deleting all its messages in the case of a collision. A simple network topology consisting of only three nodes describing a possible initial graph is depicted in Fig. 3.

This model of the gossiping protocol contains only nondeterministic behavior. During the execution, multiple nodes in the network may be able to send a message at the same time. In our model, the choice for a particular sending order is nondeterministic, which allows us to capture unknown details of the network, such as the internal behavior of the nodes and the network characteristics, e.g., varying signal strengths. Note also that the nondeterministic modeling enables us to reason about the range of possible behaviors, particularly about worst-case and best-case execution orders. However, in the used approach also the fact *whether* a node forwards a message or not has to be modeled as nondeterministic, even though this decision should be probabilistic according to the gossiping protocol. In particular, it is not possible to quantitatively specify the likelihood of the message forwarding. Similarly, it is also not possible to specify probabilities for message losses due to communication in unreliable media.

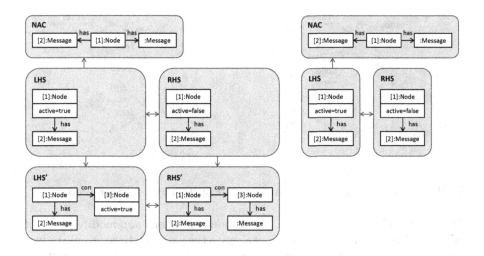

Fig. 5. Rule $send_1$ Fig. 6. Rule $send_2$

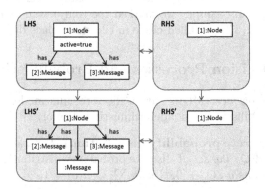

Fig. 7. Rule *reset*

Fig. 8 depicts the state space for the example with the initial configuration in Fig. 3 as a labeled transition system (LTS) in which states correspond to graphs and transitions to rule applications. For a formal definition of the derived state spaces see [11,3]. Note that both the asynchronous execution order and the local decision whether a message is forwarded is nondeterministic in this model.

To later reason about the derived state spaces, we use a specification format for graph-based atomic propositions and define a derived state labeling function. In the following, we use the notation $G \xRightarrow{p}$ to denote that there exists a graph G' and a match m such that $G \xRightarrow{p,m} G'$.

Definition 4 (Atomic propositions and labeling functions). *An* atomic proposition *is a non-modifying rule (with identical LHS and RHS). Let AP be a set of atomic propositions and Q a set of typed graphs. The* labeling function $L_Q^{AP} : Q \to 2^{AP}$ *is defined as:* $L_Q^{AP}(G) = \{\, a \in AP \mid G \xRightarrow{a} \,\}$ *for all* $G \in Q$.

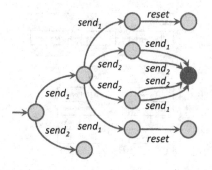
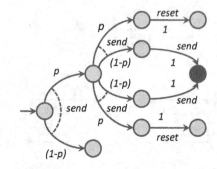

Fig. 8. Labeled transition system   **Fig. 9.** Markov decision process

Thus, a graph G satisfies an atomic proposition in form of a non-modifying rule if it is applicable to G. For our example, we define the parameterized atomic proposition $received(x)$ depicted in Fig. 4, where only the LHS of the non-modifying rule is shown. The parameter x ranges over the set of node IDs. Intuitively, a state satisfies $received(x)$ if the node x successfully received a message and became inactive. As an example, we marked the (only) state where $received(x)$ holds for all $x \in \{1, 2, 3\}$ by a filled circle in the state space depicted in Fig. 8.

3   Markov Decision Processes and Probabilistic Logic

Markov decision processes (MDPs) are a discrete-time model for systems exhibiting both probabilistic and nondeterministic behavior.

Definition 5 (Discrete probability distribution). *For a denumerable set Q, we denote with $Dist(Q)$ the set of discrete probability distributions over Q, i.e., the set of all functions $\mu : Q \to [0, 1]$ with $\sum_{q \in Q} \mu(q) = 1$.*

Definition 6 (Markov decision process). *A Markov decision process (MDP) $\mathcal{M} = (Q, q_{init}, Steps)$ consists of a denumerable set of states Q, an initial state $q_{init} \in Q$ and a probabilistic transition function $Steps : Q \to 2^{Dist(Q)}$.*

Note that $Steps$ assigns a *set* of probability distributions to states in order to incorporate nondeterministic choice. Fig. 9 depicts the required MDP for the WSN example. In contrast to the LTS in Fig. 8, the local decision whether a particular node forwards a message or not, i.e., whether $send_1$ or $send_2$ is applied for a given match, is probabilistic in this model. The intuition is that the two basic rules $send_1$ and $send_2$ are combined into one probabilistic rule $send$ which yields different results according to a given probability distribution. Specifically, the message is forwarded with a probability of p, and not forwarded with a probability of $(1 - p)$. However, the decision which of the enabled probability distributions is chosen remains nondeterministic. Thus, in contrast to the LTS, the MDP allows us to describe both the nondeterministic execution order of the message sending and the probabilistic decision whether to forward a message.

Formally, the operational semantics of an MDP can be understood as follows. A *probabilistic transition*, written as $q \xrightarrow{\mu} q'$, is made from a state $q \in Q$ by:

1. nondeterministically selecting a distribution $\mu \in Steps(q)$, and
2. making a probabilistic choice of target state q' according to μ.

A *path* of an MDP is a non-empty finite or infinite sequence of probabilistic transitions:

$$\omega = q_0 \xrightarrow{\mu_0} q_1 \xrightarrow{\mu_1} q_2 \xrightarrow{\mu_2} \dots$$

where for all $i \in \mathbb{N}$ it holds that $q_i \in Q$, $\mu_i \in Steps(q_i)$, and $\mu_i(q_{i+1}) > 0$. We denote with $\omega(i)$ the ith state of ω, and with $last(\omega)$ the last state of ω if it is finite. An *adversary* is a particular resolution of the nondeterminism in an MDP. Formally, an adversary A for $\mathcal{M}$ is a function mapping every finite path ω of $\mathcal{M}$ to a distribution $\mu \in Steps(last(\omega))$. The set of all adversaries of $\mathcal{M}$ is denoted by $Adv_{\mathcal{M}}$. For any $q \in Q$ and adversary $A \in Adv_{\mathcal{M}}$, we let $Paths_{fin}^A(q)$ and $Paths^A(q)$ be the sets of all finite and infinite paths starting in q that correspond to A, respectively. Under a given adversary, the behavior of an MDP is purely probabilistic. Formally, an adversary for an MDP induces a probability measure $Prob_q^A$ over the set of paths $Paths^A(q)$ (cf. [13] for details).

For the specification of properties of probabilistic systems, the Probabilistic Computation Tree Logic (PCTL) [14] can be used. PCTL is a branching-temporal logic based on CTL in which the existential and universal path quantifiers are replaced by a probabilistic operator which can be used to specify that the probability for a path formula meets a given lower or upper bound. Formally, the syntax of PCTL is defined as follows. A *state formula* over a set AP of atomic propositions is formed using the following grammar:

$$\Phi \quad ::= \quad \text{true} \quad | \quad a \quad | \quad \neg \Phi \quad | \quad \Phi_1 \wedge \Phi_2 \quad | \quad \mathbb{P}_{\sim \lambda}(\phi)$$

where $a \in AP$, ϕ is a path formula, $\sim \in \{<, \leq, \geq, >\}$ and $\lambda \in [0, 1]$. A *path formula* is formed using the following grammar:

$$\phi \quad ::= \quad \bigcirc \Phi \quad | \quad \Phi_1 \, \mathsf{U} \, \Phi_2 \quad | \quad \Phi_1 \, \mathsf{U}^{\leq n} \, \Phi_2$$

where Φ, Φ_1, Φ_2 are state formulas and $n \in \mathbb{N}$. The temporal operators $\bigcirc$ and U are the next- and until-operators from CTL. $\Phi_1 \, \mathsf{U}^{\leq n} \, \Phi_2$ is a step-bounded variant of the until-operator, which states that Φ_2 holds within at most n steps, while Φ_1 holds in all states visited before a Φ_2-state was reached. The eventually-operator $\Diamond$ can be derived by setting $\Diamond \Phi = \text{true} \, \mathsf{U} \, \Phi$ and analogously for a step-bounded variant of it. For example, using the graph-based atomic proposition in Fig. 4, the property 'with a probability of 0.95 or higher, node 2 correctly receives a message within 5 execution steps' can be formalized as $\mathbb{P}_{\geq 0.95}(\Diamond^{\leq 5} received(2))$.

The semantics for PCTL is defined using a satisfaction relation. Given a labeling function $L : Q \to 2^{AP}$ associating atomic propositions to states, the satisfaction relation for state formulas is defined as:

$q \models \text{true}$			$q \models \Phi_1 \wedge \Phi_2$	$\Leftrightarrow$ $q \models \Phi_1$ and $q \models \Phi_2$
$q \models a$	$\Leftrightarrow$	$a \in L(q)$	$q \models \mathbb{P}_{\geq \lambda}(\phi)$	$\Leftrightarrow$ $p_q^{\min}(\phi) \geq \lambda$
$q \models \neg \Phi$	$\Leftrightarrow$	$q \not\models \Phi$	$q \models \mathbb{P}_{\leq \lambda}(\phi)$	$\Leftrightarrow$ $p_q^{\max}(\phi) \leq \lambda$

where $p_q^{\min}(\phi)$ and $p_q^{\max}(\phi)$ are the minimum and the maximum probabilities for the set of paths starting in q and fullfiling ϕ, formally:

$$p_q^{\min}(\phi) = \inf_{A \in Adv_{\mathcal{M}}} p_q^A(\phi) \quad \text{and} \quad p_q^{\max}(\phi) = \sup_{A \in Adv_{\mathcal{M}}} p_q^A(\phi)$$

where for a given adversary A and a start state q, the probability for ϕ is:

$$p_q^A(\phi) = Prob_q^A\{\omega \in Paths(q) \mid \omega \models \phi\}$$

The satisfaction relation for path formulas is defined as follows:

$$
\begin{array}{lll}
\omega \models \bigcirc \Phi & \Leftrightarrow & \omega(1) \models \Phi \\
\omega \models \Phi_1 \cup \Phi_2 & \Leftrightarrow & \exists j \geq 0 : (\omega(j) \models \Phi_2 \wedge (\forall 0 \leq k < j : \omega(k) \models \Phi_1)) \\
\omega \models \Phi_1 \cup^{\leq n} \Phi_2 & \Leftrightarrow & \exists 0 \leq j \leq n : (\omega(j) \models \Phi_2 \wedge (\forall 0 \leq k < j : \omega(k) \models \Phi_1))
\end{array}
$$

4 Probabilistic Graph Transformation Systems

We now introduce probabilistic graph transformations systems (PGTSs), in which the format for rules is extended to incorporate probabilistic behavior.

Definition 7 (Probabilistic rule). *A probabilistic rule $\pi = \langle J, P, \mu \rangle$ consists of a typed graph J, a finite, non-empty set of rules P, such that $J = L$ for all $p = \langle L \xleftarrow{\ell} K \xrightarrow{r} R \rangle \in P$, and a probability distribution $\mu \in Dist(P)$.*

A probabilistic rule $\pi = \langle J, P, \mu \rangle$ formally consists of a finite set of basic (non-probabilistic) rules P with the same left-hand side J and a probability distribution μ over these basic rules. A probabilistic rule is interpreted as a single rule with multiple right-hand sides, which are picked randomly according to μ. Basic (non-probabilistic) rules are modeled as probabilistic rules with a single RHS and a probability distribution that assigns 1 to this RHS.

Definition 8 (Probabilistic transformation). *Let M be a typed graph, $\pi = \langle J, P, \mu \rangle$ a probabilistic rule, $m : J \to M$ a match, and $p \in P$ a basic rule. A probabilistic transformation $M \stackrel{\pi,m,p}{\Longrightarrow} N$ is defined by a basic transformation $M \stackrel{p,m}{\Longrightarrow} N$ if and only if:*

1. *for all $p' \in P$ there exists a typed graph N' such that $M \stackrel{p',m}{\Longrightarrow} N'$ and*
2. *$\mu(p) > 0$.*

Thus, a probabilistic transformation is possible if and only if (1) all its basic rules are enabled, and (2) the probability for the chosen basic rule is strictly greater than zero. Note that therefore a probabilistic rule is applicable w.r.t. a match only if the gluing condition is satisfied for all its basic rules. This is necessary because the choice for a particular basic rule is random and thus it must be ensured that all of them are enabled. Note, however, in the case of SPO graph transformation semantics, no checking of the gluing condition is required.

Definition 9 (Probabilistic graph transformation system). *A probabilistic graph transformation system (PGTS) is a tuple $\mathcal{G} = \langle T, G_{init}, \Pi \rangle$ consisting of a type graph T, an initial graph G_{init} typed over T, and a set of probabilistic rules Π typed over T.*

In a given PGTS, a probabilistic transformation $M \overset{\pi,m,p}{\Longrightarrow} N$ is made by:

1. nondeterministically selecting an applicable rule $\pi = \langle J, P, \mu \rangle \in \Pi$,
2. nondeterministically selecting a match $m : J \to M$,
3. making a probabilistic choice for a basic rule $p \in P$ according to μ,
4. transforming M into N using the basic rule p and the match m.

Thus, a probabilistic transformation $M \overset{\pi,m,p}{\Longrightarrow} N$ is a particular resolution of both the nondeterministic and the probabilistic choices in a PGTS. We denote with $G_0 \Longrightarrow_{\mathcal{G}}^* G_n$ the fact that there exists a finite sequence of consecutive probabilistic transformations using the probabilistic rules of $\mathcal{G}$:

$$G_0 \overset{\pi_0, m_0, p_0}{\Longrightarrow} G_1 \overset{\pi_1, m_1, p_1}{\Longrightarrow} \dots \overset{\pi_{(n-1)}, m_{(n-1)}, p_{(n-1)}}{\Longrightarrow} G_n$$

The operational semantics of a PGTS induces a Markov decision process.

Proposition 1 (Induced MDP). *Let $\mathcal{G} = \langle T, G_{init}, \Pi \rangle$ be a PGTS. Then $\mathcal{G}$ induces a Markov decision process $\mathcal{M}_{\mathcal{G}} = \langle Q, q_{init}, Steps \rangle$ with:*

- $Q = \{[G] \mid G_{init} \Longrightarrow_{\mathcal{G}}^* G\}$, *i.e., the set of isomorphism classes of typed graphs reachable from G_{init},*
- $q_{init} = [G_{init}]$,
- $Steps([G]) = \{ \nu \mid G \overset{\pi,m}{\Longrightarrow} \nu \}$ *where $G \overset{\pi,m}{\Longrightarrow} \nu$ with $\pi = \langle J, P, \mu \rangle \in \Pi$ denotes the fact that there exists $p \in P$ and $G' \in Q$ such that $G \overset{\pi,m,p}{\Longrightarrow} G'$, and where $\nu \in Dist(Q)$ is induced by μ as follows:*[1]

$$\nu([G']) = \sum_{p \in P: G \overset{\pi,m,p}{\Longrightarrow} G'} \mu(p) \tag{1}$$

The induced probabilistic transitions are defined in (1) by associating the probabilities of each basic rule p to the result of applying p to the current graph with the chosen match. Note that the states are defined up to graph isomorphism and that the sum in (1) is required for cases with identical or symmetric RHSs.

The concepts of graph-based atomic propositions, negative application conditions and nested rules as described in Section 2 can be directly transferred to PGTSs. NACs are defined in the usual way and restrict the applicability of a probabilistic rule as a whole. Nesting of probabilistic rules is achieved by a nesting of its basic rules, where the nested LHSs of all basic rules must be identical. For simplicity, we restrict ourselves to nested rules of depth 1 here. Note that the probabilities are associated only to the embedded rule (which is matched only once) in a nested rule. Due to lack of space, we omit the formal definition here and illustrate the concepts using an example.

---

[1] We use the convention $\sum_{\emptyset} = 0$ for sums over empty sets.

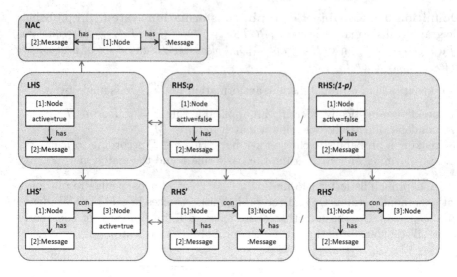

Fig. 10. Probabilistic rule *send*

5   Modeling and Comparison

In this section, we show how to use probabilistic graph transformation systems for a faithful modeling of the gossiping protocol described in Section 1 and 2 and discuss the differences to component-based models. For the PGTS model we reuse the type graph in Fig. 2, the initial graph in Fig. 3 and the rule *reset* in Fig. 7, where we trivially associate a probability of 1 to its only RHS. To specify the likelihood of the message forwarding, we combine the two basic rules $send_1$ and $send_2$ in Fig. 5 and 6 into one probabilistic rule *send* with two RHSs, depicted in Fig. 10. The first RHS models the case where the message is forwarded with a probability of p, whereas the second RHS models the case where the message is not forwarded with a probability of $(1 - p)$. Note that in both cases the node becomes inactive and that the probabilistic rule is enabled only if no collision occured. Moreover, the synchronous message passing to all neighbors is modeled again using a nested rule. To reason about this model, we reuse the atomic proposition *received*(x) in Fig. 4. As initial graphs, we consider four example topologies shown in Fig. 11, where in each network the broadcasting starts at node 1. Network 11a) is formally modeled by the typed graph in Fig. 3.

In the following, we discuss some of the advantages of using PGTSs as a modeling approach over traditional component-based modeling approaches as employed, e.g., in [4], where automata or process algebra models are used to define the behavior of the components and the system as a whole.

Modeling scalability. The first important observation is that the size of the topology has only a minor impact on the size of the PGTS model. Switching from the simple topology depicted in Fig. 11a) to the 3×3 network in Fig. 11b) only required to exchange the initial topology while the rules and the type graph remain

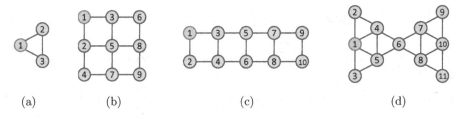

Fig. 11. Schematic example network topologies (broadcasting starts at node 1)

the same. In contrast, when using component-based models such as employed in [4], a larger network topology requires that additional components, each with its own specific local behavior and communication with its neighbors must be added to the specification. Scalability of the verification approach, however, is a separate issue and is planned to be addressed in our future work.

Changeability. When modeling different network topologies with PGTSs, this simply boils down to using different input graphs as initial states such as the four different example topologies in Fig. 11. In contrast, a modification in the network topology is a real challenge when using component-based models, since the intuitive graph structure of the network is not tangible in the specification and must be carefully encoded in the local behavior of the nodes. Moreover, in the case of a change in the protocol, only a few rules in the PGTS need to be adjusted, whereas in an component-based model the local specifications of all nodes must be altered to reflect the change in the protocol.

Expressiveness. In [4], multiple versions of the gossiping protocol are considered, which can be all modeled using PGTSs. The model in this paper corresponds to the case with nondeterministic execution order. The synchronous versions can be modeled as a PGTS by increasing the nesting depth of the rule *send* such that all active nodes with exactly one message execute the probabilistic sending at the same time. The last variant presented in [4] includes a simple, i.e., memoryless probabilistic delay for the sending of messages. This can be modeled also in a PGTS by adding a Boolean attribute which is used as an additional precondition for the *send* rule and which is switched on by a probabilistic rule.

Moreover, we argue that modeling dynamic structural changes in the network topology is (except for encodings of very simple cases) impossible using component-based models. In contrast, in the graph transformation-based approach, dynamic structural changes as needed for modeling reconfigurable and mobile systems can be expressed directly.

Simplicity. We believe that specifying the distributed protocol using a PGTS is less intricate and less error-prone because there is a clear separation between the description of the protocol modeled using the rules on the one hand, and the particular network structure on the other. This hypothesis is to some extent supported by our modeling experiment. In particular, we compared our results discussed in Section 6 for network (b) to the data presented in [4]. While

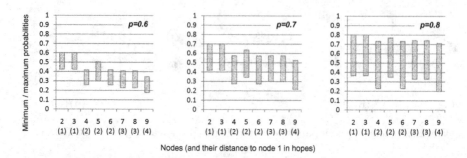

Fig. 12. Minimum / maximum probabilities for message reception for each node in network (b) and send probabilities of $p = 0.6, 0.7$ and 0.8

our experiments yielded the same maximum probabilities, we noticed that our modeling apparently predicted smaller minimum probabilities for the message receptions. Apparently, their specification contains a modeling error as the probabilistic decision whether to forward a message or not is already done at the message reception, which rules out the possibility of a collision in the case the node decides not to forward the message. However, due to the complicated encoding of the protocol this difference between their specification and the MDP induced by our model could be identified only by a detailed analysis.

6   Tool Support and Analysis

We have implemented tool support for PGTSs in version 0.9.2 of the HENSHIN [5] graph transformation tool using PRISM 4 [6] as probabilistic model checking back-end. We model probabilistic rules in HENSHIN using multiple basic graph transformation rules with the same LHS (and possible additional application conditions). Probabilities are associated to the different basic rules using annotations. We then use HENSHIN's state space generation capabilities to derive an LTS, which is subsequently converted into an MDP by (1) removing all illegal transitions where not all basic rules of a probabilistic rule are applicable for the same match, and (2) replacing the nondeterministic choice between annotated basic rules by probabilistic transitions. Our extension of HENSHIN generates an MDP in the input format of PRISM to carry out the PCTL model checking and for computing the minimum and the maximum probabilities.

Using our tool, we have modeled the gossiping protocol and ran a number of experiments. As a first setting, we fixed the send probability to $p = 0.8$ and chose the network topology (b). For these parameters, we verified using PRISM that the property $\mathbb{P}_{>0.3}(\Diamond \; received(8))$ holds, i.e., the probability that node 8 receives the message is greater than 0.3. In addition to the checking of PCTL formulas, we used PRISM to compute the minimum and the maximum probabilities for each node successfully receiving the message. Fig. 12 depicts the minimum and the maximum probabilities for each node in network (b) correctly receiving the

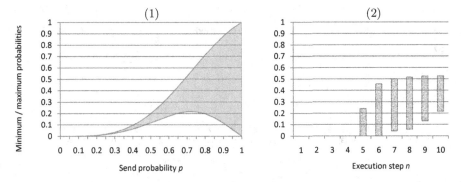

Fig. 13. Min./max. probabilities for message reception of node 9 in network (b)

message and for send probabilities of $p = 0.6, 0.7$ and 0.8. The minimum probabilities reflect worst-case, the maximum probabilities best-case execution orders for each node. We note that the minimum (and the maximum) probabilities for the different nodes vary more or less depending on the chosen send probability and the location of the node in the grid. To illustrate the impact of the send probability, we have plotted the minimum and the maximum reception probabilities for node 9 with changing p in Fig. 13.1). Note that for values of p greater than approx. 0.7, the minimum reception probability decreases again.

We further investigated how the probability for a specific node receiving the message changes over time, where time is measured as discrete execution steps. Such properties can be specified using the step-bounded until-operator in PCTL. Specifically, fixing the send probability to $p = 0.7$, we verified that the property $\mathbb{P}_{\geq 0.2}(\Diamond^{\leq 10}\ received(9))$ holds, i.e., the probability that node 9 in network (b) successfully received the message after 10 execution steps is at least 0.2. Additionally, Fig. 13.2) depicts the minimum and the maximum probabilities for node 9 having received the message after 1..10 execution steps.

Due to the graph-based approach, models with different network topologies can be easily derived. The minimum and maximum probabilities for the networks (b)-(d) are depicted in Fig. 14. The probabilities drop more for the nodes in network (c) with high indizes than in network (b) which is caused by the higher distance and the fewer number of connections. For network (d), the differences between the minimum and maximum probabilities are higher than in the other networks. This is caused by the higher connectivity of the network which increases the chance for collisions. It is also evident that node 6 is a bottleneck in the network causing the probabilities to drop abruptly for nodes 7-11.

7   Related Work

As discussed in Section 5, PGTSs compared to component-based models (as, e.g., in the PRISM specification language), provide a greater expressiveness in terms of modeling concepts, since there is a clear separation between the modeled protocols on the one hand and the used network topologies on the other.

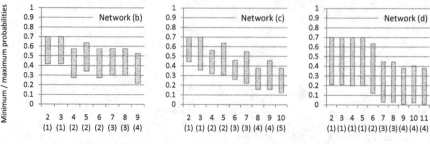

Fig. 14. Minimum / maximum probabilities for message reception for networks (b)-(d) with a fixed send probability of $p = 0.7$

Therefore, PGTSs permit to study different network topologies and to adjust protocols with minimal effort. In contrast, the component-based approaches require to encode the topology into the local behavior of the nodes, which can result in complex and erroneous specifications.

Executable term rewrite theories as used in MAUDE [15] provide similarly to GTSs natural modeling concepts for concurrent systems with structure dynamics. Probabilistic rewrite theories in PMAUDE [16] provide a combination of structure dynamics, probabilistic behavior for discrete branching, and stochastic behavior. Properties for such models can be specified using probabilistic temporal logics and checked using discrete event simulation. However, in order to simulate and analyze models in PMAUDE, all nondeterminism has to be resolved, i.e., nondeterministic choices for rules and matches as in PGTSs are not allowed.

Several extensions of GTSs with quantitative properties such as real-time [1,2] and stochastic behavior [3] exist. However, the combination of discrete probabilistic decisions and nondeterminism in PGTSs can be emulated neither by real-time nor by stochastic models. To clarify this, we discuss in detail the difference to stochastic graph transformation systems (SGTSs) [3]. While SGTSs are based on a continuous time model, PGTSs are based on a discrete one. Furthermore, SGTSs do not support nondeterminism. Instead, in any given state there is a competition between all enabled rules and their matches, which is also referred to as a *race condition*. The choice for a particular rule and match is decided probabilistically based on the rules' stochastic rates. In contrast, the choice for a particular rule and match in a PGTS is made nondeterministically whereas the effect of a rule is probabilistic. Due to the different time model and the nondeterminism, PGTSs cannot be encoded into SGTSs, nor vice versa.

8 Conclusions and Future Work

In this paper, we introduced probabilistic graph transformation systems (PGTSs), provided a sound foundation based on Markov decision processes, and presented related tool support. We further demonstrated that the modeling using PGTSs compared to existing component-based approaches as, e.g., in PRISM scale better and can be more easily adjusted to reflect changes in the topology or protocol. For future work, we plan to develop a compact visual syntax for PGTSs, to incorporate interval-valued probabilistic and real-time behavior, and to improve the scalability of the verification procedure using compositional schemes.

References

1. Gyapay, S., Varró, D., Heckel, R.: Graph transformation with time. Fundamenta Informaticae 58, 1–22 (2003)
2. Giese, H.: Modeling and Verification of Cooperative Self-adaptive Mechatronic Systems. In: Kordon, F., Sztipanovits, J. (eds.) Monterey Workshop 2005. LNCS, vol. 4322, pp. 258–280. Springer, Heidelberg (2007)
3. Heckel, R., Lajios, G., Menge, S.: Stochastic graph transformation systems. Fundamenta Informaticae 74, 63–84 (2006), doi: 1231199.1231203
4. Fehnker, A., Gao, P.: Formal Verification and Simulation for Performance Analysis for Probabilistic Broadcast Protocols. In: Kunz, T., Ravi, S.S. (eds.) ADHOC-NOW 2006. LNCS, vol. 4104, pp. 128–141. Springer, Heidelberg (2006)
5. Arendt, T., Biermann, E., Jurack, S., Krause, C., Taentzer, G.: Henshin: Advanced Concepts and Tools for In-Place EMF Model Transformations. In: Petriu, D.C., Rouquette, N., Haugen, Ø. (eds.) MODELS 2010, Part I. LNCS, vol. 6394, pp. 121–135. Springer, Heidelberg (2010)
6. Kwiatkowska, M., Norman, G., Parker, D.: PRISM 4.0: Verification of Probabilistic Real-Time Systems. In: Gopalakrishnan, G., Qadeer, S. (eds.) CAV 2011. LNCS, vol. 6806, pp. 585–591. Springer, Heidelberg (2011)
7. Corradini, A., Montanari, U., Rossi, F., Ehrig, H., Heckel, R., Löwe, M.: Algebraic approaches to graph transformation I: Basic concepts and double pushout approach. In: Handbook of Graph Grammars and Computing by Graph Transformation, pp. 163–245. World Scientific (1997)
8. Ehrig, H., Ehrig, K., Prange, U., Taentzer, G.: Fundamentals of Algebraic Graph Transformation (Monographs in Theoretical Computer Science). Springer (2006)
9. Taentzer, G.: Parallel and Distributed Graph Transformation: Formal Description and Application to Communication-Based Systems. PhD thesis, TU Berlin (1996)
10. Rensink, A., Kuperus, J.H.: Repotting the geraniums: On nested graph transformation rules. In: GT-VMT 2009. ECEASST, vol. 18 (2009)
11. Kastenberg, H., Rensink, A.: Model Checking Dynamic States in GROOVE. In: Valmari, A. (ed.) SPIN 2006. LNCS, vol. 3925, pp. 299–305. Springer, Heidelberg (2006)
12. Kwiatkowska, M., Norman, G., Parker, D.: Game-based abstraction for Markov decision processes. In: QEST 2006, pp. 157–166. IEEE Computer Society (2006), doi: 10.1109/QEST.2006.19
13. Kemeny, J., Snell, J., Knapp, A.: Denumerable Markov Chains, 2nd edn. Springer (1976)
14. Hansson, H., Jonsson, B.: A logic for reasoning about time and reliability. Formal Aspects of Computing 6, 512–535 (1994), doi: 10.1007/BF01211866
15. Clavel, M., Duran, F., Eker, S., Lincoln, P., Marti-Oliet, N., Meseguer, J., Quesada, J.: Maude: specification and programming in rewriting logic. Theoretical Computer Science 285(2), 187–243 (2002), doi: 10.1016/S0304-3975(01)00359-0
16. Agha, G., Meseguer, J., Sen, K.: PMaude: Rewrite-based specification language for probabilistic object systems. Electron. Notes Theor. Comput. Sci. 153, 213–239 (2006), doi: 10.1016/j.entcs.2005.10.040

Co-transformation of Graphs and Type Graphs with Application to Model Co-evolution[*]

Gabriele Taentzer[1], Florian Mantz[2], and Yngve Lamo[2]

[1] Philipps-Universität Marburg, Germany
taentzer@informatik.uni-marburg.de
[2] Bergen University College, Norway
{fma,yla}@hib.no

Abstract. Meta-modeling has become the key technology to define do–main-specific modeling languages in model-driven engineering. Since do–main-specific modeling languages often change quite frequently, concepts are needed for the coordinated evolution of their meta-models as well as of their models, and possibly other related artifacts. In this paper, we present a new approach to the co-transformation of graphs and type graphs and show how it can be applied to model co-evolution. This means that models are specified as graphs while model relations, especially type-instance relations, are defined by graph morphisms specifying type conformance of models to their meta-models. Hence, meta-model evolution and accompanying model migrations are formally defined by co-transformations of instance and type graphs. In our approach, we clarify the type conformance of co-transformations, the completeness of instance graph transformations wrt. their type graph modifications, and the reflection of type graph transformations by instance graph transformations. Finally, we discuss strategies for automatically deducing instance graph transformation rules from given type graph transformations.

Keywords: meta-model evolution, model migration, graph transformation.

1 Introduction

Model-driven engineering (MDE) is a software engineering discipline that uses models as the primary artifacts throughout software development processes and adopt model transformation both for their optimization as well as for model and code generation. Models in MDE describe application-specific system design which is automatically translated into code. A commonly used technique to define modeling languages is meta-modeling. In contrast to traditional software development where programming languages rarely change, domain-specific modeling languages, and therefore meta-models, often change frequently: modeling language elements may be, e.g., renamed, extended by additional attributes,

[*] This work was partially funded by NFR project 194521 (FORMGRID)

H. Ehrig et al.(Eds.): ICGT 2012, LNCS 7562, pp. 326–340, 2012.

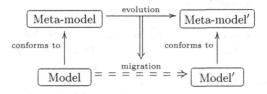

Fig. 1. Meta-model evolution and model migration

or refined by a hierarchy of sub-elements. The evolution of a meta-model requires the consistent migration of its models (See Fig. 1) which is a considerable research challenge in MDE [18].

Since graphs and graph transformations are conceptually close to models and model transformations, we consider instance and type graph co-transformations as a suitable approach to the formalization of model co-evolution. In contrast to the traditional double pushout (DPO) approach in [4] where deletion of graph parts is performed before the creation of new graph elements, we use the dual approach where creation is done before deletion. The dual approach has been introduced as co-span DPO-approach in [5] which also shows equivalence of both approaches. We choose this variant of graph transformations because they allow better synchronization of deletion and creation actions than the usual DPO approach, since the intermediate graphs contain both elements to be deleted and those to be added. Since we do not want to restrict ourselves to a specific kind of graph and graph morphism in this paper, the theoretical concepts and results are formulated at the level of (weak) adhesive categories (see e.g. [9,3]).

On this basis, we characterize co-transformations that lead to type conforming result graphs. Considering a given type graph transformation, a related instance graph transformation has to be complete, i.e., has to incorporate the whole instance graph, and has to reflect at least deletion actions of the type graph level. Furthermore, we present a strategy to deduce instance graph transformation rules from given type graph transformations.

The new graph transformation concepts and results presented are clearly motivated by our wish to develop an adequate formalization of model and meta-model co-evolution. Among existing approaches [17,20,2,8,14,16] only [16] contains fundamental results about the well-formedness of model and meta-model co-evolutions. However, the co-evolution approach presented in [16] is more restricted than ours.

The rest of the paper is organized as follows: in Section 2, we introduce a simple model co-evolution scenario. The definition of co-span transformation is recalled in Section 3 which prepares for the formalization of model and meta-model co-evolution presented in Section 4. We conclude this paper with a consideration of related work in Section 5 and final remarks in Section 6.

2   A Co-evolution Scenario

To motivate our approach, we consider a well-known Petri Net meta-model evolution scenario as presented in [2,20,14]. Figure 2 shows this scenario together with an example model migration. While the upper row presents two evolution steps of the meta-model, the lower row shows the migration of a small Petri net modeling a communication between "Alice" and "Bob". Note that the meta-model defines abstract syntax structures, while the example Petri nets are given in concrete syntax. Figure 2 shows two co-evolution steps: in the first step weights are added to outgoing and incoming arrows, in the second step the place-transitions nets are extended to coloured nets.

First step: since meta classes are needed to hold meta attributes, meta references "outArr" and "inArr" are replaced by classes "OutArr" and "InArr" as well as two additional references. In addition, attribute "weight" is added to "OutArr" first and then extracted to superclass "Arr". Second step: the token attribute of class "Place" is extracted into a new associated class "Token" as number "attribute". In addition class "Token" gets a new additional attribute "type".

The corresponding migration can be performed fully automatically. We choose to set values for new "weight" as well as "type" attributes to a default value. Afterwards a modeler may change this value. In Fig. 2, for example, the modeler changed the weight value at the outgoing transition of "to Alice" to "2" to express that Alice gets more input than Bob.

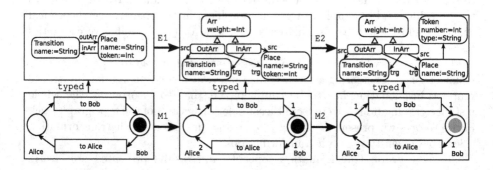

Fig. 2. A small co-evolution scenario based on Petri nets

3   Co-span Transformation

Graph transformation is the rule-based manipulation of graphs. There exists a variety of graph transformation approaches, differing mainly in the kind of transformation rules allowed and the way in which they are applied. A standard approach is the algebraic graph transformation [3] where graph parts are deleted first and then new elements are added. In the following, we consider the co-span approach [5], i.e., a variant where actions are applied in the reverse order.

Graphs are often used as an abstract representation of models. When formalizing object-oriented modeling, graph structures are used for modeling and meta-modeling leading to instance and type graphs. A fixed *type graph TG* serves as an abstract representation of a meta-model. As in object-oriented modeling, types can be structured by a generalization relation. Multiplicities and other annotations are not formalized by type graphs, but have to be expressed by additional graph constraints [6,15]. When considering meta-model conformance, we will neglect constraints for now, but we will take them into account in the future. Instance graphs define model structures and have structure-preserving mappings to their type graphs. The attribution of graph nodes and edges can be achieved by using data algebras. In that case, a type graph contains exactly one element per data type formalized by the final algebra. (For details on typed attributed graphs see [4,3].)

Considering the algebraic graph transformation approach, rules are roughly expressed by two graphs L and R, where L is the left-hand side of the rule and R is the right-hand side, which are usually overlapping in graph parts. Rule graphs may contain variables for attributes. The left-hand side L represents the pre-conditions of the rule, while the right-hand side R describes its post-conditions. The intersection $L \cap R$ (the graph part that is not changed) and the union $L \cup R$ should both form graphs, i.e., they must be compatible with source, target and type settings, in order to apply the rule. Graph $L \setminus (L \cap R)$ defines the part that is to be deleted, and graph $R \setminus (L \cap R)$ defines the part to be created. Furthermore, the application of a graph rule may be restricted by so-called *negative application conditions* (NACs) which prohibit the existence of certain graph patterns in the current instance graph. Graph elements common to L and R or common to L and a NAC, are indicated by the same name or number. (Graph inclusions are generalized to injective graph morphisms in the following definitions.)

A *direct graph transformation* $G \overset{r,m}{\Longrightarrow} H$ between two instance graphs G and H is defined by first finding a match m of the left-hand side L of rule r in an instance graph G such that m is structure-preserving, type-compatible, and satisfies the NACs (i.e., the forbidden graph patterns are not found in G). Attribute variables used in a graph object $o \in L$ are bound to concrete attribute values of graph object $m(o)$ in G. The resulting graph H is usually constructed by first removing all graph items from G that are in L but not also in R and second adding all those new graph items that are in R but not in L. This kind of graph transformation is formalized by the standard double pushout (DPO) approach as presented in [3].

However, this is not the only possible order to perform graph changes. It is possible to reverse this order which seems to better fit the needs of model co-evolution. By first adding new meta-model elements while keeping the ones to be deleted, the intermediate meta-model can be used for both, continuous typing of migrating models as well as synchronizing required migration changes. See example 4 for more details. Meta-model elements that are to be deleted, are removed in the second step. This form of transformation is called co-span DPO approach and presented in [5]. We recall the main definitions here, abstracting from the concrete graph category, and assuming graph structures and morphisms that form a (weak) adhesive category C with a selected class M of monomorphisms.

Adhesive (weak-adhesive) categories C fulfill the following properties: class M is closed under isomorphisms, composition, and decomposition. C has pushouts and pullbacks along M-morphisms, i.e., if one of the given morphisms is in M, pushouts and pullbacks over these morphisms exist. Given a pushout (PO_A) with $l \in M$ as below, then morphism $g \in M$ and pushout (PO_A) is also a pullback. Pushouts in C along M-morphisms are (weak) Van-Kampen-squares if the following holds: Consider a commutative cube like the left cube in Fig. 4: If the top is a pushout along M-morphism and the back as well as the left squares as pullbacks, the following statement holds: The bottom square is a pushout along M-morphism iff the front and the right squares are pullbacks. For weak Van-Kampen-squares, morphisms m and l or t_G, t_U, and t_I have to be in M.

Definition 1 (Co-span transformation). *Let C be a category with pushouts along class M of morphisms. An* co-span rule $p = L \xrightarrow{l} I \xleftarrow{r} R$ *consists of structures L, I and R and M-morphisms l and r which are jointly epimorphic. Given a morphism $m\colon L \to G$, called match, rule p can be applied to G if a co-span double-pushout exists as shown in the diagram below.*
$t : G \overset{p,m}{\Longrightarrow} H$ *is called a co-span transformation.*

$$
\begin{array}{ccccc}
L & \xrightarrow{l} & I & \xleftarrow{r} & R \\
\downarrow m & (PO_A) & \downarrow & (PO_B) & \downarrow m' \\
G & \xrightarrow{g} & U & \xleftarrow{h} & H
\end{array}
$$

If rule p and its match m are given, then pushout (PO_B) has to be constructed as pushout complement $R \xrightarrow{m'} H \xrightarrow{h} U$. This is possible if the co-span gluing condition is satisfied. In the category of graphs, we define each node of L whose image under m is source or target of a context edge in G as *boundary node*. The gluing condition is satisfied if boundary nodes are preserved by the rule. Formally, the co-span gluing condition is defined as follows:

Definition 2 (Co-span gluing condition). *Given morphism $m\colon L \to G$, let $b\colon B \to L$ be the boundary of m, i.e., the "smallest" morphism such that there is a pushout complement of b and m. Then, m satisfies the* co-span gluing condition *wrt. rule $p = L \xrightarrow{l} I \xleftarrow{r} R$ if there is a morphism $b'\colon B \to R$ with $r \circ b' = l \circ b$.*

$$
\begin{array}{ccccc}
 & & \overset{b'}{\frown} & & \\
 & & (=) & & \\
B & \xrightarrow{b} & L & \xrightarrow{l} & I & \xleftarrow{r} & R
\end{array}
$$

If the gluing condition is satisfied for m wrt. p, then $t : G \overset{p,m}{\Longrightarrow} H$ exists and is unique. (For more details see [5].) A co-span transformation t is *typed* over a structure TG if there are typing morphims from G and all structures belonging to the rule to TG such that they commute. In this case, also H can be typed over TG in a compatible way.

Remark 1. It is shown in [3] that attributed, typed graphs with total graph morphisms are *adhesive* with a dedicated class M of injective morphisms. Category $AGraph_{TG}$ of typed, attributed graphs with class M consisting of injective graph morphisms with isomorphisms on their data algebras, are shown to be adhesive in [4]. Considering typed graphs with node type inheritance, Theorem 1 in [7] shows that the corresponding category is *weak adhesive* if class $\mathcal{M}_{S-refl}$ contains injective morphisms being also subtype-reflecting meaning that for a mapped type, all its sub-types are also mapped. This result is extended to the category *AIGraphs* of attributed graphs with node type inheritance with class $\mathcal{A}\mathcal{M}_{S-refl}$ of S-reflecting attributed clan morphisms in Theorem 6 in [7]. (This category is used in all the following examples.)

Example 1 (Co-span graph transformation). Figure 3 shows an example for a co-span graph transformation taking up again the Petri net evolution scenario of Fig. 2. An "inArr"- edge of our example Petri net model is replaced by an "InArr"-node with weight "1" and adjacent edges. This migration step is shown in detail using the abstract syntax of Petri net models. The numbers in nodes and at edges indicate how morphisms are defined. Please note the special numbering of attributed nodes. Not only nodes but also their attributes and attribute values are numbered, since attributes are formally defined by edges and attribute values by data nodes (see [3] for more details). In the theory there are always all possible data values, however in all figures there are only those values shown that are also attribute values. The co-span gluing condition is fulfilled in this example, since edges only are deleted.

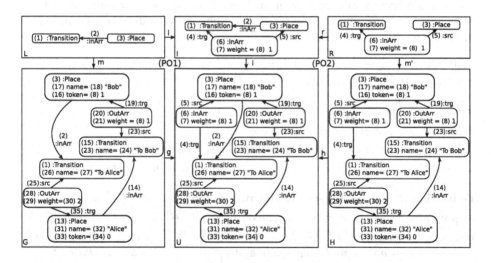

Fig. 3. Example: A co-span graph transformation

4   Formalization of Model and Meta-model Co-evolution

In this section, we formalize meta-model evolution and corresponding model migrations based on co-span transformation.

4.1 Co-evolutions by Co-transformations

Formalizing a meta-model evolution and a corresponding model migration by two coupled co-span transformations, their correspondence has to be clarified. They are type conforming if a model migration can be typed over its meta-model evolution, i.e. if all corresponding models are in consistent type-instance relations. Formally, this means that we take the category of typed attributed graphs and that there are typing graph morphisms between each pair of corresponding graphs commuting with each other.

However, this does not mean that instance graphs are always transformed such that modifications on the instance level exactly correspond to type graph modifications. To ensure a meaningful instance graph transformation, the match of an instance graph rule has to cover all graph parts typed over type graph elements taking part in the evolution. They can be formally determined by the pullback over the instance graph typing morphism and the match on type level. In this case, the migration rule match is called match complete. Furthermore, an instance rule has to reflect at least those deletions specified by its type rule. For example, an instance rule not doing anything, does not always reflect a type rule adequately, since potential deletions may not be performed and the resulting instance graph would not be well-typed over the resulting type graph, since elements with old types remain in the graph. However, creations in type graphs do not have to be (fully) reflected in corresponding instance graphs. Moreover, the action reflection of type rules in instance rules can be characterized by two pullbacks over both rules and their typing morphisms. We call co-transformations fulfilling this requirement action-reflecting.

Definition 3 (Co-transformation). *Two co-span transformations* $tt : TG \overset{tp,tm}{\Longrightarrow}$ TH *and* $t : G \overset{p,m}{\Longrightarrow} H$ *with rules* $tp = TL \overset{tl}{\longrightarrow} TI \overset{tr}{\longleftarrow} TR$ *and* $p = L \overset{l}{\longrightarrow} I \overset{r}{\longleftarrow} R$ *form a* co-transformation (tt,t), *if there are morphisms* $t_G : G \to TG$, $t_U : U \to$ TU, *and* $t_H : H \to TH$ *as well as* $t_L : L \to TL, t_I : I \to TI$, *and* $t_R : R \to TR$ *such that all squares in Fig. 4 commute.*

In such a co-transformation (tt,t), *transformation* $tt : TG \overset{tp,tm}{\Longrightarrow} TH$ *is called* an evolution *while transformation* $t : G \overset{p,m}{\Longrightarrow} H$ *is called a* migration *wrt.* tt.

Definition 4 (Evolution-reflecting migration). *A* co-transformation (tt,t) *as presented in Fig. 4 is called* match-complete *if* $G \overset{m}{\longleftarrow} L \overset{t_L}{\longrightarrow} TL$ *is a pullback and* action-reflecting *if* $TL \overset{t_L}{\longleftarrow} L \overset{l}{\longrightarrow} I$ *and* $I \overset{r}{\longleftarrow} R \overset{t_R}{\longrightarrow} TR$ *are pullbacks. We say that migration* t *is match-complete and action-reflecting wrt. evolution* tt, *or short,* t *is* evolution-reflecting *wrt.* tt.

Example 2 (Co-transformation). In this example, we consider again the meta-model evolution scenario for Petri nets introduced in Section 2 and formalize it: In Fig. 5, an excerpt of the meta-model evolution step E1 of Fig. 2 is presented as a co-span graph transformation. Before this graph transformation is applied, the "outArr" reference has already been replaced by a "OutArr" class and also weight attribute has been added and extracted to a superclass "Arr".

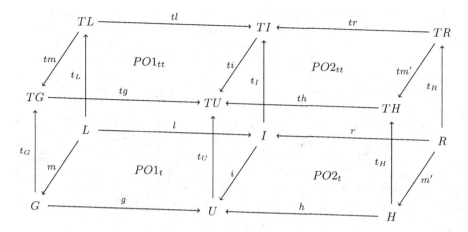

Fig. 4. Co-transformation

In the co-span rule presented in Fig. 5, reference "inArr" is replaced by a new subclass "InArr" of class "Arr". The whole transformation is typed over a simple graph representing the meta-meta-model structure. It consists of a graph node, a graph edge which is a loop on the graph node, a data node, and an attribute edge between the graph and the attribute node. The typing over this graph is not shown explicitly by morphisms but implicitly by a special graphical notation. Graph nodes are depicted by rounded rectangles, graph edges by arrows, and attribute edges and nodes by placeholders left and right to the ":=" inside graph node rectangles. In graphs TG, TU, and TH, node and edge names are depicted, since they are used as type names in corresponding migrations. Actually, these type names are not necessary, since typing is determined by graph morphisms only. However, to increase the readability we represent the typing of model graphs and migration rule graphs redundantly by type names and numbers.

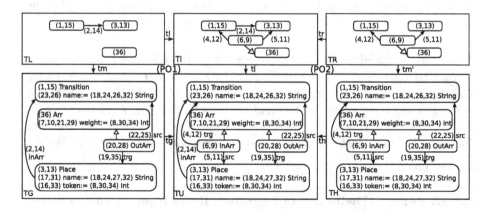

Fig. 5. Example: A meta-model evolution

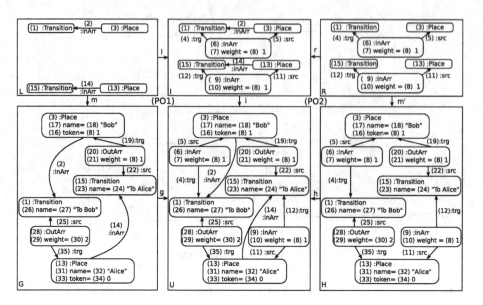

Fig. 6. Example: A match-complete, action-reflecting migration

Figure 6 shows the migration of the Petri net model triggered by the previously shown meta-model evolution rule in detail. Although similar to the one in Fig. 3, this migration is match-complete, since both "inArr" edges are matched. The migration in Fig. 3 is not match-complete, since only one "inArr" edge is matched, but still action-reflecting. In addition, the rule in Fig. 3 is not a valid migration since not all elements of deleted types are also deleted by the migration rule. Please note that the numbering in Fig. 6 is consistent with that in Fig. 5 so that both figures together form an example for the complete double-cube shown in Fig. 4. In the evolution transformation, types identified by the same "set of numbers" are related. Likewise in the migration transformation, elements identified by the same "numbers" are related. Furthermore each element of the migration is mapped to that type including the corresponding number.

Theorem 1 (Existence and uniqueness of co-transformation). *Given two co-span transformations* $tt : TG \overset{tp,tm}{\Longrightarrow} TH$ *and* $t : G \overset{p,m}{\Longrightarrow} H$ *together with typing morphisms* $t_G : G \to TG$, $t_L : L \to TL$, $t_I : I \to TI$, *and* $t_R : R \to TR$ *such that migration* t *is match-complete and action-reflecting wrt. evolution* tt, *then, structures* U *and* H *can be uniquely typed by morphisms* $t_U : U \to TU$ *and* $t_H : H \to TH$ *resp. such that* (tt, t) *forms a co-transformation. (Compare Fig. 4.)*

For the proof of this theorem see [19].

4.2   Automatic Deduction of Migration Rules from Evolution Steps

After having clarified how evolutions and migrations relate, we consider now a strategy to automatically deduce migration rules from evolution steps. Of course, we are mostly interested in deducing match-complete and action-reflecting migration rules.

The simplest case of migration rule deduction is to construct an isomorphic copy of an evolution rule. However, its application is usually not match-complete. Even worse, it might happen that the deduced rule cannot be applied to a given graph at all, since the gluing condition is not satisfied. Therefore, we present a more general deduction of migration rules taking the definition of match-complete and action-reflecting co-transformations into account. Since we will need to construct pullback complements, we recall this notion first.

Definition 5 (Pullback complement).

Let $a\colon A \to B$ and $b\colon B \to D$ be two morphisms, morphisms $c\colon A \to C$ and $d\colon C \to D$ are called a pullback complement *of a and b if a and c are the pullback of b and d.*

$$\begin{array}{ccc} B & \xrightarrow{\ b\ } & D \\ \big\uparrow a & & \big\uparrow d \\ A & \xrightarrow{\ c\ } & C \end{array}$$

Remark 2. In the category of sets and functions, a pullback can be constructed by $A = \{(x,y)|d(x) = b(y)\} \subseteq C \times D$ with morphisms $a : A \to B : (x,y) \to y$ and $c : A \to C : (x,y) \to x$. In category $AIGraphs$, pullbacks can be constructed component-wise for nodes, edges, and attributes. If at least b or d is $S-reflecting$, inheritance relations in A can be uniquely defined based on those in B and C (see also [7].) Note that there is always a pullback complement of a and b if b is in $\mathcal{M}$ and thus is S-reflecting in $AIGraphs$. (The proof is presented in [19].) A pullback complement can be constructed by C being simply equal to A and mapped to D in the same way as A is mapped to D.

Theorem 2 (Instance rule deduction). *Given a co-span transformation $tt :$ $TG \overset{tp,tm}{\Longrightarrow} TH$ and a typing morphism $t_G\colon G \to TG$ with pullback $TL \overset{t_L}{\longleftarrow} L \overset{m}{\longrightarrow} G$ of co-span $TL \overset{t_m}{\longrightarrow} TG \overset{t_G}{\longleftarrow} G$ in a (weak) adhesive category: For each pullback complement $L \overset{l}{\longrightarrow} I \overset{t_I m}{\longrightarrow} TI$ of $L \overset{t_L}{\longrightarrow} TL \overset{tl}{\longrightarrow} TI$, we obtain a match-complete, action-reflecting co-transformation and hence a co-span transformation from $G \longrightarrow TG$ to $H \longrightarrow TH$ based on $tt : TG \overset{tp,tm}{\Longrightarrow} TH$. The construction is given by the following steps:*

1. *Construct $G \overset{m}{\longleftarrow} L \overset{t_L}{\longrightarrow} TL$ as pullback of $G \overset{t_G}{\longrightarrow} TG \overset{tm}{\longleftarrow} TL$*
2. *Construct a pullback complement $L \overset{l}{\longrightarrow} I \overset{t_I}{\longrightarrow} TI$ of $L \overset{t_L}{\longrightarrow} TL \overset{tl}{\longrightarrow} TI$*
3. *Construct $I \overset{r}{\longleftarrow} R \overset{t_R}{\longrightarrow} TR$ as pullback of $I \overset{t_I}{\longrightarrow} TI \overset{t_r}{\longleftarrow} TR$*

For the proof of this theorem see [19].

Example 3 (Deduction of model migrations). Considering the example metamodel evolution in Figure 5 an edge is deleted and a node with two edges are inserted. The deduction strategy in Theorem 2 determines a skeleton migration rule with a corresponding match, but does not yield a unique result:

1. Given graph G as in Figure 3, graph L is constructed as pullback object which determines all graph elements that are affected by the evolution. This result is unique (up to isomorphism) wrt. the evolution rule and its match to the graph to be migrated.
2. The next step is a pullback complement construction which is not unique in general. Actually various outcomes are possible. We discuss some interesting ones in the following:

   (a) Nothing is added to graph L, i.e. $I = L$, however $t_I: I \to TI$ may not be surjective in this case. Non-surjectivity mean that the actions on type level are not completely reflected in the migration on the instance level which might be true if nothing is added.
   (b) One copy of each new type graph element is added to I. Consider e.g., the graph I in Figure 6: two copies of "InArr" are added here. If only one shall be added, one of these has to be selected non-deterministically. Thus, this construction seems to yield an incomplete result, although t_I would become surjective i.e., reflecting fully the creation of new types.
   (c) I contains as many copies of TI as L contains of TL: This solution is very intuitive. It would yield graph I in Figure 6 as result. Of course, it is also action-reflecting.
   (d) Copies of new type graph elements are added as often as matches of anchor nodes are found in G: considering again graph I in Figure 6, this construction would yield two more "InArr" nodes with adjacent edges combining Transition (1) with Place (13) and Transition (15) with Place (3). While this solution is action-reflecting it is not very intuitive.

3. The final deduction step is a pullback construction yielding the right-hand side R of the migration rule. This construction performs the specified type graph deletions on all occurrences found in the first deduction step. The result is unique (up to isomorphism) wrt. the evolution rule and the intermediate pullback complement chosen.

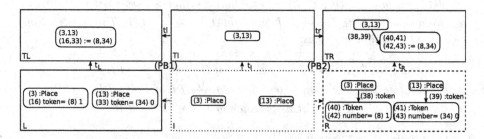

Fig. 7. Span migration and evolution rules

Example 4 (Co-span vs. span approach). As mentioned earlier we use "co-span" DPO transformations instead of "span" DPO transformations since they are more suitable for migrations. The co-span approach has a positive effect when migration rules are deduced from evolution rules. Figure 7 shows an evolution rule and a well-typed migration rule formulated as spans. Consider the evolution rule as given, the left-hand-side of the migration rule has been derived by a pullback as discussed above. The dotted and dashed parts need to be added by rule deduction. The evolution rule presented in Fig. 7 does the first part for meta-model evolution step E2: it creates a new associated class "Token" and moves attribute "token" as attribute "number" to this class. Note that since a real move does not exist in graph transformations this basically means that the attribute arrow "token" is replaced by another attribute arrow "number" which has as source node the new associated class "Token". A reasonable migration rule as presented in the figure does the same for all instances of these types. $PB1$ can be directly constructed. In the next step, pullback $PB2$ needs to be constructed as pullback complement of $I \xrightarrow{t_I} TI \xrightarrow{tr} TR$. Here, we get a problem since the values "0" and "1" in graph I are not connected to any node anymore and therefore not displayed in the figure. To find the right values for attribute "number" in rule graph R we have to consider morphism l in addition. Compare Fig. 8 now, where the same evolution rule is formulated in the co-span

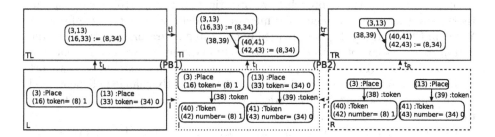

Fig. 8. Co-span migration and evolution rules

approach. This time $PB1$ has to be constructed as a suitable pullback complement. We do not run into any problem here since the values "0" and "1" are never unconnected.

Proposition 1 (Deduction of boundary). *Let* $tt : TG \xoverset{tp,tm}{\Longrightarrow} TH$ *be an evolution with rule* $tp = TL \xrightarrow{tl} TI \xleftarrow{tr} TR$ *and* $t : G \xoverset{p,m}{\Longrightarrow} H$ *with* $p = L \xrightarrow{l} I \xleftarrow{r} R$ *a migration wrt. tt deduced by the construction in Theorem 2. If tm satisfies the co-span gluing condition wrt. rule tp, then m satisfies the co-span gluing condition wrt. p.*

For the proof of this proposition see [19].

The migration rule deduction presented in Theorem 2 yields migration rules which are specific to given instance graphs. Hence, for each instance graph this construction has to be repeated. Considering a set of graphs, this deduction strategy yields a migration rule schema. It is up to future work to analyze these schemes according to certain regularities. For example, it might be possible that a basic migration rule isomorphic to its evolution rule can be identified such that the union of any number of copies yields a migration rule of the given schema.

5   Related Work

Co-evolution of structures has been considered in several areas of computer science such as database schemata, grammars, and meta-models [11,10,13,17]. Especially schema evolution has been a subject of research in the last decades. Recently, research activities have started to consider meta-model evolution and to investigate the transfer of schema evolution concepts to meta-model evolution (see e.g. [8]). In the following, we focus on meta-model evolution approaches.

In the literature, meta-model differences are most basically given by change sets. These sets can be used to deduce evolution rules [20,14] or pre-defined operations [8]. Roughly spoken, a rule is deduced from a change set by throwing away unnecessary context and by abstracting from concrete values by variables (see [1]). If pre-defined operations are given as done in COPE [8] , they can be formally defined by single rules or rule sets.

COPE [8] uses a meta-model independent representation to perform model migrations. In [16], the authors use pre-defined constructors that migrate models fully automatically. Type conformance of migrated models is mostly checked during run time in the approaches we investigated, except [16] where the authors show that their automatically migrated model is always type conforming. While in COPE and [2,16], automatically generated migrators are available, the automatic deduction of migration strategies from evolution strategies is not considered in [17,20,2,14]. However, Flock [14] (and to a minor extent [17]) support at least the automatic copying of unchanged and slightly changed elements and the automatic unsetting of deleted features wrt. a given evolution rule leading to fairly compact migration scripts.

To keep as much of the typing information as possible during evolutions, we consider an add-first-delete-then transformation approach which provides us with intermediate graphs that keep the original information and include new parts. Such intermediate graphs are able to retain typing information that is not changed during a meta-model evolution. If a corresponding migration reflects all actions of a given evolution and is match-complete wrt. to the given graph, we show that the resulting graph is uniquely typed over the resulting type graph. Thus, we do not have to check type conformance at run time as is done by nearly all approaches we considered, but can guarantee this property at design time, before migrations actually take place. A similar result can be found in [16], however our approach differs from that formalization: in [16], model migration is always fully determined, there are no possibilities to adapt migrators e.g. wrt. the reflection of new meta-model parts. We consider a general procedure to deduce migration rules from type graph evolutions and corresponding instance graphs leaving space for variants. Evolution examples in [16] are refactorings only, leaving the evaluation of other kinds of evolutions open. Furthermore, their representation of object structures is special in the sense that object slicing along class generalization relations is used to specify evolutions. Refactorings are specified as change-based requiring the specification of quite difficult folding and unfolding constructions. Our evolution specifications is more straight forward.

6   Conclusion

In this paper, we present a formal approach to model and meta-model co-evolution based on graph transformations. Meta-model evolutions and corresponding model migrations are defined by co-transformations in (weak) adhesive categories. An example is given using the category of typed, attributed graphs with inheritance. Evolutions and migrations are specified by transformation rules that can be freely designed, i.e., there are no pre-defined operations to be used. We use a kind of transformations performing the addition of new structures first and delaying the deletion of structures not needed anymore. Our main result is concerned with the type conformance of instance graph migrations wrt. their type graph evolutions. While nearly all other co-evolution approaches delay the check of type conformance to run time, i.e., when model migrations are actually performed, we developed sufficient properties and a derivation strategy to identify type conforming migrations at design time.

In addition, we consider a formal framework for automatic deductions of instance graph migrations from given type graph evolutions. While deletion actions on type graphs have to be directly reflected on the instance level to preserve type conformance, the addition of new structures usually allows for variants of migrations. The addition of type graph elements leaves the modeler with the question if new type graph elements shall be used in the instance graph and if yes, how. This is reflected by the pullback complement construction which is not unique in general. We discussed several forms of graph adaptations.

While the focus of this paper is on formalization of model and meta-model co-evolutions, the work flow of co-evolution processes as well as supporting tools are neglected here. Usually a meta-model evolution has taken place and numerous models have to be migrated accordingly. Thus, co-evolution usually does not take place synchronously as our formalization might suggest. However, interleaved evolution steps are also possible. In [12], an implementation based on the Eclipse Modeling Technology is presented where the addition of meta-model parts is performed first, models are migrated thereafter, and finally, unneeded meta-model parts are deleted. This work flow suits to our formalization in the sense that its meta-model evolutions can be directly formulated as co-span transformations. However, model migrations in [12] are less strictly typed since the intermediate meta-model is taken for typing whole migrations. It is straight forward to relax the typing of migrations being part of co- transformations as in this paper. Hence, they can be directly implemented by the approach in [12].

Acknowledgments. Many thanks to Hartmut Ehrig for his valuable comments on a previous version of this paper and to Wendy McCaull for giving valuable language feedback.

References

1. Bisztray, D., Heckel, R., Ehrig, H.: Verification of Architectural Refactorings: Rule Extraction and Tool Support. ECEASST 16 (2008)

2. Cicchetti, A., Ruscio, D.D., Eramo, R., Pierantonio, A.: Automating Co-evolution in Model-Driven Engineering. In: ECOC 2008, pp. 222–231. IEEE Computer Society (2008)

3. Ehrig, H., Ehrig, K., Prange, U., Taentzer, G.: Fundamentals of Algebraic Graph Transformation. Monographs in Theoretical Computer Science. Springer (2006)

4. Ehrig, H., Ehrig, K., Prange, U., Taentzer, G.: Fundamental Theory for Typed Attributed Graphs and Graph Transformation based on Adhesive HLR Categories. Fundam. Inform. 74(1), 31–61 (2006)

5. Ehrig, H., Hermann, F., Prange, U.: Cospan DPO Approach: An Alternative for DPO Graph Transformation. EATCS Bulletin 98, 139–149 (2009)

6. Habel, A., Pennemann, K.H.: Correctness of high-level transformation systems relative to nested conditions. Mathematical Structures in Computer Science 19(2), 245–296 (2009)

7. Hermann, F., Ehrig, H., Ermel, C.: Transformation of Type Graphs with Inheritance for Ensuring Security in E-Government Networks. In: Chechik, M., Wirsing, M. (eds.) FASE 2009. LNCS, vol. 5503, pp. 325–339. Springer, Heidelberg (2009); long version available as TR 2008-07 at TU Berlin, Germany

8. Herrmannsdoerfer, M., Benz, S., Juergens, E.: COPE - Automating Coupled Evolution of Metamodels and Models. In: Drossopoulou, S. (ed.) ECOOP 2009. LNCS, vol. 5653, pp. 52–76. Springer, Heidelberg (2009)

9. Lack, S., Sobociński, P.: Adhesive Categories. In: Walukiewicz, I. (ed.) FOSSACS 2004. LNCS, vol. 2987, pp. 273–288. Springer, Heidelberg (2004)

10. Lämmel, R.: Grammar Adaptation. In: Oliveira, J.N., Zave, P. (eds.) FME 2001. LNCS, vol. 2021, pp. 550–570. Springer, Heidelberg (2001)

11. Li, X.: A Survey of Schema Evolution in Object-Oriented Databases. In: TOOLS, pp. 362–371. IEEE Computer Society (1999)

12. Mantz, F., Jurack, S., Taentzer, G.: Graph Transformation Concepts for Meta-Model Evolution Guaranteeing Permanent Type conformance Throughout Model Migration. In: AGTIVE. LNCS, vol. 7233. Springer (2012)

13. Pizka, M., Juergens, E.: Automating Language Evolution. In: TASE 2007: Proceedings of the First Joint IEEE/IFIP Symposium on Theoretical Aspects of Software Engineering, pp. 305–315. IEEE Computer Society, Washington, DC (2007)

14. Rose, L.M., Kolovos, D.S., Paige, R.F., Polack, F.A.C.: Model Migration with Epsilon Flock. In: Tratt, L., Gogolla, M. (eds.) ICMT 2010. LNCS, vol. 6142, pp. 184–198. Springer, Heidelberg (2010)

15. Rutle, A., Rossini, A., Lamo, Y., Wolter, U.: A Formal Approach to the Specification and Transformation of Constraints in MDE. JLAP 81(4), 422–457 (2012)

16. Schulz, C., Löwe, M., König, H.: A Categorical Framework for the Transformation of Object-Oriented Systems: Models and Data. J. Symb. Comput. 46(3) (2011)

17. Sprinkle, J., Karsai, G.: A Domain-Specific Visual Language for Domain Model Evolution. J. Vis. Lang. Comput. 15(3-4), 291–307 (2004)

18. Sprinkle, J., Rumpe, B., Vangheluwe, H., Karsai, G.: Metamodelling - State of the Art and Research Challenges. In: Giese, H., Karsai, G., Lee, E., Rumpe, B., Schätz, B. (eds.) MBEERTS. LNCS, vol. 6100, pp. 57–76. Springer, Heidelberg (2010)

19. Taentzer, G., Mantz, F., Lamo, Y.: Co-Transformation of Graphs and Type Graphs with Application to Model Co-Evolution: Long Version. Tech. rep., Dep. of Mathematics and Computer Science, University of Marburg, Germany (2012), http://www.uni-marburg.de/fb12/forschung/berichte/berichteinformtk

20. Wachsmuth, G.: Metamodel Adaptation and Model Co-adaptation. In: Bateni, M. (ed.) ECOOP 2007. LNCS, vol. 4609, pp. 600–624. Springer, Heidelberg (2007)

Graph Transformations
for Evolving Domain Knowledge

Bernhard Westfechtel[1] and Manfred Nagl[2]

[1] Applied Computer Science I, University of Bayreuth
bernhard.westfechtel@uni-bayreuth.de
[2] Software Engineering, RWTH Aachen University
nagl@se.rwth-aachen.de

Abstract. Graph transformation (GraTra) systems have been used for building
tools in a wide spectrum of application domains. A GraTra system constitutes
an operational specification which may be either interpreted directly or compiled
into executable code. The specification incorporates domain knowledge concerning types of objects, operations to be performed, and patterns to be instantiated.
In many applications, domain knowledge is not fixed; rather, it evolves while
the tool based on the specification is being used. We examine and compare different approaches to support evolving domain knowledge which were developed
in several projects in different domains. Our work may be viewed as a step towards engineering of GraTra Systems for evolving domain knowledge — a topic
of practical relevance which has not gained sufficient attention so far. Although
the examples regarded in this paper have been formulated in PROGRES [26], the
arguments and results hold for other GraTra systems, as well.

1 Introduction

GraTra systems have been used for many different purposes, especially for *tool
construction* [6]: With the help of GraTras, the data and operations of a tool may be
specified at a high level of abstraction. A GraTra system constitutes an operational
specification which may be either interpreted or compiled into executable code.

Tools based on GraTras have been built for a large variety of application domains. In
order to build a domain-specific tool, the *domain knowledge* has to be formalized. The
resulting GraTra system incorporates domain knowledge concerning types of objects,
operations to be performed, and patterns to be instantiated. In many applications, domain knowledge evolves while the tool based on the specification is being used. In this
paper, we study different approaches to support *evolution* of the domain knowledge.
Our study is based on several *projects* in which GraTra systems were used internally
without exposing them to the end users of the tools: (a) *AHEAD* [11], an environment
for managing dynamic development processes, to be used by a process manager, (b)
IREEN [3], which offers tools for maintaining syntactic and semantic consistency relations between different engineering design documents, used by an engineer to maintain
consistency relations but also to install new rules for such relations, (c) *ConDes* [16],
which provides tools for the conceptual design of buildings, to be used by an experienced architect, who can also introduce new domain knowledge, and (d) *CHASID* [9],
a semantics-oriented authoring environment, to be used by a technical writer.

H. Ehrig et al.(Eds.): ICGT 2012, LNCS 7562, pp. 341–355, 2012.

The examined projects differ with respect to their requirements to evolution support, the parts of the domain knowledge which are evolved, and the ways in which evolution support is realized. In this paper, we present and compare these different approaches. In this way, we intend to prepare the grounds for the systematic *engineering* of *GraTra systems* for evolving domain knowledge.

2 Preliminaries

2.1 PROGRES

PROGRES [26] is a *specification language* for programmed GraTra systems which was used in all projects of our study. For defining the structural model (the graph schema), the language provides several advanced features such as multiple inheritance, a strati-fied type system (nodes are instances of node types which are in turn instances of node classes), both type- and instance-level attributes, and definition of derived attributes and relationships. For the behavioral model, PROGRES offers GraTra rules with both single- and set-valued nodes, use of derived data, negative application conditions, etc. Furthermore, GraTra rules may be organized into programmed methods with transac-tional behavior. The *PROGRES environment* provides a syntax-aided editor, tools for analysis and browsing, as well as both an interpreter for specification development and a compiler for producing efficient code.

2.2 Graph Transformations Used for Building Systems

As we have been following the *GraTra* approach for a long time, different ways of *using* them for building systems can be determined: (a) A GraTra specification is *manually transformed* into system code. This approach was followed at the very beginning [8] and later on, if we had to implement on an industrial platform, where our GraTra ma-chinery was not available. (b) Later, after PROGRES tools and the *code generator* were available, the translation of specifications to code was done *automatically* [26]. Quite a big number of tools were produced following this line, e.g., for software construction or maintenance [19]. After every change of the specification, the tool generation pro-cess had to be started again. (c) Distinguishing a *generic* and a *specific layer* within a specification [10] makes tool construction and modification easier if the generic layer remains unchanged. To support domain experts, we developed tools for domain-specific languages which are used to define domain knowledge and translate it into the specific layer of a PROGRES specification. (d) In many situations one even wants to have flex-ibility at the *tool's runtime*. So, an (experienced) user of a tool gets the flexibility to *modify the underlying structural and behavioral knowledge* the tool is using.

Since evolution of domain knowledge is not supported in (a) and (b), these cases are not discussed further. Modification tools are offered in case (c) to determine changes. From these changes modified specific specifications are generated. Afterwards, the tool construction cycle is started again. The case (d) is discussed here in different variations. In all those cases interpreter mechanisms are used at run time to make use of the modi-fications having been introduced at run time. So, (c) is a flexible *compilation approach*, where the tool builder is involved, (d) is an even more flexible *interpretation approach*.

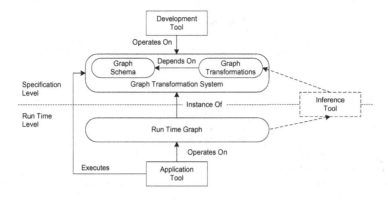

Fig. 1. GraTra systems, run time graphs, and tools

Above the term *tool* was used in quite *different meanings*: We spoke about (i) GraTra tools in order to build up and modify a GraTra specification and (ii) generator tools to translate this specification into code. Furthermore, (iii) tools were mentioned to model and modify the domain knowledge, and (iv) on one hand to generate specific specifications (case c) and on the other hand to determine some information to be interpreted at runtime (case d). Further, (v) tools will be mentioned to infer knowledge or to help to overcome the data version problem.

2.3   Classification of Evolution Support

Figure 1 illustrates the relationships among GraTra systems, run time graphs, and tools (for now, please ignore the dashed parts, which will be needed in 4.1). A *GraTra system*, which consists of a graph schema and a set of GraTra rules, is constructed with the help of a *development tool* (PROGRES). An *application tool* (AHEAD, ...) executes the GraTra system and operates on the *run time graph*, which conforms to the graph schema and is created by applying consistency-preserving GraTra rules. The horizontal dashed line in the figure separates the *specification level* from the *run time level*.

Approaches to *evolution* of domain knowledge may be *classified* with respect to the following criteria (Table 1 in Section 5): (a) *Objects of evolution*: Structural domain knowledge is represented by *types* and *object patterns*, while behavioral knowledge is modeled by *operations*. (b) *Levels of evolution*: In the case of evolution at the *specification level*, the GraTra system contains the whole domain knowledge and, therefore, has to be changed to take evolving domain knowledge into account. In contrast, evolution at the *run time level* means that domain knowledge is represented in the run time graph and is changed via the application tool. (c) *Time of evolution*: Evolution is performed at *compile time* if domain knowledge is represented in the specification and the specification is compiled into executable code. Evolution is performed at *run time* of the application tool in the following cases: (i) Domain knowledge is represented in the specification, and the specification is interpreted rather than compiled. (ii) Domain knowledge is represented in the run time graph.

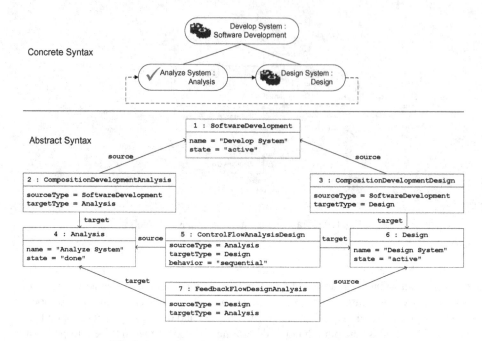

Fig. 2. Dynamic task nets (concrete and abstract syntax)

2.4   Running Example

While this paper is based on the study of multiple projects in different application domains, we discuss a single running example which is based on *dynamic task nets* [10]. A dynamic task net represents a development process to be executed. It is called "dynamic" because the task net may be changed while the process is being executed. Tasks are organized into a containment tree. Each task has a name, a type, and an execution state. Tasks are connected horizontally by control flows which resemble precedence relationships in Gantt diagrams. Feedback flows are oriented oppositely to control flows and indicate that a predecessor task receives feedback from a successor.

A simple example is given in Figure 2. The complex task Develop System of type Software Development is decomposed into two subtasks (and further subtasks not shown here): Analyze System of type Analysis and Design System of type Design. The control flow (solid arrow) indicates that analysis must be performed before design. The icons indicate the states of tasks. Analyze System is completed, while Design System is still active. The feedback flow (dashed arrow) indicates that the designer has detected a problem in the analysis document.

3   Evolution at Compile Time

3.1   Layered Specification

The AHEAD project was concerned with the management of development processes. A process is represented by a dynamic task net (Figure 2). In the case of AHEAD, the

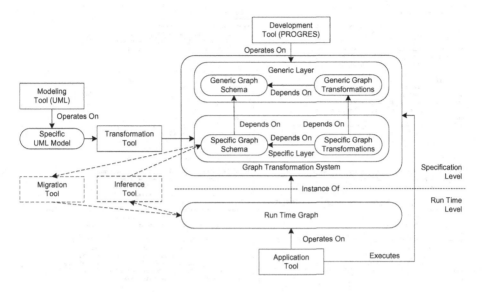

Fig. 3. Round trip evolution of the specific graph schema (AHEAD)

application tool shown in Figure 1 is provided to the process manager, who plans, analyzes, monitors, and modifies the task net. AHEAD addresses the domain "management of development processes" in general. However, in order to apply AHEAD to a specific process, it needs to be adapted. For example, the task net of Figure 2 contains tasks of specific types for software development processes. Furthermore, specific constraints have to be taken into account, e.g., concerning the order in which tasks are executed.

To support different kinds of processes, the specification was structured into two *layers*: a *generic layer*, which provides core functionality, and a *specific layer* placed below the specific layer [10] (Figure 3). The specific layer defines the types of tasks and relationships as well as customized operations being used in a specific context. Layering is realized in PROGRES with the help of *genericity*: The generic layer is parameterized with types which are instantiated in the specific layer. Parameterization is performed with the help of the stratified type system: Nodes are instances of node types which are instantiated in turn from node classes. Due to the stratified type system, types may be stored as attribute values, and they may also be passed as parameters (Figure 4): (a) *Generic graph schema*: Node classes are defined for tasks and relationships (the superclasses NAMED_NODE and PROCESS_NODE will be defined later). Each relationship is represented as a node with adjacent source and target edges. The class RELATIONSHIP is parameterized with the types sourceType and targetType, which are represented as meta attributes (i.e., class-level attributes). The subclass CONTROL_FLOW adds a meta attribute behavior, whose default value is sequential. (b) *Generic GraTra rule*: The rule for creating a control flow is parameterized with its type. The condition part checks whether source and target task are instances of the types required by controlFlowType. (c) *Specific graph schema*: Specific node types for tasks and relationships are defined as instances of the respective node classes. Furthermore, the end types of relationship

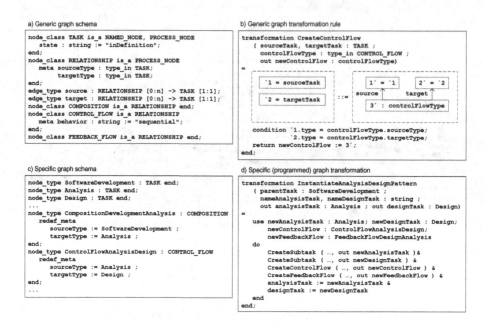

Fig. 4. Layered specification in PROGRES (AHEAD)

types are fixed by assigning values to the respective meta attributes. (d) *Specific Gra-Tra rules*: Complex operations may be defined by calling generic rules with specific parameters. For example, an analysis-design pattern may be instantiated by creating an analysis and a design task and connecting them by control and feedback flows.

The specific layer incorporates domain knowledge which has to be defined in co-operation with a domain expert. Since a domain expert usually is not familiar with the internally used specification language, a modeling tool was built which is based on UML [14]. A model is defined with the help of class diagrams (structural model) and communication diagrams (behavioral model). A transformation tool translates the model into the specific layer of the specification.

3.2   Round Trip Evolution of the Specific Graph Schema

So far, we have tacitly assumed that the specific layer is fixed. To remove this restriction, *schema versioning* was implemented on top of PROGRES [12]. The specific graph schema may consist of multiple schema versions. A new schema version is added to the specific graph schema without affecting existing schema versions. Furthermore, a task net which has been created under an old schema version may be *migrated* to the new schema version. Migration may be performed selectively; furthermore, inconsistencies may be tolerated temporarily or even permanently. Finally, an *inference tool* was implemented which constructs a proposal for a schema version by analyzing a set of instantiated task nets. Altogether, the tool set supports *round trip evolution* of the *specific graph schema* (dashed parts of Figure 3).

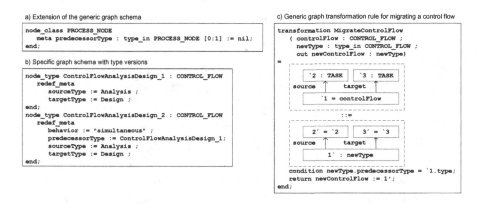

Fig. 5. PROGRES specification of schema versioning and migration (AHEAD)

Schema versioning and *migration* are realized in PROGRES as follows (Figure 5): (a) *Extension of the generic graph schema:* Versioning of types is represented by the meta attribute predecessorType of class PROCESS_NODE (which serves as the root class of the inheritance hierarchy). (b) *Specific graph schema with type versions:* For each type version, a node type is declared. The example shows two versions of control flows from analysis to design. The first version inherits the default behavior from class CONTROL_FLOW (sequential). The second version is designated as a successor of the first version. Its behavior is redefined as simultaneous, i.e., the target task may start before the source task is finished. (c) *Generic GraTra rule for migrating a control flow:* The rule receives the control flow to be migrated and its new type as parameters. The condition part checks whether the new type is a successor of the old type. Since PROGRES does not allow type changes, the old flow is deleted, and a new flow is inserted.

4 Evolution at Run Time

4.1 Round Trip Evolution of GraTra Rules

AHEAD does not specifically support the evolution of GraTra rules; the inference tool addresses only the specific graph schema. Furthermore, all changes to the specification require recompilation of the application tool. In the following, we deal with continuous evolution of GraTra rules at run time.

IREEN [3] deals with consistency control in chemical engineering. There are different documents describing different views / levels of abstraction of the overall solution (design of a chemical plant). Many increments of the internal structure of a document have a counterpart in other documents. IREEN is an interactive tool which uses *triple graph rules* [25] to assist designers in maintaining inter-document consistency. Even if the structure of all documents is fixed, the set of rules for *consistency handling* has to be *extensible*, such that a designer can add further rules (and modify existing ones).

Here, *round trip evolution* applies to *GraTra rules* (Figure 1 including the dashed parts). While the graph schemata for the documents to be integrated are well known and

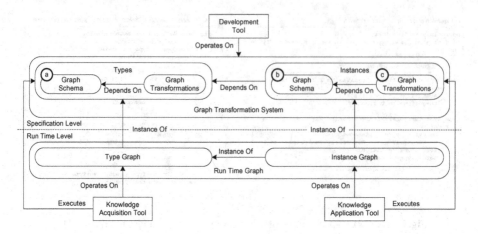

Fig. 6. Type-level domain knowledge in the run time graph (ConDes)

fixed, the domain knowledge concerning the integration of documents evolves gradually while the integration tool is in use. In Figure 1, the application tool is used to establish consistency among engineering design documents. The underlying rules are either written manually, or they are created with the help of the *inference tool* (which abstracts instance-level subgraphs connecting increments of different documents into GraTra rules). In contrast to AHEAD, the rules may be changed at *run time* of the application tool. This flexibility was achieved by a light-weight re-implementation of GraTra rules, in which rules are interpreted rather than compiled.

4.2   Type-Level Domain Knowledge in the Run Time Graph

In all approaches presented so far, domain knowledge is presented exclusively in the specification. In all approaches below, *domain knowledge* is represented in the *run time graph*. Thus, domain knowledge may be changed during the use of the application tool.

ConDes [16] is a tool for *conceptual designs* of buildings. *Semantic units*, like bath- and bedrooms and their aggregation to sleeping areas, are important and not their technical realization. These units are assembled into specific *types of buildings*, e.g., one-family houses, hospitals etc. An experienced user should be able to input / modify that domain knowledge at any time. Different tools are provided on types and instances, respectively (Figure 6). The knowledge engineer defines domain knowledge for specific types of buildings (*knowledge acquisition*). These types may be employed by architects to design actual buildings (*knowledge application*).

Therefore, domain-specific types are represented in the run time graph, which is structured into a *type graph* and an *instance graph*. In contrast to AHEAD (Figure 3), the specification is no longer structured into a generic and a specific layer. Rather, the overall specification is generic and now contains generic types and operations for defining domain-specific types. Two orthogonal kinds of *instantiation* have to be distinguished: (i) *Vertical instantiation* follows the language rules of PROGRES and concerns

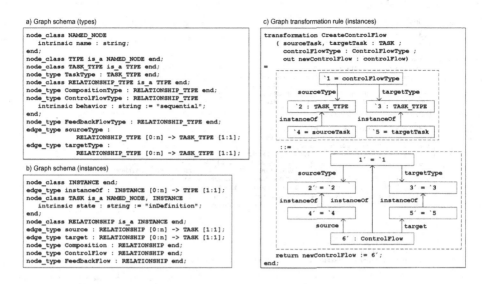

Fig. 7. Type-level domain knowledge in run time graph: PROGRES specification (ConDes)

the relationships between nodes and their types (vertical arrows in Figure 6). (ii) *Horizontal instantiation* regulates the relationships between instance nodes and type nodes of the run time graph; the rules for instantiation are defined by the modeler.

The ConDes approach is realized in PROGRES as follows (Figure 7): (a) *Graph schema (types)*: Types of tasks and relationships are represented as nodes being part of the type graph at run time. This requires the declaration of respective node classes / types, e.g., TASK_TYPE or RELATIONSHIP_TYPE. (b) *Graph schema (instances)*: Likewise, a schema for the instance graph is required (including e.g. the classes TASK and RELATIONSHIP). Instance graph nodes are connected to type graph nodes (edge type instanceOf). (c) *GraTra rule (instances)*: In contrast to the rule of Figure 4b, the current rule includes type nodes for source and target tasks to check type compatibility of the control flow. Furthermore, the control flow node is instantiated from a generic type; its specific type is represented by the instanceOf edge to the control flow type.

4.3   Instance-Level Domain Knowledge in the Run Time Graph

In ConDes, type-level domain knowledge is represented in the run time graph. In contrast, the authoring environment *CHASID* [9] maintains *instance-level domain knowledge* in the run time graph. Both approaches are complementary to each other since in CHASID domain-specific types are fixed in the graph schema.

In specific domains such as writing of scientific articles, certain *patterns* occur over and over again. For example, an article may be structured into an introduction, a description of the research method, a report on the results, and a final discussion. Also, the explanation used in the article might follow typical styles, top-down from a principle to examples, or bottom-up, starting with examples and then elaborating on the abstraction. Domain knowledge of this kind is located at the level of *abstract instances*.

a) Graph schema for pattern definitions

```
node_type Pattern : NAMED_NODE end;
node_class PATTERN_ELEMENT is_a NAMED_NODE
   elementType : type_in PROCESS_NODE;
end;
edge_type containsElement : Pattern [1:1] -> PATTERN_ELEMENT [1:n];
node_type TaskElement : PATTERN_ELEMENT end;
node_type RelationshipElement : PATTERN_ELEMENT end;
edge_type sourceElement : RelationshipElement [0:n] -> TaskElement [1:1];
edge_type targetElement : RelationshipElement [0:n] -> TaskElement [1:1];
```

b) Graph schema for pattern instances

```
node_type PatternInstance : NAMED_NODE end;
edge_type instanceOfPattern : PatternInstance [0:n] -> Pattern [1:1];
node_class PATTERN_ELEMENT_INSTANCE is_a NODE end;
edge_type instanceOfElement : PATTERN_ELEMENT_INSTANCE [0:n] -> PATTERN_ELEMENT [1:1];
edge_type references : PATTERN_ELEMENT_INSTANCE [0:n] -> PROCESS_NODE [1:1];
edge_type containsInstance : PatternInstance [1:1] -> PATTERN_ELEMENT_INSTANCE [1:n];
node_type TaskElementInstance : PATTERN_ELEMENT_INSTANCE end;
node_type RelationshipElementInstance : PATTERN_ELEMENT_INSTANCE end;
edge_type sourceInstance : RelationshipElementInstance [0:n] -> TaskElementInstance [1:1];
edge_type targetInstance : RelationshipElementInstance [0:n] -> TaskElementInstance [1:1];
```

c) Graph transformation rule for binding a pattern element

```
transformation BindRelationshipElement
   ( pattern : Pattern ; patternInstance : PatternInstance ;
     relationshipElement : RelationshipElement ;
     out relationshipElementInstance : RelationshipElementInstance)
=
```

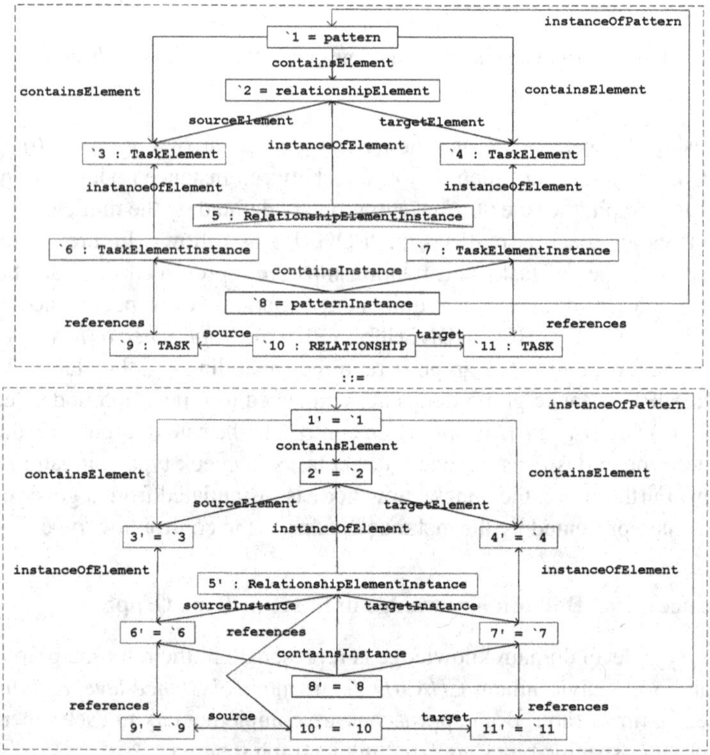


```
   condition `10.type = `2.elementType;
   return relationshipElementInstance := 5';
end;
```

Fig. 8. Definition and instantiation of patterns: PROGRES specification (CHASID)

Tool support for patterns comprises an editor for defining patterns, a tool for instantiating patterns, and an analysis tool for checking the consistency between pattern instances and pattern definitions. A pattern is defined by abstraction from a graph of instances. Thus, CHASID provides for *round trip evolution of patterns*.

The run time graph consists of a concrete *instance graph*, which is composed of "ordinary" nodes and edges, a *pattern definition graph*, which defines patterns abstracted from such nodes and edges, and a *pattern instance graph*, which contains instances (applications) of patterns. The pattern instance graph is placed in between the instance graph and the pattern definition graph to record the instantiation of patterns.

The specification of patterns is illustrated in Figure 8: (a) *Graph schema for pattern definitions*: A pattern consists of a number of pattern elements. For each element, elementType determines the type of the instance to be matched or created in the instance graph. In the case of dynamic task nets, patterns are composed of tasks and relationships. (b) *Graph schema for pattern instances*: A pattern instance is connected to a pattern definition on one end and to a graph of concrete instances on the other end. A pattern instance constitutes a copy of a pattern definition. Elements of pattern instances are represented by nodes of class PATTERN_ELEMENT_INSTANCE. Edges of type instanceOfElement and references connect these nodes to pattern elements and concrete instances, respectively. (c) *GraTra rule for binding a pattern element*: When a pattern is instantiated, all context elements are *bound* to already existing instances. For non-context elements, concrete instances are *created* and connected to instances of context elements. The role of an element (context or non-context) is determined only on pattern instantiation. The rule shown in Figure 8 handles the *binding* of a relationship element. It assumes that adjacent task elements have already been bound, creates an instance of the relationship element, and binds it to an actual, already existing relationship.

5   Comparison

Table 1 classifies the approaches presented in this paper according to Subsection 2.3.

5.1   Objects of Evolution

Evolution of Types.   In *AHEAD*, domain knowledge at the type level is represented in the graph schema. Thus, the specification developer may exploit the rich set of language features provided by PROGRES. This reduces the effort to model type-level domain knowledge considerably. On the other hand, PROGRES does not support evolution of the graph schema (apart from extensions). In AHEAD, an application-specific evolution approach was implemented on top of PROGRES (supporting schema versioning, selective migration, and toleration of inconsistencies).

The solution adopted in *ConDes* has complementary advantages and drawbacks. Flexibility is increased since the domain knowledge is part of the run time graph and may be modified at any time. Furthermore, the specification developer may design a domain-specific language for defining types freely. On the other hand, the specification effort is increased significantly: The specification developer has to provide rules for building up the type graph. Furthermore, he has to specify operations on instance graphs which control consistency with the type graph.

Table 1. Classification of evolution approaches

Criteria→	Objects			Levels		Time	
Systems ↓	Types	Operations	Patterns	Specification	Run Time	Compile Time	Run Time
AHEAD (3.1, 3.2)	x	x		x		x	
IREEN (4.1)		x		x			x
ConDes (4.2)	x				x		x
CHASID (4.3)			x		x		x

Evolution of Operations. AHEAD supports the evolution of operations by distinguishing between a generic layer and a specific layer: Domain-specific operations may be defined on top of generic operations (Figure 4d). Evolution is performed at the specification level; any change of the overall specification requires subsequent recompilation.

In *IREEN*, evolution of operations is performed at the specification level, as well. However, rules may be changed at any time since they are interpreted rather than compiled. Furthermore, rules may be derived from the run time graph with the help on an inference tool. The inference tool performs a *higher order transformation* [28], i.e., the tool executes a transformation in order to produce another transformation.

Evolution of Patterns. Patterns are supported only in *CHASID*, where they have been realized on top of PROGRES. An object pattern is a graph of abstract instances. Patterns constitute structural domain knowledge which is represented in the run time graph. Several types of operations are provided on patterns, including pattern definition, pattern instantiation, and analysis of consistency between patterns and their instantiations after either of them have been modified.

Instantiation of a *pattern* resembles the *application* of a *GraTra rule*. However, there are several crucial differences between patterns and GraTra rules in PROGRES: (a) A GraTra rule is part of the specification, while a pattern is represented in the run time graph. (b) In a GraTra rule, there is a fixed separation between context nodes and non-context nodes. In a pattern, each node may play both roles; which nodes serve as context nodes, is determined only dynamically when a pattern is instantiated. (c) In PROGRES, a GraTra rule is applied without recording its application. In contrast, when a pattern is instantiated in CHASID, its application is recorded to support traceability and consistency control.

5.2 Levels of Evolution

In *AHEAD* and *IREEN*, evolution is performed at the *specification level*. In contrast, *ConDes* and *CHASID* handle evolution at the *run time level*. Evolution at both levels may be combined, but this has not been done in any of these projects.

Evolution at both levels differ with respect to flexibility, impact, and technical realization. As mentioned above, evolution at the run time level provides for maximum *flexibility*. As a general design guideline, evolution should be performed at the run time level if domain knowledge is not stable. Nevertheless, both IREEN and AHEAD demonstrate the feasibility of specification level evolution for dynamic domain knowledge. With respect to its *impact*, evolution at the specification level is more fundamental because it may affect all existing run time graphs. Concerning the *technical realization*, evolution at the specification level has to satisfy *hard constraints* imposed by the specification language and its support environment. In particular, the consistency of run time graphs with respect to the specification must not be violated. In contrast, evolution at the run time level is concerned with *soft constraints* which are defined and (potentially) enforced by the specification developer.

5.3 Time of Evolution

Since in *ConDes* and *CHASID* domain knowledge is part of the run time graph, evolution is performed at *run time*. While domain knowledge is represented in the specification in *IREEN*, evolution is performed at run time, as well because IREEN interprets GraTra rules. Only AHEAD is constrained to evolution at *compile time*.

While the level of evolution significantly impacts the specification, the time of evolution does not impact the specification at all. If a PROGRES specification were interpreted rather than compiled, AHEAD would gain run time evolution for free.

6 Related Work

In the context of GraTra systems, the term *evolution* is frequently used with a very broad meaning: A graph is evolved by performing a graph transformation. For example, Mens [18] uses graph transformations for refactoring of object-oriented software systems; Engels et al. [7] specify consistency-preserving transformations of UML/RT models with graph transformations. In both cases, however, it is the modeled system which evolves — but not the underlying domain knowledge (see below).

Evolution of *types* has been considered for long in the context of database management systems (*schema evolution* [2]). Here, the main focus lies on the data migration problem, i.e., to migrate the data such that they conform to the new schema. More recently, the problem of migrating instances in response to type changes has been tackled also in the context of model-driven software engineering (*model migration* [22]). For example, [23] introduces the language Epsilon Flock for model migration, and [13] proposes a catalog of operators for the coupled evolution of metamodels and models in the COPE framework. Furthermore, several approaches have been developed in the GraTra field. [27] formally defines GraTra rules for co-evolution of type and instance graphs. The Model Integrated Computing (MIC) tool suite [15] includes a Model Change Language (MCL) in which GraTra rules for migration may be specified. Finally, the data migration problem is also studied in the domain of *dynamic reconfiguration* (which is performed in [29] with graph transformations).

Evolution of metamodels affects not only models, but also *model transformations*. In [24], ontology mappings are used to update model transformations after metamodel changes. These transformations constitute an example of *higher-order transformations* being surveyed in [28]. Finally, Bergman et al. [4] introduce change-based model transformations (GraTra rules trigged by the application of other GraTra rules) as a general formalism for propagating changes.

Patterns are supported e.g. in several UML tools by (usually hard-wired) commands. Levendovszky et al. [17] introduce a language and a tool for defining patterns based on arbitrary graph-based metamodels. Zhao et al. [30] define patterns with variable parts by sets of GraTra rules and use graph parsing for pattern recognition. Bottoni et al. [5] formalize patterns as graphs and describe their instantiation by triple graphs.

7 Conclusion

We have examined evolution support for evolving domain knowledge in a set of projects which were strongly application-driven. The presented evolution approaches were implemented on top of a single specification language (PROGRES), but they may be transferred to other technological spaces such as EMF-based model transformations [1] (due to space restrictions, this could not be shown in this paper). Future work should address the systematic engineering of GraTra systems for evolving domain knowledge — at the level of both languages and tools.

References

1. Arendt, T., Biermann, E., Jurack, S., Krause, C., Taentzer, G.: Henshin: Advanced Concepts and Tools for In-Place EMF Model Transformations. In: Petriu, D.C., Rouquette, N., Haugen, Ø. (eds.) MODELS 2010, Part I. LNCS, vol. 6394, pp. 121–135. Springer, Heidelberg (2010)
2. Banerjee, J., Kim, W., Kim, H.J., Korth, H.F.: Semantics and implementation of schema evolution in object-oriented databases. In: Proceedings of the 1987 ACM SIGMOD International Conference on Management of Data (SIGMOD 1987), pp. 311–322. ACM Press (1987)
3. Becker, S., Nagl, M., Westfechtel, B.: Incremental and interactive integrator tools for design product consistency. In: Nagl, Marquardt (eds.) [20], pp. 224–267
4. Bergmann, G., Ráth, I., Varró, G., Varró, D.: Change-driven model transformations: Change (in) the rule to rule the change. Software and Systems Modeling 11(3), 431–461 (2012)
5. Bottoni, P., Guerra, E., de Lara, J.: A language-independent and formal approach to pattern-based modelling with support for composition and analysis. Information and Software Technology 52, 821–844 (2010)
6. Ehrig, H., Engels, G., Kreowski, H.J., Rozenberg, G. (eds.): Handbook on Graph Grammars and Computing by Graph Transformation: Applications, Languages, and Tools, vol. 2. World Scientific, Singapore (1999)
7. Engels, G., Heckel, R., Küster, J.M., Groenewegen, L.: Consistency-Preserving Model Evolution through Transformations. In: Jézéquel, J.-M., Hussmann, H., Cook, S. (eds.) UML 2002. LNCS, vol. 2460, pp. 212–226. Springer, Heidelberg (2002)
8. Engels, G., Lewerentz, C., Nagl, M., Schäfer, W., Schürr, A.: Building integrated software development environments part I: Tool specification. ACM Transactions on Software Engineering and Methodology 1(2), 135–167 (1992)
9. Gatzemeier, F.H.: Authoring support based on user-serviceable graph transformation. In: Pfaltz, et al. (eds.) [21], pp. 170–185
10. Heimann, P., Krapp, C.A., Westfechtel, B.: Graph-based software process management. International Journal of Software Engineering and Knowledge Engineering 7(4), 431–455 (1997)
11. Heller, M., Jäger, D., Krapp, C.A., Nagl, M., Schleicher, A., Westfechtel, B., Wörzberger, R.: An adaptive and reactive management system for project coordination. In: Nagl, Marquardt (eds.) [20], pp. 300–366

12. Heller, M., Schleicher, A., Westfechtel, B.: Graph-based specification of a management system for evolving development processes. In: Pfaltz, et al. (eds.) [21], pp. 334–351

13. Herrmannsdoerfer, M., Vermolen, S.D., Wachsmuth, G.: An Extensive Catalog of Operators for the Coupled Evolution of Metamodels and Models. In: Malloy, B., Staab, S., van den Brand, M. (eds.) SLE 2010. LNCS, vol. 6563, pp. 163–182. Springer, Heidelberg (2011)

14. Jäger, D., Schleicher, A., Westfechtel, B.: Using UML for Software Process Modeling. In: Nierstrasz, O., Lemoine, M. (eds.) ESEC/FSE 1999. LNCS, vol. 1687, pp. 91–108. Springer, Heidelberg (1999)

15. Karsai, G.: Lessons Learned from Building a Graph Transformation System. In: Engels, G., Lewerentz, C., Schäfer, W., Schürr, A., Westfechtel, B. (eds.) Nagl Festschrift. LNCS, vol. 5765, pp. 202–223. Springer, Heidelberg (2010)

16. Kraft, B., Nagl, M.: Parameterized specification of conceptual design tools in civil engineering. In: Pfaltz, et al. (eds.) [21], pp. 90–105

17. Levendovszky, T., Lengyel, L., Mészáros, T.: Supporting domain-specific model patterns with metamodeling. Software and Systems Modeling 8, 501–520 (2009)

18. Mens, T.: On the Use of Graph Transformations for Model Refactoring. In: Lämmel, R., Saraiva, J., Visser, J. (eds.) GTTSE 2005. LNCS, vol. 4143, pp. 219–257. Springer, Heidelberg (2006)

19. Nagl, M. (ed.): Building Tightly Integrated Software Development Environments: The IPSEN Approach. LNCS, vol. 1170. Springer, Heidelberg (1996)

20. Nagl, M., Marquardt, W. (eds.): Collaborative and Distributed Chemical Engineering. LNCS, vol. 4970. Springer, Heidelberg (2008)

21. Pfaltz, J.L., Nagl, M., Böhlen, B. (eds.): AGTIVE 2003. LNCS, vol. 3062. Springer, Heidelberg (2004)

22. Rose, L.M., Herrmannsdoerfer, M., Williams, J.R., Kolovos, D.S., Garcés, K., Paige, R.F., Polack, F.A.C.: A Comparison of Model Migration Tools. In: Petriu, D.C., Rouquette, N., Haugen, Ø. (eds.) MODELS 2010, Part I. LNCS, vol. 6394, pp. 61–75. Springer, Heidelberg (2010)

23. Rose, L.M., Kolovos, D.S., Paige, R.F., Polack, F.A.C.: Model Migration with Epsilon Flock. In: Tratt, L., Gogolla, M. (eds.) ICMT 2010. LNCS, vol. 6142, pp. 184–198. Springer, Heidelberg (2010)

24. Roser, S., Bauer, B.: Automatic Generation and Evolution of Model Transformations Using Ontology Engineering Space. In: Spaccapietra, S., Pan, J.Z., Thiran, P., Halpin, T., Staab, S., Svatek, V., Shvaiko, P., Roddick, J. (eds.) Journal on Data Semantics XI. LNCS, vol. 5383, pp. 32–64. Springer, Heidelberg (2008)

25. Schürr, A., Klar, F.: 15 Years of Triple Graph Grammars – Research Challenges, New Contributions, Open Problems. In: Ehrig, H., Heckel, R., Rozenberg, G., Taentzer, G. (eds.) ICGT 2008. LNCS, vol. 5214, pp. 411–425. Springer, Heidelberg (2008)

26. Schürr, A., Winter, A., Zündorf, A.: The PROGRES approach: Language and environment. In: Ehrig, et al. (eds.) [6], pp. 487–550

27. Taentzer, G., Mantz, F., Lamo, Y.: Co-Transformation of Graphs and Type Graphs With Application to Model Co-Evolution. In: Ehrig, H., Engels, G., Kreowski, H.-J., Rozenberg, G. (eds.) ICGT 2012. LNCS, vol. 7562, pp. 326–340. Springer, Heidelberg (2012)

28. Tisi, M., Jouault, F., Fraternali, P., Ceri, S., Bézivin, J.: On the Use of Higher-Order Model Transformations. In: Paige, R.F., Hartman, A., Rensink, A. (eds.) ECMDA-FA 2009. LNCS, vol. 5562, pp. 18–33. Springer, Heidelberg (2009)

29. Wermelinger, M., Fiadeiro, J.L.: A graph transformation approach to software architecture reconfiguration. Science of Computer Programming 44, 133–155 (2002)

30. Zhao, C., Kong, J., Dong, J., Zhang, K.: Pattern-based design evolution using graph transformation. Journal of Visual Languages and Computing 18, 378–398 (2007)

Construction of Integrity Preserving Triple Graph Grammars

Anthony Anjorin[1,*], Andy Schürr[2], and Gabriele Taentzer[3]

[1] Technische Universität Darmstadt,
Graduate School of Computational Engineering, Germany
`anjorin@gsc.tu-darmstadt.de`
[2] Technische Universität Darmstadt,
Real-Time Systems Lab, Germany
`andy.schuerr@es.tu-darmstadt.de`
[3] Philipps-Universität Marburg,
Fachbereich Mathematik und Informatik, Germany
`taentzer@mathematik.uni-marburg.de`

Abstract. Triple Graph Grammars (TGGs) are a rule-based technique of specifying a consistency relation over a source, correspondence, and target domain, which can be used for bidirectional model transformation.

A current research challenge is increasing the *expressiveness* of TGGs by ensuring that global constraints in the involved domains are not violated by the transformation. Negative Application Conditions (NACs) can be used to enforce this property, referred to as *schema compliance*.

In previous work, we have presented a polynomial control algorithm for *integrity preserving* TGGs, using NACs *only* to ensure schema compliance, meaning that, for efficiency reasons, the usage of NACs must be restricted appropriately. In this paper, we apply the well-known translation of global constraints to application conditions for a given TGG and set of global constraints. We show that the derived set of NACs is indeed sufficient *and* necessary to ensure schema compliance, i.e., that the TGG together with the derived NACs is integrity preserving by construction.

Keywords: bidirectional transformation, triple graph grammars, schema compliance, integrity preservation, negative application conditions.

1 Introduction and Motivation

Model-Driven Engineering (MDE) is an established, viable means of coping with the increasing complexity of modern software systems, promising an increase in productivity, interoperability and a reduced gap between solution and problem domains. *Model transformation* plays a central role in any MDE approach and, as industrial applications demand efficiency and other suitable properties, the need for formal verification arises. This is especially the case for *bidirectional* transformation, a current research focus in various communities.

---

* Supported by the 'Excellence Initiative' of the German Federal and State Governments and the Graduate School of Computational Engineering at TU Darmstadt.

Triple Graph Grammars (TGGs) [13] are a declarative, rule-based technique of specifying a consistency relation between models in a source, correspondence and target domain, from which forward and backward unidirectional operational rules can be automatically derived. Coupled with a *control algorithm* to guide the overall transformation process, the derived operational rules can be used to realize bidirectional model transformation with useful formal properties such as correctness, completeness and polynomial runtime [2,11,13].

A current challenge is to increase the *expressiveness* of TGGs by ensuring *schema compliance*, i.e., that the derived transformation produces triples that are not only consistent with respect to the TGG, but also fulfil a set of *global constraints* in the involved domains. Although *Negative Application Conditions* (NACs) can clearly be used to restrain a TGG and control the derived transformation appropriately [6,11], due to efficiency reasons, TGG implementations typically have to restrict the usage of NACs appropriately. In [11], a polynomial algorithm is presented for *integrity preserving TGGs*, which use NACs *only* to ensure schema compliance. All formal properties are proven for integrity preserving TGGs, but a *constructive* definition of the exact supported class of NACs and a formal static *verification* of integrity preservation is left to future work.

Inspired by the construction of application conditions from constraints presented in [5], we close this gap and characterize the class of allowed NACs for integrity preserving TGGs as exactly those NACs that can be derived automatically from negative constraints according to the construction in [5].

Our contribution in this paper is to (i) extend results from [3] to show that existing graph transformation theory from [5] can be applied to TGGs, (ii) use these results to give a precise and constructive definition of the exact class of supported NACs used by integrity preserving TGGs, and (iii) prove that this class is indeed necessary *and* sufficient to guarantee schema compliance.

The paper is structured as follows: Sect. 2 discusses our running example, used to introduce TGGs, constraints, NACs and our formalization for these concepts. Our main contribution is presented in Sect. 3, while Sect.4 discusses related approaches. Sect. 5 concludes with a summary and future work.

2   Running Example and Formalization

Our running example is a *tool adapter scenario* inspired by [12], and is depicted schematically in Fig. 1. A Commercial Off-The-Shelf (COTS) tool is used by a certain group of domain experts to specify dependencies in a system via component/block diagrams (1). After using the tool for a period of time, a series of domain-specific rules and conventions are bound to evolve and the need to manipulate the diagrams to enforce these rules in an automatized but high-level manner arises. COTS tools are, however, typically closed source and often only offer an import and export format (2) as a means of data exchange [12]. This tool-specific textual exchange format (3), can be parsed via string grammars to

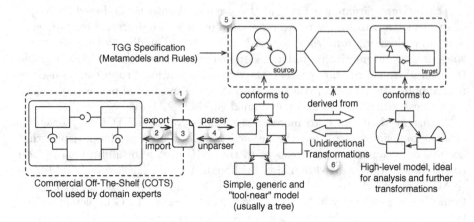

Fig. 1. Overview of the bidirectional transformation for our running example

a simple, generic tree and unparsed via a set of templates (4). From a TGG specification (5), describing how a model can be extracted from the (parse) tree, a pair of unidirectional transformations can be automatically derived (6), which can be used to forward transform the tree to a model, and after enforcing rules and possibly applying refactorings on the model, to backward transform the model to a tree that can be unparsed to text.

In such a scenario, bidirectionality is a natural requirement and TGGs, as a bidirectional model transformation language, can increase productivity by supporting a *single* specification of the bidirectional transformation and by guaranteeing that derived forward and backward transformations always produce consistent triples with respect to the consistency relation defined by the TGG.

In the following, we use this example to introduce TGGs, constraints, NACs and our formalization for these concepts. Due to space limitations, we assume a basic knowledge of category theory and fundamental concepts from algebraic graph transformation as presented in detail in [3].

2.1   TGGs as an Adhesive HLR Category

Rule-based transformation systems can be treated in a very general manner so that fundamental results can be applied to a large class of algebraic structures. As presented in [3], the minimal requirement on such a structure to build up a theory of rule-based transformation systems is the existence of generalized forms of a union of structures called a *pushout*, and an intersection of structures called a *pullback*. In addition, these constructions need to show a certain notion of compatibility. We refer the interested reader to [3] for further details.

Any category that fulfils these minimal requirements is called an *adhesive High-Level-Replacement (HLR)* category allowing the general theory on adhesive HLR categories to be instantiated by a particular category. The following definitions are adapted from [3,5].

Definition 1 (Adhesive High-Level-Replacement Category).

A category **C** *with a morphism class* $\mathcal{M}$ *is called an adhesive HLR category if:*

1. $\mathcal{M}$ *is a class of monomorphisms closed under isomorphisms, composition, and decomposition.*
2. **C** *has pushouts and pullbacks along* $\mathcal{M}$*-morphisms, and* $\mathcal{M}$*-morphisms are closed under pushouts and pullbacks.*
3. *Pushouts in* **C** *along* $\mathcal{M}$*-morphisms are Van Kampen squares [3,5].*

Definition 2 (Rule and Direct Derivation).

Given a rule $r = (L \leftarrow K \rightarrow R)$ *consisting of two morphisms in* $\mathcal{M}$ *with a common domain* K*, a direct derivation consists of two pushouts (1) and (2) in the category* **C** *(cf. diagram to the right), denoted as*

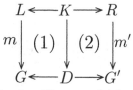

$G \overset{r@m}{\leadsto} G'$ *(or simply* $G \overset{r}{\leadsto} G'$*), where* $m : L \rightarrow G$ *is referred to as the* match *and* $m' : R \rightarrow G'$ *as the* comatch *of* r*. Given* m*, (1) has to be constructed as a pushout-complement, a generalized subtraction of graphs, which exists if* m *fulfills a* gluing condition *in* **C** *(cf. [3] for details).*

The following definitions instantiate this theory for the case of TGGs:

Definition 3 (Graph and Graph Morphism).

A graph $G = (V, E, s, t)$ *consists of finite sets* V *of nodes and* E *of edges, and two functions* $s, t : E \rightarrow V$ *that assign each edge source and target nodes.*
A graph morphism $h : G \rightarrow G'$*, with* $G' = (V', E', s', t')$*, is a pair of functions* $h := (h_V, h_E)$ *where* $h_V : V \rightarrow V'$*,* $h_E : E \rightarrow E'$ *and* $\forall e \in E : h_V(s(e)) = s'(h_E(e)) \wedge h_V(t(e)) = t'(h_E(e))$.

Graphs and graph morphisms form a category called **Graphs** (cf. [3] for details).

Definition 4 (Typed Graph and Typed Graph Morphism).

A type graph is a graph $TG = (V_{TG}, E_{TG}, s_{TG}, t_{TG})$.
A typed graph is a pair $(G, type)$ *of a graph* G *together with a graph morphism* type: $G \rightarrow TG$*. Given* $(G, type)$ *and* $(G', type')$*,* $g : G \rightarrow G'$ *is a typed graph morphism iff* $type = type' \circ g$.
$\mathcal{L}(TG) := \{ G \mid \exists\, type : type(G) = TG \}$ *denotes the set of all graphs of type* TG.

In the following, we denote *triple graphs* with single letters (e.g., G), which consist of graphs denoted with an index $X \in \{S, C, T\}$ (e.g, G_S, G_C, G_T).

Definition 5 (Typed Triple Graph and Typed Triple Morphism).

A triple graph $G := G_S \overset{\gamma_S}{\leftarrow} G_C \overset{\gamma_T}{\rightarrow} G_T$ *consists of typed graphs* $G_X \in \mathcal{L}(TG_X)$*,* $X \in \{S, C, T\}$*, and two morphisms* $\gamma_S : G_C \rightarrow G_S$ *and* $\gamma_T : G_C \rightarrow G_T$.

A typed triple morphism $h := (h_S, h_C, h_T) : G \rightarrow G'$*,* $G' = G'_S \overset{\gamma'_S}{\leftarrow} G'_C \overset{\gamma'_T}{\rightarrow} G'_T$ *is a triple of typed morphisms* $h_X : G_X \rightarrow G'_X, X \in \{S, C, T\}$*, s.t.* $h_S \circ \gamma_S = \gamma'_S \circ h_C$ *and* $h_T \circ \gamma_T = \gamma'_T \circ h_C$.

A type triple graph is a triple graph $TG = TG_S \overset{\Gamma_S}{\leftarrow} TG_C \overset{\Gamma_T}{\rightarrow} TG_T$*. A typed triple graph is a pair* $(G, type)$ *of a triple graph* G *and triple morphism* type : $G \rightarrow TG$*. Analogously to Def. 4,* $\mathcal{L}(TG)$ *denotes the set of all triple graphs of type* TG.

Remark 1 (Attributed Typed Graphs with Inheritance). *Practical applications require* attributed *typed graphs with* node type inheritance *and the usage of* abstract types *to allow for concise rules. Due to space limitations, we omit the corresponding formalization and refer to [8,3] for details.*

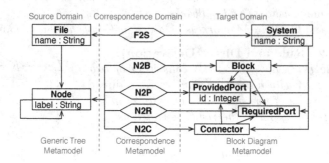

Fig. 2. Metamodels for trees, links and corresponding block diagrams

Example: Figure 2 depicts the type triple graph for our running example. On the left, the type graph for the source domain defines concepts such as a `File`, which can contain `Nodes` with labels. On the right, the type graph for the target domain defines a `System`, which consists of `Blocks` with `ProvidedPorts` and `RequiredPorts`. `Blocks` can depend on other `Blocks` via `Connectors` that assign a `RequiredPort` to a `ProvidedPort`. The type graph for the correspondence domain defines which source elements correspond to which target elements via correspondence types (visually distinguish as hexagons) such as `F2S` (`Files` with `Systems`). A typed triple graph for our example is depicted in Fig. 3.

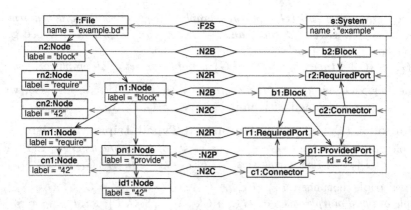

Fig. 3. Model triple of tree, links and corresponding block diagram

Fact 1 (TriGraphs is an adhesive HLR category).
The category **TriGraphs** *of typed triple graphs and triple morphisms, together with the class* $\mathcal{M}$ *of injective triple morphisms is an adhesive HLR category.* For the proof we refer the interested reader to Fact. 4.18 in [3].

Definition 6 (Triple Graph Grammar).
Let TG be a type triple graph. A TGG rule $r = (L \leftarrow K \rightarrow R)$ *is a monotonic creating rule, where* $L = K \subseteq R$ *are typed triple graphs over TG.*
TGG rules, denoted simply as $r = (L, R)$, *are applied at injective matches only. A triple graph grammar* $TGG := (TG, \mathcal{R})$ *is a pair consisting of a type triple graph TG and a finite set* $\mathcal{R}$ *of TGG rules. The generated language is denoted as* $\mathcal{L}(TGG) := \{G \in \mathcal{L}(TG) \mid \exists \, r_1, r_2, \ldots, r_n \in \mathcal{R} : \; G_{\emptyset} \overset{r_1}{\leadsto} G_1 \overset{r_2}{\leadsto} \ldots \overset{r_n}{\leadsto} G_n = G\}$, *where* $G_{\emptyset}$ *denotes the empty triple graph.*

Example: The TGG consisting of five rules for our example is depicted in Fig. 4. Each rule $r = (L, R)$ is depicted in a concise manner by merging L and R in a single diagram: Black elements without any markup are context elements of the rule (L), while green elements with a "++" markup are created by the rule $(R \setminus L)$. Rule (I) creates a file **F** and a corresponding system **S**, ensuring that the name of the system corresponds to the name of the file minus its extension. Rule (II) requires a file and a corresponding system as context, and creates a node with a "block" label and a corresponding block. Rule (III) creates required ports, while Rule (IV) creates provided ports, ensuring that the id of the provided port is equal to the label of the id node of the corresponding provider node. Children nodes of require nodes are created with corresponding connectors in Rule (V).

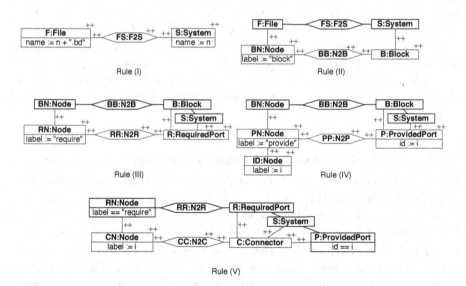

Fig. 4. TGG rules for simultaneous evolution of trees and corresponding block diagrams

2.2   Bidirectional Model Transformation with TGGs

TGGs can be used for bidirectional model transformation by automatically deriving forward and backward unidirectional rules. These *operational* rules form the atomic steps of a complete transformation sequence determined by a *control algorithm*. A TGG implementation, therefore, faces the challenge of (i) determining an appropriate *operationalization* of the TGG rules, and (ii) providing an efficient (polynomial) control algorithm to determine and control the sequence in which nodes of the input graph are transformed with operational rules in an efficient manner. In addition to the restriction to non-deleting TGG rules in Def. 6, the control algorithm presented in [11] poses further restrictions on the class of supported TGGs. As we aim to close the gap between formal results and an actual TGG algorithm and implementation, we shall present the general theory from [5] already simplified and reformulated with these restrictions. We believe that this does not only improve readability but also allows for a closer correlation with our running example. We refer the interested reader to [5] for a more general formulation of Def. 7, 8 and Theorem 1.

A metamodel does not only specify the allowed structure of models (abstract syntax) but also restricts the class of valid models via some form of constraints such as multiplicities or OCL (static semantics). In our formalization, we complement type graphs with negative constraints[1] defined in the following:

Definition 7 (Negative Constraints and Schema Compliance).
Let $\mathcal{M}$ denote the class of monomorphisms in **TriGraphs**. *A schema is a pair $(TG, \mathcal{NC})$ of a type triple graph TG and a set $\mathcal{NC} \subseteq \mathcal{L}(TG)$ of negative constraints. A typed triple graph $G \in \mathcal{L}(TG)$ satisfies a negative constraint $NC \in \mathcal{NC}$, denoted as $G \models NC$, if $\nexists nc \in \mathcal{M} : NC \to G$. A negative source constraint is of the form $NC_S \xleftarrow{\emptyset} \emptyset \xrightarrow{\emptyset} \emptyset$ (negative target constraints are defined analogously). Let $\mathcal{L}(TG, \mathcal{NC}) := \{G \in \mathcal{L}(TG) \mid \forall NC \in \mathcal{NC}, G \models NC\}$ denote the set of all schema compliant triple graphs. $TGG = (TG, \mathcal{R})$ is schema compliant, iff $\mathcal{L}(TGG) \subseteq \mathcal{L}(TG, \mathcal{NC})$, i.e., the language generated by TGG consists only of schema compliant triple graphs.*

Example: Fig. 5 depicts two negative target constraints[2] for our example: $\emptyset \xleftarrow{\emptyset} \emptyset \xrightarrow{\emptyset}$ NC1 forbids a block from satisfying its own dependencies, while $\emptyset \xleftarrow{\emptyset} \emptyset \xrightarrow{\emptyset}$ NC2 prohibits connecting a required port to more than one provider. Now consider Rule (V) (Fig. 4) and $\emptyset \xleftarrow{\emptyset} \emptyset \xrightarrow{\emptyset}$ NC1 (Fig. 5). Note that the TGG consisting of the five rules is *not* schema compliant as, by creating connectors between ports of the *same* block via Rule (V), it can create triples graphs that violate $\emptyset \xleftarrow{\emptyset} \emptyset \xrightarrow{\emptyset}$ NC1. The question is how to restrain TGG rules in order to ensure schema compliance. Assuming that NACs are sufficient for this task (we prove this in Sect. 3), the following definition formalizes NACs for TGG rules.

---

[1] The negative constraints we use are powerful enough to enforce upper bounds on multiplicities, lower bounds can be enforced by *positive constraints* as handled in [5,15].

[2] Only the non-trivial target components of the negative constraints are shown.

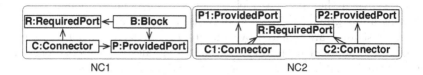

Fig. 5. Negative constraints for block diagrams

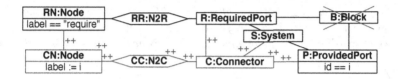

Fig. 6. Rule (VI) which uses a NAC to ensure that NC1 is not violated

Definition 8 (Negative Application Conditions (NACs)).

Let TG be a type triple graph. A negative application condition over a typed triple graph $L \in \mathcal{L}(TG)$ is a typed triple graph $N \in \mathcal{L}(TG)$ such that $L \subseteq N$.
A triple morphism $m : L \to G$ satisfies N, i.e., $m \models N$, if $\nexists n : N \to G, n|_N = m$.
A neg. application condition $N(r)$ for a TGG rule $r = (L, R)$ is a NAC over L.
A direct derivation $G \overset{r@m}{\rightsquigarrow} G'$ satisfies $N(r)$, if $m \models N(r)$.
A source NAC is of the form $N_S \overset{q_L}{\leftarrow} L_C \overset{r_L}{\to} L_T$ (target NACs are defined analogously). Every TGG rule r is equipped with $\mathcal{N}$, a set of NACs, i.e., $r = (L, R, \mathcal{N})$.

Example: Taking $\emptyset \overset{\emptyset}{\leftarrow} \emptyset \overset{\emptyset}{\to}$ NC1 (Fig. 5) as the negative target constraint that must not be violated by the TGG for our running example, Rule (V) can be extended by a target NAC N that appropriately restricts matches to ensure schema compliance. This new Rule (VI) $= (L, R, N)$ is shown in Fig. 6 with nodes[3] in $N \setminus L$ depicted as *crossed out* elements (i.e., block B).

For bidirectional model transformation, each TGG rule must be decomposed into a *source rule* that only changes the source graph of a graph triple and a *forward rule* that retains the source graph and changes the correspondence and target graphs. For NACs that can be separated into source and target components, formal results [6,11] guarantee that this is always uniquely possible, and that the sequence of derived source and forward rules can be reordered appropriately so that the source rules first build up the source graph completely and the sequence of forward rules perform a forward transformation. Handling arbitrary (negative) application conditions efficiently is an open challenge[4], but if the usage of source and target NACs as defined above is restricted to ensuring schema

---

[3] Note that all links incident to B are also elements in N.

[4] To ensure efficiency, we employ a context-driven recursive algorithm and a dangling edge check to avoid backtracking. This strategy, however, does not work for arbitrary NACs that can influence rule application with respect to the source domain.

compliance, then they can be safely ignored in the source domain (assuming consistent input graph triples) and only used in the target domain to ensure that no constraints are violated by applying the forward rule [11]. The following definition, adapted from [6,11], formalizes the decomposition process for a forward transformation. As TGGs are symmetric, all definitions and arguments can be formulated analogously for a backward transformation.

Definition 9 (Operationalization for Rules with Source/Target NACs).
*Let $TGT = (TGT, \mathcal{R})$ and $r = (L_S \xleftarrow{q_L} L_C \xrightarrow{\tau_L} L_T, R_S \xleftarrow{q_R} R_C \xrightarrow{\tau_R} R_T, \mathcal{N}) \in \mathcal{R}$.
If $\mathcal{N} = \mathcal{N}_S \cup \mathcal{N}_T$, where $\mathcal{N}_S$ is a set of source NACs, and $\mathcal{N}_T$ a set of target NACs, then* source rule $r_S = (SL, SR, \mathcal{N}_S)$ *and* forward rule $r_F = (FL, FR, \mathcal{N}_T)$ *can be derived according to the following diagram (triangles represent sets of NACs).*

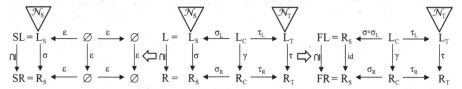

Example: The derived forward and backward rules for Rule (VI) (Fig. 6) are depicted in Fig. 7. As the required NAC is a target NAC, it does not restrain the backward rule r_B in any way, but restrains the forward rule r_F to ensure that the negative target constraint $\emptyset \xleftarrow{\emptyset} \emptyset \xrightarrow{\emptyset} NC1$ is not violated.

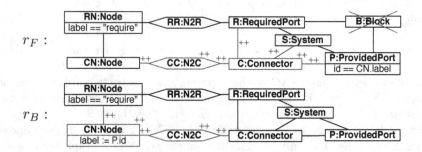

Fig. 7. Derived operational rules from Rule (V) of our running example

For *integrity preserving* TGGs that fulfil certain restrictions, including using NACs only to ensure schema compliance, [11] has shown that a polynomial, context-driven TGG control algorithm can guarantee all important formal properties (correctness, completeness and efficiency). Due to space limitations, we refer to [11] for further details concerning the context-driven TGG control algorithm, definitions and proofs of formal properties, and exactly *why* these restrictions are necessary. The following definition, taken from [11], formalizes the concept of integrity preservation.

Definition 10 (Integrity Preserving TGGs).
Given a schema $(TG, \mathcal{NC})$ and a TGG rule $r = (L, R, \mathcal{N})$, where $r^- := (L, R, \emptyset)$ is the same rule as r but without NACs. The TGG rule r is integrity preserving iff $\forall\, G, G' \in \mathcal{L}(TG):$

1. *$\mathcal{NC}$ and $\mathcal{N}$ consist of only negative source/target constraints and source/target negative application conditions.*
2. *r preserves schema compliance for triple graphs:*
   For every $G \overset{r@m}{\rightsquigarrow} G' : G \in \mathcal{L}(TG, \mathcal{NC}) \Rightarrow G' \in \mathcal{L}(TG, \mathcal{NC})$,
3. *r^- preserves violations of schema compliance for triple graphs:*
   For every $G \overset{r^-@m}{\rightsquigarrow} G' : G \notin \mathcal{L}(TG, \mathcal{NC}) \Rightarrow G' \notin \mathcal{L}(TG, \mathcal{NC})$
4. *NACs are only used to ensure schema compliance in the following manner:*
   For every $G \overset{r^-@m}{\rightsquigarrow} G' : \exists\, N = N_S \leftarrow L_C \rightarrow L_T \in \mathcal{N}$ with $m \not\models N \Rightarrow$
   $\exists\, NC = NC_S \overset{\emptyset}{\leftarrow} \emptyset \overset{\emptyset}{\rightarrow} \emptyset \in \mathcal{NC} : G' \not\models NC$ (analogously for target NACs)

A TGG is integrity preserving if it consists only of integrity preserving rules.

3   From Global Constraints to Integrity Preserving TGGs

Given a TGG and a set of global constraints, specifying the *right* NACs, i.e., NACs that are just strong enough to ensure schema compliance but not stronger, is quite a challenge even for experienced users.

In this section, we show that the required NACs can be automatically derived from the set of negative constraints and that the set of NACs derived according to this construction is indeed sufficient *and* necessary to ensure schema compliance, i.e., that the TGG together with the derived set of NACs is integrity preserving.

In order to apply results from [5], we have to show two additional properties for **TriGraphs**: the existence of a generalized disjoint union (binary coproduct), and a generalized factorization in surjective and injective parts for every triple morphism (epi-mono factorization).

Definition 11 (Binary Coproducts).
A category $C = (Ob_C, Mor_C, \circ, id)$ has binary coproducts iff $\forall A, B \in Ob_C$ the binary coproduct of A and B, $(A + B, i_A : A \rightarrow A + B, i_B : B \rightarrow A + B)$, can be constructed such that $\forall X \in Ob_C, f : A \rightarrow X, g : B \rightarrow X,$

$$\exists\, [f, g] \in Mor_C : A + B \rightarrow X \text{ with } [f, g] \circ i_A = f \text{ and } [f, g] \circ i_B = g.$$

Lemma 1. *Category **TriGraphs** has binary coproducts.*

Proof. This is a direct consequence of the facts that category **Graphs** has binary coproducts and coproducts in **TriGraphs** are constructed componentwise (cf. Facts 4.15 and 4.16 in [5]).

Given $(A_S \overset{\alpha_S}{\leftarrow} A_C \overset{\alpha_T}{\rightarrow} A_T), (B_S \overset{\beta_S}{\leftarrow} B_C \overset{\beta_T}{\rightarrow} B_T)$, the binary coproduct can be constructed as follows:

$$A + B := (A_S + B_S) \overset{[i_{A_S} \circ \alpha_S, i_{B_S} \circ \beta_S]}{\longleftarrow} (A_C + B_C) \overset{[i_{A_T} \circ \alpha_T, i_{B_T} \circ \beta_T]}{\longrightarrow} (A_T + B_T),$$
$$i_A := (i_{A_S}, i_{A_C}, i_{A_T}), \ i_B := (i_{B_S}, i_{B_C}, i_{B_T})$$

i.e., by constructing the binary coproduct componentwise in the category of graphs **Graphs** which exists as shown in [3].

Definition 12 (Epi-Mono Factorization). *A category is said to have an epi-mono factorization if* $\forall f : A \rightarrow B$, $\exists e : A \rightarrow C$, $m : C \rightarrow B$ *such that* e *is an epimorphism,* m *is a monomorphism and* $m \circ e = f$.

Lemma 2 (Category TriGraphs has an epi-mono factorization).

Proof. Given $(f_S, f_C, f_T) : (A_S \leftarrow A_C \rightarrow A_T) \rightarrow (B_S \leftarrow B_C \rightarrow B_T)$ the diagram below shows that $C = C_S \overset{\sigma_S}{\leftarrow} C_C \overset{\sigma_T}{\rightarrow} C_T$ can be determined appropriately, such that $e := (e_S, e_C, e_T) : A \rightarrow C$ and $m := (m_S, m_C, m_T) : C \rightarrow B$, composed componentwise using composition in **TriGraphs** and the property of the epi-mono factorization in **Graphs** [3], fulfil $m \circ e = (m_S \circ e_S, m_C \circ e_C, m_T \circ e_T) = f$. $\sigma_S := m_S^{-1} \circ \beta_S \circ m_C$, is well-defined as m_S is injective and thus reversible. Furthermore, we have $f_S \circ \alpha_S = m_S \circ e_S \circ \alpha_S = \beta_S \circ f_C$, thus the images of $m_S \circ e_S \circ \alpha_S$ and $\beta_S \circ f_C$ are equal. It remains to show that the left two new squares commute (proof for σ_T analogously):

$m_S \circ \sigma_S = m_S \circ m_S^{-1} \circ \beta_S \circ m_C$
$= \beta_S \circ m_C$
$m_S \circ e_S \circ \alpha_S = f_S \circ \alpha_S$
$= \beta_S \circ f_C = \beta_S \circ m_C \circ e_C$
$= m_S \circ \sigma_S \circ e_C$
As m_S is mono, it follows:
$e_S \circ \alpha_S = \sigma_S \circ e_C$

$$
\begin{array}{ccc}
A_S & \overset{\alpha_S}{\leftarrow} A_C & \overset{\alpha_T}{\rightarrow} A_T \\
\downarrow{\scriptstyle e_S} & \downarrow{\scriptstyle e_C} & \downarrow{\scriptstyle e_T} \\
C_S & \overset{\sigma_S}{\leftarrow} C_C & \overset{\sigma_T}{\rightarrow} C_T \\
\downarrow{\scriptstyle m_S\, f_S} & \downarrow{\scriptstyle m_C\, f_C} & \downarrow{\scriptstyle m_T\, f_T} \\
B_S & \overset{\beta_S}{\leftarrow} B_C & \overset{\beta_T}{\rightarrow} B_T
\end{array}
$$

$\sigma_S := m_S^{-1} \circ \beta_S \circ m_C$
$\sigma_T := m_T^{-1} \circ \beta_T \circ m_C$

After proving these additional properties for **TriGraphs**, we can now apply results from [5] to construct NACs from negative constraints:

Theorem 1 (Constructing NACs from Negative Constraints).
Given a $TGG = (TG, \mathcal{R}^-)$ *with rules without NACs, and a set of global constraints* $\mathcal{NC} \subseteq \mathcal{L}(TG)$. *For every rule* $r^- = (L, R, \emptyset) \in \mathcal{R}^-$, *there is a construction* A *for NACs such that:*
$$\forall G \overset{r@m}{\rightsquigarrow} G' \text{ with } r = (L, R, A(\mathcal{NC})) \in \mathcal{R}: G \in \mathcal{L}(TG, \mathcal{NC}) \Rightarrow G' \in \mathcal{L}(TG, \mathcal{NC})$$

Proof. We refer to [5] for a detailed proof that NACs derived according to this construction are necessary and sufficient to ensure schema compliance.

In the following, we present a version of the general construction from [5], simplified for negative constraints (Def. 7) and NACs (Def. 8) as required for our running example.

Let $r^- = (L, R, \emptyset)$, and $NC \in \mathcal{NC}$:

1. Construct the disjoint union $R + NC$ as a binary coproduct with $i_R : R \to R + NC$ and $i_{NC} : NC \to R + NC$.
2. Given a graph $G' \not\models NC$, i.e., $\exists nc' : NC \to G' \in \mathcal{M}$, for any co-match $m' : R \to G'$ there is a unique morphism $f : R + NC \to G'$, such that $f \circ i_R = m'$ and $f \circ i_{NC} = nc'$.
3. Separate f in two parts, such that the first part glues graph elements and the second part embeds this gluing into a larger context, i.e., let $f = e \circ p'$ be an epi-mono factorization, where $e : R + NC \to N'$.
4. Then the post-condition $n' : R \to N'$ is defined by $n' = e \circ i_R$ and there is morphism $p' : N' \to G'$ such that $p' \circ n' = m'$. Thus, $m' \not\models N'$.

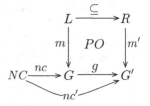

5. Try to construct the pre-condition N as part of the pushout complement over rule $r = (L, R)$ and post-condition N'. If it exists, the assumption that $m' \not\models N'$ yields p' with $p' \circ n' = m'$. Due to pushout decomposition, we get $p : N \to G$ and $n : L \to N$ with $p \circ n = m$. Thus, $m \not\models N$.

This is repeated for all constraints $NC \in \mathcal{NC}$ to construct the rule $r = (L, R, \mathcal{N})$ and for all rules $r^- \in \mathcal{R}^-$ to obtain the set R of rules with required NACs.

Proposition 1.
Given a TGG $= (TG, \mathcal{R}^-)$ with rules without NACs, and a set of negative constraints $\mathcal{NC} \subseteq \mathcal{L}(TG)$. For every $G \overset{r^- @m}{\rightsquigarrow} G'$ with rule $r^- = (L, R, \emptyset) \in \mathcal{R}^-$:
$$G' \in \mathcal{L}(TG, \mathcal{NC}) \Rightarrow G \in \mathcal{L}(TG, \mathcal{NC})$$

Proof. Given $G \notin \mathcal{L}(TG, \mathcal{NC}) \wedge G \overset{r^- @m}{\rightsquigarrow} G'$ there is an $NC \in \mathcal{NC}$ with $G \not\models NC$. Due to $G \overset{r^- @m}{\rightsquigarrow} G'$ with $g : G \to G'$, there is a morphism $nc' = g \circ nc$ as depicted in the diagram to the right. Hence, $G' \notin \mathcal{L}(TG, \mathcal{NC})$.

Proposition 2. *Given a TGG with rules without NACs, and a set of negative source constraints, the construction of NACs as specified in Theorem 1 yields source NACs only (analogously for target constraints and target NACs).*

Proof. Since all steps in the NAC construction are defined componentwise, this case leads back to the NAC construction on simple graphs in the source and the target case. For correspondence graphs we have the following situation: As $NC_C = \emptyset$, $(i_R)_C = (id_R)_C$ and $f_C = m'_C$, N'_C depends only on m'_C. As m'_C is a monomorphism, $N'_C = R_C$. Furthermore, we get $N_C = L_C$ due to pushout properties. Correspondingly, it is straight forward to show that $N_S = L_S$ if $NC_S = \emptyset$ in case of negative source constraints and that $N_T = L_T$ if $NC_T = \emptyset$ in case of negative target constraints.

Corollary 1.
TGGs with NACs constructed according to Theom. 1 only employ NACs to en-
force schema compliance, i.e., for TGG = (TG, R), and G, G' as in Prop. 1:
$$\forall r = (L, R, \mathcal{N}) \in \mathcal{R} \text{ and } G \overset{r^- @m}{\rightsquigarrow} G' : \exists N \in \mathcal{N} \text{ with } m \not\models N \Rightarrow G' \notin \mathcal{L}(TG, \mathcal{NC}).$$

Proof. Given $N \in \mathcal{N} : G \overset{r^- @m}{\rightsquigarrow} G' \wedge m \not\models N$, Theom. 1 can be reformulated as:
$$\forall G \in \mathcal{L}(TG, \mathcal{NC}), G \overset{r^- @m}{\rightsquigarrow} G' : m \not\models N \Leftrightarrow G' \notin \mathcal{L}(TG, \mathcal{NC}).$$

Corollary 2.
TGGs with NACs as constructed according to Theom. 1 are integrity preserving

Proof. This follows directly from Def. 10, Theom. 1, Props. 1 and 2, and Coroll. 1.

Example: In the following, we shall use the construction process to derive an
appropriate NAC for Rule (V) of our example (Fig. 4), regarding $\emptyset \overset{\emptyset}{\leftarrow} \emptyset \overset{\emptyset}{\rightarrow} NC1$
(Fig. 5) as the constraint that should not be violated.

To allow for a compact notation, a schematic
representation of the target component (L_T, R_T)
of Rule (V) and the target component $NC1$ of
the negative constraint is depicted in the diagram to the right, displaying L_T and R_T sepa-
rately and e.g., B:Block simply as B without its corresponding type. The task is
to derive $r = (L, R, \mathcal{N})$ from $r^- = (L, R, \emptyset)$ according to our construction. There
are in sum eight different intersections $D_T^1 - D_T^8$ of R_T and NC1, i.e., eight different ways of gluing elements in $R + NC$ to construct the post-condition N'.
The resulting post-conditions and pre-conditions for all cases are depicted in
Fig. 8. Assuming consistent input, D_T^1 only leads to the NAC N_T^1 being added
as $L_S \leftarrow L_C \rightarrow N_T^1$ to $\mathcal{N}$. It corresponds exactly to the NAC used by Rule (VI)
(Fig. 6) to ensure schema compliance. In general, a *set* of NACs is derived for
each rule, which can be further optimized using a weaker-than relationship between NACs [5]. In all other cases, the pushout complement does not exist or
the resulting precondition already violates the constraint and is thus discarded.

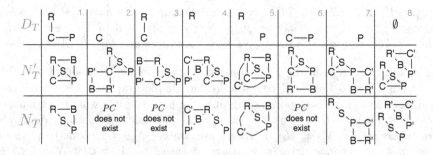

Fig. 8. Results of construction for all possible intersections of R_T and NC1

4   Related Work

In the following, we consider three different groups of approaches that introduce and use (negative) application conditions in the context of TGGs:

NACs to Guarantee Schema Compliance: An "on-the-fly" technique of determining a sequence of forward rules via a context-driven algorithm with polynomial runtime is presented in [14]. This algorithm supports NACs, but a proof of completeness is left to future work. In [11], an extended algorithm is presented for integrity preserving TGGs, appropriately restricting the usage of NACs, so that completeness, correctness and polynomial runtime can be proven.

Filter NACs to Guarantee Polynomial Runtime: The Decomposition and Composition Theorem in [2] is extended in [6] for NACs with a similar treatment as in [14], showing correctness and completeness for a backtracking algorithm. The "on-the-fly" technique of determining a sequence of forward rules employed in [14] is formalized in [4] and, although concepts of parallel independence are introduced and the possibility of employing a critical pair analysis are mentioned, the presented approach is still exponential in general. This basis provided by [4] is extended in [9,10], and a critical pair analysis is used to enforce functional behaviour. Efficiency (polynomial runtime) is guaranteed by the construction of *filter NACs* that cut off possible backtracking paths of the algorithm and eliminate critical pairs. Local completeness is, however, weaker than functional behaviour and the authors of [11], based on experience with industrial case studies [12], regard explicit support for non-functional TGGs as an important requirement.

General Application conditions: In [7], a larger class of general application conditions (positive, negative, complex and nested) are introduced for TGGs. An extension of the algorithm introduced in [10] is, however, left to future work. An integration of OCL with TGGs and corresponding tool support is presented in [1]. It is, however, unclear exactly how and to what extent the arbitrary OCL constraints must be restricted to ensure correctness and completeness of the derived translators.

5   Summary and Conclusion

We have shown that formal techniques from [5] can be appropriately applied to a given TGG and negative constraints to derive a set of NACs, which is both necessary and sufficient to ensure schema compliance, i.e., the TGG together with the derived set of NACs is integrity preserving as required by [11].

As the category of triple graphs has binary coproducts and an epi-mono factorization, the general definitions of constraints and application conditions, and the construction from [5] can be directly used for TGGs, i.e., without the restrictions to negative constraints and NACs as used in this paper. However, it remains an open challenge to extend our control algorithm to handle such general constraints and application conditions efficiently, and we leave this to future work.

References

1. Dang, D.-H., Gogolla, M.: On Integrating OCL and Triple Graph Grammars. In: Chaudron, M.R.V. (ed.) MODELS 2008 Workshop. LNCS, vol. 5421, pp. 124–137. Springer, Heidelberg (2009)
2. Ehrig, H., Ehrig, K., Ermel, C., Hermann, F., Taentzer, G.: Information Preserving Bidirectional Model Transformations. In: Dwyer, M.B., Lopes, A. (eds.) FASE 2007. LNCS, vol. 4422, pp. 72–86. Springer, Heidelberg (2007)
3. Ehrig, H., Ehrig, K., Prange, U., Taentzer, G.: Fundamentals of Algebraic Graph Transformation. Springer, Berlin (2006)
4. Ehrig, H., Ermel, C., Hermann, F., Prange, U.: On-the-Fly Construction, Correctness and Completeness of Model Transformations Based on Triple Graph Grammars. In: Schürr, A., Selic, B. (eds.) MODELS 2009. LNCS, vol. 5795, pp. 241–255. Springer, Heidelberg (2009)
5. Ehrig, H., Habel, A., Ehrig, K., Pennemann, K.H.: Theory of Constraints and Application Conditions: From Graphs to High-Level Structures. Fundamenta Informaticae 74(1), 135–166 (2006)
6. Ehrig, H., Hermann, F., Sartorius, C.: Completeness and Correctness of Model Transformations based on Triple Graph Grammars with Negative Application Conditions. In: Boronat, A., Heckel, R. (eds.) GTVMT 2009. ECEASST, vol. 18. EASST (2009)
7. Golas, U., Ehrig, H., Hermann, F.: Formal Specification of Model Transformations by Triple Graph Grammars with Application Conditions. In: Echahed, R., Habel, A., Mosbah, M. (eds.) GCM 2010. ECEASST, vol. 39. EASST (2011)
8. Guerra, E., de Lara, J.: Attributed Typed Triple Graph Transformation with Inheritance in the Double Pushout Approach. Tech. rep., Universidad Carlos III, Madrid (2006)
9. Hermann, F., Ehrig, H., Orejas, F., Golas, U.: Formal Analysis of Functional Behaviour for Model Transformations Based on Triple Graph Grammars. In: Ehrig, H., Rensink, A., Rozenberg, G., Schürr, A. (eds.) ICGT 2010. LNCS, vol. 6372, pp. 155–170. Springer, Heidelberg (2010)
10. Hermann, F., Golas, U., Orejas, F.: Efficient Analysis and Execution of Correct and Complete Model Transformations Based on Triple Graph Grammars. In: Bézivin, J., Soley, M.R., Vallecillo, A. (eds.) MDI 2010. ICPS, vol. 482, pp. 22–31. ACM, New York (2010)
11. Klar, F., Lauder, M., Königs, A., Schürr, A.: Extended Triple Graph Grammars with Efficient and Compatible Graph Translators. In: Engels, G., Lewerentz, C., Schäfer, W., Schürr, A., Westfechtel, B. (eds.) Nagl Festschrift. LNCS, vol. 5765, pp. 141–174. Springer, Heidelberg (2010)
12. Rose, S., Lauder, M., Schlereth, M., Schürr, A.: A Multidimensional Approach for Concurrent Model Driven Automation Engineering. In: Osis, J., Asnina, E. (eds.) Model-Driven Domain Analysis and Software Development: Architectures and Functions, pp. 90–113. IGI Publishing (2011)
13. Schürr, A.: Specification of Graph Translators with Triple Graph Grammars. In: Mayr, E.W., Schmidt, G., Tinhofer, G. (eds.) WG 1994. LNCS, vol. 903, pp. 151–163. Springer, Heidelberg (1995)
14. Schürr, A., Klar, F.: 15 Years of Triple Graph Grammars. In: Ehrig, H., Heckel, R., Rozenberg, G., Taentzer, G. (eds.) ICGT 2008. LNCS, vol. 5214, pp. 411–425. Springer, Heidelberg (2008)
15. Taentzer, G., Rensink, A.: Ensuring Structural Constraints in Graph-Based Models with Type Inheritance. In: Cerioli, M. (ed.) FASE 2005. LNCS, vol. 3442, pp. 64–79. Springer, Heidelberg (2005)

Applying Incremental Graph Transformation to Existing Models in Relational Databases*

Gábor Bergmann, Dóra Horváth, and Ákos Horváth

Budapest University of Technology and Economics,
Department of Measurement and Information Systems,
1117 Budapest, Magyar tudósok krt. 2
horvathdora@sch.bme.hu, {ahorvath,bergmann}@mit.bme.hu

Abstract. As industrial practice demands larger and larger system models, the *efficient execution* of graph transformation remains an important challenge. Additionally, for real-world applications, compatibility and integration with already well-established technologies is highly desirable. Therefore, *relational databases* have been investigated before as off-the-shelf environments for graph transformation, since they are already widely used for storing, processing and querying large graphs.

The *graph pattern matching* phase of graph transformation typically dominates in cost due to its combinatorial complexity. Therefore significant attempts have been made to improve this process; *incremental pattern matching* is an approach that has been shown to exert favorable performance characteristics in many practical use cases. To this day, however, no solutions are available for applying incremental techniques side by side with *already deployed systems* built over relational databases.

In the current paper, we propose an approach that translates graph patterns and transformation rules into event-driven (trigger-based) SQL programs that seamlessly integrate with existing relational databases to perform incremental pattern matching. Additionally, we provide experimental evaluation of the performance of our approach.

1 Introduction

1.1 Motivation

Nowadays *model-driven systems development* is being increasingly supported by a wide range of conceptually different *model transformation tools*. Several such model transformation approaches rely on the rule-based formalism of *graph transformation* (GT) [1] for specifying model transformations. Informally, a graph transformation rule performs local manipulation on graph models by finding a match of its left-hand side (LHS) graph pattern in the model, and replacing it with the right-hand side (RHS) graph.

As industrial practice demands ever larger system models, the scalability of storage, query and manipulation of complex graph-based model structures, and

---

* This work was partially supported by SecureChange (ICT-FET-231101), CertiMoT (ERC_HU-09-01-2010-0003) and TÁMOP-4.2.1/B-09/1/KMR-2010-0002.

H. Ehrig et al.(Eds.): ICGT 2012, LNCS 7562, pp. 371–385, 2012.

thus the *efficient execution* of graph transformation, gains importance. *Graph pattern matching* is the phase of graph transformation where matches of the LHS are identified in the graph; due to its dominant combinatorial complexity, significant attempts have been made to improve this process [2,3]. *Incremental pattern matching* is an approach that has been shown [4,5] to exert favorable performance characteristics in many practical use cases.

For industrial applications, compatibility and integration with already well-established technologies is preferred to custom solutions. *Relational Databases* (RDBs) have successfully served as the storage medium for business critical data for large companies. As explored in [6], RDBs offer a promising implementation environment for large graph models and graph transformation.

Regarding GT execution performance, however, RDBs have had mixed success [7]. Incremental pattern matching in RDBs has been proposed in [4]. This approach guarantees the consistency of the incremental store only if the system evolves along the specific GT rules. However, this solution is not compatible with already deployed (legacy) software, which may manipulate the underlying database in an arbitrary way. In fact, in many industrial scenarios, the underlying relational database (where the graph model is stored) is accessed in multiple ways (client programs, server side scripts), which are unaware of the incremental caches, hence they do not update them properly. For consistent behavior, these programs would have to be re-engineered with unrealistic effort.

In the paper, we propose to extend incremental pattern matching over RDBs in order to obtain an efficient system in an industrial environment that can exist side-by-side with already deployed software. With our current approach, incrementality will be guaranteed regardless of any external changes to the underlying database. Our proposal is complemented with a performance evaluation on a prototype implementation.

1.2   Goals

To summarize, the proposed solution will keep the beneficial properties of [4], including *Declarativity* (automatic execution based on GT specification, without requiring manually written code) and *Incrementality* (incremental evaluation techniques for the graph pattern matching phase, to improve performance on certain types of tasks). Additionally, we will also address the new requirement of *Compatibility*, permitting side-by-side operation with any existing legacy scripts and programs already reading or writing the database contents.

An implementation of this solution will be evaluated by measuring its performance on some known GT benchmarks. Results will be contrasted with the performance of the non-incremental execution and a non-RDB implementation.

1.3   Structure of the Paper

The rest of the paper is structured as follows. Sec. 2 introduces the concepts of graph transformation rules, incremental pattern matching and relational databases. Sec. 3 gives an overview on our approach, while Sec. 4 highlights the key phases that enable incrementality. Performance evaluation on different well-known case studies are conducted in Sec. 5 and finally Sec. 6 concludes the paper.

2   Background and Related Work

2.1   Graph Transformation

Graph patterns represent conditions (or *constraints*) controlling how *pattern variables* can be mapped to a part of the instance *graph model*. The pattern variables here are the nodes and edges of the graph pattern. The most important constraints express the graph structure of the pattern and types of variables. A *match* of the pattern is a mapping of variables to model elements so that all constraints are satisfied (analogously to a morphism into the model graph).

Extensions to the formalism include attribute checks, injectivity, composition and disjunction. A *negative application condition* (NAC) constraint expresses non-existence of certain elements in the form of a negated subpattern.

Graph transformation (GT) [1] is a rule-based manipulation language for graph models. Graph transformation rules can be specified by using a left-hand side – LHS (or precondition) graph pattern determining the applicability of the rule, and a right-hand side – RHS (postcondition) pattern which declaratively specifies the result model after rule application. Elements that are present only in (the image of) the LHS are deleted, elements that are present only in the RHS are created, and other model elements remain unchanged.

Example 1. An example GT rule taken from the Sierpinski Triangles case study [8] of the AGTIVE 2007 Tool Contest [9] is depicted in Fig. 1(b). In this example, all edges are of the same type, while several node types are used (with subtyping relationships) according to the metamodel in Fig. 1(a). For easier readability, the names of pattern variables (nodes and edges) are excluded from the figure.

The GT rule describes how triangles are generated for the Sierpinski fractal. The LHS of the rule captures a simple triangle that has three different types of nodes A, B, C. As the result of the application of the rule on a match of the LHS, the original three nodes (appearing both in the LHS and RHS) are preserved, the original three edges will be deleted (absent from the RHS), and three new nodes will be created along with nine new edges (exclusive to the RHS).

Tool support. Many GT tools emerged during the years for various purposes. The rest of the paper uses the GT language of the VIATRA2 model transformation framework [10,11] to demonstrate the technicalities of our approach; VIATRA2 will also be used as the platform for the prototype evaluation.

2.2   Incremental Pattern Matching

Most graph transformation systems use local search-based (LS) pattern matching, meaning that each time the pattern matching is initiated, the model graph

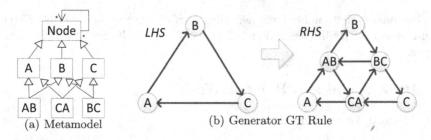

Fig. 1. Sierpinsky Case Study

will be searched for matches of the pattern. As an alternate approach, *incremental pattern matching* (INC) [4,12,13,14,15] relies on a *cache* in which the matches of a pattern are stored explicitly. The match set is readily available from the cache at any time without searching, and the cache is incrementally updated whenever changes are made to the model. The result can be retrieved in constant time – excluding the linear cost induced by the size of the match set itself –, making pattern matching extremely fast. The trade-off is space consumption of the match set caches, model manipulation performance overhead related to cache maintenance, and possibly the initialization cost of the cache.

Several experiments [4,5,15] have observed better performance of INC than conventional LS in various scenarios. A performance advantage is to be expected when complex patterns with moderately sized match sets are matched frequently, without too much model manipulation in-between. This is typical of the "as long as possible" style of execution, that involves repeated cycles of matching the LHS pattern and applying the rule on one match. [16] has found that under certain conditions the best results are achieved by hybridizing INC and LS.

There are several INC approaches. The TREAT [17] algorithm (or LEAPS [18]) will only construct caches for the match set of graph patterns; whereas RETE nets [19] (or [12]) use extensive caching and create internal caches for sub-patterns as well. Several GT systems [13,15] have already adopted a pattern matcher based on RETE. This paper, however, will focus on implementing GT on an RDB; therefore we have chosen TREAT as it allows us to delegate more work to the (presumably well-optimized) built-in query engine of the RDB.

2.3 Relational Databases

Relational databases manage large quantities of information structured according to the well-known *relational data model* (schema, table/relation, row/tuple, column, key). Conventional RDBs store all data on disk (using memory as cache), while others are entirely *in-memory* for better performance.

The most common interface language for RDBs by far is *SQL*, capable of schema definition, data manipulation and querying alike. A reusable data manipulation program segment is called a *stored procedure*, while a reusable query expression defines a *view*. A *trigger* on a table is a special stored procedure that is invoked upon each row manipulation affecting the table (receiving the description of the change as input), thereby facilitating event-driven programming.

2.4   Related Work over Databases

The idea of representing graphs in RDBs and executing graph transformation over the database has been explored before in [6]. A related paper [4] is one of the first to suggest the idea of incremental evaluation in GT, and describes a proof-on-concept experiment implemented in a RDB. This is clearly a predecessor to our work; but since its main focus was the feasability of incremental pattern matching in general, the particulatities of the RDB-based implementation and the automated mapping were not elaborated in detail. Furthermore, the consistency of the incremental cache is only guaranteed if the system evolves along the specified GT rules, while external programs manipulating the underlying database may cause inconsistency of the match results. This is a consequence of making the incremental maintenance of the results an explicit part of the manipulation phase of GT rule execution, which is not invoked when pre-existing programs manipulate the model. The main conceptual extension over [4] is that our solution meets the Compatibility goal (see Sec. 1.2) through automated maintenance of pattern match sets even upon model manipulations carried out by unmodified legacy programs.

The incremental matching algorithm Rete is integrated into an RDB in [20], but the user formulates queries in SQL as opposed to declarative graph patterns; also, Rete maintenance is performed periodically, not by event-driven triggers.

In the context of relational databases, the cached result of a query is called *materialized view*. Some commercial database engines provide this feature along with an option of automatic and incremental maintenance. However, in mainstream databases this non-standard feature is typically restricted to a subset of SQL queries which is insufficient to express complex graph patterns (especially NACs); therefore our approach could not simply rely on it. For example, Flexviews for MySQL and Indexed Views in MS SQL do not support outer joins (or existence checking) that will be required for NAC enforcement (see Sec. 4), while Oracle's Materialized Views do not even support top-level inner joins, and finally there is no built-in incremental maintenance at all in PostGreSQL.

There are also significantly more powerful approaches [21,22] for incremental query evaluation over databases, that support the highly expressive *Datalog* language. The greatest challenge in terms of correctness is the handling of recursive Datalog queries (especially when combined with negation), thus algorithms in this field sacrifice performance and simplicity to address this issue[1] which is not relevant in the context of graph patterns. Even though there exists a rarely used recursive extension [23,24] to the language of graph patterns (even supported

---

[1] With the exception of the simple Counting algorithm [21] (largely analogous to TREAT discussed above), which is quite fast (also confirmed by our limited experiments), but incompatible with recursion.

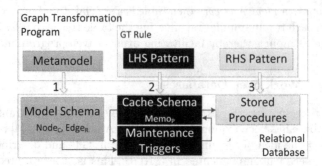

Fig. 2. Overview of the mapping process

by our implementation platform VIATRA2), our current investigation focuses on conventional graph patterns (incl. NACs), therefore the benefits and challenges of recursive queries are out of scope for this paper. Since unlike SQL, Datalog is not supported by mainstream commercial databases, we opted to target SQL.

3   Overview of the Approach

Our approach aims to conduct graph transformation over models represented in relational databases. The most important difference to prior work [6] is the application of INC to improve the performance of the graph pattern matching phase (see Incrementality in Sec. 1.2); while [4] is extended by (i) the detailed description of a universal procedure (inspired by TREAT [17]) that achieves incrementality for any GT program (see Declarativity in Sec. 1.2), (ii) the non-interference with existing programs that manipulate the graph model (see Compatibility in Sec. 1.2), and (iii) some pattern language features (see Sec. 4).

Incremental pattern matching requires (a) caches preserving previously computed results and (b) mechanisms to update said caches upon changes of the input. The first is achieved by using additional database tables to store cached relations. One possible solution to the second problem could be a global policy stating that all operations writing to the graph model must be suffixed by cache maintenance routines that propagate the changes to the pattern match results [4]. However, in order to satisfy the goal of Compatibility (see Sec. 1.2), our solution *does not require the modification of any existing programs manipulating the graph model.* Our approach employs database triggers instead to refresh the contents of the cache tables in an event-driven fashion, after arbitrary transactions manipulating the model.

We provide an algorithm to generate SQL code from the graph transformation program, in accordance with the Declarativity goal (see Sec. 1.2). The proposed approach, depicted in Fig. 2, has three main phases:

1. **Mapping between the graph metamodel and the relational schema**
   is the well-known problem of Object-Relational Mapping (ORM) [25] executed in either direction. To retain Compatibility (see Sec. 1.2) with systems

already deployed in RDB, a relational schema can be used as input if available. In the particular ORM strategy used in the example, each class (node type) C in the metamodel corresponds to a table $Node_C(ID, \ldots)$, with a separate column for each data attribute, and a primary key column as unique identifier. Model elements appear as a row in the table of their class and those of the superclasses, logically connected by the common identifier. An edge type R from node type $SrcT$ to type $TrgT$ corresponds to a separate table $Edge_R(SrcID, TrgID)$. Each row of $Edge_R$ represents one edge instance, and the two columns reference the identifiers of the source and target node (rows in $Node_{SrcT}$ and $Node_{TrgT}$, respectively). An edge R with multiplicity of 1 could map to a column of $Node_{SrcT}$ referencing the $TrgT.ID$ key of the single target node. Note that there are several other possible ORM methods, and the approach outlined in this paper is applicable to all of them.

2. **Cache tables and maintenance triggers for patterns** are our main contribution. A database table $Memo_P$ is created for each pattern P to preserve its match set. Unlike [4], incremental maintenance of the contents of these tables is performed by database triggers, that are automatically generated from the declarative pattern description. The triggers are activated by updates to the database tables representing the graph elements, and potentially the other cache tables as well. The solution is described in detail in Sec. 4.

3. **Mapping GT rules to stored procedures** is performed according to [6], no modifications are required. The main idea is that the application of the GT rule is decomposed into individual node and edge operations (creation, deletion), which are then simply transcribed into SQL manipulation commands. The resulting sequence is then automatically assembled into an SQL stored procedure that takes the LHS match as input.

4   Incremental Pattern Matching Using Cache Tables and Triggers

Most of the following ideas are applicable to any GT tool, not just VIATRA2.

4.1   Basic Pattern Matching

We assume that the metamodel of the graph has already been mapped into a relational schema (see Sec. 3). For a simple graph pattern P consisting of node and edge constraints, the SQL view definition $View_P$ is a *relational join operation* that yields the matches of pattern P. This solution from [6] is not yet incremental, as each evaluation of $View_P$ will re-execute the join.

With the ORM mapping in Sec. 3, for each node constraint $n\langle V, C\rangle$ that restricts the pattern variable V to node type C, $View_P$ will involve the table $Node_C$ as a participant in the join operation. Likewise for each edge constraint $e\langle V_{Src}, V_{Trg}, R\rangle$ expressing that there is an edge of type R from pattern variable V_{Src} to V_{Trg}, the join in $View_P$ will include the table $Edge_R$. The incidence of

Listing 1. SQL View Definition to return matches of the LHS of the Sierpinski rule (subscripts indicated by underscores)

```
1 CREATE VIEW View_LHS AS
2 SELECT Node_A.ID, Node_B.ID, Node_C.ID
3 FROM Edge_e AS eAB, Edge_e AS eBC, Edge_e AS eCA, Node_A, Node_B, Node_C
4 WHERE eAB.TrgID=eBC.SrcID AND eBC.TrgID=eCA.SrcID AND eCA.TrgID=eAB.SrcID
5   AND Node_A.ID=eAB.SrcID AND Node_B.ID=eBC.SrcID AND Node_C.ID=eCA.SrcID
```

nodes and edges are enforced by join conditions. *Injectivity* constraints (if a table appears several times in the join) are easily checked by comparing identifiers.

Example 2. The LHS of the Sierpinski GT rule introduced in Sec. 2.1 is mapped into the SQL view definition in Lst. 1. As the edge type was unnamed in the example, let's assume it was mapped into the table $Edge_e$. The pattern expresses three edge constraints, the unnamed edge type appears in the pattern three times; therefore the join expression involves $Edge_e$ three times, distinguished by aliases eAB, eBC and eCA. Additionally, three node constraints assert a type for each of the node variables, thus $Node_A$, $Node_B$ and $Node_C$ are also included in the join (there is currently no need to create aliases).

4.2   Achieving Incrementality

For each pattern P, a table $Memo_P$ will be created that caches the matches of P. The previously defined view $View_P$ can be used to initialize the table. For each node type C referenced by P, the match set of the pattern may change whenever nodes of type C are created or deleted (or retyped). The match set may also change when edges of type R are created or deleted, provided R is mentioned in P. In the core pattern formalism, the match set is invariant to all other manipulations of the graph model. All of the changes listed here are observable as row insertions / deletions in tables $Node_C$ or $Edge_R$, therefore database triggers can be tasked with refreshing the match set. Triggers for row insertion and deletion are registered for each node table $Node_C$ or edge table $Edge_R$ that P depends on. The trigger will compute the *delta* in the match set of the pattern, and update $Memo_P$ accordingly.

Let's consider the case where there is an edge constraint $e\langle V_{\text{Src}}, V_{\text{Trg}}, R\rangle$ and the change is the creation of a new edge of type R, appearing as a new row $\langle ID_{\text{Src}}, ID_{\text{Trg}}\rangle$ in table $Edge_R$. The delta will be the set of matches that contain the newly created edge. Therefore the trigger will insert into $Memo_P$ the result of query Δ_e^+, which is a modified ("seeded") version of $View_P$, restricted to the new matches that are produced by this edge insertion. Δ_e^+ is formed by omitting from the join the $Edge_R$ operand that corresponds to the edge constraint e, and substituting its source and target identifier values respectively with the triggering ID_{Src} and ID_{Trg}. These input values reduce the cardinality of the result relation significantly, making incremental maintenance efficient.

If the pattern contains k edge constraints for the type R, then $View_P$ is seeded similarly for each of them, and the delta is the union of the results. If the pattern

Listing 2. Seeded SQL query for computing the delta (edge insertion, eAB case)

```
1 SELECT Node_A.ID, Node_B.ID, Node_C.ID
2 FROM Edge_e AS eBC, Edge_e AS eCA, Node_A, Node_B, Node_C
3 WHERE ID_Trg=eBC.SrcID AND eBC.TrgID=eCA.SrcID AND eCA.TrgID=ID_Src
4   AND Node_A.ID=ID_Src AND Node_B.ID=ID_Trg AND Node_C.ID=eCA.SrcID
```

is not required to be *injective*, then it is also possible that the new edge produces a match by simultaneously satisfying several of these k constraints; the delta will be the union of $2^k - 1$ branches (at least one of the k is the new edge).

With deletion, there are two basic options in variants of TREAT. The straightforward solution is to implement deletion triggers that are symmetric to creation triggers, evaluate seeded Δ_e^- queries, and remove the results from $Memo_P$. A potentially faster (though never by more than 50%) solution would be to directly scan $Memo_P$ and remove all matches that were produced by the deleted edge.

Node creations and deletions (more properly, the assignment or removal of a type label to a node) are processed analogously to edge operations.

Example 3. Let's consider the Sierpinski GT rule again. Assuming node retyping is allowed in the system, triggers will have to be registered for the insertion (and, symetrically, deletion) of rows into tables $Node_A$, $Node_B$, $Node_C$ and $Edge_e$. Because $Edge_e$ was involved in the original join expression three times (eAB, eBC and eCA from Lst. 1), each row insertion into $Edge_e$ has to be considered in three different ways. First, to contribute to the formation of a new match as the $A \rightarrow B$ edge eAB, then as eBC, and finally as eCA. The union of the deltas obtained in these cases will be used to update the cache table $Memo_{\text{LHS}}$.

Focusing now on the edge constraint eAB in case of the insertion of the row $\langle ID_{\text{Src}}, ID_{\text{Trg}} \rangle$ into $Edge_e$, the trigger will evaluate a seeded query to obtain the delta relation Δ_{eAB}^+, and add the contents of the delta to the cached match set $Memo_{LHS}$. The contents of the seeded query are listed in Lst. 2.

4.3 Advanced Pattern Language Features

Attribute Checks. Attributes are columns of the $Node_C$ tables, and attribute constraints are translated into simple attribute checks in the WHERE SQL clause. Polymorphism (type inheritance) may require the query to first join $Node_C$ against $Node_{C'}$ if C' is the supertype of C that defines the particular attribute.

Composition. The pattern language of VIATRA2 has a *pattern composition* feature, not supported in [4], that makes a *called* graph pattern reusable within a *caller* pattern. Our approach considers compositions as a pattern constraint similar to edge constraints, and the $Memo_{Called}$ table will participate in the join operation computing the deltas of $Memo_{Caller}$. As columns in $Memo_{Called}$ will be used as join keys, SQL commands are issued that build index structures on these columns to improve performance. Triggers on $Memo_{Called}$ will also be registered to propagate the changes between the match sets of the patterns.

NAC Support. In the presence of a NAC, the join operation computing the match set will involve an *outer join* of $Memo_{NAC}$ which checks that there are no corresponding matches of the NAC pattern. Insertion triggers on $Memo_{NAC}$ will delete matches of the positive pattern; and deletion triggers will produce new matches if no remaining rows of $Memo_{NAC}$ inhibit it.

Parameters and Disjunction. Some of the pattern variables used in a VIA-TRA2 graph pattern may not be exposed as *parameters*, resulting in information hiding and existential quantification of these *local/hidden variables*. Furthermore, a *disjunctive pattern* has several alternative *pattern bodies*, the union of their match sets constituting the match set of the disjunctive pattern. These are not handled in [4]. Our approach creates a separate internal cache table for each of the pattern bodies, which will be updated as before. An externally visible cache table will also be created for the pattern itself, that contains the union of the individual match sets (projected onto the set of parameter variables). This table will be maintained by triggers defined on the internal cache tables.

5   Performance Evaluation

The solution proposed in Sec. 3 was evaluated on a prototype implementation. Performance of the prototype as well as VIATRA2, both with and without incremental matching, were measured against established GT benchmarks selected to compare the relative merit of the approaches under different contexts. A synthetic benchmark is conducted to assess the processing of large models with a simple GT rule. A model simulation benchmark is demonstrating nearly ideal conditions for incremental optimization. A model-to-model synchronization example reflects conditions that are closer to real-world applications.

The prototype is a plug-in of the VIATRA2 model transformation tool, implemented in Java. Metamodels defined in VIATRA2 are mapped into a database schema expressed in the SQL dialect of MySQL, and then GT programs defined in VIATRA2 are transformed into SQL code that executes the GT rules with incremental pattern matching. The SQL output is fed into the RDB engine MySQL, where a small, manually written script will initiate the benchmark[2].

5.1   Sierpinski Triangles Benchmark

Our first GT benchmark is the synthetic *Sierpinski Triangles* case study [8] of the AGTIVE 2007 Tool Contest [9], which uses the GT rule described in Sec. 2.1.

The initial state consists of a triangle of three nodes (matching the LHS of the rule), and each step (generation) of the benchmark applies the GT rule simultaneously at all of its current matches. The size of the graph grows exponentially with each step, therefore this benchmark mainly focuses on the ability of tools

---

[2] Measurement environment: Intel Core 2 Duo P8600 laptop (2*2,4GHz) with 4GB RAM, running Microsoft Windows 7 64-bit. Oracle Java 1.6.0_25 (32 bit) was used with 1.5G heap space for VIATRA2; in-memory MySQL 5.1.53 served as the RDB.

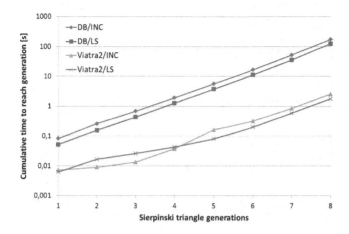

Fig. 3. Sierpinski triangles benchmark up to the 8th generation

to store large graphs. The LHS pattern is easy to match, since it is just a triangle, and each graph node has a small, bounded degree. As pattern matching is performed only once in each step, and practically the whole model is replaced between steps, we do not expect INC to have an advantage in this case.

See Fig. 3 for results (note that the chart is logarithmically scaled). The measurements show that, in accordance with expectations, the overheads of INC could not be offset by its benefits in this case. It also apparent that such volumes of raw model manipulation is significantly more efficient in an object-oriented model representation, than in a relational data model. On the other hand, the DB solution has much better memory characteristics even in the incremental case ($\approx$ 10MB instead of the $\approx$ 95MB of VIATRA2).

5.2   Petri Net Firing Benchmark

Petri nets [26] is a graph-based behavioral formalism with explicit representation of current state (in the form of token markings), and concise definition of possible state transitions. [5] uses simulation of Petri nets as a GT benchmark. Each step of the simulation is the firing of a Petri transition; first a graph pattern identifies the set of *fireable* transitions, then a single selected transition is fired, leading to a new state, with a different set of fireable transitions. The graph pattern that captures fireable transitions has a NAC that in turn has a nested NAC itself; and this complex pattern has to be re-evaluated at each step. Since there are frequent complex queries with only small model change in-between, and a long simulation sequence can offset any initialization costs, this is a scenario where INC has a lot of potential. The measurement was carried out on synthetical Petri nets generated using the process described in [5]. The "size" parameter of such a net is proportional to the number of places, transitions, outgoing and incoming edges; e.g. the net of size 100000 has 75098 places, 75177 transitions, 122948 outgoing and 123287 incoming edges.

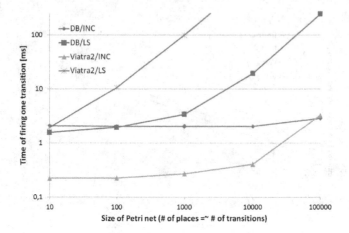

Fig. 4. Petri net firing benchmark, average from 1000 transition firings

Due to the mentioned characteristics, INC can have great advantage over conventional LS in context of such model simulation tasks. Fig. 4 confirms this intuition by showing that INC is several orders of magnitude quicker to execute firings both in a DB context and in VIATRA2, and has a better complexity characteristic when measured against nets of increasing size. Actually, INC takes roughly constant time for an individual transition firing irrespective of the net size. The execution time of VIATRA2/INC increases towards the largest nets only because the significantly inflated memory usage of the RETE-based INC implementation ($\approx$ 1GB of JVM heap size vs. $\approx$ 17 MB of RDB) leads to thrashing effects starting at sizes around 100000. The DB/INC solution does not suffer from this, due to the more compact memory management of the DB solution, and due to the TREAT algorithm doing less caching than the RETE variant implemented in VIATRA2.

5.3   Object-Relational Mapping Synchronization Benchmark

The model transformation case study is a GT-based *Object-Relational Mapping* (ORM) [25], similar to the ORM in phase 1 of Sec. 3. To demostrate advantages of INC at model-to-model synchronization, our previous work [5] extended the benchmark by two additional steps: (i) after the initial mapping, the *source models are modified* by a "legacy" program unaware of the transformation; then (ii) the system has to *synchronize the changes to the target model* (i.e. find the changes in the source and alter the target accordingly).

Our measurements examined the total execution time of the incremental synchronization (including the modification that triggers the synchronization). The experiments were carried out with the particular rules, input model synthesis method and source model modifications that were used in [5]. Each input model is a package with N classes, N attributes for each class, and one association for each pair of classes with two association ends. The largest input model at

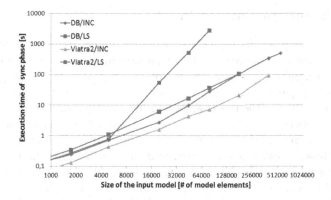

Fig. 5. ORM benchmark, execution time of model synchronization after modification

$N = 250$ had 560751 model elements and links altogether. The modification of the source model deleted every third class (together with associations and attributes), also deleted every fifth remaining association, and finally renamed every second remaining attribute.

In many model transformation programs, the GT rules cannot be applied in any order, because some rules have the application of others as prerequisites. For example, the mapping of an attribute must be predated by the mapping of its class, etc. INC is expected to efficiently re-evaluate the applicability of rules after each rule application, gaining significant advantage for these kinds of transformations. Such re-evaluation is not necessary for this ORM case, as the dependencies can be satisfied by imposing an order on the GT rule definitions (e.g. package mapping before class mapping before attribute mapping), and simply applying the rules on all of their LHS matches in this order. This solution may not always be possible; e.g. if a class can only be mapped if all of its superclasses have already been mapped, then repeatedly re-evaluating the applicability of the class mapping rule is inevitable. But even in the ORM case, incremental change synchronization itself creates the need for efficient re-evaluation of rule applicability conditions. Aside from rule dependencies, INC may also pay off e.g. by pre-caching the result set of complex NACs or called patterns.

The ORM measurement results (see Fig. 5) reveal that it is possible to scale to model sizes with over half a million model elements where VIATRA2 fails (32-bit Java heap exhausted at 1.5GB); while RDB (though slower) takes merely 102 MB of memory, suggesting that it can scale up even further. This justifies choosing RDBs as a scalable platform of model-driven engineering and GT.

Also, INC has an advantage in execution time both in VIATRA2 and over the RDB. The latter difference is more pronounced over the RDB, which we attribute to two factors: (a) instead of a generic SQL query engine, the local search based pattern matcher of VIATRA2 was built to handle pattern composition (especially NACs) reasonably well, and (b) the TREAT-based incremental matcher has fewer caches (making it slower but more memory-efficient) than its RETE-based counterpart in VIATRA2.

6   Conclusion and Future Work

In the paper, we proposed a graph transformation engine based on traditional relational databases using incremental pattern matching techniques for increased performance. Extending [4], the essence of our approach is to formulate the matches of graph patterns as database tables and incrementally update them using triggers that monitor the changes (including those caused by other programs) in the underlying model. The solution meets all goals stated in Sec. 1.2.

The main conclusion that can be drawn from our experiments is that relational databases provide a promising, scalable platform for the execution of graph transformation rules. Additionally, incremental techniques can be effectively applied in relational databases for graph pattern matching and it provides good runtime characteristics for certain workloads including behavioural model simulation or model-to-model synhcronization.

As a main direction for future work, we plan to (i) further optimize our translation as in certain cases the generated SQL queries were using inefficient join structures and (ii) implement the complete translation of the VIATRA2 language to support large model transformations with complex control structures.

References

1. Ehrig, H., Engels, G., Kreowski, H.J., Rozenberg, G. (eds.): Handbook on Graph Grammars and Computing by Graph Transformation. Applications, Languages and Tools, vol. 2. World Scientific (1999)
2. Geiß, R., Batz, G.V., Grund, D., Hack, S., Szalkowski, A.: GrGen: A Fast SPO-Based Graph Rewriting Tool. In: Corradini, A., Ehrig, H., Montanari, U., Ribeiro, L., Rozenberg, G. (eds.) ICGT 2006. LNCS, vol. 4178, pp. 383–397. Springer, Heidelberg (2006)
3. Zündorf, A.: Graph Pattern Matching in PROGRES. In: Cuny, J., Engels, G., Ehrig, H., Rozenberg, G. (eds.) Graph Grammars 1994. LNCS, vol. 1073, pp. 454–468. Springer, Heidelberg (1996)
4. Varró, G., Varró, D.: Graph transformation with incremental updates. In: Proc. GT-VMT 2004, International Workshop on Graph Transformation and Visual Modelling Techniques. ENTCS, vol. 109, pp. 71–83. Elsevier (2004)
5. Bergmann, G., Horváth, Á., Ráth, I., Varró, D.: A Benchmark Evaluation of Incremental Pattern Matching in Graph Transformation. In: Ehrig, H., Heckel, R., Rozenberg, G., Taentzer, G. (eds.) ICGT 2008. LNCS, vol. 5214, pp. 396–410. Springer, Heidelberg (2008)
6. Varró, G., Friedl, K., Varró, D.: Implementing a graph transformation engine in relational databases. Journal of Software and Systems Modelling 5(3), 313–341 (2006)
7. Varró, G., Schürr, A., Varró, D.: Benchmarking for graph transformation. In: Proc. of the 2005 IEEE Symposium on Visual Languages and Human-Centric Computing, Dallas, Texas, USA, pp. 79–88 (September 2005)
8. Taentzer, G., Biermann, E., Bisztray, D., Bohnet, B., Boneva, I., Boronat, A., Geiger, L., Geiß, R., Horvath, Á., Kniemeyer, O., Mens, T., Ness, B., Plump, D., Vajk, T.: Generation of Sierpinski Triangles: A Case Study for Graph Transformation Tools. In: Schürr, A., Nagl, M., Zündorf, A. (eds.) AGTIVE 2007. LNCS, vol. 5088, pp. 514–539. Springer, Heidelberg (2008)

9. Schürr, A., Nagl, M., Zündorf, A. (eds.): AGTIVE 2007. LNCS, vol. 5088. Springer, Heidelberg (2008)

10. Varró, D., Balogh, A.: The model transformation language of the VIATRA2 framework. Science of Computer Programming 68(3), 214–234 (2007)

11. Varró, D., Pataricza, A.: VPM: A visual, precise and multilevel metamodeling framework for describing mathematical domains and UML. Software and Systems Modeling 2(3), 187–210

12. Varró, G., Varró, D., Schürr, A.: Incremental Graph Pattern Matching: Data Structures and Initial Experiments. In: Graph and Model Transformation (GraMoT 2006). Electronic Communications of the EASST, vol. 4. EASST (2006)

13. Bergmann, G., Ökrös, A., Ráth, I., Varró, D., Varró, G.: Incremental pattern matching in the VIATRA transformation system. In: GRaMoT 2008, 30th International Conference on Software Engineering (2008)

14. Hearnden, D., Lawley, M., Raymond, K.: Incremental Model Transformation for the Evolution of Model-Driven Systems. In: Wang, J., Whittle, J., Harel, D., Reggio, G. (eds.) MoDELS 2006. LNCS, vol. 4199, pp. 321–335. Springer, Heidelberg (2006)

15. Ghamarian, A.H., Jalali, A., Rensink, A.: Incremental pattern matching in graph-based state space exploration. Electronic Communications of the EASST (2010) GraBaTs 2010, Enschede

16. Bergmann, G., Horváth, Á., Ráth, I., Varró, D.: Efficient Model Transformations by Combining Pattern Matching Strategies. In: Paige, R.F. (ed.) ICMT 2009. LNCS, vol. 5563, pp. 20–34. Springer, Heidelberg (2009)

17. Miranker, D.P., Lofaso, B.J.: The organization and performance of a TREAT-based production system compiler. IEEE Transactions on Knowledge and Data Engineering 3(1), 3–10 (1991)

18. Batory, D.: The LEAPS algorithm. Technical Report CS-TR-94-28 (January 1994)

19. Forgy, C.L.: Rete: A fast algorithm for the many pattern/many object pattern match problem. Artificial Intelligence 19(1), 17–37 (1982)

20. Jin, C., Carbonell, J., Hayes, P.: ARGUS: Rete + DBMS = Efficient Persistent Profile Matching on Large-Volume Data Streams. In: Hacid, M.-S., Murray, N.V., Raś, Z.W., Tsumoto, S. (eds.) ISMIS 2005. LNCS (LNAI), vol. 3488, pp. 142–151. Springer, Heidelberg (2005)

21. Gupta, A., Mumick, I.S., Subrahmanian, V.S.: Maintaining views incrementally. In: Proc. Int. Conf. on Management of Data, pp. 157–166. ACM (1993)

22. Dong, G., Su, J., Topor, R.: First-order incremental evaluation of datalog queries. Annals of Mathematics and Artificial Intelligence, 282–296 (1993)

23. Varró, G., Horváth, Á., Varró, D.: Recursive Graph Pattern Matching: With Magic Sets and Global Search Plans. In: Schürr, A., Nagl, M., Zündorf, A. (eds.) AGTIVE 2007. LNCS, vol. 5088, pp. 456–470. Springer, Heidelberg (2008)

24. Hoffmann, B., Jakumeit, E., Geiss, R.: Graph rewrite rules with structural recursion. In: Workshop on Graph Computation Models, Leicester, UK (2008)

25. Ullman, J.D., Garcia-Molina, H., Widom, J.: Database Systems: The Complete Book, 1st edn. Prentice Hall PTR, Upper Saddle River (2001)

26. Murata, T.: Petri nets: Properties, analysis and applications. Proceedings of the IEEE, 541–580 (April 1989)

Incremental Pattern Matching for the Efficient Computation of Transitive Closure*

Gábor Bergmann, István Ráth, Tamás Szabó, Paolo Torrini, and Dániel Varró

Budapest University of Technology and Economics,
Department of Measurement and Information Systems,
1117 Budapest, Magyar tudósok krt. 2
{bergmann,rath,varro}@mit.bme.hu, {szabta89,ptorrx}@gmail.com

Abstract. Pattern matching plays a central role in graph transformations as a key technology for computing local contexts in which transformation rules are to be applied. Incremental matching techniques offer a performance advantage over the search-based approach, in a number of scenarios including on-the-fly model synchronization, model simulation, view maintenance, well-formedness checking and state space traversal [1,2]. However, the incremental computation of *transitive closure* in graph pattern matching has started to be investigated only recently [3]. In this paper, we propose multiple algorithms for the efficient computation of *generalized transitive closures*. As such, our solutions are capable of computing reachability regions defined by *simple graph edges as well as complex binary relationships defined by graph patterns*, that may be used in a wide spectrum of modeling problems. We also report on experimental evaluation of our prototypical implementation, carried out within the context of a stochastic system simulation case study.

1 Introduction

In model-driven software engineering, queries and transformations are nowadays core techniques to process models used to design complex embedded or business systems. Unfortunately, many modeling tools used in practice today have scalability issues when deployed in large-scale modeling scenarios, motivating research and development efforts to continue improving performance for essential use-cases such as model management, transformations, design-time analysis and code generation.

Transitive closure is generally needed to express model properties which are recursively defined, often used in reasoning about partial orders, and thus widely found in modeling applications, e.g. to compute model partitions or reachability regions in traceability model management [4] and business process model analysis [5]. In graph transformations, recursive graph patterns are most frequently used to specify transitive closure for processing recursive model structures [6].

---

* This work was partially supported by the CERTIMOT (ERC_HU-09-01-2010-0003) project and the János Bolyai Scholarship and the grant TÁMOP-4.2.1/B-09/1/KMR-2010-0002.

At the meta-level, they may provide the underpinnings for n-level metamodeling hierarchies where transitive type-subtype-instance relationships need to be maintained [7], or for maintaining order structures such as those proposed in [8] for spatially aware stochastic simulation.

Incremental graph pattern matching has already demonstrated scalability in a number of such scenarios [1,9,10], especially when pattern matching operations are dominating at runtime (e.g. view maintenance, model synchronization, well-formedness checking and state space traversal [2]). However, recursive incremental pattern matching was, up to now, supported only for acyclic subgraphs. Therefore, as it has been recently recognized in [3], the efficent integration of transitive closure computation algorithms for graphs would provide a crucial extension to the current capabilities of incremental pattern matchers.

Challenges. By analyzing related work, we observed that in order to efficiently adapt transitive closure computation for the specific needs of graph pattern matching in graph transformations, three key challenges need to be addressed. First, the Rete algorithm (used e.g. in VIATRA2 [6], EMF-INCQUERY [10], GROOVE [2], JBoss Drools [11] and other tools) does not *handle cyclic closure correctly*, i.e. in the presence of graph cycles, incremental updates of recursive patterns may yield false matching results. Second, for functionally complete pattern matching it is important to support *generic transitive closure*, i.e. the ability to compute the closure of not only simple graph edges (edge types), but also derived edges defined by binary graph patterns that establish a complex logical link between a source and a target vertex. Finally, the adaptation should *align with the general performance characteristics of incremental pattern matching* to impose a low computational overhead on model manipulation operations and minimize runtime memory overhead.

Contributions of the paper. To address the above challenges, we adapted different general purpose graph transitive closure algorithms [12,13] to the specific needs of incremental graph pattern matching. After analyzing common characteristics of several modeling scenarios, we developed a novel version of *IncSCC* [12], the incremental transitive computation algorithm based on the maintenance of strongly connected components. We demonstrate the feasibility of our approach by extending the high-level pattern language of the VIATRA2 framework to support the correct computation of transitive closure. In order to evaluate experimentally the performance of the extended pattern matcher, we relied on the GRaTS stochastic simulator [14] built on VIATRA2 that was used to run a simple structured network model scenario, specifically tailoring the simulation to compare the characteristics of these algorithms.

Structure. The rest of the paper is structured as follows. Sec. 2 introduces (meta)modeling and graph transformation as preliminaries necessary to understand the rest of the discussion, and describes a case study – the stochastic simulation of a structured network – that illustrates the technical details of our approach. Sec. 3 elaborates the transitive closure problem, and describes the

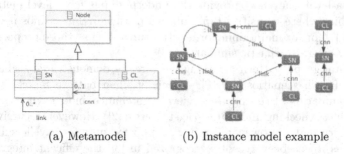

(a) Metamodel   (b) Instance model example

Fig. 1. Meta- and instance models of the case study

novel adaptation of the *IncSCC* algorithm in detail. Sec. 4 reports on our performance benchmark findings. The paper is concluded by a discussion of related work in Sec. 5 and the conclusion in Sec. 6.

2   Preliminaries

2.1   Metamodeling and Graph Patterns

We rely on the VIATRA2 model transformation tool [6] as the technological framework for our approach. However, all metamodels will be presented in a traditional EMF syntax to stress that all the main concepts presented could be transferred to other modeling environments as well, e.g. by using the EMF-INCQUERY framework [10] with EMF. VIATRA2 uses a canonical metamodeling approach for its model repository, with three core concepts (entities, properties and relations) that correspond to vertices, attributes (labels) and edges of a typed graph. All metalevels (types and instances) are uniformly represented in a containment hierarchy that makes up the VIATRA2 *model space*.

Case study example (meta- and instance models). Theoretically, here we rely on a typed single pushout graph transformation approach with attributes and negative application conditions [15]. We consider a structured network evolving according to some rules. The wider network is formed by an overlay on supernodes that represent external ports of local area networks, and we may query the existence of connections between any pair of nodes. A simple VIATRA2 metamodel is shown in Fig. 1a. We model networks as graphs that may consist of two kinds of Nodes: they may either be *LAN* clients (instances of the type CL) or *LAN* supernodes (SN) to which clients may connect (through connections of type cnn). Supernodes can connect to each other through connections of type link (see [16] for technical details). A sample instance model (as visualized in VIATRA2) is shown in Fig. 1b.

Case study example (graph patterns). A sample graph pattern is shown in Fig. 2, using a graphical concrete syntax for illustration (on the left) and also the textual representation from the actual VIATRA2 transformation (on the right). This pattern represents the linked relation between any two supernodes $S1$ and

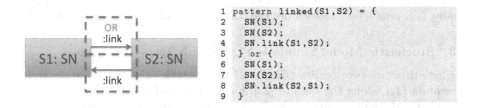

```
1  pattern linked(S1,S2) = {
2    SN(S1);
3    SN(S2);
4    SN.link(S1,S2);
5  } or {
6    SN(S1);
7    SN(S2);
8    SN.link(S2,S1);
9  }
```

Fig. 2. Graph pattern to capture linked supernodes

$S2$ (both required to be of type SN, as expressed in lines 2–3 and 6–7) that are connected by a relation of type link in either direction (as expressed by means of the *or* construct).

2.2 Incremental Pattern Matching in Graph Transformations

Graph transformation systems use pattern matching algorithms to determine the parts of the model that correspond to the *match set* of a graph pattern. *Incremental pattern matching engines* (INC) [17,18,19,2] rely on a *cache* in which the matches of a pattern are stored explicitly. The match set is readily available from the cache at any time without searching, and the cache is incrementally updated whenever changes are made to the model. The result can be retrieved in constant time – excluding the linear cost induced by the size of the match set itself –, making pattern matching extremely fast. The trade-off is space consumption of the match set caches, model manipulation performance overhead related to cache maintenance, and possibly the initialization cost of the cache.

In terms of transformation performance, INC has been observed by several experiments [17,1,2] to be highly scalable in a number of scenarios, particularly when complex patterns with moderately sized match sets are matched frequently, without excessive changes to the model in-between. This is typical of the *as-long-as-possible* style of transformation execution, which is frequently used for model simulation purposes.

Overview of Rete-based incremental pattern matching. Rete [20] is a well-known incremental pattern matching technique from the field of rule-based expert systems. A Rete net consists of *Rete nodes* (not to be confused with the vertices of the model graph), each storing a relation corresponding to the *match set* of a partial pattern, i.e. the set of model element tuples that satisfy a given subset of pattern constraints. Rete nodes are connected by *Rete edges* so that the content of a Rete node can be derived from its parent nodes. The Rete edges propagate incremental updates of the match sets, i.e. whenever the contents of a Rete node is changed, child nodes are also updated using the difference (inserted or deleted tuples). There are three types of nodes in the Rete net: (i) *input nodes* serve as the knowledge base of the underlying model, e.g. there is a separate node for each entity or relation type, enumerating the set of instances as tuples;

(ii) *intermediate nodes* perform operations to derive a set of partial matches; finally, (iii) *production nodes* store the complete match set of a given pattern.

2.3   Stochastic Model Simulation by Graph Transformations

A simulation framework called *GRaTS* [21] for *generalized stochastic graph transformation* [22], along the lines of [23], has been introduced in [14] and further extended [21], built on top of the Rete-based pattern matching infrastructure of VIATRA2, to support the design-time analysis of discrete event systems and to validate stochastic properties of the system-under-design. A model in GraTS consists of a graph transformation system in which each transformation *action rule* (see Fig. 3), is augmented with a probability distribution governing the delay of its application (in our simple case study, we use exponential distributions that are characterised by a *weight* parameter – a higher weight will result in the rule being executed more frequently). Additionally, each valid action rule match represents a possible event. A stochastic experiment consists of a model together with a set of transformation rules, each used as a *probe rule*, allowing to aggregate user-defined statistics on simulation runs.

Simulation in GraTS procedes by discrete steps, each determined by the execution of an action rule, leading from one state to another, where each state is characterised by the set of enabled events — i.e. all the valid rule matches, maintained as a priority queue. Statistics are also collected step-wise, by computing valid matches of probe rules and aggregating data. The simulation engine relies heavily on the incremental pattern matcher to keep track efficiently of valid rule matches and especially, in the case of events, of their enabling time [14,8,21].

Discrete event stochastic simulation can be characterised as a semi-Markov process [24,25], as – aside from exponential distributions found in most stochastic tools – *generalised probability distributions are supported*. Even though this capability is not used in the simple model of this paper, it allows for modelling of complex network scenarios [22] involving hybrid features, such as jitter and bandwidth in realistic modelling of VoIP [22,16], in which transitive closures can typically arise quite often [14].

Case study example (action rule). In our case study, a sample action rule is shown in Fig. 3. Here, the AddLink operation is defined, whereby redundant overlay links can be added to a pair of *LAN* supernodes $S1, S2$ that are not directly connected, as expressed by the negative application condition (line 5) referring to the linked pattern of Fig. 2. By an execution of this rule for a given $S1, S2$ pair, a new link will be added to the model (line 7).

3   Transitive Closure in a Rete-Based Matcher

A brief overview about transitive closure in graph transformations is given in Sec. 3.1. The applied solution - a special Rete node capable of efficient incremental transitive closure calculations - is discussed in Sec. 3.2. Then Sec. 3.3 presents the general purpose incremental graph transitive closure algorithms that we we adapted and evaluated.

```
1  gtrule AddLink() = {
2    precondition pattern lhr(S1,S2) = {
3      SN(S1);
4      SN(S2);
5      neg find linked(S1,S2);
6    } action {
7      new(SN.link(S1,S2));
8    }
9  }
```

Fig. 3. Graph transformation rule to add redundant overlay links between disconnected supernodes

3.1 Transitive Closure

Generic and Irreflexive Transitive Closure. For a binary relation E over a domain D, the *irreflexive transitive closure* E^+ consists of $\langle u, v \rangle$ pairs of elements for which there is a *non-empty* finite linked sequence $\langle u = w_0, w_1 \rangle, \langle w_1, w_2 \rangle, \ldots,$ $\langle w_{k-1}, w_k = v \rangle$ of pairs in E.

In case of *generic transitive closure*, the base relation E is a "derived edge", not restricted to simple graph edges, but defined by any two-parameter graph pattern (e.g. with path expressions, attribute checks). We focus on the most general approach: generic, irreflexive transitive closure.

Transitive Closure Operations. Any program computing the transitive closure E^+ of a binary relation E is required to expose a subroutine Construct(E) that builds a data structure for storing the result and possibly auxiliary information as well. Afterwards, the following reachability queries can be issued: Query(Src,Trg) returns whether Trg is reachable from Src; Query(Src,?) returns all targets reachable from Src, while Query(?,Trg) returns all sources from where Trg can be reached; finally Query(?,?) enumerates the whole E^+.

In case of incremental computation, the following additional subroutines have to be exposed: Insert(Src,Trg) updates the data structures after the insertion of the $\langle Srg, Trg \rangle$ edge to reflect the change, while Delete(Src,Trg) analogously maintains the data structures upon an edge deletion. To support further incremental processing, both of these methods return the delta of E^+, i.e. the set of source-target pairs that became (un)reachable due to the change.

Strongly Connected Components (SCC), Condensed Graph. A graph is strongly connected iff all pairs of its vertices are mutually transitively reachable. An SCC of a graph is a maximal subset of vertices within a graph that is strongly connected. As the SCC of a vertex v is the intersection of the set of ancestors and descendants of the vertex, each graph has a unique decomposition S into disjoint SCCs. For a graph $G(V, E)$, the SCCs form the *condensed graph* $G_c(S, E_c)$, where two SCCs are connected iff any of their vertices are connected: $E_c = \{\langle s_i, s_j \rangle \mid s_i, s_j \in S \land \exists u \in s_i, v \in s_j : \langle u, v \rangle \in E\}$. It follows from the definitions that a condensed graph is always acyclic.

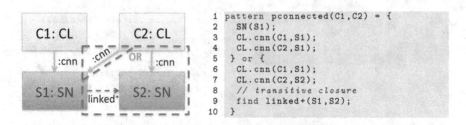

```
1  pattern pconnected(C1,C2) = {
2    SN(S1);
3    CL.cnn(C1,S1);
4    CL.cnn(C2,S1);
5  } or {
6    CL.cnn(C1,S1);
7    CL.cnn(C2,S2);
8    // transitive closure
9    find linked+(S1,S2);
10 }
```

Fig. 4. Transitive closure within graph pattern to capture overlay-connected clients

Case study example (transitive closure in graph patterns). The example in Fig. 4 demonstrates transitive closure features in graph pattern matching. A transitive closure over the overlay network of supernodes is specified by the pattern pconnected that defines the relationship between any two client nodes $C1, C2$ which are (i) either sharing a common supernode to which they are both directly connected along cnn edges (lines 2–4), or (ii) their "pconnection" is indirect in the sense that their supernodes $S1, S2$ are reachable each other through a transitive linked+ relationship (lines 6–9). The latter is the generic transitive closure of the derived edge defined by binary pattern linked (see Fig. 2).

3.2   Integration of Graph Transitive Closure into Rete

A transitive closure result will be represented by a Rete node, like any other pattern. We integrate dynamic transitive closure algorithms into Rete nodes by exploiting the operations specified in Sec. 3.1. Generic transitive closure (see Sec. 3.1) is achieved by attaching such a Rete node to a parent node that matches a graph edge or an arbitrary binary graph pattern (derived edge).

Fig. 5 (a) shows the transitive closure node in the Rete network. It is an intermediate node which receives updates from a binary graph pattern (here denoted as binary relation E) and forms a two-way interface between Rete and a transitive closure maintenance algorithm. Whenever the Rete node for E^+ receives an insertion / deletion update from its parent node E, the Insert()/Delete() subroutine is invoked. The subroutine computes the necessary updates to E^+, and returns these delta pairs, which will then be propagated along the outgoing edge(s) of the Rete node. Queries are invoked when initializing the child nodes, and later as a quick lookup to speed up join operations over the node contents.

Alternatively, transitive closure can be expressed as a recursive graph pattern. This solution was rejected, as Rete, having first-order semantics without fixpoint operators, might incorrectly yield a (still transitive) superset of the transitive closure: in graph models containing cycles, obsolete reachabilities could cyclically justify each other after their original justification was deleted.

Case study example (transitive closure Rete node). Here we demonstrate the behaviour of a Rete node that computes the transitive closure E^+ of the binary graph pattern E, e.g. linked+ for the overlay network linked between super

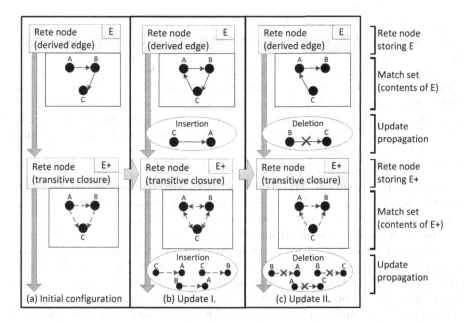

Fig. 5. Transitive closure Rete node during insertion of $\langle C, A \rangle$ and deletion of $\langle B, C \rangle$

nodes. Initially, as seen in Fig. 5 (a), the parent node E stores linked, i.e. the binary relation $\{\langle A, B \rangle, \langle B, C \rangle\}$. Its child node E^+ contains the set of reachable pairs: $\{\langle A, B \rangle, \langle A, C \rangle, \langle B, C \rangle\}$.

Fig. 5 (b) shows the insertion of edge $\langle C, A \rangle$ into E. Rete propagates this update from the E to E^+, where the operation Insert(C,A) is invoked to adjust the transitive closure relation to $\{\langle A, B \rangle, \langle A, C \rangle, \langle B, A \rangle, \langle B, C \rangle, \langle C, A \rangle, \langle C, B \rangle\}$, i.e. the whole graph becomes strongly connected. The computed difference (delta) is the insertion of $\{\langle B, A \rangle, \langle C, A \rangle, \langle C, B \rangle\}$ into E^+, which is propagated in the Rete network to child nodes of E^+.

Finally, Fig. 5 (c) shows an edge deletion. E^+ is notified of the deletion of $\langle B, C \rangle$ from E, and invokes Delete(B,C). Thus E^+ becomes $\{\langle A, B \rangle, \langle C, A \rangle, \langle C, B \rangle\}$, and the propagated delta is the deletion of $\{\langle A, C \rangle, \langle B, A \rangle, \langle B, C \rangle\}$.

3.3   Incremental Graph Transitive Closure Maintenance Algorithms

An incremental transitive closure algorithm is required to operate the Rete node proposed in Sec. 3.2. From the rich literature (see Sec. 5), we selected and adapted two such algorithms. Here we provide an overview of their core ideas.

DRed - Delete and REDerive. This simple algorithm explicitly stores E^+. Construct() initializes the closure relation using a standard non-incremental algorithm, and Query() is directly answered based on E^+. The update operations are derived from the DRed [26] algorithm for recursive Datalog queries.

Insert(Src,Trg) computes the newly reachable pairs as $E^* \circ \{\langle Src, Trg \rangle\} \circ E^*$, and adds them to E^+ (unless already reachable), where $A \circ B := \{\langle u, v \rangle \mid \exists w : \langle u, w \rangle \in A \land \langle w, v \rangle \in B\}$.

Delete(Src,Trg) computes an overestimation of the delta as $E_D^+ = (E^* \circ \{\langle Src, Trg \rangle\} \circ E^*) \setminus E$, and marks these pairs for deletion. Then it attempts to derive again these marked reachability pairs using unaffected ones as $E_D^+ \cap (E \circ (E^+ \setminus E_D^+))$; successfully rederived pairs are removed from E_D^+, allowing further ones to be rederived until a fixpoint is reached. The final contents of E_D^+ are the deleted reachability pairs removed from E^+.

IncSCC - Incremental Maintenance of Strongly Connected Components.

We have also implemented the transitive closure algorithm *IncSCC*, where the name *IncSCC* stands for Incremental SCC maintenance.

The main idea of the algorithm, from [12], is to reduce update time and memory usage by eliminating unnecessary reachability information, namely, that each vertex is reachable from every other vertex within the same SCC. Thus, the two concerns of the algorithm are maintaining (i) a decomposition S of the graph into SCCs, and (ii) transitive reachability within the condensed graph. The latter is a simpler problem with several efficient solutions, as the condensed graph is acyclic; our implementation relies on the "basic algorithm" from the original paper [12], that will be called the *Counting Algorithm*, as it simply keeps track of the number of derivations of each transitive reachability pair.

In the following, we give a brief overview of (our implementation of) IncSCC. For details and analysis, refer to [12].

Implementing Construct(E). The SCC partitioning of the initial graph are computed using *Tarjan's algorithm* [27] based on depth-first search. Afterwards, the condensed graph is constructed, and the Counting Algorithm is initialized to provide reachability information between SCCs.

Implementing Query() operations. As the most significant improvement over [12], the transitive closure relation E^+ is not stored explicitly in our IncSCC solution to reduce the memory footprint. However, reachability in graph $G(V, E)$ can be reconstructed from the partitioning S of SCCs and the reachability relation E_c^+ of condensed graph $G_c(S, E_c)$, since for $s_1, s_2 \in S, u \in s_1, v \in s_2 : \langle s1, s2 \rangle \in E_c^*$ iff $\langle u, v \rangle \in E^*$. Therefore when receiving a reachability query, the parameter vertices are mapped to SCCs, where reachability information in the condensed graph is provided by the Counting Algorithm. Vertices enumerated in the answer are obtained by tracing back the SCCs to vertices.

Implementing Insert(Source,Target). First, a lookup in S maps the vertices to SCCs. Afterwards, there are three possible cases to distinguish. If (i) $\langle Source, Target \rangle$ are in different SCCs, the new edge of the condensed graph is handled by the Counting Algorithm, which can confirm that no cycle is created in the condensed graph. If, however, (ii) the inserted edge caused a cycle in the condensed graph, then the cycle is collapsed into a single SCC. Finally, if (iii) $\langle Source, Target \rangle$ are in the same SCC, there is no required action. Computation details of the delta relation is omitted here for space considerations.

Implementing `Delete(Source,Target)`. The algorithm first performs a lookup in S to map the vertices to SCCs; afterwards, we once again distinguish three possible cases. (1) If $\langle Source, Target \rangle$ are in the same SCC but $Target$ remains reachable from $Source$ after the edge deletion (as confirmed by a depth-first-search), no further actions are required. (2) If $\langle Source, Target \rangle$ are in the same SCC but $Target$ is no longer reachable from $Source$ after the edge deletion, then the SCC is broken up (using Tarjan's algorithm) into smaller SCCs, because it is no longer strongly connected. Finally, (3) if $\langle Source, Target \rangle$ are in different SCCs, then the edge is deleted from the condensed graph, which is in turn is handled by the Counting Algorithm.

4   Benchmarking

4.1   Measurement Scenario

To find out the performance differences between various pattern matching algorithms for transitive closure, we ran a series of measurements[1] on simplified stochastic model simulation process, used to analyse the probability of the network being (fully) connected (so that each client can communicate with every other one, through their direct supernode connections and the transitive overlay links between supernodes). The connectivity measure was registered through a *probe* of the match set of the `pconnected` pattern (Fig. 4), reporting the size of the match set after each simulation step.

A simulation run consisted of 2000 steps (rule applications), and along with the total execution time of the run, we also registered the wall times for various sub-phases – such as the time it took to propagate updates through the transitive closure Rete node – using code instrumentation. The experiments were carried out with three different strategies of evaluating graph patterns and transitive closure: (a) local search pattern matching as implemented in VIATRA2, (b) Rete-based incremental matching with the DRed algorithm for transitive closure, (c) Rete with IncSCC for transitive closure. We have investigated the performance of these solutions in two series of experiments.

The first series considered various model structures induced by different probability weight parameterizations of the `addLink` rule (i.e. increasingly frequent applications of the rule). It was run on an initial model of 2000 vertices in 20 isolated components, each containing 10 supernodes and 9 clients per supernode. The second series settled on a fixed value 2 of `addLink` weight (thus keeping the frequency of the rule application roughly constant), and considered increasingly

---

[1] Performed on Intel Core i5 2,5 GHz, 8 GB RAM, Java Hotspot Server vm build 1.7.0_02-b13 on 64-bit Windows, with 4 Rete threads. The entire corpus is available at `http://viatra.inf.mit.bme.hu/publications/inctc`

Table 1. Graph properties and simulation performance, depending on `addLink` weight

addLink	Graph properties (avg.)			Local search	DRed		IncSCC	
	#SCCs	Net size	Overlay connectivity	Total time [ms]	Total time [ms]	Tc time	Total time [ms]	Tc time
0,005	15,30	428,55	107,56	54833	17236	13,9 %	18833	1,3 %
0,01	15,22	420,44	111,72	51681	16875	14,8 %	16461	1,9 %
0,05	16,74	417,13	133,70	50533	22295	15,3 %	19228	1,8 %
0,1	18,65	415,53	149,70	55562	21297	17,4 %	18736	1,9 %
1	11,45	459,25	1663,45	151913	47211	59,9 %	20707	3,3 %
2	5,27	509,01	4543,02	309476	67718	70,5 %	21008	3,9 %
5	2,63	594,35	7480,20	579774	97755	78,2 %	26643	3,6 %

larger models sizes (from 1000 to 10000 vertices), initially divided into 10 to 100 components similarly to the first series.

4.2 Results and Analysis

Table 1 shows the results of the first experiment series. For each value of `addLink` *weight*, we have displayed (i) the values of the probes (as well as the number of strongly connected components) averaged over an entire simulation run; (ii) for each of the three solutions the total execution time and, in case of the incremental algorithms, (iii) the time spent initializing and updating the transitive closure node (expressed as a percentage of total time).

The first series of experiments reveals that as the application frequency of `addLink` increases, the frequent rule executions make the graph more and more connected. DRed performance significantly degrades for more connected graphs (e.g. as larger and larger number of pairs have to be rederived after deletion), to the point that transitive closure maintenance dominates the total execution time of the simulation. IncSCC however takes advantage of SCCs and runs efficiently in all cases, having a negligible impact on the overall runtime of the simulation and Rete maintenance. Local search in VIATRA2 is orders of magnitudes slower than either of the incremental approaches.

Fig. 6 shows the results of the second experiment series. For each model size on the horizontal axis, we have displayed the average number of SCCs in the model, and on the logarithmic vertical axis the total simulation execution times in case of the three solutions. The second measurement series demonstrates that IncSCC has a better complexity characteristic on large models than DRed, while both scale significantly better than LS.

5   Related Work

Dynamic computation of transitive closure. While there are several classical algorithms (depth-first search, etc.) for computing transitive reachability in graphs, efficient incremental maintenance of transitive closure is a more challenging task. As transitive closure can be defined as a recursive Datalog query, incremental Datalog view maintenance algorithms such as DRed [26] can be applied as a

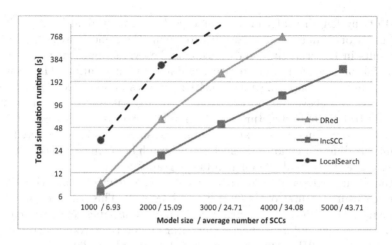

Fig. 6. Performance on increasing model sizes (`addLink` weight is 2)

generic solution. There is also a wide variety [28] of algorithms that are specifically tailored for the fully dynamic[2] transitive reachability problem. Some of these algorithms provide additional information (shortest path, transitive reduction), others may be randomized algorithms (typically with one-sided error); the majority focuses on worst-case charactersitics in case of dense graphs. The spectrum of solutions offers various trade-offs between the cost of operations specified in Sec. 3.1.

Even if the original graph has a moderate amount of edges (*sparse graph*), the size of the transitive closure relation can easily be a quadratic function of the number of vertices, raising the relative cost of maintenance. A key observation, however, is that in many typical cases vertices will form large SCCs. This is exploited in a family of algorithms [12,13] including IncSCC that maintain (a) the set of SCC using a dynamic algorithm, and also (b) the transitive reachability relationship between SCCs. Choosing such an algorithm is justified by simplicity of implementation, the sparse property of typical graph models and the practical observation that large SCCs tend to form.

Incremental pattern matching and transitive closure in graph and model transformation. Apart from VIATRA2, GROOVE [2] also features a Rete-based incremental pattern matcher, and is therefore the most closely related work. In fact, the Rete implementation in GROOVE has recently been extended [3] by the capability of incrementally maintaining transitive closure relations. They also introduced a new type of Rete node that accepts a binary relationship as input and emits its transitive closure as output. The transitive closure node in GROOVE implements a simple algorithm that maintains the set of all paths

---

[2] Note that the graph algorithms community uses the term "fully dynamic" instead of "incremental", as the latter has a secondary, more restrictive meaning in context of the transitive closure maintenance problem.

(walks) of any length that can be composed from the original binary relation, even if many of them are redundant due to having the same sources and targets. This results in factorial time and space complexity, as opposed to the various polynomial solutions found in literature and also in our solution. Their solution is only capable of computing the transitive closures of so called *regular (path) expressions*; we believe our notion of "derived edge" is more general, as it includes arbitrary graph structures (e.g. circular patterns as context, attribute restrictions, etc.). Finally, the experimental assessment in [3] is conducted under substantially different conditions, such as the graph being linear; in contrast, our solution proves to be scalable for non-linear graphs as well.

In the future, we would like to carry out experimental comparison of the transitive closure features of GROOVE and VIATRA2. This will need significant additional effort, as the running example of our current paper relies on a complex peer-to-peer model and a stochastic simulator engine that would be difficult to replicate on GROOVE, while the case study example in [3] relies on model checking capabilities that are not supported in VIATRA2.

Some other graph transformation tools [29,30] feature path expressions, including transitive closure, without maintaining the result incrementally. In a graph with a low branching factor, they can still be feasible in practice. There are other model transformation tools that offer incremental evaluation. The incremental tranformation solution in ATL [31] relies on impact analysis of OCL expressions, meaning that the entire OCL expression will be re-evaluated whenever a relevant element in the model changes; however standard OCL cannot express transitive closure in arbitrary graphs. There is an incremental evaluation technique for Tefkat [19] that maintains an SLD resolution tree of the pattern expression; but without special handling of transitive closure, the SLD tree expands all possible paths from source to target, leading to factorial complexity similarly to GROOVE.

6   Conclusion

We have presented the extension of the incremental pattern matcher of VIATRA2 with a dedicated capability for maintaining generic transitive closure built on a fully dynamic transitive closure maintenance strategy. The results were evaluated in terms of performance on a P2P stochastic system simulation case study.

Our measurements have shown the performance impact of incrementally evaluating generalized transitive closure to be affordable. This implies that the inclusion of transitive closure based probes and rule guard conditions is feasible and scalable in stochastic model simulation, even in case of dynamically changing graph structures. As for the performance of transitive closure algorithms, our investigation demonstrated the overall superiority of *IncSCC* in a wide range of model structures.

As future work, we plan to (i) conduct more detailed benchmarking in other scenarios, (ii) integrate transitive closure maintenance into the EMF-based EMF-INCQUERY [10], and (iii) investigate additional transitive closure algorithms.

References

1. Bergmann, G., Horváth, Á., Ráth, I., Varró, D.: A Benchmark Evaluation of Incremental Pattern Matching in Graph Transformation. In: Ehrig, H., Heckel, R., Rozenberg, G., Taentzer, G. (eds.) ICGT 2008. LNCS, vol. 5214, pp. 396–410. Springer, Heidelberg (2008)
2. Ghamarian, A.H., Jalali, A., Rensink, A.: Incremental pattern matching in graph-based state space exploration. Electronic Communications of the EASST (2010), GraBaTs 2010, Enschede
3. Jalali, A., Ghamarian, A.H., Rensink, A.: Incremental pattern matching for regular expressions. In: 11th International Workshop on Graph Transformation and Visual Modeling Techniques (GT-VMT 2012) (2012)
4. Anquetil, N., Kulesza, U., Mitschke, R., Moreira, A., Royer, J.C., Rummler, A., Sousa, A.: A model-driven traceability framework for software product lines. Software and Systems Modeling 9, 427–451 (2010), doi:10.1007/s10270-009-0120-9
5. Kovács, M., Gönczy, L., Varró, D.: Formal analysis of bpel workflows with compensation by model checking. International Journal of Computer Systems and Engineering 23(5) (2008)
6. Varró, D., Balogh, A.: The model transformation language of the VIATRA2 framework. Sci. Comput. Program. 68(3), 214–234 (2007)
7. Madari, I., Lengyel, L., Mezei, G.: Incremental model synchronization by bi-directional model transformations. In: IEEE International Conference on Computational Cybernetics, ICCC 2008, pp. 215–218. IEEE (2008)
8. Torrini, P., Heckel, R., Ráth, I., Bergmann, G.: Stochastic graph transformation with regions. ECEASST 29 (2010)
9. Horváth, Á., Bergmann, G., Ráth, I., Varró, D.: Experimental assessment of combining pattern matching strategies with VIATRA2. International Journal on Software Tools for Technology Transfer (STTT) 12, 211–230 (2010), doi:10.1007/s10009-010-0149-7
10. Bergmann, G., Horváth, Á., Ráth, I., Varró, D., Balogh, A., Balogh, Z., Ökrös, A.: Incremental Evaluation of Model Queries over EMF Models. In: Petriu, D.C., Rouquette, N., Haugen, Ø. (eds.) MODELS 2010, Part I. LNCS, vol. 6394, pp. 76–90. Springer, Heidelberg (2010)
11. Proctor, M., et al.: Drools Documentation. JBoss, http://labs.jboss.com/drools/documentation.html
12. La Poutré, J.A., van Leeuwen, J.: Maintenance of Transitive Closures and Transitive Reductions of Graphs. In: Göttler, H., Schneider, H.-J. (eds.) WG 1987. LNCS, vol. 314, pp. 106–120. Springer, Heidelberg (1988)
13. Frigioni, D., Miller, T., Nanni, U., Zaroliagis, C.: An experimental study of dynamic algorithms for transitive closure. ACM Journal of Experimental Algorithmics 6, 2001 (2000)
14. Torrini, P., Heckel, R., Ráth, I.: Stochastic Simulation of Graph Transformation Systems. In: Rosenblum, D.S., Taentzer, G. (eds.) FASE 2010. LNCS, vol. 6013, pp. 154–157. Springer, Heidelberg (2010)
15. Ehrig, H., Ehrig, K., Prange, U., Taentzer, G.: Fundamentals of algebraic graph transformation. Springer (2006)
16. Khan, A., Heckel, R., Torrini, P., Ráth, I.: Model-Based Stochastic Simulation of P2P VoIP Using Graph Transformation System. In: Al-Begain, K., Fiems, D., Knottenbelt, W.J. (eds.) ASMTA 2010. LNCS, vol. 6148, pp. 204–217. Springer, Heidelberg (2010)

17. Varró, G., Varró, D.: Graph transformation with incremental updates. In: Proc. GT-VMT 2004, International Workshop on Graph Transformation and Visual Modelling Techniques. ENTCS, vol. 109, pp. 71–83. Elsevier (2004)
18. Varró, G., Varró, D., Schürr, A.: Incremental Graph Pattern Matching: Data Structures and Initial Experiments. In: Graph and Model Transformation (GraMoT 2006). Electronic Communications of the EASST, vol. 4. EASST (2006)
19. Hearnden, D., Lawley, M., Raymond, K.: Incremental Model Transformation for the Evolution of Model-Driven Systems. In: Wang, J., Whittle, J., Harel, D., Reggio, G. (eds.) MoDELS 2006. LNCS, vol. 4199, pp. 321–335. Springer, Heidelberg (2006)
20. Forgy, C.L.: Rete: A fast algorithm for the many pattern/many object pattern match problem. Artificial Intelligence 19(1), 17–37 (1982)
21. Torrini, P., Ráth, I.: GraTS: graph transformation-based stochastic simulation — Documentation (2012), http://viatra.inf.mit.bme.hu/grats
22. Khan, A., Torrini, P., Heckel, R.: Model-based simulation of VoIP network reconfigurations using graph transformation systems. In: Corradini, A., Tuosto, E. (eds.) Intl. Conf. on Graph Transformation (ICGT) 2008 - Doctoral Symposium. Electronic Communications of the EASST, vol. 16 (2009), http://eceasst.cs.tu-berlin.de/index.php/eceasst/issue/view/26
23. Kosiuczenko, P., Lajios, G.: Simulation of generalised semi-Markov processes based on graph transformation systems. Electronic Notes in Theoretical Computer Science 175, 73–86 (2007)
24. Cassandras, C.G., Lafortune, S.: Introduction to discrete event systems. Kluwer (2008)
25. D'Argenio, P.R., Katoen, J.P.: A theory of stochastic systems part I: Stochastic automata. Inf. Comput. 203(1), 1–38 (2005)
26. Gupta, A., Mumick, I.S., Subrahmanian, V.S.: Maintaining views incrementally (extended abstract). In: Proc. of the Int. Conf. on Management of Data, pp. 157–166. ACM (1993)
27. Tarjan, R.: Depth-first search and linear graph algorithms. SIAM Journal on Computing 1(2), 146–160 (1972)
28. Demetrescu, C., Italiano, G.F.: Dynamic shortest paths and transitive closure: algorithmic techniques and data structures. J. Discr. Algor. 4, 353–383 (2006)
29. Schürr, A.: Introduction to PROGRES, an Attributed Graph Grammar Based Specification Language. In: Nagl, M. (ed.) WG 1989. LNCS, vol. 411, pp. 151–165. Springer, Heidelberg (1990)
30. Nickel, U., Niere, J., Zündorf, A.: Tool demonstration: The FUJABA environment. In: The 22nd International Conference on Software Engineering (ICSE), Limerick, Ireland. ACM Press (2000)
31. Jouault, F., Tisi, M.: Towards Incremental Execution of ATL Transformations. In: Tratt, L., Gogolla, M. (eds.) ICMT 2010. LNCS, vol. 6142, pp. 123–137. Springer, Heidelberg (2010)

Efficient Model Synchronization
with Precedence Triple Graph Grammars

Marius Lauder*, Anthony Anjorin*, Gergely Varró**, and Andy Schürr

Technische Universität Darmstadt, Real-Time Systems Lab,
Merckstr. 25, 64283 Darmstadt, Germany
`name.surname@es.tu-darmstadt.de`

Abstract. Triple Graph Grammars (TGGs) are a rule-based technique
with a formal background for specifying bidirectional and incremental model transformation. In practical scenarios, unidirectional rules for
incremental forward and backward transformation are automatically derived from the TGG rules in the specification, and the overall transformation process is governed by a control algorithm. Current incremental
implementations either have a runtime complexity that depends on the
size of related models and not on the number of changes and their affected elements, or do not pursue formalization to give reliable predic-
tions regarding the expected results. In this paper, a novel incremental
model synchronization algorithm for TGGs is introduced, which employs
a static analysis of TGG specifications to efficiently determine the range
of influence of model changes, while retaining all formal properties.

Keywords: triple graph grammars, model synchronization, control algorithm of incremental transformations, node precedence analysis.

1   Introduction

Model-Driven Engineering (MDE) established itself as a promising means of
coping with the increasing complexity of modern software systems and, in this
context, *model transformation* plays a central role. As industrial applications
require reliability and efficiency, the need for formal frameworks that guarantee useful properties of model transformation arises. Especially for *bidirectional*
model transformation, it is challenging to define precise semantics for the manipulation and synchronization of models with efficient tool support. The *Triple
Graph Grammar (TGG)* approach has not only solid formal foundations [3,11]
but also various tool implementations [1,6,10]. TGGs provide a declarative, rule-
based means of specifying the consistency of two models in their respective domains, and tracking inter-domain relationships between elements explicitly by
using a correspondence model. Although TGGs describe how *triples* consisting

---

* Supported by the 'Excellence Initiative' of the German Federal and State Governments and the Graduate School of Computational Engineering at TU Darmstadt.
** Supported by the Postdoctoral Fellowship of the Alexander von Humboldt Founda-
tion and associated with the Center for Advanced Security Research Darmstadt.

H. Ehrig et al.(Eds.): ICGT 2012, LNCS 7562, pp. 401–415, 2012.

of source, correspondence, and target models are derived in parallel, most practical scenarios involve existing models and require unidirectional transformation. Consequently, TGG tools support model transformation based on unidirectional forward and backward operational rules, automatically derived from a single TGG specification, as basic transformation steps, and use an algorithm to control which rule is to be applied on which part of the input graph. Such a *batch transformation* is the standard scenario for model transformation, where existing models are transformed (completely) from scratch.

In contrast, *incremental* model transformation supports changing already related models and propagating deltas appropriately. The challenge is to perform the update in an *efficient* manner and to avoid information loss by retaining unaffected elements of the models. Determining such an update sequence is a difficult task because transformations of deleted elements and their dependencies, as well as transformations of potential dependencies of newly added elements must be revoked [9]. The challenge is to identify such dependent elements in the model and to undo their previous transformation taking all changes into account.

Current incremental TGG approaches guarantee either the formal properties of *correctness* meaning that only consistent graph triples are produced, and *completeness* meaning that all possible consistent triples, which can be derived from a source or a target graph, can actually be produced, but are inefficient (scale with the size of the overall models) [9], or are efficient (scale with the number of changes and affected elements), but do not consider formal aspects [5,7].

In this paper, we introduce a novel incremental TGG control algorithm for model synchronization and prove its correctness, completeness, and efficiency. Based on our *precedence*-driven TGG batch algorithm presented in [12], a static *precedence analysis* is used to retrieve information, which allows for deciding which elements may be affected by deletions and additions of elements.

Section 2 introduces fundamentals and our running example. Section 3 presents our node precedence analysis, used by the incremental TGG algorithm presented in Sect. 4. Section 5 discusses related approaches, while Sect. 6 concludes with a summary and future work.

2   Fundamentals and Running Example

In this section, all concepts required to formalize and present our contribution are introduced and explained using our running example.

2.1   Type Graphs, Typed Graphs and Triples

We introduce the concept of a *graph*, and formalize *models* as *typed graphs*.

Definition 1 (Graph and Graph Morphism). *A graph $G = (V, E, s, t)$ consists of finite sets V of nodes, and E of edges, and two functions $s, t : E \to V$*

that assign each edge source and target nodes. A graph morphism $h : G \rightarrow G'$, *with* $G' = (V', E', s', t')$, *is a pair of functions* $h := (h_V, h_E)$ *where* $h_V : V \rightarrow V'$, $h_E : E \rightarrow E'$ *and* $\forall e \in E : h_V(s(e)) = s'(h_E(e)) \wedge h_V(t(e)) = t'(h_E(e))$.

Definition 2 (Typed Graph and Typed Graph Morphisms).

A type graph is a graph $TG = (V_{TG}, E_{TG}, s_{TG}, t_{TG})$.
A typed graph $(G, type)$ *consists of a graph* G *together with a graph morphism* type: $G \rightarrow TG$.
Given typed graphs $(G, type)$ *and* $(G', type')$, $g : G \rightarrow G'$ *is a typed graph morphism iff the depicted diagram commutes.*

These concepts can be lifted in a straightforward manner to *triples* of connected graphs denoted as $G = G_S \xleftarrow{h_S} G_C \xrightarrow{h_T} G_T$ as shown by [4,11]. In the following, we work with *typed graph triples* and corresponding morphisms.

Example: Our running example specifies the integration of *class diagrams* and corresponding *database schemata*. The *TGG schema* depicted in Fig. 1(a) is the type graph triple for our running example. In the *source domain*, class diagrams consist of **Packages**, **Classes**, and inheritance between **Classes**. In the *target domain*, a database schema consists of **Databases** and **Tables**. The *correspondence domain* specifies links between elements in the different domains, in this case **P2D** relating packages with databases, and **C2T** relating classes with tables. In Fig. 1(b), a schema conform (typed graph) triple is depicted: a package **p**

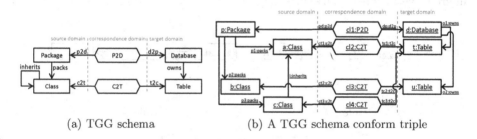

(a) TGG schema   (b) A TGG schema conform triple

Fig. 1. TGG schema for the running example and a conform triple

consists of three classes **a**, **b**, and **c**, while the corresponding database schema **d** contains two tables **t** and **u**.

2.2   Triple Graph Grammars and Rules

The simultaneous evolution of typed graph triples can be described by a *triple graph grammar* consisting of *transformation rules*. In general, transformation rules can be formalized via a double-pushout to allow for creating and deleting elements in a graph [4]. As TGG rules are restricted to the creation of elements, we simplify the definition in the following:

Definition 3 (Graph Triple Rewriting for Monotonic Creating Rules).

A monotonic creating rule $r := (L = K, R)$, *is a pair of typed graph triples s.t.* $L \subseteq R$. *A rule* r *rewrites (via adding elements) a graph triple* G *into a graph triple* G' *via a match* $m : L \to G$, *denoted as* $G \overset{r@m}{\rightsquigarrow} G'$, *iff* $m' : R \to G'$ *can be defined by building the pushout* G' *as denoted in the diagram to the right.*

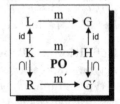

Elements in L denote the precondition of a rule and are referred to as *context elements*, while elements in $R \setminus L$ are referred to as *created elements*.

Definition 4 (Triple Graph Grammar).

A triple graph grammar $TGG := (TG, \mathcal{R})$ *is a pair consisting of a type graph triple* TG *and a finite set of monotonic creating rules* $\mathcal{R}$. *The generated language is* $\mathcal{L}(TGG) := \{G \mid \exists r_1, r_2, \ldots, r_n \in \mathcal{R} : G_{\emptyset} \overset{r_1@m_1}{\rightsquigarrow} G_1 \overset{r_2@m_2}{\rightsquigarrow} \ldots \overset{r_n@m_n}{\rightsquigarrow} G_n = G\}$, *where* $G_{\emptyset}$ *denotes the empty graph triple.*

Example: In Fig. 2, Rules (a)–(c) declare how an integrated class diagram and a database schema are created simultaneously. Rule (a) creates the root elements (a `Package` and a corresponding `Database`), while Rule (b) appends a `Class` and a `Table`, and Rule (c) extends the models with an inheriting `Class`, which is related to the same `Table`. We use a concise notation (merging L and R) depicting context elements in black without any markup, and created elements in green with a "++" markup.

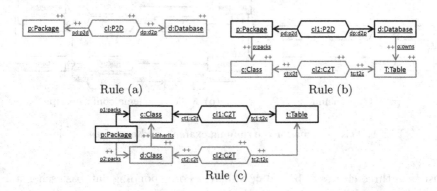

Rule (a)   Rule (b)

Rule (c)

Fig. 2. TGG Rules (a)–(c) for the integration

2.3   Derived Operational Rules

The real potential of TGGs as a bidirectional transformation language lies in the automatic derivation of *operational rules*. Such operational rules can be used to transform a given source domain model to a corresponding target domain model, and vice versa. Although we focus in the following sections only on forward transformations, all concepts and arguments are symmetric and can be applied analogously for the case of backward transformations.

As shown in [3], a sequence of TGG rules, which describes a simultaneous evolution, can be uniquely decomposed into (and conversely composed from) a sequence of *source rules* that only evolve the source model and *forward rules* that retain the source model and evolve the correspondence and target models. In addition, *inverse forward rules* revoke the effects of forward rules. These operational rules serve as the building blocks used by our control algorithm. As inverse forward rules only delete elements, we define monotonic deleting rules:

Definition 5 (Graph Triple Rewriting for Monotonic Deleting Rules).

A monotonic deleting rule $r := (L, K = R)$, is a pair of typed graph triples s.t. $L \supseteq R$. A rule r rewrites (via deleting elements) a graph triple G into a graph triple G' via a match $m : L \to G$, denoted as $G \overset{r@m}{\leadsto} G'$, iff $m' : R \to G'$ can be defined by building the pushout complement $H = G'$ as denoted in the diagram to the right.

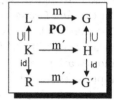

The elements in $L \setminus R$ of a monotonic deleting rule are referred to as *deleted elements*. Using this definition, operational rule derivation is formalized as follows:

Definition 6 (Derived Operational Rules). *Given a $TGG = (TG, \mathcal{R})$ and a rule $r = (L, R) \in \mathcal{R}$, a source rule $r_S = (SL, SR)$, a forward rule $r_F = (FL, FR)$ and an inverse forward rule $r_{F-1} = (FR, FL)$ are derived as follows:*

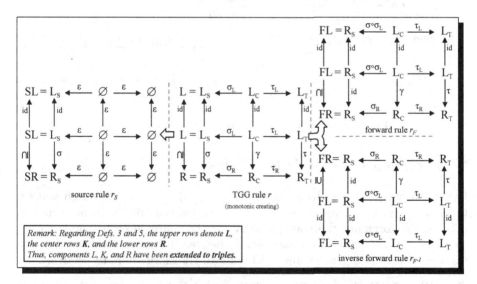

The forward rule r_F can be applied according to Def. 3, i.e., this involves building a pushout to create the required elements, while the inverse forward rule r_{F-1} involves building a pushout complement to delete the required elements according

to Def. 5. *Given a forward rule r_F, the existence of rule r_{F-1}, which reverses an application of r_F up to isomorphism, can be shown according to Fact 3.3 in [4].*

Although forward and inverse forward rules retain all source elements, the control algorithm keeps track of which source elements are transformed by a rule application. This can be done by introducing marking attributes [9], or maintaining a bookkeeping data structure in the control algorithm [6]. In concrete syntax, we equip every *transformed* element with a *checked box*, and every *untransformed* elements with an *unchecked box* (cf. Fig. 4) as introduced in [11].

Example: From Rule (b) (Fig. 2), the operational rules r_S, r_F, and r_F^{-1} depicted in Fig. 3 are derived. The source rule extends the source graph by adding a Class to an existing Package, while the forward rule r_F transforms (denoted as $\square \rightarrow \boxtimes$) an existing Class by *creating* a new C2T link and Table in the corresponding Database. The inverse forward rule *untransforms* (denoted as $\boxtimes \rightarrow \square$) a Class in a Package by *deleting* the corresponding link and Table, i.e., revoking the modifications of the forward rule. In addition to the already introduced merged representation of L and R of a rule, we further indicate deleted elements by a "$--$" markup and red color. Forward and inverse forward rules match the same context element and retain the checked box (denoted as $\boxtimes \rightarrow \boxtimes$).

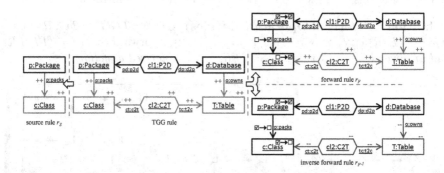

Fig. 3. Source and forward rules derived from Rule (b)

3   Precedence Analysis for TGGs

In the following, we introduce a path-based precedence analysis, which is used to partially sort the nodes in a source graph and thus control the transformation process. We formalize the concepts only for the source domain and a corresponding forward transformation, but, as before, all concepts can be directly transferred to the target domain and backward transformation, respectively.

Definition 7 (Paths and Type Paths). *Let G be a typed graph with type graph TG. A path p between two nodes $n_1, n_k \in V_G$ is an alternating sequence of nodes and edges in V_G and E_G, respectively, denoted as $p := n_1 \cdot e_1^{\alpha_1} \cdot n_2 \cdot \ldots \cdot n_{k-1} \cdot e_{k-1}^{\alpha_{k-1}} \cdot n_k$, where $\alpha_i \in \{+, -\}$ specifies if an edge e_i is traversed from* source $s(e_i) = n_i$ to target $t(e_i) = n_{i+1}$ *(+), or in a reverse direction (-). A type path is a path between node types and edge types in* V_{TG} *and* E_{TG}, *respectively. Given a path p, its type (path) is defined as* $type_p(p) := type_V(n_1) \cdot type_E(e_1)^{\alpha_1} \cdot type_V(n_2) \cdot type_E(e_2)^{\alpha_2} \cdot \ldots \cdot type_V(n_{k-1}) \cdot type_E(e_{k-1})^{\alpha_{k-1}} \cdot type_V(n_k)$.

For our analysis we are only interested in paths that are induced by certain *patterns* present in the *TGG rules*:

Definition 8 (Relevant Node Creation Patterns). *For a TGG* $= (TG, \mathcal{R})$ *and all rules* $r \in \mathcal{R}$, *where* $r = (L, R) = (L_S \leftarrow L_C \rightarrow L_T, R_S \leftarrow R_C \rightarrow R_T)$, *the set* $Paths_S$ *denotes all paths in* R_S *(note that* $L_S \subseteq R_S$). *The predicates* $context_S : Paths_S \rightarrow \{true, false\}$ *and* $create_S : Paths_S \rightarrow \{true, false\}$ *in the source domain are defined as follows:*
$context_S(p_r) := \exists \, r \in \mathcal{R}$ *s.t.* p_r *is a path between two nodes* $n_r, n'_r \in R_S$:
$(n_r \in L_S) \land (n'_r \in R_S \setminus L_S)$, *i.e., a rule r in* $\mathcal{R}$ *contains a path* p_r *which is isomorphic to the node creation pattern depicted in the diagram to the right.*

$create_S(p_r) := \exists \, r \in \mathcal{R}$ *s.t.* p_r *is a path between two nodes* $n_r, n'_r \in R_S$:
$(n_r \in R_S \setminus L_S) \land (n'_R \in R_S \setminus L_S)$, *i.e., a rule in* $\mathcal{R}$ *contains a path* p_r *which is isomorphic to the node creation pattern depicted in the diagram to the right.*

We can now define the set of interesting type paths, relevant for our analysis.

Definition 9 (Type Path Sets). *The set* $TPaths_S$ *denotes all type paths of paths in* $Paths_S$ *(cf. Def. 8), i.e.* $TPaths_S := \{tp \mid \exists \, p \in Paths_S \text{ s.t. } type_p(p) = tp\}$. *Thus, we define the restricted* create *type path set for the source domain as*
$$TP_S^{create} := \{tp \in TPaths_S \mid \exists \, p \in Paths_S \land type_p(p) = tp \land create_S(p)\},$$
and the restricted context *type path set for the source domain as*
$$TP_S^{context} := \{tp \in TPaths_S \mid \exists \, p \in Paths_S \land type_p(p) = tp \land context_S(p)\}.$$

In the following, we formalize the concept of *precedence between nodes*, reflecting that one node could be used as context for transforming another node.

Definition 10 (Precedence Function $\mathcal{PF}_S$). *Let* $\mathcal{P} := \{<, \doteq, \cdot \not\succ \cdot\}$ *be the set of precedence relation symbols. Given a TGG* $= (TG, \mathcal{R})$ *and the restricted type path sets for the source domain* TP_S^{create}, $TP_S^{context}$. *The precedence function for the source domain* $\mathcal{PF}_S : \{TP_S^{create} \cup TP_S^{context}\} \rightarrow \mathcal{P}$ *is computed as follows:*
$$\mathcal{PF}_S(tp) := \begin{cases} < & \text{iff } tp \in \{TP_S^{context} \setminus TP_S^{create}\} \\ \doteq & \text{iff } tp \in \{TP_S^{create} \setminus TP_S^{context}\} \\ \cdot \not\succ \cdot & \text{otherwise} \end{cases}$$

Example: For our running example, $\mathcal{PF}_S$ consists of the following entries:
Rule (a): $\emptyset$.   Rule (b): $\mathcal{PF}_S(\texttt{Package} \cdot \texttt{packs}^+ \cdot \texttt{Class}) = \,<$.
Rule (c): $\mathcal{PF}_S(\texttt{Package} \cdot \texttt{packs}^+ \cdot \texttt{Class}) = \,<$, $\mathcal{PF}_S(\texttt{Class} \cdot \texttt{inherits}^- \cdot \texttt{Class}) = \,<$, $\mathcal{PF}_S(\texttt{Class} \cdot \texttt{packs}^- \cdot \texttt{Package} \cdot \texttt{packs}^+ \cdot \texttt{Class}) = \,<$.
Note that regarding our running example, path $\texttt{Class} \cdot \texttt{packs}^- \cdot \texttt{Package}$ is not in $\mathcal{PF}_S$ as this path is neither in TP_S^{create} nor in $TP_S^{context}$.

Restriction: As our precedence analysis depends on paths in rules of a given TGG, the presented approach only works for TGG rules that are *(weakly) connected* in each domain. Hence, considering the source domain, the following must hold: $\forall\, r \in \mathcal{R} : \forall\, n, n' \in R_S : \exists\, p \in \text{Paths}_S$ between n, n'.

Based on the precedence function $\mathcal{PF}_S$, we now analyze typed graphs with two relations $\lessdot_S$ and $\doteq_S^*$. These are used to topologically sort a given source input graph and determine the sets of affected elements due to changes.

Definition 11 (Source Path Set). *For a given typed source graph G_S, the source path set for the source domain is defined as follows:*
$P_S := \{p \mid p \text{ is a path between } n, n' \in V_{G_S} \wedge type_p(p) \in \{TP_S^{create} \cup TP_S^{context}\}\}.$

Definition 12 (Precedence Relation $\lessdot_S$). *Given $\mathcal{PF}_S$, the precedence function for a given TGG, and a typed source graph G_S. The precedence relation $\lessdot_S \subseteq V_{G_S} \times V_{G_S}$ for the source domain is defined as follows: $n \lessdot_S n'$ if there exists a path $p \in P_S$ between nodes n and n', such that $\mathcal{PF}_S(type_p(p)) = \lessdot$.*

Example: For our example triple (Fig. 1(b)), the following pairs constitute $\lessdot_S$: $(\text{p} \lessdot_S \text{a}), (\text{p} \lessdot_S \text{b}), (\text{p} \lessdot_S \text{c}), (\text{a} \lessdot_S \text{c})$.

Definition 13 (Relation $\doteq_S$). *Given $\mathcal{PF}_S$, the precedence function for a given TGG, and a typed source graph G_S. The symmetric relation $\doteq_S \subseteq V_{G_S} \times V_{G_S}$ for the source domain is defined as follows: $n \doteq_S n'$ if there exists a path $p \in P_S$ between nodes n and n' such that $\mathcal{PF}_S(type_p(p)) = \doteq$.*

Definition 14 (Equivalence Relation $\doteq_S^*$). *The equivalence relation $\doteq_S^*$ is the transitive and reflexive closure of the symmetric relation $\doteq_S$.*

Example: For our example triple (Fig. 1(b)), relation $\doteq_S^*$ partitions the nodes of the source graph into the following equivalence classes: $\{p\}, \{a\}, \{b\}$, and $\{c\}$. For a more complex example with non-trivial equivalence classes we refer to [12].

We now define the concept of a *precedence graph* based on our relations $\doteq_S^*$, $\lessdot_S$ to sort a given graph according to its precedences, which is used by the incremental algorithm to determine if an element is available for transformation.

Definition 15 (Precedence Graph $\mathcal{PG}_S$). *The precedence graph for a given source graph G_S is a graph $\mathcal{PG}_S$ constructed as follows:*
(i) The equivalence relation $\doteq_S^*$ is used to partition V_{G_S} into equivalence classes
    $EQ_1, \ldots EQ_n$ which serve as the nodes of $\mathcal{PG}_S$, i.e., $V_{\mathcal{PG}_S} := \{EQ_1, \ldots, EQ_n\}$.
(ii) The edges in $\mathcal{PG}_S$ are defined as follows:
    $E_{\mathcal{PG}_S} := \{e \mid s(e) = EQ_i, \ t(e) = EQ_j : \exists\, n_i \in EQ_i, n_j \in EQ_j \text{ with } n_i \lessdot_S n_j\}$.

Example: The corresponding $\mathcal{PG}_S$ constructed from our example triple is depicted in Fig. 5(a) in Sect. 4.

Remark: $\mathcal{PG}_S$ defines a partial order over equivalence classes. This is a direct consequence of Def. 15.

Finally, we define the class of typed graph triples that do not introduce contradicting precedence relations between connected source and target domain elements. This is important as the synchronization control algorithm presented in Sect. 4 relies *only* on the source domain when applying appropriate changes to the correspondence and target domain.

Definition 16 (Forward Precedence Preserving Graph Triples). *Given a graph triple* $G = G_S \xleftarrow{h_S} G_C \xrightarrow{h_T} G_T$ *and two corresponding precedence graphs* $\mathcal{PG}_S$ *and* $\mathcal{PG}_T$*. For* $EQ_S \in V_{\mathcal{PG}_S}$ *and* $EQ_T \in V_{\mathcal{PG}_T}$*, the predicate* cross-domain-connected *on pairs of equivalence classes in precedence graphs of different domains is defined as follows:* cross-domain-connected$(EQ_S, EQ_T) := \exists\, n_C \in V_{G_C}$ *s.t.* $h_S(n_C) \in EQ_S \wedge h_T(n_C) \in EQ_T$.
Given $EQ_S, EQ'_S \in V_{\mathcal{PG}_S}, EQ_S \neq EQ'_S$ *and* $EQ_T, EQ'_T \in V_{\mathcal{PG}_T}, EQ_T \neq EQ'_T$ *s.t.* cross-domain-connected$(EQ_S, EQ_T) \wedge$ cross-domain-connected(EQ'_S, EQ'_T). *The graph triple* G *is* forward precedence preserving *iff*
$$\exists\ path\ p_T(EQ_T, EQ'_T) = EQ_T \cdot e_{T_1}^{\alpha_{T_1}} \cdot \ldots \cdot e_{T_n}^{\alpha_{T_n}} \cdot EQ'_T\ s.t.\ \alpha_{T_i} = +\ \forall\ i \in \{1, \ldots, n\}$$
$$\Rightarrow$$
$$\exists\ path\ p_S(EQ_S, EQ'_S) = EQ_S \cdot e_{S_1}^{\alpha_{S_1}} \cdot \ldots \cdot e_{S_n}^{\alpha_{S_n}} \cdot EQ'_S\ s.t.\ \alpha_{S_i} = +\ \forall\ i \in \{1, \ldots, n\}$$

Example: The running example (Fig. 1(b)) satisfies this property.

4   Incremental Precedence TGG Algorithm

To realize bidirectional incremental model synchronization with TGGs, a *control algorithm* is required that accepts a triple $G = G_S \leftarrow G_C \rightarrow G_T \in \mathcal{L}(TGG)$, an update graph triple [9] for the source domain $\Delta_S = G_S \leftarrow D \rightarrow G'_S$, the pre-compiled precedence function for the source domain $\mathcal{PF}_S$, and precedence graph $\mathcal{PG}_S$ used in a previous batch or incremental transformation, and returns a consistent graph triple $G' = G'_S \leftarrow G'_C \rightarrow G'_T$ with all changes propagated to the correspondence and target domain. Therefore, this algorithm (i) untransforms deleted elements and their dependencies in a valid order, (ii) untransforms elements (potentially) dependent on additions in a valid order, and (iii) transforms all untransformed and newly created elements by using the precedence-driven batch algorithm of [12]. Regarding the valid order, the algorithm has to find a way to delete elements in the opposite domain without compromising the transformation of existing elements. As a (fomal) restriction, edges can only be added (deleted) together with adjacent nodes, hence we focus on nodes only. In practice, Ecore for example assigns all edges to nodes, which overcomes this restriction.

Example: Using our example, we describe the incremental forward propagation of the following changes in the source domain (Fig. 4(a)): class a is deleted (indicated by «del») and a new class d is added (indicated by «add»). Parameters passed to the algorithm (line 1) are the original graph triple G (Fig. 1(b)), its source domain precedence graph $\mathcal{PG}_S$ (Fig. 5(a)), update Δ_S with deleted nodes $\Delta^- := V_{G_S} \setminus V_D$ and added nodes $\Delta^+ := V_{G'_S} \setminus V_D$, and the pre-compiled source domain precedence function $\mathcal{PF}_S$ (cf. example for Def. 10).

The algorithm returns a consistent graph triple with all changes propagated (Fig. 4(d)) on line 13.

Algorithm 1. Incremental Precedence TGG Algorithm

```
 1: procedure PROPAGATECHANGES(G, Δ_S, PF_S, PG_S)
 2:     for (node n⁻ ∈ Δ⁻) do
 3:         UNTRANSFORM(n⁻, PG_S)
 4:     end for
 5:     (G_S⁻, PG_S⁻) ← remove all n⁻ in Δ⁻ from G_S and PG_S
 6:     (G_S⁺, PG_S⁺) ← insert all n⁺ in Δ⁺ to G_S⁻ and PG_S⁻
 7:     if PG_S⁺ is cyclic then
 8:         terminate with error                    ▷ Additions invalidated G'_S
 9:     end if
10:     for (node n⁺ ∈ Δ⁺) do
11:         UNTRANSFORM(n⁺, PG_S⁺)
12:     end for            ▷ At this point G has changed to G* = G'_S ← G_C* → G_T*
13:     return (G'_S ← G'_C → G'_T) ← TRANSFORM(G*, PF_S)    ▷ Call batch algo [12]
14: end procedure
15: procedure UNTRANSFORM(n, PG_S)
16:     deps ← all nodes in all equiv. classes in PG_S with incoming edges from EQ(n)
17:     for node dep in deps do
18:         if dep is transformed then
19:             UNTRANSFORM(dep, PG_S)
20:         end if
21:     end for
22:     neighbors ← all nodes in EQ(n)
23:     for node neighbor in neighbors do
24:         if n is transformed then
25:             APPLYINVERSERULE(n)        ▷ Throw exception if Def. 16 is violated
26:         end if
27:     end for
28: end procedure
```

A for-loop (line 2) untransforms every deleted node in Δ^- (in our case class **a**) by calling method UNTRANSFORM. Line 16 places **c** in deps as this is dependent on $EQ(\mathbf{a})$ ($EQ(x)$ returns the appropriate equivalence class of node x) and calls UNTRANSFORM recursively on line 19. The equivalence class of **c** has no dependent elements in PG_S and on line 25, calling APPLYINVERSERULE untransforms **c** by applying the inverse forward rule of Rule (c) (Fig. 2). Note that with an appropriate bookkeeping data structure (not explained here) this method is aware of all previous rule applications and applies the correct inverse forward rule to the same match used previously by the forward transformation. The rule application can only fail if building the pushout complement was not possible due to dependencies in G_T which would violate the forward precedence preserving property for graph triples (Def. 16). In this case, an appropriate exception is thrown. After returning from the recursive call, **a** is untransformed by using the inverse forward rule of Rule (b). The resulting graph triple is depicted in Fig. 4(b). Next, all changes in Δ_S are used to update G_S and PG_S on lines 5 and 6. Adding elements may result in a cyclic precedence graph indicating

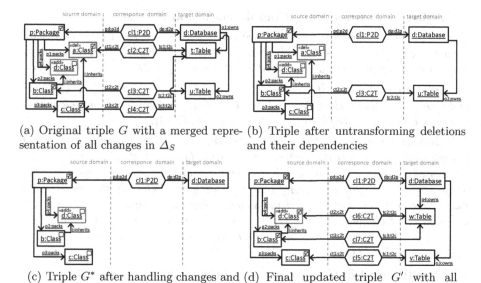

(a) Original triple G with a merged representation of all changes in Δ_S

(b) Triple after untransforming deletions and their dependencies

(c) Triple G^* after handling changes and untransforming their dependencies

(d) Final updated triple G' with all changes propagated

Fig. 4. Consistent change propagation from source to target domain

cyclic context dependencies and the algorithm would terminate with an error on line 8. For our running example, the updated precedence graph $\mathcal{PG}_S$ is acyclic (Fig. 5(b)), so the algorithm continues untransforming all elements that potentially depend on newly added elements as context. The only dependent element of d, which is b, is untransformed by calling UNTRANSFORM on line 11 which results in the triple G^* (Fig. 4(c)). Finally, on line 13 the intermediate triple G^* is passed to the TGG batch transformation algorithm of [12], which transforms all untransformed elements (with empty checkboxes) and returns the integrated and updated graph triple $G'_S \leftarrow G'_C \rightarrow G'_T$ depicted in Fig. 4(d).

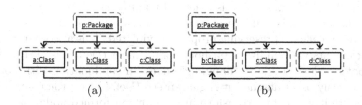

Fig. 5. $\mathcal{PG}_S$ for the original (left) and $\mathcal{PG}_S^+$ for the updated source graph (right)

Formal Properties of the Incremental Precedence TGG Algorithm

In this section, we prove that our algorithm retains all formal properties proposed in [13] and proved for the precedence-driven TGG batch algorithm of [12].

Definition 17 (Correctness, Completeness and Efficiency).

Correctness: Given an input graph triple $G_S \leftarrow G_C \rightarrow G_T \in \mathcal{L}(TGG)$ and an update $\Delta_S = G_S \leftarrow D \rightarrow G'_S$, the transformation algorithm either terminates with an error or produces a consistent graph triple $G'_S \leftarrow G'_C \rightarrow G'_T \in \mathcal{L}(TGG)$.

Completeness: $\forall\, G_S \leftarrow G_C \rightarrow G_T \in \mathcal{L}(TGG), G'_S \leftarrow G'_C \rightarrow G'_T \in \mathcal{L}(TGG)$ and a corresponding update $\Delta_S = G_S \leftarrow D \rightarrow G'_S$, the transformation algorithm produces a consistent triple $G'_S \leftarrow G^*_C \rightarrow G^*_T \in \mathcal{L}(TGG)$.

Efficiency: According to [13], a TGG batch transformation algorithm is efficient (polynomial runtime) if its runtime complexity class is $O(n^k_S)$, where n_S is the number of nodes in the source graph to be transformed and k is the largest number of elements to be matched by any rule r of the given TGG. In the incremental case, the algorithm is efficient if the synchronization runtime effort scales with the number of changes $(|\Delta^-| + |\Delta^+|)$ and (potentially) dependent elements n_δ and not with the size of the updated graph triple, i.e., the incremental algorithm runs in the order of $O(n^k_\delta)$.

All properties are defined analogously for backward transformations.

Theorem. *Algorithm 1 is correct, complete, and efficient for any source-local complete TGG (due to space restrictions we refer to Def. 13 in [11]) and forward precedence preserving graph triples (Def. 16).*

Proof.

Correctness: Lines 2 – 12 of the algorithm only invert previous rule applications. The order of rule applications is directed by the precedence graph (Def. 15), which represents potential dependencies between nodes, i.e., a node x has as dependencies all other nodes y, which may be transformed by applying a rule that matches x as context. These dependencies are potential dependencies as actual rule applications may select other nodes in place of x. Nevertheless, y potentially depends on x. The algorithm traverses to the very last dependency of every deleted/added node and applies the inverse of the rule used in a previous transformation. Demanding precedence preserving graph triples (Def. 16) guarantees that $\mathcal{PG}_S$ is sufficient to correctly revoke forward rules in a valid order. If an element on the target side is deleted by applying an inverse forward rule, although this element is still in use as context for another element, we know that the forward precedence preserving property is violated. This also guarantees that deleting elements via building a pushout complement (Def. 5) is always possible and cannot be blocked due to "dangling" edges. In combination with bookkeeping of previously used matches, it is guaranteed (Def. 6) that the resulting triple is in the state it was before transforming the untransformed node.

It directly follows that if the triple G was consistent, the remaining integrated part of G remains consistent. Since UNTRANSFORM inverts rule applications of a previous transformation, we know that the graph triple after line 12 is a valid intermediate graph triple produced by the batch transformation algorithm. As shown in [12], the precedence-driven TGG batch algorithm is correct (produces only correct graph triples or terminates with an error if no correct graph triple can be produced), so it directly follows that Algorithm 1 is also correct.   □

Completeness: The correctness proof shows that the incremental update produces a triple via a sequence of rule applications that the batch algorithm could have chosen for a forward transformation of G'_S. Completeness arguments from [12] for the batch algorithm can, hence, be transferred to this algorithm.□

Efficiency: Efficiency is influenced mainly by the cost of (i) untransforming dependent elements of a deleted or added node (lines 2–4 and 10–12), (ii) updating the precedence graph and graph triple itself (lines 5 and 6), and (iii) transforming all untransformed elements via our precedence-driven TGG batch algorithm (line 13). The number of deleted/added nodes ($|\Delta^-|+|\Delta^+|$) and their dependencies is denoted by n_δ. Regarding UNTRANSFORM, a recursive depth-first search on the precedence graph $\mathcal{PG}_S$ is invoked starting at a certain node. Depth-first search has a worst-case complexity of $O(|V_{\mathcal{PG}_S}| + |E_{\mathcal{PG}_S}|)$ if the changed node was an (indirect) dependency of all other equivalence classes in $\mathcal{PG}_S$. If the algorithm encounters an already untransformed element on line 18, we know for sure that all subsequent elements are already untransformed and, therefore, can safely terminate recursion. Independent of the position of the changed element, UNTRANSFORM traverses every dependent element exactly once. Finally, applying the inverse operational rule (line 25) is (at most) of the same complexity as the appropriate previous rule application since the rule and match are already known. Considering both untransformation runs together, we know that n_δ elements are untransformed, and that every element is treated exactly once. Updating G_S on line 5 (6) involves deleting (inserting) $|\Delta^-|$ ($|\Delta^+|$) elements $m \in \Delta^-(\Delta^+)$ and updating, each time, a number of adjacent nodes (degree(m)). Updating $\mathcal{PG}_S$ has similar costs since elements have to be deleted (added) and updating the edge set of $\mathcal{PG}_S$ means to traverse all adjacent nodes of a deletion or addition in G_S and retrieve appropriate entries from $\mathcal{PF}_S$. Thus, the complexity of line 5 and 6 can be estimated with $O(|\Delta_S|)$, as Δ_S contains all nodes and edges that have been changed and, therefore, need to be revised. Finally, transforming the rest of the prepared graph (line 13) has $O(n_\delta^k)$ complexity [12]. Because only added elements, their dependencies, and the dependencies of removed elements have been untransformed, n_δ refers to these elements only, and not to all elements in G_S. The algorithm, therefore, scales with the number of changes and their dependencies and not with the size of the graph triple: $n_\delta \leq n$.   □

5   Related Work

This section complements the discussion from Sect. 1 on related incremental synchronization approaches grouped according to their strengths.

Formality: Providing formal aspects for incremental updates that guarantee well-behavedness according to a set of laws or properties is challenging. Algebraic approaches such as lenses [2] and the framework introduced by Stevens [14] provide a solid basis for formalizing concrete implementations that support incremental model synchronization. Inspired by [2], a TGG model synchronization framework was presented in [9] that is correct and complete. The proposed algorithm, however, requires a complete remarking of the entire graph triple and depends, therefore, on the size of the related graphs and not on the size of the update and affected elements. This is infeasible for an *efficient* implementation and the need for an improved strategy is stated as future work in [9].

Efficiency: In contrast to this formal framework, an incremental TGG transformation algorithm has been presented in [5], which exploits the correspondence model to determine an efficient update strategy. Although the batch mode of this algorithm has been formally presented in [6], the incremental version has not been fully formalized and it is unclear how the update propagation order is determined correctly for changes to elements that are not linked via the correspondence model to other elements. The authors describe an event-handling mechanism and so it can be assumed that model changes are instantly propagated. This allows for reduced complexity regarding dependencies between changes, but forbids the option of collecting a set of changes before propagating. This is, however, a requirement for scenarios in which changes are applied to models offline (i.e., without access to the related model) and the actual model synchronization must be performed later. The TGG interpreter described in [7] employs basically the same approach as [5], but additionally attempts to *reuse* elements instead of deleting and creating them. This is important as it prevents a loss of information that cannot be recovered by (re-)creating an element (user added contents). Unfortunately, this approach has also not been formalized and it is unclear whether the algorithm guarantees correctness and completeness. Nonetheless, this concept of reuse is crucial for industrial relevance and should be further investigated.

Concurrency: The challenge of dealing with *concurrent* changes to both domains has been discussed and investigated in [8,15]. A cascade of *propagate*, *calculate diff*, and *merge* steps is proposed that finally results in a consistent model. Extending our TGG algorithm based upon these ideas but retaining efficiency is also an important task of future research.

6   Conclusion and Future Work

A novel incremental algorithm for TGG has been presented that employs a precedence analysis to determine the effects of model changes. This involves not only determining which elements rely on deletions and, hence, must be untransformed, but also includes finding all elements that may rely on additions and also have to be untransformed. This must be achieved without compromising formal properties (i.e., correctness and completeness) while scaling efficiently with the size of the changes and their dependencies and not with the size of the overall graph.

Current restrictions include the lack of support for concurrent change propagation, which we plan to handle according to [15], and the formal requirement that edges can only be deleted or added together with adjacent nodes. Last but not least, we shall implement the presented incremental algorithm as an extension of our current implementation in our metamodelling tool eMoflon[1][1] and perform empirical performance assessments and comparisons with other implementations.

---

[1] http:\\www.moflon.org

References

1. Anjorin, A., Lauder, M., Patzina, S., Schürr, A.: eMoflon: Leveraging EMF and Professional CASE Tools. In: Heiß, H.U., Pepper, P., Schlingloff, H., Schneider, J. (eds.) Informatik 2011. LNI, vol. 192, p. 281. GI, Bonn (2011)
2. Diskin, Z., Xiong, Y., Czarnecki, K., Ehrig, H., Hermann, F., Orejas, F.: From State- to Delta-Based Bidirectional Model Transformations: The Symmetric Case. In: Whittle, J., Clark, T., Kühne, T. (eds.) MODELS 2011. LNCS, vol. 6981, pp. 304–318. Springer, Heidelberg (2011)
3. Ehrig, H., Ehrig, K., Ermel, C., Hermann, F., Taentzer, G.: Information Preserving Bidirectional Model Transformations. In: Dwyer, M.B., Lopes, A. (eds.) FASE 2007. LNCS, vol. 4422, pp. 72–86. Springer, Heidelberg (2007)
4. Ehrig, H., Ehrig, K., Prange, U., Taentzer, G.: Fundamentals of Algebraic Graph Transformation. Springer, Berlin (2006)
5. Giese, H., Hildebrandt, S.: Efficient Model Synchronization of Large-Scale Models. Tech. Rep. 28, Universitätsverlag Potsdam (2009)
6. Giese, H., Hildebrandt, S., Lambers, L.: Toward Bridging the Gap Between Formal Semantics and Implementation of Triple Graph Grammars. In: MoDeVVA 2010, pp. 19–24. IEEE, New York (2010)
7. Greenyer, J., Pook, S., Rieke, J.: Preventing Information Loss in Incremental Model Synchronization by Reusing Elements. In: France, R.B., Kuester, J.M., Bordbar, B., Paige, R.F. (eds.) ECMFA 2011. LNCS, vol. 6698, pp. 144–159. Springer, Heidelberg (2011)
8. Hermann, F., Ehrig, H., Ermel, C., Orejas, F.: Concurrent Model Synchronization with Conflict Resolution Based on Triple Graph Grammars. In: de Lara, J., Zisman, A. (eds.) FASE 2012. LNCS, vol. 7212, pp. 178–193. Springer, Heidelberg (2012)
9. Hermann, F., Ehrig, H., Orejas, F., Czarnecki, K., Diskin, Z., Xiong, Y.: Correctness of Model Synchronization Based on Triple Graph Grammars. In: Whittle, J., Clark, T., Kühne, T. (eds.) MODELS 2011. LNCS, vol. 6981, pp. 668–682. Springer, Heidelberg (2011)
10. Kindler, E., Rubin, V., Wagner, R.: An Adaptable TGG Interpreter for In-Memory Model Transformations. In: Schürr, A., Zündorf, A. (eds.) Fujaba Days 2004, pp. 35–38. Paderborn (2004)
11. Klar, F., Lauder, M., Königs, A., Schürr, A.: Extended Triple Graph Grammars with Efficient and Compatible Graph Translators. In: Engels, G., Lewerentz, C., Schäfer, W., Schürr, A., Westfechtel, B. (eds.) Nagl Festschrift. LNCS, vol. 5765, pp. 141–174. Springer, Heidelberg (2010)
12. Lauder, M., Anjorin, A., Varró, G., Schürr, A.: Bidirectional Model Transformation with Precedence Triple Graph Grammars. In: Vallecillo, A., Tolvanen, J.-P., Kindler, E., Störrle, H., Kolovos, D. (eds.) ECMFA 2012. LNCS, vol. 7349, pp. 287–302. Springer, Heidelberg (2012)
13. Schürr, A., Klar, F.: 15 Years of Triple Graph Grammars. In: Ehrig, H., Heckel, R., Rozenberg, G., Taentzer, G. (eds.) ICGT 2008. LNCS, vol. 5214, pp. 411–425. Springer, Heidelberg (2008)
14. Stevens, P.: Towards an Algebraic Theory of Bidirectional Transformations. In: Ehrig, H., Heckel, R., Rozenberg, G., Taentzer, G. (eds.) ICGT 2008. LNCS, vol. 5214, pp. 1–17. Springer, Heidelberg (2008)
15. Xiong, Y., Song, H., Hu, Z., Takeichi, M.: Synchronizing Concurrent Model Updates Based on Bidirectional Transformation. SoSyM, 1–16 (2011) (Online First^{TM}, January 4, 2011)

ICGT 2012 Doctoral Symposium

Andrea Corradini[1] and Gabriele Taentzer[2]

[1] Dipartimento di Informatica, Università di Pisa, Italy
andrea@di.unipi.it
[2] University of Marburg, Germany
taentzer@mathematik.uni-marburg.de

Given the success of the previous Doctoral Symposia of ICGT held in 2008 in Leicester and in 2010 in Twente, also this year a specific event of the International Conference on Graph Transformation was explicitly dedicated to Ph.D. students. The Doctoral Symposium consisted of a technical session dedicated to presentations by doctoral students, held during the main conference, giving them a unique opportunity to present their research project and to interact with established researchers of the graph transformation community and with other students.

Among the submissions that we received, the following three-page abstracts were accepted for presentation at the conference and are included in the ICGT 2012 proceedings:

1. Yongzhi Ong, *Multi-scale Rule-based Graph Transformation using the Programming Language XL*
2. Christopher M. Poskitt, *Towards the Analysis and Verification of Graph Programs*
3. Andrea Vandin, *Towards the Specification and Verification of Modal Properties for Structured Systems*

These contributions were selected among the submitted abstracts according to their originality, significance and general interest, by the members of the following Program Committee:

- Paolo Baldan
- Luciano Baresi
- Andrea Corradini (co-chair)
- Juan De Lara
- Gregor Engels
- Claudia Ermel
- Holger Giese
- Annegret Habel
- Reiko Heckel
- Dirk Janssens
- Hans-Jörg Kreowski
- Barbara König
- Mark Minas
- Fernando Orejas
- Detlef Plump
- Andy Schürr
- Gabriele Taentzer (co-chair)
- Dániel Varró
- Bernhard Westfechtel

H. Ehrig et al.(Eds.): ICGT 2012, LNCS 7562, p. 416, 2012.

Multi-scale Rule-Based Graph Transformation Using the Programming Language XL

Yongzhi Ong

Department of Ecoinformatics, Biometrics & Forest Growth,
Georg-August University of Göttingen, Germany
yong@gwdg.de

1 Introduction and State-of-the-Art

The XL (eXtended L-System) programming language is an extension of the Java programming language by parallel rule-based graph rewriting features [10]. XL is primarily used in functional structural plant modelling where L(Lindenmayer)-systems [13,16] are used. Other main L-system implementations used in plant modelling are cpfg [15,8], lpfg [9], L-Py [1] and GROGRA [12]. The relational growth grammar (RGG) formalism [10] implemented by XL provides a connection mechanism emulating L-System string re-writing on graphs. An extension of such a rewriting formalism to multiscale structures is interesting in various domains from systems biology [14] to simulations of crop plants competing for resources [2].

2 Problems

Godin and Caraglio introduced the multiscale tree graph (MTG) to represent multi-scale plant topological structures in 1998[7]. Existing L-System implementations did not provide means to specify rules on such multi-scalar structures until the advent of L-Py. L-Py is introduced with an L-system feature that operates on multiscale L-system strings [1] converted from MTGs. While this allows L-system re-writing on MTGs, graph models are suitable for a wider variety of modelling requirements [3,6,11] as compared to strings.

The use of typed attributed graphs with inheritance [4,10] allows RGG to define multi-scalar graph models resembling MTGs. However, the current L-system-style connection mechanism does not take into account edges relating nodes from different scales. The other, single pushout [5] based embedding mechanism in XL results in lengthy and non-intuitive rule statements. It is the aim of the current project to design a more elegant and generic solution, to implement it as part of XL and to combine it flexibly with visualization tools.

3 Proposed Solutions

3.1 Multi-scale RGG Graph Model

An extension of the existing RGG graph model is introduced. A structure of scales defined as a partially ordered set is appended to the previous formalism

H. Ehrig et al.(Eds.): ICGT 2012, LNCS 7562, pp. 417–419, 2012.
© Springer-Verlag Berlin Heidelberg 2012

of Type Graph [4,10] in RGG. The extended type graph contains a unique edge label for inter-scale relationships between nodes. A multi-scale typed graph is then formalized over the multi-scale type graph.

The new graph model does not enforce the representation of an entity across all scales, i.e. is not restricted to an axial tree [7] structure.

3.2 Multi-scale L-System Style Grammar in XL

An L-System style connection mechanism is introduced coupled with a compact XL syntax for specifying rules using it. The syntax references the multi-scale type graph and implicitly represents inter-scale relationships with few newly introduced symbols (see Fig. 1).

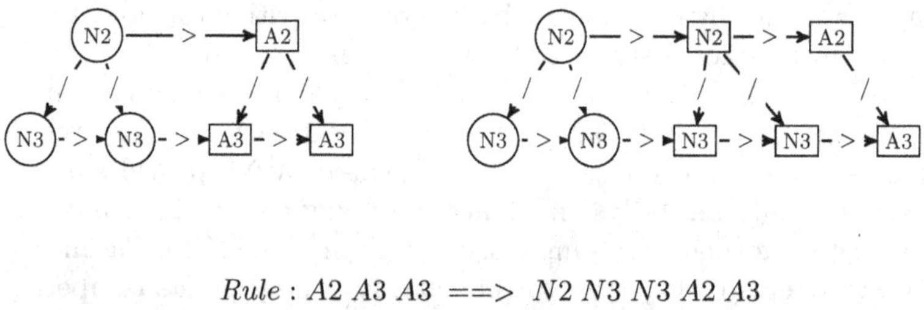

Rule : A2 A3 A3 ==> N2 N3 N3 A2 A3

Fig. 1. Multi-scale rule and embedding using XL. 'A' and 'N' are node types while digits are in ascending order representing coarse to fine scales respectively. Graph before and after the rule-based transformation is shown on the left and right respectively. In a biological application, the node types could, e.g., stand for a branch axis of a tree, an annual shoot, an internode and an apical bud producing (at the finest scale) new internodes. Circled nodes are not affected by the rule. Rectangular nodes are removed and inserted by the rule.

4 Project Status

The project has started 9 months ago. In a first step, GroIMP, the software platform for XL programming has been enabled to decode and visualize static MTG multi-scale graphs. The implementation of the new graph grammar formalism, including possibilities for visualization and analysis, is in progress.

References

1. Boudon, F., Pradal, C., Cokelaer, T., Prusinkiewicz, P., Godin, C.: L-Py: an L-System simulation framework for modeling plant development based on a dynamic language. Frontiers in Plant Science 3(00076) (2012)
2. Buck-Sorlin, G., Hemmerling, R., Kniemeyer, O., Burema, B., Kurth, W.: A rule-based model of barley morphogenesis, with special respect to shading and gibberellic acid signal transduction. Annals of Botany 101(8), 1109–1123 (2008)

3. Culik, K., Lindenmayer, A.: Parallel graph generating and graph recurrence systems for multicellular development. International Journal of General Systems 3(1), 53–66 (1976)

4. Ehrig, H., Ehrig, K., Prange, U., Taentzer, G.: Fundamentals of Algebraic Graph Transformation, 1st edn. Monographs in Theoretical Computer Science. An EATCS Series. Springer (March 2006)

5. Ehrig, H., Heckel, R., Korff, M., Löwe, M., Ribeiro, L., Wagner, A., Corradini, A.: Algebraic approaches to graph transformation. Part II: Single pushout approach and comparison with double pushout approach. In: Handbook of Graph Grammars and Computing by Graph Transformation. Foundations, vol. I, ch. 4, pp. 247–312. World Scientific Publishing Co., Inc. (1997)

6. Gabriele, T., Beyer, M.: Amalgamated Graph Transformations and Their Use for Specifying AGG – an Algebraic Graph Grammar System. In: Ehrig, H., Schneider, H.-J. (eds.) Dagstuhl Seminar 1993. LNCS, vol. 776, pp. 380–394. Springer, Heidelberg (1994)

7. Godin, C., Caraglio, Y.: A multiscale model of plant topological structures. Journal of Theoretical Biology 191, 1–46 (1998)

8. Hanan, J.: Parametric L-systems and their application to the modelling and visualization of plants. Ph.D. thesis, The University of Regina, Canada (1992)

9. Karwowski, R., Prusinkiewicz, P.: Design and Implementation of the L+C Modeling Language. Electronic Notes in Theoretical Computer Science 86(2), 1–19 (2003)

10. Kniemeyer, O.: Design and implementation of a graph grammar based language for functional-structural plant modelling. Ph.D. thesis, Brandenburg University of Technology Cottbus (2008)

11. Kniemeyer, O., Kurth, W.: The Modelling Platform GroIMP and the Programming Language XL. In: Schürr, A., Nagl, M., Zündorf, A. (eds.) AGTIVE 2007. LNCS, vol. 5088, pp. 570–572. Springer, Heidelberg (2008)

12. Kurth, W.: Growth Grammar Interpreter GROGRA 2.4 - a software tool for the 3-dimensional interpretation of stochastic, sensitive growth grammars in the context of plant modelling. In: Berichte des Forschungszentrums Waldökosysteme der Universität Göttingen, Ser. B, vol. 38 (1994)

13. Lindenmayer, A.: Mathematical models for cellular interactions in development I & II. Journal of Theoretical Biology 18(3), 280–315 (1968)

14. Maus, C., Rybacki, S., Uhrmacher, A.: Rule-based multi-level modeling of cell biological systems. BMC Systems Biology 5(1), 166 (2011)

15. Prusinkiewicz, P., Hanan, J., Měch, R.: An L-System-Based Plant Modeling Language. In: Nagl, M., Schürr, A., Münch, M. (eds.) AGTIVE 1999. LNCS, vol. 1779, pp. 395–410. Springer, Heidelberg (2000)

16. Prusinkiewicz, P., Lindenmayer, A.: The algorithmic beauty of plants. Springer (1996)

Verification of Graph Programs

Christopher M. Poskitt

Department of Computer Science, The University of York, UK
cposkitt@cs.york.ac.uk

1 Introduction

GP (for Graph Programs) is an experimental nondeterministic programming language which allows for the manipulation of graphs at a high level of abstraction [11]. The program states of GP are directed labelled graphs. These are manipulated directly via the application of (conditional) rule schemata, which generalise double-pushout rules with expressions over labels and relabelling. In contrast with graph grammars, the application of these rule schemata is directed by a number of simple control constructs including sequential composition, conditionals, and as-long-as-possible iteration. GP shields programmers at all times from low-level implementation issues (e.g. graph representation), and with its nondeterministic semantics, allows one to solve graph-like problems in a declarative and natural way.

An important question to ask of any program is whether it is correct with respect to its specification. For more traditional programming languages, verification techniques to help answer this have been studied for many years [1]. But a number of issues prevent these techniques being used for graph programs "out of the box" (e.g. the state we must reason about is a graph, not a mapping from variables to values). Fortunately, research into verifying graph transformations is gaining momentum, with numerous verification approaches emerging in recent years [15,2,9,3,8] (though typically focusing on sets of rules or graph grammars). Recent work by Habel, Pennemann, and Rensink [5,6] contributed a weakest precondition based verification framework for a language similar to GP, although this language lacks important features like expressions as graph labels in rules.

2 Research Aims and Progress

Our research programme is concerned with the challenge of verifying graph programs using a Hoare-style approach, especially from a theoretical viewpoint so as to provide the groundwork for later development of e.g. tool support, and formalisations in theorem provers. The particular contributions we aim to make in our thesis are discussed below.

Nested conditions with expressions. In [5,6], nested conditions are studied as an appropriate graphical formalism for expressing and reasoning about structural properties of graphs. However, in the context of GP, where graphs are labelled

H. Ehrig et al.(Eds.): ICGT 2012, LNCS 7562, pp. 420–422, 2012.

over an infinite label alphabet and graph labels in rules contain expressions, nested conditions are insufficient. For example, to express that a graph contains an integer-labelled node, one would need the infinite condition ∃(⓪)∨∃(①)∨ ∃(⓪) ∨ ∃(②) ∨ ∃(②) ∨ · · · .

In [13,12], we added expressions and assignment constraints to yield nested conditions with expressions (short E-conditions). E-conditions can be thought of as finite representations of (usually) infinite nested conditions, and are shown to be appropriate for reasoning about first-order properties of structure and labels in the graphs of GP. For example, an E-condition equivalent to the infinite nested condition earlier is ∃(ⓧ | type(x) = int), expressing that the variable x must be instantiated with integer values. A similar approach was used earlier by Orejas [10] for attributed graph constraints, but without e.g. the nesting allowed in E-conditions. Despite the graphical nature of E-conditions, they are precise (the formal definition is based on graph morphisms), and thus suitable for use as an assertion language for GP.

Many-sorted predicate logic. In [14] we defined a many-sorted first-order predicate logic for graphs, as an alternative assertion language to E-conditions. This formalism avoids the need for graph morphisms and nesting, and is more familiar to classical logic users. It is similar to Courcelle's two-sorted graph logic [4] in having sorts (types) for nodes and edges, but additionally has sorts for labels (the semantic domain of which is infinite): these are organised into a hierarchy of sorts corresponding to GP's label subtypes. This hierarchy is used, for example, to allow predicates such as equality to compare labels of any subtype, while restricting operations such as addition to expressions that are of type integer. We have shown that this logic is equivalent in power to E-conditions, and have constructed translations from E-conditions to many-sorted formulae and vice versa.

Hoare Logic. In [13,12] we proposed a Hoare-style calculus for partial correctness proofs of graph programs, using E-conditions as the assertion language. We demonstrated its use by proving properties of graph programs computing colourings. In proving ⊢ {c} P {d} where P is a program and c, d are E-conditions, from our soundness result, if P is executed on a graph satisfying c, then if a graph results, it will satisfy d. Currently we are extending the proof rules to allow one to reason about both termination and freedom of failure. We require the termination of loops to be shown outside of the calculus, by defining termination functions # mapping graphs to naturals, and showing that executing loop bodies (rule schemata sets) yields graphs for which # returns strictly smaller numbers.

Case studies and further work. We will demonstrate our techniques on larger graph programs in potential application areas, e.g. in modelling pointer manipulations as graph programs and verifying properties of them. Also, the challenges involved in formalising our Hoare logic in an interactive theorem prover like Isabelle will be explored. Finally, we will discuss how our calculus could be extended to integrate a stronger assertion language such as the HR conditions of [7], which can express non-local properties.

Acknowledgements. The author is grateful to Detlef Plump and the anonymous referees for their helpful comments, which helped to improve this paper.

References

1. Apt, K.R., de Boer, F.S., Olderog, E.-R.: Verification of Sequential and Concurrent Programs, 3rd edn. Springer (2009)
2. Baldan, P., Corradini, A., König, B.: A framework for the verification of infinite-state graph transformation systems. Information and Computation 206(7), 869–907 (2008)
3. Bisztray, D., Heckel, R., Ehrig, H.: Compositional Verification of Architectural Refactorings. In: de Lemos, R., Fabre, J.-C., Gacek, C., Gadducci, F., ter Beek, M. (eds.) Architecting Dependable Systems VI. LNCS, vol. 5835, pp. 308–333. Springer, Heidelberg (2009)
4. Courcelle, B.: Graph rewriting: An algebraic and logic approach. In: van Leeuwen, J. (ed.) Handbook of Theoretical Computer Science, vol. B, ch. 5. Elsevier (1990)
5. Habel, A., Pennemann, K.-H.: Correctness of high-level transformation systems relative to nested conditions. Mathematical Structures in Computer Science 19(2), 245–296 (2009)
6. Habel, A., Pennemann, K.-H., Rensink, A.: Weakest Preconditions for High-Level Programs. In: Corradini, A., Ehrig, H., Montanari, U., Ribeiro, L., Rozenberg, G. (eds.) ICGT 2006. LNCS, vol. 4178, pp. 445–460. Springer, Heidelberg (2006)
7. Habel, A., Radke, H.: Expressiveness of graph conditions with variables. In: Proc. Colloquium on Graph and Model Transformation on the Occasion of the 65th Birthday of Hartmut Ehrig. Electronic Communications of the EASST, vol. 30 (2010)
8. König, B., Esparza, J.: Verification of Graph Transformation Systems with Context-Free Specifications. In: Ehrig, H., Rensink, A., Rozenberg, G., Schürr, A. (eds.) ICGT 2010. LNCS, vol. 6372, pp. 107–122. Springer, Heidelberg (2010)
9. König, B., Kozioura, V.: Towards the Verification of Attributed Graph Transformation Systems. In: Ehrig, H., Heckel, R., Rozenberg, G., Taentzer, G. (eds.) ICGT 2008. LNCS, vol. 5214, pp. 305–320. Springer, Heidelberg (2008)
10. Orejas, F.: Attributed Graph Constraints. In: Ehrig, H., Heckel, R., Rozenberg, G., Taentzer, G. (eds.) ICGT 2008. LNCS, vol. 5214, pp. 274–288. Springer, Heidelberg (2008)
11. Plump, D.: The Graph Programming Language GP. In: Bozapalidis, S., Rahonis, G. (eds.) CAI 2009. LNCS, vol. 5725, pp. 99–122. Springer, Heidelberg (2009)
12. Poskitt, C.M., Plump, D.: A Hoare Calculus for Graph Programs. In: Ehrig, H., Rensink, A., Rozenberg, G., Schürr, A. (eds.) ICGT 2010. LNCS, vol. 6372, pp. 139–154. Springer, Heidelberg (2010)
13. Poskitt, C.M., Plump, D.: Hoare-style verification of graph programs. Fundamenta Informaticae 118(1-2), 135–175 (2012)
14. Poskitt, C.M., Plump, D., Habel, A.: A many-sorted logic for graph programs (2012) (submitted for publication)
15. Rensink, A., Schmidt, Á., Varró, D.: Model Checking Graph Transformations: A Comparison of Two Approaches. In: Ehrig, H., Engels, G., Parisi-Presicce, F., Rozenberg, G. (eds.) ICGT 2004. LNCS, vol. 3256, pp. 226–241. Springer, Heidelberg (2004)

Specification and Verification
of Modal Properties for Structured Systems*

Andrea Vandin

IMT Institute for Advanced Studies Lucca, Italy
{andrea.vandin}@imtlucca.it

1 Problem Statement

System specification formalisms should come with suitable property specification languages and effective verification tools. We sketch a framework for the verification of quantified temporal properties of systems with dynamically evolving structure. We consider visual specification formalisms like graph transformation systems (GTS) where program states are modelled as graphs, and the program behaviour is specified by graph transformation rules. The state space of a GTS can be represented as a graph transition system (GTrS), i.e. a transition system with states and transitions labelled, respectively, with a graph, and with a partial morphism representing the evolution of state components. Unfortunately, GTrSs are prohibitively large or infinite even for simple systems, making verification intractable and hence calling for appropriate abstraction techniques.

2 State-of-the-Art in GTS Logics

After the pioneering works on monadic second-order logic (MSO) [7], various graph logics have been proposed and their connection with topological properties of graphs investigated [8]. The need to reason about the evolution of graph topologies has then led to combining temporal and graph logics in propositional temporal logics using graph formulae as state observations (e.g. [4]). However, due to the impossibility to interleave the graphical and temporal dimensions it was not possible to reason on the evolution of single graph components. To overcome this limitation, predicate temporal logics were proposed (e.g. [2, 16]), where edge and node quantifiers can be interleaved with temporal operators.

More recent approaches [2] propose quantified μ-calculi combining the fixpoint and modal operators with MSO. These logics fit at the right level of abstraction for GTSs, allowing to reason on the topological structure of a state, and on the evolution of its components. We refer to § 8 of [11] for a more complete discussion. Unfortunately, the semantical models for such logics are less clearly cut. Current solutions are not perfectly suited to model systems with dynamic structure, where components might get merged [2, 16], or (re)allocated [2]. These problems are often solved by restricting the class of admissible models or by reformulating the state transition relation, hampering the meaning of the logic.

---

* Partly supported by the EU FP7-ICT IP ASCENS and by the MIUR PRIN SisteR.

H. Ehrig et al.(Eds.): ICGT 2012, LNCS 7562, pp. 423–425, 2012.

3   State-of-the-Art in GTS Verification

Various approaches have been proposed for the verification of GTSs, often adopting traditional techniques (e.g. model checking) to the area of graph transformation. We mention two research lines that have integrated the proposed techniques in verification tools, namely GROOVE [5, 12, 15, 17] and AUGUR [1–4, 13][1].

The model checking problem for GTSs is in general not decidable for reasonably expressive logics, since GTSs are Turing complete languages. Pragmatically, GTSs are often infinite-state and it is well known that only some infinite-state model checking problems are decidable. Several approximation techniques inspired to abstract interpretation have thus been proposed, the main idea being to consider a *finite-state* system approximating an *infinite-state* one. In order to be meaningful, those approximations may be related with the original systems via behavioural relations. The above mentioned research lines developed approximation techniques: namely *neighbourhood abstractions* [5], and *unfoldings* [1–4].

4   Current Contributions

In [10, 11] we introduced a novel semantics for quantified μ-calculi. We defined *counterpart models*, generalizing GTrSs, where states are algebras and the evolution relation is given by a family of partial morphisms. One of the main characteristics of our approach is that formulae are interpreted over sets of pairs (w, σ_w), for w a state and σ_w an assignment from formula variables to components of w. This allows for a straightforward interpretation of fixed points and for their smooth integration with quantifiers, which often asked for a restriction of the class of admissible models. Our proposal avoids the limitations of existing approaches, in particular in what regards merging and name reuse. Moreover it dispenses with the reformulation of the transition relation, obtaining a streamlined and intuitive semantics, yet general enough to cover the alternatives we are aware of.

In [14] we presented a first step towards a tool support for our approach, preparing the ground for an efficient tool framework. We first presented a Maude [6] implementation of graph rewriting as conditional rewrite rules on object multisets. Then we introduced a prototypal model checker for finite counterpart models. Our tool allows to analyze the evolution of individual components, and, as far as we know, it is one of the few model checkers for quantified μ-calculi.

Finally, [9] proposes a general formalization of similarity-based counterpart model approximations, and a technique for approximated verification exploiting them. We extended and generalized in several directions the type system of [4], proposed within the unfolding technique to classify formulae as preserved or reflected by a given approximation: (i) our type system is *technique-agnostic*, meaning that it does not require a particular approximation technique; (ii) we consider counterpart models, a generalization of GTrSs; (iii) our type system is parametric with respect to a given simulation relation (while the original one

---

[1] See groove.cs.utwente.nl and www.ti.inf.uni-due.de/research/tools/augur2.

considers only those with certain properties); (iv) we use the type system to reason on all formulae (rather than just on closed ones); and (v) we propose a technique that exploits approximations to estimate properties more precisely, handling also part of the untyped formulae.

References

1. Baldan, P., Corradini, A., König, B.: A Static Analysis Technique for Graph Transformation Systems. In: Larsen, K.G., Nielsen, M. (eds.) CONCUR 2001. LNCS, vol. 2154, pp. 381–395. Springer, Heidelberg (2001)
2. Baldan, P., Corradini, A., König, B., Lluch Lafuente, A.: A Temporal Graph Logic for Verification of Graph Transformation Systems. In: Fiadeiro, J.L., Schobbens, P.-Y. (eds.) WADT 2006. LNCS, vol. 4409, pp. 1–20. Springer, Heidelberg (2007)
3. Baldan, P., König, B.: Approximating the Behaviour of Graph Transformation Systems. In: Corradini, A., Ehrig, H., Kreowski, H.-J., Rozenberg, G. (eds.) ICGT 2002. LNCS, vol. 2505, pp. 14–29. Springer, Heidelberg (2002)
4. Baldan, P., König, B., König, B.: A Logic for Analyzing Abstractions of Graph Transformation Systems. In: Cousot, R. (ed.) SAS 2003. LNCS, vol. 2694, pp. 255–272. Springer, Heidelberg (2003)
5. Bauer, J., Boneva, I., Kurbán, M.E., Rensink, A.: A Modal Logic Based Graph Abstraction. In: Ehrig, H., Heckel, R., Rozenberg, G., Taentzer, G. (eds.) ICGT 2008. LNCS, vol. 5214, pp. 321–335. Springer, Heidelberg (2008)
6. Clavel, M., Durán, F., Eker, S., Lincoln, P., Martí-Oliet, N., Meseguer, J., Talcott, C.: All About Maude. LNCS, vol. 4350. Springer, Heidelberg (2007)
7. Courcelle, B.: The Expression of Graph Properties and Graph Transformations in Monadic Second-Order Logic. In: Rozenberg, G. (ed.) Handbook of Graph Grammars and Computing by Graph Transformation, pp. 313–400. World Scientific (1997)
8. Dawar, A., Gardner, P., Ghelli, G.: Expressiveness and Complexity of Graph Logic. Information and Computation 205(3), 263–310 (2007)
9. Gadducci, F., Lluch Lafuente, A., Vandin, A.: Exploiting Over- and Under-Approximations for Infinite-State Counterpart Models. In: Ehrig, H., Engels, G., Kreowski, H.-J., Rozenberg, G. (eds.) ICGT 2012. LNCS, vol. 7562, pp. 51–65. Springer, Heidelberg (2012)
10. Gadducci, F., Lluch Lafuente, A., Vandin, A.: Counterpart Semantics for a Second-Order μ-Calculus. In: Ehrig, H., Rensink, A., Rozenberg, G., Schürr, A. (eds.) ICGT 2010. LNCS, vol. 6372, pp. 282–297. Springer, Heidelberg (2010)
11. Gadducci, F., Lluch Lafuente, A., Vandin, A.: Counterpart Semantics for a Second-Order μ-Calculus. Fundamenta Informaticae 118(1-2) (2012)
12. Ghamarian, A.H., de Mol, M., Rensink, A., Zambon, E., Zimakova, M.: Modelling and Analysis Using GROOVE. STTT 14(1), 15–40 (2012)
13. König, B., Kozioura, V.: Counterexample-Guided Abstraction Refinement for the Analysis of Graph Transformation Systems. In: Hermanns, H., Palsberg, J. (eds.) TACAS 2006. LNCS, vol. 3920, pp. 197–211. Springer, Heidelberg (2006)
14. Lluch Lafuente, A., Vandin, A.: Towards a Maude Tool for Model Checking Temporal Graph Properties. In: Gadducci, F., Mariani, L. (eds.) GT-VMT. ECEASST, vol. 42. EAAST (2011)
15. Rensink, A.: Towards Model Checking Graph Grammars. In: Leuschel, M., Gruner, S., Lo Presti, S. (eds.) AVOCS. DSSE-TR, vol. 2003-2. University of Twente
16. Rensink, A.: Model Checking Quantified Computation Tree Logic. In: Baier, C., Hermanns, H. (eds.) CONCUR 2006. LNCS, vol. 4137, pp. 110–125. Springer, Heidelberg (2006)
17. Rensink, A., Zambon, E.: Neighbourhood Abstraction in GROOVE. In: de Lara, J., Varró, D. (eds.) GraBaTs. ECEASST, vol. 32. EAAST (2010)

Author Index